The Geological Society of America, Inc.
Memoir 159

Circum-Pacific Plutonic Terranes

Edited by
J. A. Roddick

1983

This volume is dedicated to Paul C. Bateman who created the Circum-Pacific Plutonic Project and held the fragile creation together for the decade required for its completion.

Published by The Geological Society of America, Inc.
3300 Penrose Place, P.O. Box 9140, Boulder, Colorado 80301
Printed in U.S.A.

Library of Congress Cataloging in Publication Data
Main entry under title:

Circum-Pacific plutonic terranes.

(Memoir / The Geological Society of America ; 159)
Includes bibliographies.
1. Rocks, Igneous. 2. Petrology—Pacific Ocean.
I. Roddick, J. A.   II. Series: Memoir (Geological Society
of America) ; 159.
QE461.C49 1983   552'.3'091823   83-1557
ISBN 0-8137-1159-2

Contents

Geological Society of America
Memoir 159
1983

Preface

Paul C. Bateman
Project Leader
U.S. Geological Survey
345 Middle Field Road
Menlo Park, California 94025

This Memoir is one of three final products of IGCP Project 30, Circum-Pacific Plutonism. The other two are an "Annotated Bibliography of Circum-Pacific Plutonism," which is being published in the G.S.A. Microform Series, and a "Map of Circum-Pacific Magmatism," being compiled and published in the Soviet Union.

The Circum-Pacific Plutonism Project was one of the first to be organized under IGCP, which is sponsored jointly by the International Union of Geological Sciences and UNESCO. The idea of such a project was conceived in 1969 during a preliminary organizational meeting of IGCP in Budapest. A hypothetical Circum-Pacific Plutonism Project was cited in the final report of the meeting as an example of the kind of project that would be suitable for sponsorship under the category "Geologic Patterns in Space and Time."

A year later, in 1970, Brian Harland, then Secretary of the Coordinating Panel for IGCP, suggested that I organize an actual Circum-Pacific Plutonism Project to assist in getting IGCP underway. After considering the difficulties involved, I did submit a proposal, and it was quickly approved by the U.S. National Committee on Geology and the IGCP Coordinating Panel. I then began a search for interested and qualified scientists in the other Pacific nations. This was a slow and uncertain task, because at that time the almost complete lack of communication between scientists studying plutonic rocks in the different Pacific nations made it difficult to identify the principal investigators in each country.

The first project meeting was held in Santa Cruz, California, in September 1972, and was followed by an excursion into the Sierra Nevada to examine features of the Sierra Nevada batholith. This meeting was attended by scientists from Canada, Japan, Australia, Chile, and the United States. Later, project membership grew to include scientists from the USSR, Korea, Thailand, Malaysia, Indonesia, New Zealand, Mexico, and Papua-New Guinea. At the first meeting I was confirmed as project leader, and it was decided that project meetings should be held at yearly intervals in a succession of Pacific nations. It was also decided that each meeting would have a theme and that the

formal meeting would be followed by a field excursion to allow the participants to examine the plutonic rocks of the area and to exchange ideas in an informal environment and to develop friendships.

Subsequent project meetings were as follows (year-place-theme-excursion):

1973   Santiago, Chile. The setting of batholiths. Chile, Argentina, and Peru.

1974   San Diego, California. Mechanics of emplacement of Circum-Pacific batholiths. The Peninsula Range batholith of Baja California and adjacent parts of southern California.

1975   Vancouver, Canada; part of the 13th Pacific Science Congress. Timing of events during plutonism. The Coast Plutonic Complex of British Columbia.

1975   Kuala Lumpur, Malaysia. The relation between granitoids and associated ore deposits of the Circum-Pacific region. Malaysia and Thailand.

1976   Sydney, Australia; held in connection with the 25th International Congress. Symposium on S- and I-type granites. The Lachlan belt of southeastern Australia.

1977   Toyama, Japan. Plutonism in relation to volcanism and metamorphism. Japan and South Korea.

1978   Bangkok, Thailand; mini-meeting held in connection with the Third Regional Conference on Geology and Mineral Resources of Southeast Asia. Thailand.

1979   Khabarovsk, USSR; held in connection with the 14th Pacific Science Congress. Intrusive magmatism and its relation to tectonics. The upper Kolyma Region of the Soviet Far East.

1981   Yosemite National Park, California. Business meeting to review final products of the project. Local field trips.

The tangible products of the project are the publications that resulted from its activities. The IGCP Catalogue for 1973-1979 lists 95 publications that were prepared in response to project activities. Other papers have been published more recently. The papers presented at two meetings are published in separate volumes. Pacific Geology, v. 8, contains the papers presented at the 1973 meeting in Santiago, Chile on the theme "The setting of batholiths," and Bulletin 9 (The 10th Anniversary volume) of the Geological Society of Malaysia contains the papers presented at the 1975 meeting in Kuala Lumpur on the theme "The relations between granitoids and associated ore deposits." There can be little doubt that the publications of project members have increased understanding of not only Circum-Pacific plutonism but also plutonism in all parts of the world.

However, the most important products of the project are intangible. They are friendship, good will, and increased communication and collaboration among scientists from different nations. Reciprocal visits and collaboration are common now and should increase if the world can manage its affairs so as to remain at peace.

Geological Society of America
Memoir 159
1983

Circum-Pacific plutonic terranes: *An overview*

J. A. Roddick
Geological Survey of Canada
100 West Pender Street
Vancouver, B. C., V6B 1R8

Plutonic terranes wherever situated are enormously complex, and our understanding of them is barely more than incipient. Whether more data will increase our understanding is, perhaps, debatable, but the process of reviewing what we know focusses our attention on what we do not know and should lead eventually to a more comprehensive picture of what in these enigmatic terranes remains unsatisfactorily explained. The articles in this volume are mainly reviews of the available information concerning plutonic terranes of countries or regions adjacent to the Pacific Ocean, but the articles also contain new information or insights. Although it was our original intention that each major review paper be supported by a number of short papers on current research, the difficulties in preparing the review papers indicated that the inclusion of many shorter papers would cause unacceptable further delays in an already much-delayed volume. Four topical papers, however, do support the review paper for Japan by Nozawa.

The articles are arranged so that the regions described form a clockwise sequence around the Pacific Ocean, beginning with New Zealand and ending with Chile. Information on circum-Pacific terranes is very uneven, and for certain major segments no coverage could be obtained in the time available. Especially missed are descriptions of the granitoid rocks of northeastern China and northwestern South America.

The obvious question "Is there a common thread other than their circum-Pacific location that links the plutonic terranes?" must be answered, "Perhaps." At present the variety and differences found in the terranes are more striking than their similarities.

Possibly a pattern exists only on the east side of the Pacific Ocean basin, where in both North and South America, the plutonic rocks are commonly more basic and older on the oceanic side, becoming more silicic and potassic, and younger inland. That, however, is a widebrush, high-tolerance generalization, and cross sections that show wide deviations from the model are easily found. The vaguely perceived norm is probably best exhibited by the west to east, gabbro-tonalite-granite transition in Baja California.

The validity of the generalization also depends on whether a single belt or the total width of the Cordillera is considered. In the Coast Plutonic Complex of British Columbia, for example, there is an eastward younging but it seems to be roughly repeated (through different ranges) in the flanking belts, Vancouver Island on the west and the Intermontane Belt on the east. Also, the rocks become more acid eastward if the whole width of the Cordillera is taken into account, but the most silicic part of the Coast Plutonic Complex is its axial part, although it does have more mafic and denser rocks on the western margin. In Peru the "normal" pattern is sustained in a crude way. The mainly Late Cretaceous Coastal Batholith is more mafic and less siliceous than the younger (Miocene to Pliocene) Cordillera Blanca Batholith which lies to the east. Such variations, however, are not found along most cross sections within the confines of the Coastal Batholith.

The arcuate masses of diorite, quartz diorite, granodiorite, and granite that form the centered ring complexes in the Peruvian Coastal Batholith are of special interest. Similar ring complexes are known elsewhere, but none is so well exposed as, nor exhibits the regular 35- to 50-km spacing of those in Peru. Whether these intriguing structures actually hold the key to the mechanism of batholithic emplacement, as workers there suggest, is debatable, mainly because the complexes are small in comparison to the enormous bulk of the batholith. Yet they must be accounted for in any acceptable theory of plutonic genesis.

On the west side of the Pacific Ocean, the pattern is complicated by the intersection of Asiatic structures with the Pacific basin. The tin granites of Malaysia and Thailand have no major equivalent in the circum-Pacific belt, although plutons as siliceous and otherwise chemically similar are known (such as the Cathedral Peak granodiorite of the Sierra Nevada Batholith). With about 10 ppm Sn in granite away from the greisenized areas, the Malaysian granites are exceptional not in their mean tin content, only in their spectacular tin deposits.

Japan forms the major island-arc model for the Pacific; yet plutonism there seems to have migrated oceanward in zones parallel with the Honshu arc. Also unusual for island arcs is the high silica content of Japanese granites, considerably above world averages.

1

The settings of circum-Pacific plutonic terranes have some common aspects but differ in others. Most are fairly close to plate boundaries, except for the Paleozoic granites of eastern Australia and the Cretaceous granties of the Yana-Kalyma foldbelt of northeastern USSR. All the Mesozoic batholiths were emplaced in continental crust or are simply mobilized parts of continental crust. Near most batholiths the adjoining strata are strongly deformed and metamorphosed to greenschist or amphibolite facies. Regional metamorphism and deformation generally preceded batholithic emplacement. In Peru, however, the adjoining strata are not metamorphosed and are only gently deformed; the level of exposure may account for such differences. Whether any structure that we can perceive at the surface exercised control on granite emplacement cannot be definitely established, although some evidence suggests that the Coastal Batholith of Peru may be fault-controlled, and the Sierra Nevada Batholith may have been emplaced within a complex faulted synclinorium.

Most of the granitoid rocks exposed in the circum-Pacific terranes were emplaced in the Cretaceous or Early Tertiary. Precambrian ages have been determined in many places, however, and large volumes of Paleozoic plutonic rocks are found in eastern Australia. Plutonic activity appears to have been episodic rather than periodic. At nearly all times plutonism was taking place somewhere, but at no time can we confirm that all parts of a large region were simultaneously undergoing plutonism. Little, however, should be drawn from this, because our record of ancient deep-seated processes will always be incomplete.

In overall composition the granites of Malaysia and Thailand are the most silicic, and the Coast Plutonic Complex of British Columbia is the most basic. The other terranes are intermediate, but all contain a wide range of plutonic rock types.

In this volume most of the papers employ the classification of plutonic rocks recommended by the IUGS Subcommission on the Systematics of Igneous Rocks (Streckeisen, 1976). Although we could perhaps classify plutonic rocks more intelligently if their mode of origin were definitely known, it is unlikely that any classification could solve the problem of their genesis. Obviously, however, many advantages result from the wide use of a single classification. The IUGS classification has been adopted by a number of national geological surveys and has rapidly become the most widely used system of nomenclature for plutonic rocks.

Apart from nomenclature, it can be useful to distinguish among broad groups of plutonic rocks. In this matter the Australian participants in this project have made a valuable contribution with their concept of I- and S-type granitoids. Each type has a number of characteristics from which it is postulated their source rock can be determined as either igneous or sedimentary. The Japanese division of plutonic rocks into ilmenite- and magnetite-types is similar in

philosophy but not entirely equivalent. By either method, difficulties are encountered with plutonic rocks which do not clearly belong to one or the other group.

Most mineral deposits in the circum-Pacific region are associated with plutonic rocks. Tin, tungsten, molybdenum, and porphyry copper deposits are found within the granitic rocks, but most mineral deposits occur in adjacent or included rocks. As the interiors of large batholiths are essentially devoid of mineral deposits, the relation between the two may be more structural than genetic. Possibly the deposits we call primary are in another sense secondary, with the primary concentration taking place over broad regions (metallogenic provinces) deep in the crust by metasomatic processes (it being doubtful whether any other mechanism is capable of concentrating a wide variety of minerals). Tapping of the "primary" mineral concentrations could be a simple matter of connecting them with near-surface strata by hydrous fluid in deep faults commonly associated with pluton emplacement (in a solid or plastic state).

Although "solved" many times, the problem of granite genesis remains a problem, in fact, the most fundamental problem in geology. All of the papers in this volume except for the author's are based on the assumption of a magmatic origin of one type or another for the plutonic terranes. That is a ratio of 20:1 and is probably fairly representative of world geological opinion. The prevailing but not unanimous scenario of magmatic genesis derives granitic magmas by partial melting of the lower crust to account for plutonic rocks with high initial strontium ratios, and by partial melting of upper mantle material or "underplated continent" for plutonic rocks with low strontium ratios. The magma generation is related by most to subduction. Although held responsible for a variety of characteristics of plutonic terranes, subduction can logically be held responsible for the transfer of heat to the lower crust via basic magma generated near the Benioff zone.

In northeastern USSR the subduction model might apply to the Okhotsk-Chukchi coastal belt which exhibits a sodic series of Cretaceous plutonic rocks on the ocean side and a potassic series on the continental side, but not to the granites (also Cretaceous) in the Yana-Kolyma foldbelt which intersects the coastal belt at a high angle and extends several thousand kilometers inland.

In Chile and possibly in New Zealand there appears to be a close relationship between rhyolite and granite, possibly a magmatic relationship. But whether the rhyolite issued from the top or from the base of plutons has not been established, although clearly, radically different concepts are involved.

From a cosmic viewpoint granitic rocks seem to be the grey hair of medium-sized planets that inhabit water-window orbits. This, of course, assumes that planets are not born with sialic crusts. Rather, they develop sialic crusts as a secondary result of the reaction between the basaltic lithosphere and the hydrosphere. Whether by metasomatism or by fusion (or both) granitoid terrane can be produced eventually from the products of weathering, for in that process is the major differentiation of basaltic crust. In some complex way the amount and siliceousness of sialic crust represent time — not necessarily the age of the planet, but the length of residency in the water window. The relationship is complex because of variable rates of crustal churning (volcanism, orogeny, and subduction) and the gradual slowing of the basalt-hydrosphere reaction due to encroachment of sialic crust over oceanic crust. Although the subject draws one up many curious paths, such as consideration of the final product of this evolution, it would be inappropriate to pursue the matter in this less lofty overview. One aspect, however, should be raised; that is, the amount of new sialic material that has been generated since the Precambrian seems startlingly small. Much of what appears at first to be new crust is almost certainly mobilized Precambrian sialic crust, low strontium initial ratios in certain cases notwithstanding. This implies that the earth may be much older than the approximate 5-billion-year age often attributed to it, although evidently, some event of considerable importance to the solar system took place about that time.

We are in the stage of developing geologic models to account for plutonic terranes. They are characterized by a style which in decades to come will be identified with our time, probably as the "magmatic plate-tectonic style." If most of the models are unconvincing, it may be attributed perhaps to a sort of impenetrable denseness, but more probably to incomplete information which in turn forces the models to be based on apparently independent events. The models often resemble a series of overlapping silhouettes devoid of the necessary dimensions of cause and inevitable effect. They are, however, improving, even though growth in our understanding of plutonic terranes lags far behind the mushrooming database.

REFERENCES CITED

Streckeisen, A., 1976, To each plutonic rock its proper name: Earth Science Reviews, v. 12, p. 1–33.

MANUSCRIPT ACCEPTED BY THE SOCIETY JULY 12, 1982

Geological Society of America
Memoir 159
1983

Granitoid rocks of New Zealand — A brief review

A. J. Tulloch
New Zealand Geological Survey
P.O.B. 30368
Lower Hutt, New Zealand

ABSTRACT

Granitoid rocks in New Zealand occur in two areas of the South Island, separated by 450 km displacement on the Alpine Fault. Biotite and biotite-muscovite granitoids predominate, with granites, granodiorites, and tonalites occurring in subequal proportions.

In west Nelson-Westland granitoid rocks dominate the basement and intrude upper Proterozoic gneisses, and lower to middle Paleozoic sedimentary rocks which generally do not exceed greenschist facies metamorphic grade, producing andalusite, cordierite, biotite, sillimanite and wollastonite, and possibly garnet, staurolite, chloritoid-bearing contact aureoles. Two N-NNE-trending batholiths are distinguished: the eastern Separation Point Batholith of Cretaceous age is relatively sodic and commonly contains hornblende in addition to biotite. The larger Karamea Batholith is composed chiefly of Paleozoic (430-280 Ma) biotite and/or muscovite granites, granodiorites, and tonalites of the Karamea Suite. Plutons are large (some 100-200 km^2) and are commonly foliated. The Karamea Suite is discordantly intruded by small ($<$ 30 km^2), massive, high-level upper Mesozoic biotite granodiorite to muscovite granite plutons (Rahu Suite), which were locally extruded onto the surface. Minor but widespread Mo (Cu) mineralization is associated with this latter suite. Hornblende-bearing granitoid rocks are a minor constituent of the batholith; minor but widespread quartz diorites are intermediate in age between the Karamea and Rahu Suites.

Karamea Suite granitoids are relatively potassic and enriched in Fe, Mg, Ti, and Rb; have Sr$_0$ 0.709-10; and conform to S-type criteria of Chappell and White (1974); Separation Point Batholith rocks, on the other hand, are relatively sodic and enriched in Ca and Sr, with Sr$_0$ 0.704-5 (I-type). Rahu Suite plutons have Na/K approximately equal to 1 with Sr$_0$ 0.706-7 and are ambiguous in terms of I/S criteria. Reconnaissance dating indicates that Separation Point and Rahu granitoids are approximately coeval and are possibly related to the same plutonic event; the Rahu magmas possibly contain a greater contribution from the sub-Karamea Suite continental basement.

In Fiordland/Stewart Island, correlatives of west Nelson granitoids border an amphibolite-granulite facies metamorphic complex. In eastern Fiordland, sodic granites are correlated with the Separation Point Batholith of west Nelson; in southwest Fiordland correlatives of the Karamea Batholith occur in an area of relatively low metamorphic grade similar to that of the Paleozoic host rocks of west Nelson. In central and southern Stewart Island relatively sodic granitoids predominate. Less abundant potassic granites are associated with Sn-mineralization in metamorphic rocks similar to those of the Fiordland complex. Syntectonic granitoids also occur within central Fiordland.

Granitoid rocks are also associated with several mafic complexes that occur along the junction between the granitoid-bearing Tasman metamorphic belt in the west, and the Wakatipu belt to the east, which lacks granitoid rocks.

Samples from outlying islands and drillholes show that granitoid rocks are extensively distributed throughout the continental shelf to the south and north of the South Island of New Zealand, while a Cretaceous rhyolite has been dredged from the Lord Howe Rise, some 600 km west-northwest of west Nelson.

Granitoid plutonism is probably associated with Quaternary voluminous rhyolitic volcanism in the central North Island of New Zealand, related to the Tonga-Kermadec-New Zealand island arc, and minor quartz diorite-granodiorite plutons are associated with copper mineralization along the east coast of the North Island from the Coromandel Peninsula northwards.

INTRODUCTION

The general distribution of granitoid rocks in New Zealand was summarized from early mapping on 1:250,000-scale regional maps published by the New Zealand Geological Survey in the 1960s. Early work on these rocks was reviewed by Reed (1958), Grindley (1978, p. 93-99 and p. 108-110), and Nathan (1978c, p. 75-79). However, early geologists did not attempt to map or subdivide plutonic terranes, at least partly because of heavy forest cover and inaccessibility. More recently detailed mapping has been undertaken in local areas by Nathan (1975, 1976, 1978a) and Tulloch (1979a). It must be emphasized, however, that this paper is only a reconnaissance survey of generally scanty data.

The nomenclature used herein follows that of Streckeisen (1976).

DISTRIBUTION

With the exception of granitoid plutonism presumed to be associated (Ewart and Cole, 1967) with the voluminous acid lavas of the presently active Taupo Volcanic Zone (Ewart and others, 1977) and minor granitoid rocks in the Coromandel area (Skinner, 1976), granitoid rocks in New Zealand are restricted to two areas of the west coast of the South Island, and Stewart Island (Figs. 1 and 2). However, the presence of granitoid rocks on the Bounty, Snares, and Auckland Islands, and in offshore drillholes, demonstrates the plutonic nature of much of the continental shelf to the west (Wodzicki, 1974) and south of New Zealand. Cretaceous rhyolites occur in the central South Island (Wood, 1974) and on the Lord Howe Rise northwest of New Zealand (McDougall and van der Lingen, 1974). The two areas in the South Island are separated by 450-km displacement on the Alpine Fault (Fig. 2): (1) West Nelson-Westland, which includes approximately 7000 km² of granitoid rocks; (2) Fiordland and Stewart Island, each of which includes approximately 1000 km² of granitoid rocks.

This study will concentrate on the west Nelson-Westland region for which more information is available.

REGIONAL SETTING

The west Nelson-Westland and Fiordland-Stewart Island regions together form the high-temperature/low-pressure Tasman metamorphic belt of Landis and Coombs (1967), which they contrasted with the low-temperature/high-pressure Wakatipu belt to the east (cf. Shelley, 1975). The structural grain in both regions trends north to north-northeast but deeper structural levels and consequent higher metamorphic grades are generally exposed in Fiordland.

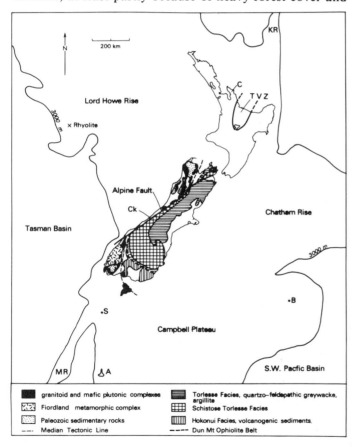

Figure 1. Part of the New Zealand continental shelf, showing the location of Paleozoic-Mesozoic granitoid rocks in the South Island; offshore islands on which granitoid rocks occur: Bounty (B), Auckland (A), and Snares (S); the currently active Taupo Volcanic Zone (TVZ), and the Coromandel Peninsula (C). Other North Island geology omitted.

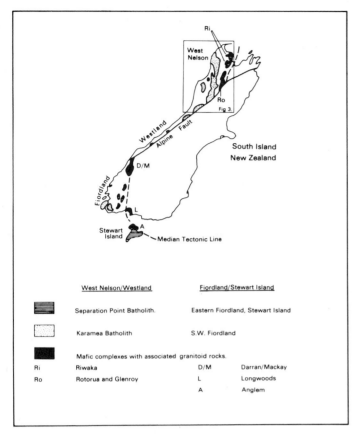

Figure 2. South Island, New Zealand, showing the distribution of the Karamea and Separation Point Batholiths in West Nelson and Westland, their suggested correlatives in Fiordland/Stewart Island, mafic complexes with which minor granitoid rocks are associated.

Details of the regional geology of these areas may be found in Suggate and others (1978) and Cooper (1979) — only a brief summary is given here.

West Nelson-Westland

In west Nelson-Westland, granitoid rocks intrude late Precambrian gneisses and lower to middle Paleozoic sediments (Fig. 3). The metamorphic grade of the latter generally does not exceed greenschist facies. Intruded rocks are generally psammitic-pelitic except in the east where marble host rocks also occur (Cooper, 1975; Grindley, 1971, 1980). Both sediments and major faults strike NNE (Laird, 1968; Grindley, 1978, Fig. 2.13), and the major batholith segments are subparallel with this trend. Rangitata (Cretaceous) and Kaikoura (Late Tertiary-Recent) orogenies have deformed and obscured many of the original plutonic features. Although fault contacts are common, contact aureoles up to 2 km wide (but generally less than 1 km) occur in a number of areas. Assemblages in Karamea Batholith (see Granitoid Phases below) aureoles include biotite, muscovite, andalusite, cordierite, and rare sillimanite—the

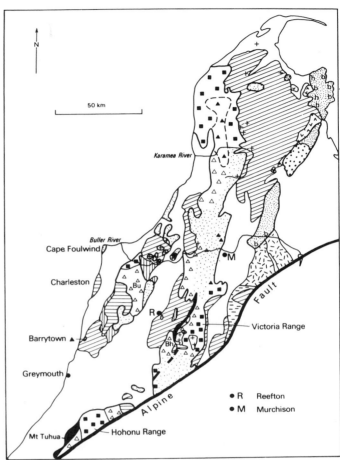

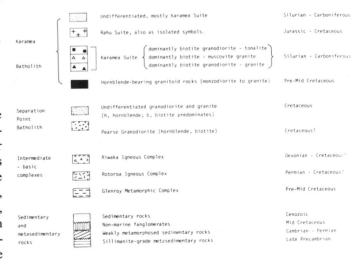

Figure 3. Sketch-map of the pre-Tertiary geology of West Nelson-Westland. Sources: Grindley, Fig. 2.13 in Suggate and others 1978, 1980; Nathan 1978a; Tulloch 1979a; unpublished data (unpublished mapping of the Glenroy Metamorphic Complex with J. K. Campbell), and P. J. Oliver, personal communication. Bu and Bh are Buckland and Bald Hill Granites, respectively.

presence of cordierite and absence of garnet indicate a relatively low pressure emplacement of the magmas. Contact metamorphism adjacent to the Separation Point Batholith has produced grossularite, diopside, sphene, idocrase, hornblende, and wollastonite-bearing skarns. Biotite-muscovite-almandine-staurolite-clinozoisite-chloritoid assemblages in pelitic rocks of the Pikikiruna and Onekaka Schists (Ghent, 1968; Grindley, 1980) may also have resulted from contact metamorphism but may in part have developed during earlier regional metamorphism (Shelley, 1975). Many aureole rocks are weakly schistose indicating a degree of forcible emplacement of some plutons. In the Victoria Range a dominant biotite-muscovite schistosity is mimetic upon bedding and is parallel to the batholith contact (Tulloch, 1979a), whereas the Olympus Granite pluton in the northern Karamea Batholith (Brathwaite, 1968) deflected the strike of the host rock and shows more convincing evidence for forcible emplacement.

Areas of late Precambrian pelitic paragneiss and granitic orthogneiss (Charleston Metamorphic Group) are mapped separately in the western part of this region (Bowen, 1964; Nathan, 1975, 1978a; Laird, 1967; Hume, 1977), but in many other areas similar metamorphic rocks were previously mapped as granitoid rocks by early geologists (e.g., Mt. Bonar and Granite Hill massifs). Their metamorphic grade is upper amphibolite facies, and they are the oldest known rocks in New Zealand (700 Ma, Adams, 1975). Granulitic rocks with high-pressure assemblages comparable with those commonly observed in Fiordland (see the following section) occur in one small area adjacent to the Alpine Fault in southeast Nelson.

Tertiary and Quaternary sedimentary rocks obscure much of the basement in west and southwest Nelson, and Westland.

Fiordland-Stewart Island

In Fiordland the metamorphic grade of metasediments and metaigneous rocks generally ranges from upper amphibolite in the eastern and central regions to granulite facies in western Fiordland (Oliver and Coggan, 1979; Oliver, 1980; Blattner, 1978; Gibson, 1979). High-grade metamorphism and a strong positive gravity anomaly in western Fiordland indicate the proximity of the mantle to the surface. Syntectonic granitoids occur in central and western Fiordland, but granitoid rocks of batholithic proportions are restricted to the east and southwest margins of the region. In the southwest granitoids intrude lower Paleozoic metasediments similar in composition and metamorphic grade to the Paleozoic sequence of west Nelson (Grindley, 1978, p. 100-105; Ward, 1980).

On Stewart Island isolated roof pendants of schist (Williams, 1934) are lithologically similar to rocks of the Fiordland complex; and the same can probably be said of

schists from the Snares Island farther south (Watters and Fleming, 1975).

GRANITOID PHASES OF WEST NELSON-WESTLAND

The reconnaissance Rb-Sr studies of Aronson (1968) suggested at least two plutonic events in the "foreland province" of New Zealand. Subsequent mapping and dating, albeit limited, revealed, not surprisingly, that the elongate (5-30 km wide and up to 200 km long) granitoid bodies are individually complex and were produced by at least four significant plutonic events. Reed (1958) subdivided the granitoid rocks geographically into three approximately north-trending belts. While Reed's eastern belt is petrographically, chemically, and geographically distinct from the other granitoid rocks in the region, the western and central belts, and those granitoid rocks adjacent to the Alpine Fault in Westland, all contain a similarly diverse range of lithologies and ages and are not readily geographically separable. Thus all granitoid rocks in west Nelson and Westland, west and south of the eastern belt, are now referred to as the Karamea Batholith (Nathan, 1978a; cf. Grindley, 1961, 1971). It should be noted, however, that the Karamea Batholith as so defined includes plutonic rocks of at least two, and possibly four, distinct ages. Thus the Tuhua Intrusive Group (Bowen, 1964) comprises an eastern Separation Point Batholith and a larger, western Karamea Batholith. The term Tuhua Intrusive Group, or Tuhua Granite, has been loosely used in the past to cover all granitoid and metamorphic rocks of the region; it is recommended here that it be abandoned. A plutonic map of the west Nelson-Westland region (Fig. 3) can be regarded only as a progress report. Off-shore drillhole data (Wodzicki, 1974) show that the major lithologies of west Nelson extend northwards along strike, below the Cenozoic cover to the west of the North Island.

Karamea Batholith

Biotite and biotite-muscovite granitoid rocks greatly predominate. Granite (sensu stricto), granodiorite, and tonalite are estimated to occur in subequal proportions, in contrast to some other segments of the circum-Pacific plutonic belt where tonalites predominate (Cobbing and Pitcher, 1972). Numerous faults and shear zones are parallel with, and commonly form boundaries of, the batholith. These structural features range in age from at least pre-Cretaceous to presently active faults. The oldest structures include the Fraser Fault in Westland (Reed, 1964; Young, 1968) and a major shear zone to the west of the Victoria Range (Tulloch, 1979a).

Two major suites (Karamea and Rahu) comprise most of the batholith.

Karamea Suite. The term Karamea Suite is proposed here for the calc-alkali to alkali, mica- and quartz-rich granitoids (Fig. 4) of mid-Paleozoic age which form the bulk of the Karamea Batholith — the Karamea "Granite" of Grindley (1961, 1971) and the Tarn Summit Suite of Tulloch (1979a). Several characteristic but often gradational varieties occur widely (Fig. 3) and tend to form relatively large (commonly exceeding 150 km²) plutons.

Biotite granodiorites and tonalites — coarse-grained, inequigranular to porphyritic, and commonly foliated —are widespread. These rocks subconcordantly intrude late-Precambrian paragneiss in the Victoria Range, but the foliation (marked by clots of decussate biotite) is strongly discordant with the Lower Paleozoic-sediment contacts. Biotite-rich xenoliths are petrographically and Sr-isotopically similar to the paragneiss. In thin section the texture is relatively simple with similarities to a metamorphic, granular-polygonal, texture.

Biotite-muscovite granites are also abundant, particularly in the Buller River area. They also may show a foliation, defined either by an alignment of K-feldspar megacrysts, as in the Windy Point, Dunphy, and Foulwind Granites (Nathan, 1976, 1978a) or alignment of micas, as in the Buckland Granite (Nathan, 1978a). Tentatively included here are the Buckland and Bald Hill biotite-muscovite granites (Fig. 3) that structurally and petrographically assemble the Karamea Suite, but which have geochemical characteristics more akin to those of the Rahu Suite.

Fine- to medium-grained, biotite-rich granodiorites and granites, some weakly porphyritic, occur on the west coast at Barrytown and Meybelle Bay (20 km north of Barrytown), in the Matiri Valley northwest of Murchison, and in a large belt centered on a northern tributary of the Karamea River in northwest Nelson (Grindley, 1978, p. 95). In the small, zoned Barrytown pluton, red-brown biotite is more abundant in a marginal zone (Tulloch, 1973).

Within the Karamea Suite potassium feldspar is generally maximum microcline, although twinning may often be patchily developed within a single grain. Plagioclase ranges in composition from andesine to albite and is characterized by a general lack of complex zoning or twinning. Quartz occasionally exhibits a pale blueish tinge in hand specimens of the more mafic varieties. Biotite is a deep orange- or red-brown, or a dark brown in some of the more evolved rocks. Biotite has low Fe^{3+}/Fe^{2+} and high TiO_2 (Table 2). In Karamea Suite rocks from the Victoria Range Fe/Fe+Mg ranges from 46 to 70 and ΣAl falls within the narrow range of 2.95-3.13, although with ΣAl at 3.5 the Bald Hill Granite is once again clearly anomalous within the Karamea Suite. Hornblende is a rare phase in some more mafic variants, while almandine garnet is common in some more leucocratic phases. A single cordierite xenocryst pseudomorph has been observed in one rock. Apatite, ilmenite, and one or more zircon populations are abundant as accessory phases, and

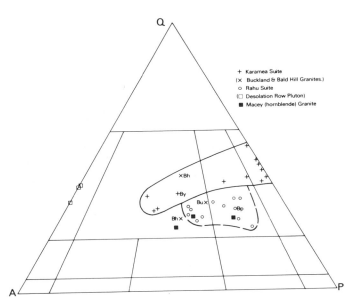

Figure 4. Modal QAP plot of Karamea and Rahu Suite rocks from the Victoria Range segment of the Karamea Batholith. Also plotted are the Barrytown Granite (By), average holocrystalline Berlins Porphyry (Bp), Buckland Granite (Bu), and Bald Hill Granite (Bh). The relatively quartz-rich compositions of Karamea Suite rocks can be at least partially attributed to the biotite-rich nature of this suite. Classification fields are from Streckeisen (1976).

monazite occurs in inferred correlatives of the Karamea Suite in Southwest Fiordland, while sphene, epidote, and allanite are much less common (Tulloch, 1979a; Smale and Nathan, 1980).

Rahu Suite. High-level, massive plutons of medium-grained calc-alkali granodiorite and granite that discordantly cut the Karamea Suite in the Victoria Range were named the Rahu Suite by Tulloch (1979a), and the term is extended here to cover all such plutons in the Karamea Batholith. Only in the two areas where such plutons are known to occur in any significant volume — the Victoria Range and the lower Buller Valley — do stratigraphic and Rb-Sr whole-rock studies unambiguously demonstrate a late Mesozoic age. A number of other structurally and petrographically similar plutons which are included in this suite have been dated by K-Ar only.

In the Victoria Range two relatively large plutons of about 30 km² each and two smaller plutons discordantly intrude paragneiss and Karamea Suite basement. Contacts are steep, commonly chilled, and generally show a similar orientation to the regional NNE and NW fracture pattern, suggesting emplacement by uplift of fault-bounded blocks. A circular topographic structure, about 15 km across, appears to coincide with this locus of Rahu plutonism; it may have resulted from regional doming and concentric fracturing during such plutonism (Tulloch, 1979a, 1981a; Eggers, 1979). The plutons are homogeneous and largely equigranular and range from biotite granodiorite to muscovite-biotite granite, except for a single pluton of garnetiferous muscovite alkali-feldspar granite (Desolation Row pluton). Xenoliths are not abundant, but comprise two varieties: gneissic xenoliths derived from subjacent paragneiss country rock and more abundant, but still volumetrically insignificant, fine-grained ovoid xenoliths, mostly 10-30 cm across, which are petrographically and chemically mafic variants of their host granites. The latter are interpreted as early crystallization products from the host magma. Modal quartz is less abundant than in the Karamea Suite (Fig. 4). Biotite is more evenly distributed throughout the rock compared to the Karamea Suite, although microscopic textures are often more irregular due to late stage activity of K-feldspar and quartz. Chilled margins contain phenocrysts of quartz, K-feldspar, plagioclase, and biotite in a microgranitic groundmass. Perthitic microcline commonly exhibits core zones, and plagioclase (calcic oligoclase to albite) ehibits complex twinning, with both normal and oscillatory zoning. Synneusis aggregates of truncated plagioclase crystals are common, and sodic rims are generally corroded by microcline. Biotite is the sole mafic phase, with the exception of muscovite in more evolved members of the suite, and is green-brown due to relatively high Fe^{3+}/Fe^{2+} and Fe^{3+}/Ti (Table 2). Compared to Karamea Suite biotites, those of the Rahu Suite in the Victoria Range have a more restricted Fe/Fe+Mg range (50-60) but a wider range of ΣAl (2.8-3.13). Manganese is relatively enriched in Rahu Suite biotites. Victoria Range biotite compositions plotted in the system $Fe^{3+}:Fe^{2+}:Mg$ (Wones and Eugster, 1965) indicate oxygen fugacities close to that of the hematite-magnetite buffer for the Rahu Suite (fO_2 approximately 10^{15} bars at 650°C and 3kb PH_2O), compared to FMQ-NNO conditions for Karamea Suite biotites. High fO_2 is also indicated by the characteristic accessory suite of magnetite, manganiferous ilmenite, epidote, allanite, and sphene, in addition to apatite and zircon. However, accessory minerals in the Desolation Row pluton include almandine-spessartine garnet, columbite, gahnite, monazite, and rare magnetite (Tulloch, 1981b).

In the lower Buller Valley a number of small plutons (1-6 km across), all included in the Berlins Porphyry (Bowen, 1964; Nathan, 1978a), intrude Greenland Group sedimentary rocks to relatively high levels (to within 2 km of the surface). Margins of the granodiorite plutons are composed of black glassy dacite, which grades, with increase in grain size but without compositional change, into granodiorite in the cores of the larger plutons (Nathan, 1974). Acid tuffaceous material in lake beds underlying Albian fanglomerates indicates local volcanism. Biotite and plagioclase in Berlins Porphyry plutons show similar textural and compositional features to those phases in the Victoria Range area, but K-feldspar is absent from chilled margins. This is possibly due to the relatively low-pressure crystallization and corre-

lates with a slightly higher proportion of modal biotite compared to the Rahu Suite plutons in the Victoria Range. Biotite and quartz phenocrysts, euhedral in the chilled margins, exhibit strong resorption in the cores of the pluton.

The large (260 km²) Buckland Granite pluton immediately southwest of the Berlins Porphyry has been indirectly dated as Cretaceous (Adams and Nathan, 1978) but is included in the Karamea Suite here. The two known concentrations of Rahu plutonism, in the Victoria Range and lower Buller Valley, and the Bald Hill and Buckland Granites which lie southwest of each of these areas, respectively (Fig. 3), lie along the NNW structural trend which, together with a conjugate NNE trend, appears to have controlled the emplacement of Rahu Suite plutons in the Victoria Range (Tulloch, 1979a).

A number of other small, high-level plutons associated with molybdenum prospects occur within and adjacent to the Karamea Batholith, and many of these have been dated (K-Ar, mica) as Cretaceous (Eggers and Adams, 1979, and references therein). Although they are included here in the Rahu Suite on structural and geographic grounds, their generally more sodic nature relates them more to the Separation Point Batholith.

Hornblende-bearing granitoid rocks. Granitoid rocks in which hornblende is a major mafic phase are relatively minor constituents of the Karamea Batholith. Their ages are generally unknown except that they are probably pre-Late Cretaceous.

The Macey Granite forms a 1-2 km wide, fault-bounded sliver, associated with a major high-angle reverse shear zone, parallel to the western margin of the Victoria Range (Fig. 3). It is characterized by abundant, large, stubby, euhedral, pink K-feldspar megacrysts, a general predominance of green hornblende (ferro-edenite) over biotite, and lenticular dioritic xenoliths. Epidote, magnetite, sphene, and apatite are abundant accessory minerals.

The small MacIntosh monzodiorite pluton (Nathan, 1978a) in the lower Buller Valley varies considerably in composition, from monzodiorite to gabbro, but has a common mineral assemblage of calcic plagioclase, hornblende, and magnetite (Table 1).

An equigranular, uniform body of quartz monzonite to quartz monzodiorite outcrops over approximately 40-50 km² in the vicinity of Mt. Tuhua, close to the Alpine and Fraser Faults in North Westland. Plagioclase is strongly zoned, biotite is dark brown, and dark green hornblende (after pyroxene in some instances) is abundant. Sphene and magnetite are abundant accessories, and microcline twinning is weak to absent in K-feldspar of more mafic varieties.

In South Westland two small, isolated outliers of monzodiorite to quartz monzodiorite occur at Canavans Knob (Fig. 1) and in the nearby Omoeroa Range. Plagioclase is oscillatorily zoned andesine to oligoclase and the mafic minerals include augite, hypersthene, orange-brown biotite,

TABLE 1. CHEMICAL AND MODAL ANALYSES OF REPRESENTATIVE GRANITOID ROCKS FROM WEST NELSON AND WESTLAND, NEW ZEALAND

| | Karamea Batholith | | | | | | | | | | | | | | Separation Point Batholith | | | | |
| | Karamea Suite | | | | | | Rahu Suite | | | | | | | | | | | | |
	1	2	3	4	5	6	7	8	9	10	11	12	13	14	15	16	17	18	19
SiO_2	62.19	66.29	71.81	72.0	73.38	74.89	69.4	69.7	70.05	73.23	75.22	67.37	52.29	60.30	65.40	66.61	67.63	71.30	71.66
Al_2O_3	15.29	14.95	13.98	13.98	13.88	14.13	14.8	15.9	15.16	14.69	14.38	15.22	17.02	16.48	16.57	17.70	15.99	15.79	15.60
TiO_2	1.21	0.85	0.39	0.24	0.16	0.24	0.47	0.41	0.31	0.10	0.05	0.49	1.18	0.64	0.47	0.33	0.37	0.21	0.19
Fe_2O_3	1.83	0.84	0.10	0.53	} 1.26	0.31	0.77	1.04	1.13	0.16	0.33	1.41	2.09	1.34	0.18	1.16	1.42	0.47	0.85
FeO	5.25	4.22	2.57	2.12		1.10	2.10	1.79	1.69	0.75	0.39	2.18	7.02	4.77	2.23	0.95	1.20	0.71	0.59
MnO	0.11	0.09	0.07	0.05	0.00	0.01	-	0.04	0.13	0.08	0.45	0.13	0.15	-	0.05	0.04	0.26	0.02	0.04
MgO	2.32	1.80	0.77	0.48	0.28	0.59	1.40	1.20	0.80	0.20	0.12	1.14	4.97	1.80	0.91	0.64	0.92	0.40	0.35
CaO	3.24	3.05	0.85	1.27	0.92	0.85	2.73	1.90	3.23	1.28	0.10	4.34	7.69	6.08	3.09	3.05	2.67	1.89	1.83
Na_2O	2.85	2.90	2.90	2.99	2.47	2.82	3.73	3.78	3.61	4.12	4.43	3.79	3.28	3.88	5.37	6.33	4.89	4.90	5.27
K_2O	3.51	2.86	4.91	4.90	6.21	4.75	3.31	3.55	3.06	4.64	3.65	2.44	1.59	2.41	3.36	1.95	3.41	3.76	2.90
P_2O_5	0.36	0.25	0.14	0.22	0.08	0.19	0.13	0.15	0.19	0.02	0.01	0.22	0.36	0.38	0.18	0.17	0.21	0.06	0.04
H_2O	1.01	1.08	1.46	0.41	0.61	0.90	1.36	0.65	0.39	0.20	0.54	0.68	1.38	1.39	0.70	0.70	0.50	0.71	0.05
	99.17	99.18	99.95	99.19	99.25	100.78	100.2	100.1	99.75	99.47	99.67	99.41	99.02	99.47	98.51	99.63	99.47	100.22	99.37
Rb	194	174	279	-	199	299	-	-	115	115	488	79	65	79	102	-	115	-	77
Sr	177	170	75	-	9	62	-	-	622	258	<10	1023	459	1300	1087	-	864	-	873
Ba	414	504	276	-	410	272	-	-	859	652	13	671	371	838	-	-	-	-	-
Zr	347	238	171	-	2	-	-	-	177	75	43	140	118		-	-	-	-	-
Quartz	29.8	34.4	33.8	-	33.9	-	27.0	8.9	26.9	25.8	36.9	26.0	5.0	-					
K-spar	tr	5.5	32.7	-	39.6	-	16.0	0.0	14.2	26.4	19.1	29.1	0.0	-					
Plagioclase	56.1	41.9	24.8	-	21.6	-	47.4	34.8	48.0	41.8	36.3	37.6	48.7	-			Not available		
Hornblende	0.0	0.0	0.0	-	0.0	-	0.0	0.0	0.0	0.0	0.0	3.5	32.0	-					
Biotite	12.0	15.7	8.9	-	4.5	-	9.2	9.7	7.8	3.6	0.0	0.7	13.2	-					
Muscovite	tr	0.6	0.0	-	0.4	-	0.0	0.0	0.8	1.6	6.9	-	-	-					

Analysis 12 Macey Granite, 13 Tobin Quartz diorite, 14 McIntosh Monzodorite.
Sources: Challis (1971), Harrison and McDougall (1980), Nathan (1974, 1976, and personal communication, Tulloch (1979).

and green hornblende. These rocks are unusual not only in their composition but also in that their greywacke host rocks are chemically and mineralogically (especially in the abundant presence of K-feldspar) atypical of the monotonous lower Paleozoic Greenland Group of Westland. The only known similar igneous rocks occur in the Riwaka Complex of Nelson and the Darran Complex of Fiordland, which are now separated more than 450 km by the Alpine Fault (Blattner, 1980), and at Gulches Head, southwest Fiordland.

Tobin Quartz Diorite Suite. In the Victoria Range a number of small (generally less than 1 km²) bodies of tholeiitic quartz diorite (Table 1) intrude Karamea Suite granitoid rocks but not the younger Rahu Suite, and were named the Tobin Epidiorite Suite by Tulloch (1979a). They are relatively homogeneous massive bodies, composed of an equigranular to granular-polygonal (metamorphic?) assemblage of intermediate to calcic plagioclase, green magnesio-hornblende, and minor orange-brown biotite (Table 2) and quartz. Apatite, sphene, and ilmenite are abundant accessory minerals. K-Ar dates on the hornblende are Triassic.

Reconnaisance studies suggest similar rocks are widespread but minor components of the Karamea Batholith.

Pre-Silurian granitoid rocks. The Okari Granite-gneiss associated with the Charleston Metamorphic Group is assumed to be of late Precambrian age (Nathan, 1978a). Granitoid cobbles are present in the Cambrian Lockett conglomerate (Grindley, 1980); a lack of metasedimentary cobbles suggests they were *not* derived from the Okari Granite-gneiss. The source rocks for the Greenland Group, from which 1170-1370 Ma zircons have been recovered (Aronson, 1968), are presumed to have been an acid plutonic terrane.

Dike rocks associated with the Karamea Batholith. Lamprophyre dikes occur throughout the Karamea Batholith often forming swarms, the most conspicuous of which is a lithologically diverse swarm centered on the Hohonu Range in North Westland (Wellman and Cooper, 1971; Adams and Nathan, 1978; Hunt and Nathan, 1976). Camptonitic varieties of Late Cretaceous age predominate, and in some areas have produced minor fenitization of their granitoid host

TABLE 2. ANALYSES OF BIOTITE FROM THE KARAMEA BATHOLITH, WEST NELSON-WESTLAND, NEW ZEALAND

	Karamea Suite			Rahu Suite		Bald Hill Granite	Macey Granite	Tobin Quartz Diorite
	1	2	3	4	5	6	7	8
SiO_2	35.26	34.91	34.11	35.35	36.37	35.08	37.32	37.48
Al_2O_3	16.76	16.90	18.25	15.65	14.62	19.15	17.29	16.45
TiO_2	3.24	2.85	3.42	2.97	4.24	2.53	2.60	1.88
Fe_2O_3	1.07	2.78	-	5.47	3.28	3.51	-	1.31
FeO	20.13	22.80	22.72*	16.10	14.72	20.34	16.31*	15.95
MnO	0.21	0.12	0.39	0.93	0.05	0.52	0.30	0.15
MgO	8.80	6.79	6.46	9.21	13.36	5.77	10.94	12.67
CaO	0.03	0.00	0.02	0.01	0.02	0.01	0.02	0.05
Na_2O	0.04	0.05	0.09	0.18	0.19	0.13	0.04	0.20
K_2O	9.77	9.29	9.38	9.28	9.01	8.83	9.22	8.93
	95.31	96.49	94.84	95.15	95.86	95.87	94.04	95.07
Si	5.44	5.39	5.33	5.44	5.44	5.37	5.65	5.62
Al^{iv}	2.56	2.61	2.67	2.56	2.56	2.63	2.35	2.38
Al^{vi}	0.48	0.47	0.69	0.28	0.02	0.83	0.74	0.53
Ti	0.38	0.33	0.40	0.34	0.48	0.29	0.30	0.21
Fe^{3+}	0.12	0.32	-	0.63	0.37	0.41	-	0.15
Fe^{2+}	2.59	2.95	2.97	2.07	1.84	2.61	2.07	2.00
Mn	0.03	0.02	0.05	0.09	0.01	0.07	0.04	0.02
Mg	2.02	1.56	1.51	2.11	3.00	1.32	2.47	2.83
Ca	0.01	0.00	0.00	0.00	0.01	0.00	0.00	0.01
Na	0.01	0.02	0.03	0.06	0.06	0.04	0.01	0.06
K	1.93	1.83	1.87	1.83	1.72	1.73	1.78	1.71
$\frac{100\Sigma Fe}{\Sigma Fe+Mg}$	57.3	67.7	66.3	56.1	42.4	69.6	45.6	43.2

All analyses are from Tulloch 1979, except 5 (Nathan, 1974) and 3 (Tulloch, unpublished).
* Total Fe as FeO, atomic proportions on basis of (O,OH) = 22.

rock. In the Victoria Range appinitic and spessartitic varieties appear to be slightly older than the camptonites (Tulloch, 1979a). Alkali basalts and to a lesser extent trachyandesites and trachytes are also widespread (Nathan, 1978b; Suggate, 1957; Tulloch, 1979a).

Composite dykes (emulsion-textured basaltic andesite/trondhjemite) occur within Karamea Suite rocks in the Victoria Range. The acid phase is considered to represent melted granitoid wall rock, subsequently modified by reaction with the basic host.

Separation Point Batholith

The Separation Point Batholith is composed of three elongate segments which average 10 km in width and total 120 km in length (Fig. 3). Possible correlatives intrude high-grade metamorphic rocks 40 km south of the mapped batholith and occur in offshore wells 90 km to the north of the South Island (Wodzicki, 1974). The bulk of the batholith is relatively homogeneous, composed of medium-grained leucocratic (hornblende) biotite granite. In thin section, textures tend to be more equigranular than those of the Rahu Suite, and biotite commonly has a yellowish green-brown pleochroism.

The northern segment of the batholith is zoned from equigranular biotite-hornblende granodiorite and quartz diorite in the west (Grindley, 1971; Reid, 1972) to porphyritic (hornblende) biotite granite in the east (Henderson, 1950). Accessory minerals include magnetite, epidote, apatite, and conspicuously abundant sphene. A small pluton of aplitic garnetiferous muscovite granite occurs within the southern end of this segment, and a garnetiferous biotite-muscovite pluton (Onahau Granite, Grindley, 1971; Wodzicki, 1972) lies 15 km west of this segment. Satellite stocks of hornblende-biotite granodiorite (Canaan Granodiorite, Grindley, 1971, 1980) to the west of the bounding shear zone are both older and younger than the main phase of the batholith (Harrison and McDougall, 1980).

The remainder of the batholith to the south comprises a uniform (hornblende) biotite granodiorite/granite except for the Pearse Granodiorite (Grindley, 1980) which forms the southern two-thirds of the central segment. This body is composed of plagioclase, hornblende (plus or minus pyroxene), biotite, minor quartz and K-feldspar, and abundant accessory epidote and sphene. It is more mafic and has a distinctive, more granular-polygonal texture than the rest of the batholith.

GEOCHEMISTRY: WEST NELSON-WESTLAND

In the following discussion the data for the Rahu and Karamea Suites are largely taken from Tulloch (1979a) due to a dearth of published geochemical data. Whole rock analyses of material from the Karamea and Separation Point Batholiths are presented in Table 1.

Reed (1958) chemically distinguished the Separation Point Batholith from what is referred to here as the Karamea Batholith on the basis of Na/K ratios. Taking Ca into account (Fig. 5) it is also possible to distinguish between the Rahu and Karamea Suites in the Victoria Range segment of the Karamea Batholith, at least in the case of the more evolved members of the suites (but note that the Bald Hill Granite provides at least one exception). The Karamea Suite is relatively potassic and the Separation Point Batholith relatively sodic, while the Rahu Suite has intermediate K/Na ratios of approximately 1:1. Thus the Mesozoic granites increase in K/Na away from the Pacific Basin, from the Separation Point Batholith to the Rahu Suite. Less evolved rocks of the Karamea and Rahu Suites can be distinguished on major and trace element plots against a differentiation index (Fig. 6). Rocks of the Rahu Suite are relatively rich in Ca, Mn, Sr, and Ba, while Karamea Suite rocks are relatively rich in total iron, Mg, Ti, P, and Rb. Rahu Suite rocks have higher Fe_2O_3/FeO and lower Ca/Sr (Fig. 7) than Karamea Suite rocks. Values of Al/(Na+K+Ca) range from 1.0 to 1.45 for *both* the Karamea and Rahu Suites; normative corundum is less than 1.2 for the Rahu Suite (excluding Desolation Row pluton) and greater than 1.38 for the Karamea Suite.

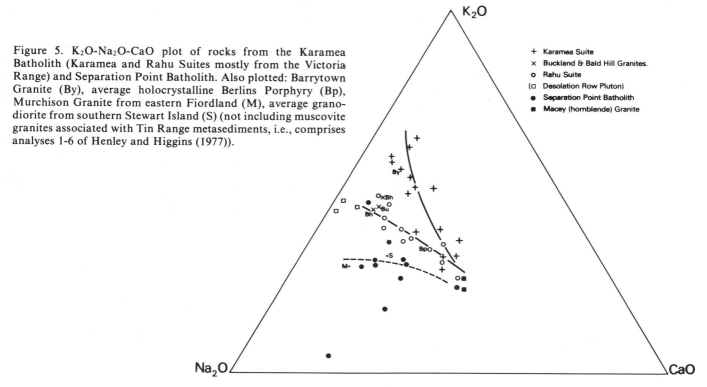

Figure 5. K₂O-Na₂O-CaO plot of rocks from the Karamea Batholith (Karamea and Rahu Suites mostly from the Victoria Range) and Separation Point Batholith. Also plotted: Barrytown Granite (By), average holocrystalline Berlins Porphyry (Bp), Murchison Granite from eastern Fiordland (M), average grano-diorite from southern Stewart Island (S) (not including muscovite granites associated with Tin Range metasediments, i.e., comprises analyses 1-6 of Henley and Higgins (1977)).

In these characteristics Separation Point rocks are similar to those of the Rahu Suite. Figure 6 shows ΣFe and CaO values are similar in Rahu and Separation Point rocks, while MgO and TiO₂ values are both higher in Separation Point rocks and intermediate in the Rahu and Karamea Suites. Rocks of the Separation Point Batholith, however, appear to be distinctly enriched in Sr relative to the Rahu Suite, although data are scanty. Preliminary isotope data (Table 3) suggest that each of the three groups have characteristic initial stontium isotope ratios and that the Rahu Suite has slightly lower $\delta^{18}O$ than the Karamea Suite.

In the "granite" system (Q-Ab-Or) rocks of the Rahu Suite from the Victoria Range plot close to the ternary minimum at about 2 kb, while Karamea Suite rocks form a more widely spread group about the 4 kb minimum. Relationships in the "feldspar" system (Ab-Or-An) are compatible with a single Rahu lineage following the computed fractionation trends of Presnall and Bateman (1973), possibly resulting from plagioclase fractionation. Mixing models, however, suggest that in general variation in the Rahu Suite results from progressive partial melting rather than simple fractional crystallization. Karamea Suite rocks plot widely on both sides of the thermal trough, negating crystal fractionation models.

Together with the field and petrographic data presented previously, these data suggest that the Separation Point Batholith (which occurs on the Pacific Ocean side of the west Nelson-Westland plutonic terrane) is dominantly I-type, whereas the Karamea Suite (continental side) is dominantly S-type (Chappell and White, 1974; O'Neil and Chappell, 1978). Classification of the Rahu Suite is not quite so clear, however. While the Rahu Suite is geographically associated with the Karamea Suite in the Karamea Batholith, in age and geochemistry it is more akin to the Separation Point Batholith. The differences that are observed between the Rahu Suite and the Separation Point Batholith, such as Na₂O/K₂O ratios, the presence or absence of hornblende, and the abundance of sphene, are consistent with ocean-continent variations observed in other circum-Pacific plutonic terranes (e.g., White and others, 1947; Kistler, 1974). Thus the Rahu Suite may be a modification of the approximately coeval Separation Point magmas and may possibly involve a greater component of the sub-Karamea Suite continental basement. In composition, structure and tectonic setting (post-closure, late orogenic uplift regime) the Rahu Suite has much in common with the Caledonian granites of the United Kingdom and Ireland, and thus with the I-Caledonian grouping of W.S. Pitcher (personal communication). The felsic Desolation Row pluton of the Rahu Suite, the peralkaline granite at Lake Monowai in eastern Fiordland, and the peralkaline rhyolites of the North Island (Smith and others, 1977) may correspond to the A-type granites of White and Chappell (this volume).

GEOPHYSICS

Few geophysical studies directly concerned with granitoid rocks have been undertaken, although 1:250,000-scale

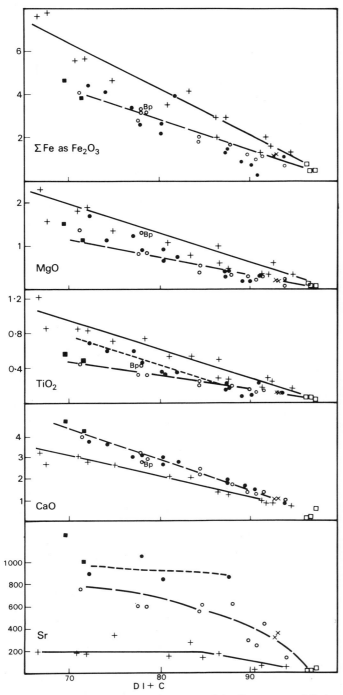

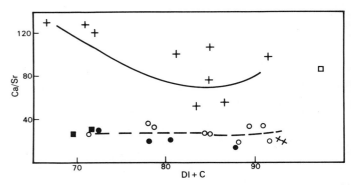

Figure 7. Distinction of Karamea Suite rocks from those of the Rahu Suite and Separation Point Batholith on a plot of Ca/Sr vs. DI+C.

enabled Tulloch (1979a) to clearly distinguish Rahu Suite plutons in a traverse of the Victoria Range. One particularly large magnetic anomaly in the lower Buller Valley may indicate a lamprophyre source body (Hunt and Nathan, 1976).

Gravity maps (1:250,000) are also becoming available (Hunt and others, 1977; Bennie and Ferry, 1977) and although some granitoid bodies are associated with gravity anomalies (e.g., a positive anomaly over the Canavans Knob monzodiorite and a negative anomaly over the Olympus Granite pluton), interpretive work awaits regional geological mapping of the plutonic basement.

GRANITOID PHASES OF FIORDLAND-STEWART ISLAND

Granitoid rocks in Fiordland can be divided into three geographical groups (Oliver and Coggan, 1979): 1. Eastern Fiordland, 2. Southwest Fiordland, and 3. Central Fiordland. Groups 1 and 2 together largely comprise the Post Tectonic Intrusives of Grindley (1978, p. 108-110), whereas 3 approximates his Syntectonic Intrusives. The first two groups can be correlated with granitoid rocks in the West Nelson-Westland region.

Eastern Fiordland

Granitoid rocks are mostly medium-grained, equigranular to porphyritic, biotite granodiorite, and granite. Included are the Murchison and Pomona Granites of Wood (1962,

Figure 6. Variation diagrams for rocks of the Karamea and Rahu Suites (Karamea Batholith), and Separation Point Batholith. Symbols as for Fig. 5, DI+C is Thornton and Tuttle Differentiation Index, plus normative C. Trend lines are visual estimates only.

aeromagnetic maps are now available for most of west Nelson-Westland (Hunt, 1978, and references to 1:250,000 aeromagnetic maps therein). These show a highly irregular pattern. Magnetite in the Rahu Suite, and its general absence in the Karamea Suite and basement paragneiss,

TABLE 3. PRELIMINARY ISOTOPE DATA[1] FOR GRANITOIDS OF WEST NELSON-WESTLAND, NEW ZEALAND

| | Karamea Batholith | | Separation Point Batholith |
	Karamea Suite	Rahu Suite[2]	
$^{87}Sr/^{86}Sr_o$	0.709 - 0.710	0.706 - 0.707	0.704 - 0.705
$\delta^{18}O$	10.7	9.6	

[1]Sr data for Separation Point Batholith is from Harrison and McDougall, 1980. All other data is unpublished data of A. J. Tulloch, C. J. Adams, and P. Blattner.
[2]Excludes Desolation Row Pluton

1966) and Turner (1937), and several plutons further south, including the Green Lake Granodiorite of Higgins and Kawachi (1977). Sodic granitoids of the Mackay Suite in northern Fiordland (Williams and Harper, 1978) may belong to this group. Few published geochemical data are available, but the Green Lake Granodiorite of Higgins and Kawachi is also clearly sodic and an aegirine-riebeckite granite occurs at Lake Monowai (Challis, 1971; Wood, 1969). The Murchison granite, with Na_2O/K_2O of 2.0, is petrographically similar to biotite granites from the Separation Point Batholith, and the apparent similarity of at least some of the associated Eastern Fiordland diorites (Oliver and Coggan, 1979) with the Pearse Granodiorite supports a general correlation of these Eastern Fiordland granitoid rocks with the Separation Point Batholith of Nelson.

Southwest Fiordland

Granitoid rocks in this area intrude rocks of relatively low metamorphic grade, and are dominated by the potassic Kakapo Granite (Wood, 1960; Benson and Bartrum, 1935), which comprises inequigranular to porphyritic biotite granitoids similar to those of the Karamea Suite. However, a concordant biotite-muscovite Rb-Sr date (Aronson, 1968) and the presence of a granite similar to the Desolation Row pluton suggest that some correlatives of the Rahu Suite may also be present. Quartz monzonites occur at Gulches Head in the southwest of this area (D. G. Bishop personal communication) and appear to be similar to those found at Canavans Knob in Westland.

Central Fiordland granitoids

Migmatitic and foliated granitoids (the Syntectonic Intrusives of Grindley (1978, p. 108) in central Fiordland generally lack contact aureoles and include the Western Manapouri Injection Complex of Turner (1937). Similar but less common granitoids in western Fiordland are included in this group. A Paleozoic age is probable (Oliver, 1980), and these rocks could be regarded as deeper equivalents of the Karamea Batholith.

Stewart Island

Most of the southern two-thirds of Stewart Island consists of medium-grained, leucocratic biotite granite and granodiorite — the Rakeahua Granite of Watters (1978a, p. 111). The Rakeahua Granite is remarkably homogeneous although the northern half tends to be more coarse grained and more strongly foliated than the southern half (W. A. Watters, personal communication). Petrographic features such as a relatively granular texture, the yellowish-green colour of biotite, the sodic nature of analyses from the

southern end of the island (Henley and Higgins, 1977), and the occurrences of rocks similar to the Pearse Granodiorite, suggest correlation with the eastern Fiordland granitoids and the Separation Point Batholith of Nelson. However, more potassic garnetiferous muscovite granites are present, in association with metasedimentary rocks in the Tin Range (Williams, 1934).

GRANITOID ROCKS OF THE CAMPBELL PLATEAU

Granitoid rocks occur on several offshore islands and in drillholes on the Campbell Plateau, south and southeast of the South Island (Fig. 1). The Snares Islands (Fleming and others, 1953; Watters and Fleming, 1975) consist of gneissic garnetiferous muscovite granite similar to the Tin Range muscovite granite of Stewart Island, and the association with metamorphic rocks comparable to the Pegasus Formation supports this correlation.

The Bounty Islands consist of (reddish-brown) biotite granodiorite (Wasserburg and others, 1963), from which biotite yields a concordant Rb-Sr and K-Ar date of 189 Ma. Biotite granite forms a small outcrop in the core of a basaltic shield volcano on the Auckland Islands (Speight and Finlayson, 1909). A Late Cretaceous K-Ar date on biotite was obtained by Denison and Coombs (1975).

GRANITOIDS ASSOCIATED WITH MAFIC COMPLEXES

Granitoid rocks, predominantly quartz dioritic, also occur in association with several ultramafic complexes along or near the boundary between the granitoid-rich Tasman metamorphic belt, and the Wakatipu belt to the east, which lacks granitoid rocks (Figs. 2, 3).

In west Nelson several discrete mafic to ultramafic bodies which form the Riwaka Complex (Grindley, 1980; Harrison and McDougall, 1980) are exposed immediately west of the Separation Point Batholith. The Brooklyn diorite (augite, hypersthene, hornblende, biotite, andesine, quartz, orthoclase, magnetite) forms two small plutons intrusive into peridotite and pyroxenite phases of the complex.

Farther south adjacent to the Alpine Fault, the Rotoroa Complex consists of a layered basic intrusion, which has been intruded and marginally assimilated by Separation Point granitoids to produce hybrid diorites (G. A. Challis, personal communication).

Blattner (1978) and Williams and Harper (1978) described rocks southeast of the Alpine Fault which range from gabbro to leucoquartz diorite (Darran Basic Complex), and from gabbronorite through abundant quartz diorite to leucogranite (Mackay Intrusives), respectively. On the basis of mineral halogen contents, Blattner (1980) correlated the Darran Basic Complex with the Riwaka Complex of Nelson.

Farther south, minor granitoid rocks are associated with basic and ultrabasic rocks of the Longwood Complex. Challis and Lauder (1977) concluded that the granitoid rocks intruded and assimilated mafic host rocks to produce hybrid rocks similar to parts of the Rotoroa Complex. However, Price and Sinton (1978) considered that some of the quartz diorites resulted from crystal fractionation of high-Al gabbro, and the remaining xenolith-rich (restite?) quartz diorites from lower crustal partial melts.

The Anglem Group (Watters, 1978b; 1978a, p. 171) underlies the northern third of Stewart Island, and is separated from the Rakeahua Granite of southern Stewart Island by a tectonic break. Most of the Anglem Group consists of a complex of gabbroic diorite and amphibolitic rock which are commonly gneissic, but microcline granite, aplite, and pegmatite veins are also present.

RADIOMETRIC DATING

Radiometric dating of intrusive events is still at a reconnaissance stage in New Zealand; most available dates are from minerals by K-Ar and Rb-Sr methods.

At least four periods of plutonism/metamorphism are now known to have occurred in the New Zealand region since the late Precambrian. Evidence for pre-Silurian granites has been presented above, and the reconnaissance Rb-Sr work of Aronson (1965, 1968) clearly established the existence of thermal events in the Carboniferous to Devonian (Tuhua Orogeny) and in the Cretaceous (Rangitata Orogeny). More recently additional determinations have established another event, in the Triassic.

The most voluminous plutonism (Karamea Suite granitoids) occurred in the mid-Paleozoic. The 280-380 Ma range of the Tuhua granites determined by Aronson now appears to extend back at least into the Silurian (Adams, 1974), with a number of rocks, particularly on the western side of the Karamea Batholith, yielding ages in the 390-430 Ma range. The 280-380 Ma thermal events may have obscured the older group by resetting the mineral ages of Aronson (1968). K-Ar ages of Lower Ordovician slates fall into two groups of 298-370 Ma and 395-438 Ma (Adams and others, 1975), which may correlate with this subdivision of the Tuhua Orogeny. A whole rock Rb-Sr isochron for the Karamea Suite in the Victoria Range yields an age of 380 Ma (Tulloch, 1979a).

A number of Triassic to Early Jurassic K-Ar and Rb-Sr mineral dates have been recorded from the New Zealand region, generally from intermediate rather than acid rocks. These include the Mistake Diorite from the Mackay Intrusives, 208-180 Ma (Williams and Harper, 1978); Bounty Islands, 189 Ma (Wasserburg and others, 1963); Ruapuke Island, 217-211 Ma, and Pahia Point 188 Ma — both in the Foveaux Strait region between the Longwood and Anglem Complexes (Devereux and others, 1968); Charleston, 210

Ma (Aronson, 1968); Upper Buller River, 222 Ma on hornfelsed metasediments adjacent to the Karamea Batholith (Eggers and Adams, 1979); and quartz diorites from the Victoria Range, 232-213 Ma (Tulloch, 1979a).

Unfortunately Aronson's Rb-Sr determinations were mostly mineral isochrons which include at best a single whole-rock analysis and as such are susceptible to post-emplacement metamorphism/uplift events, although concordant mineral ages and constraints imposed by maximum initial strontium ratios did confirm the presence of Cretaceous plutonism. Nevertheless, as a result of widespread metamorphism and uplift in the Cretaceous, most granitoid rocks yield Cretaceous K-Ar mineral dates, although in some cases plutonism closely preceded this uplift (Adams and Nathan, 1978). In the Victoria Range, K-Ar biotite ages from the Rahu Suite, in the range 110-90 Ma, decrease with increasing host rock differentiation and biotite composition — possible compositional control on biotite Ar-closure temperatures suggests an uplift rate of .3 m/1000 years (Tulloch, 1979a). Stratigraphic evidence in the Buller Gorge (Nathan, 1978a) confirms that some volcanism and associated plutonism did occur in Early to mid-Cretaceous times.

A general problem with Rb-Sr dating of Cretaceous plutonism is that Rahu Suite and Separation Point granitoid rocks have low $^{87}Rb/^{86}Sr$, generally less than, or equal to one. An isochron for the Rahu Suite in the Victoria Range (Tulloch, 1979a) gives an approximate age of 125 ± 25 Ma, which compares well with a 140-Ma zircon determination (Aronson, 1968).

A Cretaceous age has recently been confirmed for the Separation Point Batholith by a combined mineral K-Ar, ^{40}Ar-^{39}Ar, Rb-Sr, U-Pb and fission-track study (Harrison and McDougall, 1980). Their data yielded an age of 114 Ma for intrusion of the Separation Point Batholith and 367 Ma for an adjacent phase of the Riwaka Complex. However, the metamorphic texture of the Pearse Granodiorite may indicate an older age for this phase.

In the North Island minor plutons ranging in composition from quartz diorite to granodiorite (Skinner, 1976) are of mid-Miocene age. Quaternary eruption of some 16,000 km³ of rhyolitic magma, which includes individual ignimbrite flows with volumes of the order of 100-1000 km³, suggests associated plutonism to be on a similar scale to that now exposed in the West Nelson-Westland region of the South Island.

MINERALIZATION

Few minerals have been added to the list of economic minerals which Reed (1958) considered to be related to granitoid rocks. However, a number of new occurrences of these minerals, particularly molybdenite, have been located recently.

Reed considered that several Cu-Mo-Au-Ag-Pb-Zn and

W-Sn lodes were *directly* related to granite. Of the former, the only significant occurrence was at Mt. Radiant in the Karamea Batholith, the only such lode that has been worked. Wolframite and cassiterite-bearing lodes occur in greisen within biotite schists of the Tin Range metasediments on Stewart Island. Although the associated granites and granodiorites are potash-rich muscovite granites, the predominant granite variety on Stewart Island is relatively sodic, unlike the S-type granitoids with which tin is normally associated. Cassiterite has also been observed in central Fiordland (Turner, 1937) and as a common constituent of some heavy mineral suites on the west coast of the South Island, notably those recovered from gold-dredge concentrates (Hutton, 1950; McDonald, 1965; Bradley and others, 1978). Reed lists two other minor quartz-wolframite lodes in west Nelson-Westland; scheelite is also known in association with the Barrytown Pluton, exposed just under its tourmalinised roof rock (Tulloch, 1973), and at Canaan within granite and diorite and adjacent marble near the margin of the Separation Point Batholith.

At the present time molybdenite is the most important of these minerals in west Nelson, and at least 14 occurrences have been recently, or are at present, under exploration (Williams, 1974; Eggers and Adams, 1979). All prospects are associated with small plutons and stocks of sodic, fine-to medium-grained quartz; feldspar porphyritic granites; leucogranodiorites; and trondhjemites (of the Rahu Suite?). They generally occur close to the margin of the Karamea Suite/lower Paleozoic sediment contact, and K-Ar mineral ages range from 116 to 95 Ma. Pyrite, chalcopyrite, bismuth minerals, galena, and sphalerite are commonly associated with molybdenite; and bornite, tetrahedrite, cubanite, and argentite also occur. Most molybdenite mineralization occurs in quartz-vein stockworks but it is also locally disseminated within the host granitoid. At Questa Creek an outer chlorite halo gives way to biotite, sericite, and quartz-orthoclase zones progressively, towards the centre of mineralization (Smale, 1977). In a detailed study of the Copperstain Creek prospect, Wodzicki (1972) described an outer alteration halo (in noncalcareous rocks) characterized by muscovite ± albite, and an inner halo characterized by microcline and muscovite, together with the addition of large amounts of K, S, and minor Mo.

Although Mo-mineralization is clearly associated with these minor sodic granitoids which have much in common with Separation Point rocks, little Mo has been found within the Separation Point Batholith itself. Possibly the Separation Point Batholith is too deeply eroded, or the Mo has been derived from the sub-Karamea Suite basement (Wallace and others, 1978).

Alteration of granitoids in which plagioclase and biotite break down to form Ca-Al silicates is a common and widespread feature, but one which does not appear to have any direct relationship to mineralization (Tulloch, 1979b).

Reed considered that the auriferous quartz lodes of the Reefton district, extensively worked between 1871 and 1951, were *probably* related to granites. Pyrite, arsenopyrite, and stibnite-bearing lodes occur in prehnite-actinolite grade (Nathan, 1978b) rocks of the Greenland Group (Williams, 1974). Subeconomic uranium mineralization is widespread within the granitoid-bearing facies of the Cretaceous nonmarine fanglomerates referred to earlier and is also present in a soda porphyry dike (Williams, 1974).

In the North Island minor porphyry-type Cu-mineralization is associated with Miocene quartz diorites at Knuckle Point, Coppermine Island, and Paritu on the Northland and Coromandel Peninsulas (Williams, 1974; Hay, 1975). Gold-silver mineralization associated with andesitic propylitization on Coromandel Peninsula is being intensively prospected.

Gold, silver, mercury, antimony, sphalerite, chalcopyrite, and galena mineralization is associated with active geothermal fields of the Taupo Volcanic Zone.

ACKNOWLEDGMENTS

Thanks are due to S. Nathan, G. W. Grindley, R. L. Brathwaite, J. A. Roddick, and L. Aguirre for critically reading the manuscript, and to P. J. Oliver and W. A. Watters for helpful discussion. The author's Ph.D. project on the Victoria Range was supervised by A. F. Cooper at the University of Otago. However, responsibility for inaccuracies and generalization rests with the author alone. Unpublished isotope analyses of Victoria Range rocks were done in conjunction with C. J. Adams, J. Gabites, and P. Blattner. N. Orr, F. Tonks, P. K. Hodgson and M. Haronga were responsible for fast and competent thin section preparation, typing, and draughting, respectively. Funds to attend the final meeting of Circum-Pacific Plutonism (IGCP Project 30) were awarded by the Royal Society of New Zealand, the New Zealand National Commission for UNESCO, the New Zealand Department of Scientific and Industrial Research, and IGCP Project 30.

REFERENCES CITED

Adams, C. J. D., 1974, Rb-Sr age of the Greenland Group and Constant Gneiss; Nelson, Buller and Westland, New Zealand: Institute of Nuclear Sciences Internal Report, INS-R-140, N.Z. Department of Scientific and Industrial Research.

——,1975, Discovery of Precambrian rocks in New Zealand. Age relations of the Greenland Group and Constant Gneiss, West Coast, South Island: Earth and Planetary Science Letters, v. 28, p. 98–104.

Adams, C. J. D., and Nathan, S., 1978, Cretaceous chronology of the Lower Buller Valley, South Island, New Zealand: N.Z. Journal of Geology and Geophysics, v. 21, p. 455–462.

Adams, C. J. D., Harper, C. T., and Laird, M. G., 1975, K-Ar ages of low-grade metasediments of the Greenland and Waiuta Groups in Westland and Buller, New Zealand: N.Z. Journal of Geology and Geophysics, v. 18, p. 39–48.

Aronson, J. L., 1965, Reconnaissance rubidium-strontium geochronology of New Zealand plutonic and metamorphic rocks: N.Z. Journal of Geology and Geophysics, v. 8, p. 401–423.

——, 1968, Regional geochronology of New Zealand: Geochimica et Cosmochimica Acta, v. 32, p. 669–697.

Bennie, S. L., and Ferry, L. M., 1977, Sheet 17, Hokitika "Gravity Map of New Zealand, 1:250,000": N.Z. Department of Scientific and Industrial Research.

Benson, W. N., and Bartrum, J. A., 1935, The geology of the region about Preservation and Chalky Inlets, south-west Fiordland, N.Z.: Transactions of the Royal Society of New Zealand, v. 65, p. 108–152.

Blattner, P., 1978, Geology of the crystalline basement between Milford Sound and the Hollyford Valley, New Zealand: N.Z. Journal of Geology and Geophysics, v. 21, p. 33–47.

——, 1980, Chlorine-enriched leucogabbro in Nelson and Fiordland, New Zealand: Contributions to Mineralogy and Petrology, v. 72, p. 291–296.

Bowen, F. E., 1964, Sheet 15, Buller "Geological Map of New Zealand 1:250,000": N.Z. Department of Scientific and Industrial Research.

Bradley, J. P., Wilkins, C. J, Oldershaw, W., and Smale, D., 1979, A study of detrital heavy minerals in the Taramakau catchment: Journal of the Royal Society of New Zealand, v. 9, p. 233–251.

Brathwaite, R. L., 1968, The geology of the Boulder Lake area, North-west Nelson, Part 2 — The Mount Olympus Granite Pluton: N.Z. Journal of Geology and Geophysics, v. 11, p. 92–122.

Challis, G. A., 1971, Chemical analyses of New Zealand rocks and minerals with CIPW norms and petrographic descriptions, 1917–57, Part I: Igneous and pyroclastic rocks: N.Z. Geological Survey Bulletin 84.

Challis, G. A., and Lauder, W. R., 1977, Pre-Tertiary geology of the Longwood Range 1:50,000: N.Z. Geological Survey Miscellaneous Series, Map 11.

Chappell, B. W., and White, A. J.R., 1974, Two contrasting granite types: Pacific Geology, v. 8, p. 173–174.

Cobbing, E. J., and Pitcher, W. S., 1972, The Coastal Batholith of Central Peru: Journal of the Geological Society of London, v. 128, p. 421–460.

Cooper, R. A., 1975, New Zealand and South-East Australia in the Early Paleozoic: N.Z. Journal of Geology and Geophysics, v. 18, p. 1–20.

——, 1979, Lower Paleozoic rocks of New Zealand: Journal of the Royal Society of New Zealand, v. 9, p. 29–84.

Denison, R. E., and Coombs, D. S., 1977, Radiometric ages for some rocks from the Snares and Auckland Islands, Campbell Plateau: Earth and Planetary Science Letters, v. 34, p. 23–29.

Devereux, I., McDougall, I., and Watters, W. A., 1968, Potassium-argon mineral dates on intrusive rocks from the Foveaux Strait area: N.Z. Journal of Geology and Geophysics, v. 11, p. 1230–1235.

Eggers, A. J., 1979, Large-scale circular features in North Westland and West Nelson, New Zealand; possible structural control for porphyry molybdenum-copper mineralization: Economic Geology, v. 74, p. 1490–1494.

Eggers, A. J., and Adams, C. J. D., 1979, Potassium-argon ages of molybdenum mineralization and associated granites at Bald Hill and correlation with other molybdenum occurrences in the South Island, New Zealand: Economic Geology, v. 74, p. 628–637.

Ewart, A., and Cole, J. W., 1967, Textural and mineralogical significance of the granitic xenoliths from the central volcanic region, North Island, New Zealand: N.Z. Journal of Geology and Geophysics, v. 10, p. 31–54.

Ewart, A., Brothers, R. N., and Mateen, A., 1977, An outline of the geology and geochemistry, and the possible petrogenetic evolution of the volcanic rocks of the Tonga-Kermadec-New Zealand Island Arc: Journal of Volcanology and Geothermal Research, v. 2, p. 205–250.

Fleming, C. A., Reed, J. J., and Harris, W. F., 1953, Geology of the Snares Islands: Cape Expedition Series Bulletin 13, N.Z. Department of Scientific and Industrial Research.

Ghent, E. D., 1968, Petrology of metamorphosed pelitic rocks and quartzites, Pikikiruna Range, north-west Nelson, New Zealand:

Transactions of the Royal Society of New Zealand (Geology), v. 5, p. 193–213.

Gibson, G. M., 1979, Margarite in kyanite- and corundum-bearing anorthosite, amphibolite and hornblende from Central Fiordland, New Zealand: Contributions to Mineralogy and Petrology, v. 68, p. 171–179.

Grindley, G. W., 1961, Sheet 13, Golden Bay "Geological Map of New Zealand, 1:250,000": N.Z. Department of Scientific and Industrial Research.

——, 1971, Sheet S8, Takaka "Geological Map of New Zealand 1:63,360": N.Z. Department of Scientific and Industrial Research.

——, 1978, West Nelson: *in* Suggate, R. P., Stevens, G. R., and Te Punga, M. T., The Geology of New Zealand. Government Printer, 2 vols., 820 p.

——, 1980, Sheet S13, Cobb "Geological Map of New Zealand 1:63,360": N.Z. Department of Scientific and Industrial Research.

Harrison, T. M., and McDougall, I., 1980, Investigations of an intrusive contact, northwest Nelson, New Zealand-I. Thermal, chronological and isotopic constraints: Geochimica et Cosmochimica Acta, v. 44, p. 1985–2003.

Hay, R. F., 1975, Sheet N7, Doubtless Bay "Geological Map of New Zealand 1:63,360": N.Z. Department of Scientific and Industrial Research.

Henderson, J., 1950, Cornish stone and feldspar in New Zealand: N.Z. Journal of Science and Technology, V. B31, p. 25–44.

Henley, R. W., and Higgins, N. C., 1977, Geology of the granitic terrain, South-west Stewart Island: N.Z. Journal of Geology and Geophysics, v. 20, p. 779–796.

Higgins, N. C., and Kawachi, Y., 1977, Microcline megacrysts from the Green Lake Granodiorite, Eastern Fiordland, New Zealand: N.Z. Journal of Geology and Geophysics, v. 20, p. 273–286.

Hume, B. J., 1977, The relationship between the Charleston Metamorphic Group and the Greenland Group in the Central Paparoa Range, South Island, New Zealand: Journal of the Royal Society of New Zealand, v. 7, p. 379–392.

Hunt, T. M., 1978, Stokes Magnetic Anomaly System: N.Z. Journal of Geology and Geophysics, v. 21, p. 595–606.

Hunt, T., and Nathan, S., 1976, Inangahua magnetic anomaly, New Zealand. N.Z. Journal of Geology and Geophysics, v. 19, p. 395–406.

Hunt, T. M., Doone, A., and Mathews, S., 1977, Sheet 13, Golden Bay "Gravity Map of New Zealand 1:250,000": N.Z. Department of Scientific and Industrial Research.

Hutton, C. O., 1950, Studies of heavy detrital minerals: Geological Society of America Bulletin, v. 61, p. 635–716.

Kistler, R. W., 1974, Phanerozoic batholiths in western North America. A summary of some recent work on variations in time, space, chemistry and isotopic compositions: Annual Review of Earth and Planetary Science, v. 2, p. 403–418.

Laird, M. G., 1967, Field relations of the Constant Gneiss and Greenland Group in the Central Paparoa Range, West Coast, South Island. N.Z. Journal of Geology and Geophysics, v. 10, p. 247–256.

——, 1968, The Paparoa Tectonic Zone: N.Z. Journal of Geology and Geophysics, v. 11, p. 435–454.

Landis, C. A., and Coombs, D. S., 1967, Metamorphic belts, and orogenesis in southern New Zealand: Tectonophysics, v. 4, p. 501–518.

MacDonald, S., 1965, Tin-ore potential of the South Island, New Zealand: N.Z. Journal of Geology and Geophysics, v. 8, p. 440–452.

McDougall, I., and van der Lingen, G. J., 1974, Age of the rhyolites of the Lord Howe Rise, and the evolution of the south west Pacific Ocean: Earth and Planetary Science Letters, v. 21, p. 117–126.

Nathan, S., 1974, Petrology of the Berlins Porphyry: A study of the crystallization of granitic magma: Journal of the Royal Society of New Zealand, v. 4, p. 463–483.

——, 1975, Sheets S23 and S30, Foulwind and Charleston "Geological Map of New Zealand 1:63,360": N.Z. Department of Scientific and Industrial Research.

——,1976, Sheets S23/9 and S24/7, Foulwind and Westport "Geological Map of New Zealand 1:25,000": N.Z. Department of Scientific and Industrial Research.

——,1978a, Sheets S31 and part S32, Buller-Lyell "Geological Map of New Zealand 1:63,360": N.Z. Department of Scientific and Industrial Research.

——,1978b, Sheet S44, Greymouth "Geological Map of New Zealand 1:63,360": N.Z. Department of Scientific and Industrial Research.

——,1978c, Tuhua Intrusive Group: *in* Suggate, R. P., Stevens, G. R., and Te Punga, M. T., The Geology of New Zealand. Government Printer, 2 vols., 820 p.

Oliver, G. J. H., 1980, Geology of the granulite and amphibolite facies gneisses of Doubtful Sound, Fiordland, New Zealand: N.Z. Journal of Geology and Geophysics, v. 23, p. 27–41.

Oliver, G. J. H., and Coggan, J. H., 1979, Crustal structure of Fiordland, New Zealand: Tectonophysics, v. 54, p. 253–292.

O'Neil, J. R., and Chappell, B. W., 1977, Oxygen and hydrogen isotope relations in the Berridale batholith: Journal of the Geological Society of London, v. 133, p. 559–571.

Presnall, D. C., and Bateman, P. C., 1973, Fusion relations in the system $NaAlSi_3O_8\text{-}CaAl_2Si_2O_8\text{-}KAlSi_3O_8\text{-}SiO_2\text{-}H_2O$, and generation of granitic magmas in the Sierra Nevada Batholith. Geological Society of America Bulletin, v. 84, p. 3181–3202.

Price, R. C., and Sinton, J. M., 1978, Geochemical variations in a suite of granitoids and gabbros from Southland, New Zealand: Contributions to Mineralogy and Petrology, v. 67, p. 267–278.

Reed, J. J., 1958, Granites and mineralization in New Zealand: N.Z. Journal of Geology and Geophysics, v. 1, p. 47–64.

——,1964, Mylonites, cataclasites, and associated rocks along the Alpine Fault, South Island, New Zealand: N.Z. Journal of Geology and Geophysics, v. 7, p. 645–684.

Reid, D. L., 1972, Thermal metamorphism and assimilation of schists by a dioritic magma, Ligar Bay (S8) North-West Nelson: N.Z. Journal of Geology and Geophysics, v. 15, p. 632–642.

Shelley, D., 1975, Metamorphic belt and volcanic arc migration in New Zealand: Nature, v. 258, p. 668–672.

Skinner, D. N. B., 1976, Sheet N40 and part sheets N35, N36 and N39 Northern Coromandel "Geological Map of New Zealand 1:63,360": N.Z. Department of Scientific and Industrial Research.

Smale, D., 1977, Hydrothermal alteration around younger intrusives near Karamea Bend, North-west Nelson, New Zealand: Australasian Institute of Mining and Metallurgy, Proceedings, v. 260, p. 53–58.

Smale, D., and Nathan, N., 1980, Heavy minerals in basement rocks of the Paparoa Range, Westland: unpublished New Zealand Geological Survey Report, NZGS G36.

Smith, I. E. M., Chappell, B. W., Ward, G. K., and Freeman, R. S., 1977, Peralkaline rhyolites associated with andesitic arcs of the southwest Pacific: Earth and Planetary Science Letters, v. 37, p. 230–236.

Speight, R., and Finlayson, A. M., 1909, Physiography and geology of the Auckland Bounty and Antipodes Islands: *in* Chilton, C., ed., "The Subantarctic Islands of New Zealand": Philosophical Institute of Canterbury, v. 2, p. 705–744.

Streckeisen, A., 1976, To each plutonic rock its proper name: Earth-Science Reviews, v. 12, p. 1–33.

Suggate, R. P., 1957, The Geology of the Reefton subdivision: N.Z. Geological Survey Bulletin 56.

Suggate, R. P., Stevens, G. R., and Te Punga, M. T., 1978, The Geology of New Zealand. Government Printer, 2 vols., 820 p.

Tulloch, A. J., 1973, The Barrytown adamellite stock and its contact aureole: Unpublished B.Sc. (Hons.) thesis, University of Canterbury.

——,1979a, Plutonic and metamorphic rocks of the Victoria Range segment of the Karamea Batholith, Southwest Nelson, New Zealand: Unpublished Ph.D. thesis, University of Otago.

——,1979b, Secondary Ca-Al silicates as low-grade alteration products of granitoid biotite: Contributions to Mineralogy and Petrology, v. 69, p. 105–117.

——,1981a, Large-scale circular features in North Westland and West Nelson, New Zealand: possible structural control for porphyry molybdenum-copper mineralization: discussion: Economic Geology, v. 76, p. 2061–2063.

——,1981b, Gahnite and columbite in an alkali-feldspar granite from New Zealand: Mineralogical Magazine, v. 44, p. 275–278.

Turner, F. J., 1937, The metamorphic and plutonic rocks of Lake Manapouri, Fiordland, New Zealand, Part I: Transactions of the Royal Society of New Zealand, v. 67, p. 83–100.

Wallace, S. R., MacKenzie, W. B., Blair, R. G., and Muncaster, N. K., 1978, Geology of the Urad and Henderson molybdenite deposits, Clear Creek County, Colorado, with a section on a comparison of these deposits with those at Climax, Colorado: Economic Geology, v. 73, p. 325–368.

Ward, C. M., 1980, Lithostratigraphy of an area south of Dusky Sound, Fiordland, and its correlation with the Nelson Lower Paleozoic: Programme and Abstracts, N.Z. Geological Society Conference, Christchurch 1980, p. 96.

Wasserburg, G. J., Craig, M., Menard, H. W., Engle, A. E. J., and Engel, G., 1963, Age and composition of a Bounty Islands granite and age of a Seychelles Islands granite: Journal of Geology, v. 71, p. 785–789.

Watters, W. A., 1978a, Stewart, Snares and Bounty Islands: *in* Suggate, R. P., Stevens, G. R., and Te Punga, M. T., The Geology of New Zealand. Government Printer, 2 vols., 820 p.

——,1978b, Diorite and associated intrusive and metamorphic rocks between Port William and Paterson Inlet, Stewart Island, and on Ruapuke Island: N.Z. Journal of Geology and Geophysics, v. 21, p. 423–442.

Watters, W. A., and Fleming, C. A., 1975, Petrography of rocks from the Western Chain of the Snares Islands: N.Z. Journal of Geology and Geophysics, v. 18, p. 491–498.

Wellman, P., and Cooper, A. F., 1971, Potassium-argon ages of some New Zealand lamprophyre dykes near the Alpine Fault: N.Z. Journal of Geology and Geophysics, v. 14, p. 341–350.

White, A. J. R., Chappell, B. W., and Cleary, J. R., 1974, Geologic setting and emplacement of some Australian Paleozoic batholiths and implications for intrusive mechanisms: Pacific Geology, v. 8, p. 159–171.

Williams, I. G., 1934, A granite-schist contact in Stewart Island, New Zealand: Quarterly Journal of the Geological Society of London, v. 90, p. 322–350.

——,1974, Economic Geology of New Zealand: Australasian Institute of Mining and Metallurgy Monograph Series, no. 4, 490 p.

Williams, J. G., and Harper, C. T., 1978, Age and status of the Mackay Intrusives in the Eglinton-upper Hollyford area: N.Z. Journal of Geology and Geophysics, v. 21, p. 733–742.

Wodzicki, A., 1972, Mineralogy, geochemistry and origin of hydrothermal alteration and sulphide mineralization in the disseminated molybdenite and skarn-type copper sulphide deposit at Copperstain Creek, Takaka, New Zealand: N.Z. Journal of Geology and Geophysics, v. 15, p. 599–631.

——,1974, Geology of the pre-Cenozoic basement of the Taranaki-Cook Strait-Westland area, New Zealand, based on recent drillhole data: N.Z. Journal of Geology and Geophysics, v. 17, p. 747–758.

Wones, D. R., and Eugster, H. P., 1965, Stability of biotite: experiment, theory and application: American Mineralogist, v. 50, p. 1228–1272.

Wood, B. L., 1960, Sheet 27, Fiordland "Geological Map of New Zealand 1:250,000": N.Z. Department of Scientific and Industrial Research.

——,1962, Sheet 22, Wakatipu "Geological Map of New Zealand 1:250,000": N.Z. Department of Scientific and Industrial Research.

——,1966, Sheet 24, Invercargill "Geological Map of New Zealand 1:250,000": N.Z. Department of Scientific and Industrial Research.

——,1969, Geology of the Tuatapere Subdivision, Western Southland, New Zealand: Geological Survey Bulletin 79.

Wood, C. P. 1974, Petrogenesis of garnet-bearing rhyolites from Canterbury, New Zealand: N.Z. Journal of Geology and Geophysics, v. 17, p. 759–788.

Young, D. J., 1968, The Frazer Fault in Central Westland and its associated rocks: N.Z. Journal of Geology and Geophysics, v. 11, p. 291–311.

MANUSCRIPT ACCEPTED BY THE SOCIETY JULY 12, 1982

Geological Society of America
Memoir 159
1983

Granitoid types and their distribution in the Lachlan Fold Belt, southeastern Australia

A. J. R. White
La Trobe University
Bundoora, Victoria
Australia

B. W. Chappell
Australian National University
Canberra, ACT
Australia

ABSTRACT

The Lachlan Fold Belt in southeastern Australia comprises rocks ranging in age from Cambrian to Devonian. Granitoid emplacement and related volcanic activity occurred in Silurian and Devonian times, with minor development of Carboniferous plutons in the most easterly part of the belt. The belt is at least 800 km wide, which is much wider than the Mesozoic and Cenozoic fold belts of the circum-Pacific. Granitoids are extensively developed in the Lachlan belt and make up 36 percent of exposed Paleozoic rocks in the relatively well-exposed easternmost part east of longitude 148° E, a strip up to 200 km wide.

Granitoids in the Lachlan Fold Belt can be grouped into suites, where each suite has a distinctive chemical character, consistent with its having been derived from source rocks of unique composition. Most of the variation within suites can be ascribed to varying degrees of separation of material residual from partial melting, or restite, from melt. The differences between suites result from differences in source rock composition. The first-order subdivision between suites is between those granitoids derived from sedimentary and from igneous source rocks, the S- and I-types. These two types have chemical, mineralogical, and isotopic characters reflecting the distinctive features of their sources, specifically the fact that the S-type source rocks have been through at least one cycle of chemical weathering at the earth's surface. There is an eastern limit to the occurrence of S-type granitoids, called the I-S line, which is thought to represent the eastern limit of thick crystalline basement. A late-formed group of felsic granitoids, the A-types, are thought to have been derived from crust that had previously produced I-type magmas so that the source rocks were residual from that prior melting event.

REGIONAL SETTING

The dominantly Paleozoic Tasman Fold Belt flanks eastern Australia and is divided into a northern New England Fold Belt and a southern Lachlan Fold Belt, by the Permian to Mesozoic Sydney-Bowen Basin (Fig. 1). The younger New England Belt is intruded by a variety of granitoids, ranging in age from Carboniferous to Early Triassic. Recent summaries of the granitoids at the northern and southern ends of the New England Belt have been provided by Richards (1980) and Shaw and Flood (1981), respectively.

In the Lachlan Fold Belt sedimentation of turbidite type (greywackes and shales) occurred mainly during the

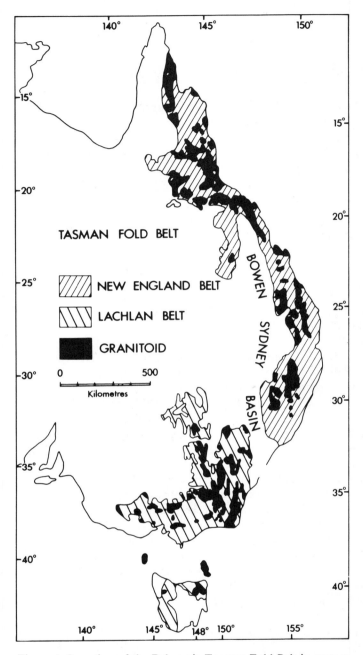

Figure 1. Location of the Paleozoic Tasman Fold Belt in eastern Australia. The northern New England Fold Belt (mid- to late-Paleozoic) is younger than the Lachlan Fold Belt (early- to mid-Paleozoic) to the south. Granitoids are extensively developed in both belts.

Ordovician. By latest Silurian and into Devonian time, sedimentation included shelf facies coralline limestones. Granitoids were mainly emplaced during the Late Silurian and Early Devonian, with restricted occurrences of Carboniferous age on the eastern side of the belt.

The present exposures of Lachlan Fold Belt rocks are virtually confined to the eastern mountain region, which is a series of plateaus up to 2100 m high uplifted since the Cretaceous (Wellman, 1979). These mountains represent only the eastern part of the Lachlan belt which from east to west, perpendicular to the dominantly meridionally trending structures, is at least 800 km wide. It is thus much wider than the Mesozoic to Cenozoic fold belts of the circum-Pacific region. We are here particularly concerned with those granitoids outcropping east of longitude 148° E, a strip 200 km wide, that we refer to as the eastern Lachlan Fold Belt.

BASEMENT ROCKS

No rocks older than Cambrian are known to outcrop in the Lachlan belt. Some altered basaltic rocks are found in strips a few kilometers wide and are considered to have ocean ridge basalt affinities; some of these are Cambrian. Mostly based on the presence of these basalts low in the sequence, structural and stratigraphic arguments have been made to suggest that the Ordovician turbidites were deposited on oceanic crust (Scheibner, 1974; Crook, 1980). We do not accept this hypothesis since basaltic rocks and mafic belts resembling ophiolites are very insignificant compared with the vast areas of granitoids, amounting to more than one-third of the eastern Lachlan belt. Granitoids are thought to "image" their source rocks in the lower crust (Chappell, 1979), and hence data from a study of the granitoids give more significant information about the deep crust. Such data indicate that the Paleozoic sediments are probably underlain by felsic Proterozoic rocks, consistent with arguments by Rutland (1976) that Proterozoic basement extends eastwards beneath the Lachlan Fold Belt.

Seismic investigations (Finlayson and others, 1979) indicate that the crust of the belt is complex with no clearly defined layers but with gradational changes. The crust-mantle boundary is also transitional. A depth to the Moho of between 40 and 52 km is indicated. These data are not consistent with the simple model of Phanerozoic sediments resting on basaltic oceanic crust. It is also inconceivable that the Ordovician sediments could be 40 km thick even if they are now a series of thrust wedges. Some evidence (Pinchin, 1979) that the crust thins east from Canberra is consistent with the suggestion from our granitoid studies that the crust is different near Canberra than farther east.

Heat flow measurements over the area of Australia underlain by Lachlan belt rocks show that present heat flow is well above world average. Heat flow values are as high as 110 mWm^{-2} and in the immediate study area, values range between 70 and 90 mWm^{-2} (Cull, 1979; Sass and Lachenbruch, 1979). Estimates of P-T conditions of formation of inclusions from breccia pipes carrying upper-mantle and lower-crustal inclusions, give conclusive evidence that thermal gradients were also high in the region during Mesozoic and Cenozoic times. These pipes include the Jurassic nephelinitic breccia at Delegate (Lovering and

White, 1969; White and Chappell, 1982), the Jurassic Meredith kimberlitic breccia (Day and others, 1979), and the Late Cenozoic(?) Jugiong kimberlitic breccia (Ferguson and others, 1977, 1979). These pipes all intrude the southeastern part of the Lachlan belt. The abundance of granitoids through the very wide Lachlan belt (Fig. 2), as well as the presence of several large belts of low pressure-high temperature metamorphic rocks of middle Paleozoic age, indicates that the thermal gradients were also high at that time.

Thermal gradients depend on concentrations of radioactive elements in the crust and mantle, conductivity, and mass transfers of heat by magmas. The fragmental evidence listed above suggests that heat flow might always have been high in the Lachlan Fold Belt, at least since the mid-Paleozoic. If this is so, then relatively high concentrations of radioactive elements may be the main cause of the high heat flow which may also be the reason for the extensive magmatism of the region. Alternatively, there may have been periods of thermal highs resulting from the introduction of magma into the upper crust.

GRANITE PLUTONS, GRANITE COMPLEXES, AND BATHOLITHS

Over four hundred granitoid plutons occur in the eastern part of the Lachlan Fold Belt. These range in exposed area from less than one to a maximum of 970 km². The largest

pluton, the Bemboka Adamellite, is partly covered by younger sedimentary rocks; if these were removed its area would exceed, but not greatly, 1000 km². Data on the total and relative abundances of granitoids in the eastern Lachlan Fold Belt are:

Total area east of longitude 148° E,
south of the Great Artesian Basin,
south and west of the Sydney Basin .... 108,400 km²
Area of younger Paleozoic strata 5,800
Area of Cenozoic volcanic rocks 3,600
Area of Cenozoic sedimentary deposits 6,800
Net total exposure of pregranitoid and
granitoid rocks 92,300
Area of exposed granitoid 32,950
Percent. area of exposed granitoid 36%

The figure of 36 percent exposed granitoid is a remarkably high one and indicates a massive magmatic recycling of the southeast Australian crust in the mid-Paleozoic.

The numerous granitoid plutons are grouped into granite complexes ("Granites") or batholiths (Table 1 and Fig. 2). We use the term "Batholith" for a group of plutons that are contiguous or nearly so, with a total exposed area generally in excess of 500 km². Granite complexes are smaller units,

TABLE 1. GRANITOID BATHOLITHS AND COMPLEXES OF THE EASTERN LACHLAN FOLD BELT (EAST OF LONGITUDE 148°E)

	Batholith or Complex	Area (km²)	% of type I	S	A
1	Gulgong	795	100	-	-
2	Bathurst	1680	100	-	-
3	Oberon	475	100	-	-
4	Marulan	138	100	-	-
5	Moruya	263	100	-	-
6	Gabo	55	3	-	97
7	Bega	8620	99	-	1
8	Wologorong	725	-	100	-
9	Wyangala	3170	~40	~60	~1
10	Murrumbidgee	1470	<1	>99	-
11	Cooma	14	-	100	-
12	Gingera	238	14	86	-
13	Berridale	1650	46	54	-
14	Bonang	430	66	14	20
15	Kosciusko	3940	9	91	-
16	Yeoval	1490	?100	?	?
17	Grenfell	610	?	?100	?
18	Young	3990	1	99	-
19	Tumut	380	100	-	-
20	Maragle	3950*	12	88	-
	TOTAL	34100*	52	47	1

* includes 1150 km² west of longitude 148° E.

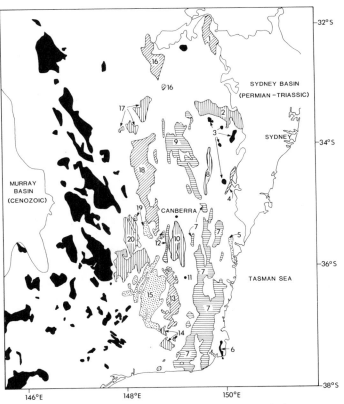

Figure 2. Granitoid batholiths and complexes of the eastern Lachlan Fold Belt. Numbers correspond to those listed in Table 1, and the patterns indicate the extent of each grouping but are not otherwise significant.

sometimes good "mini-batholiths," such as Marulan, Moruya, and Gingera (Table 1); the term "Batholith" is sometimes used for these contiguous units. The term "Granites" is also used for an area of dispersed plutons, e.g., Oberon and Bonang. It may also be used for an area of dispersed plutons with a total area in excess of 500 km², of which the Gulgong Granites comprise the only area in the eastern Lachlan belt.

Subdivision into batholiths and granite complexes is sometimes arbitrary, and historical usage must be considered. For example, the Cooma Granodiorite could be assigned to the Murrumbidgee Batholith. However, it has for over 50 years been a focus of attention as the core rock of the Cooma Metamorphic Complex and this separate identity is retained.

The batholiths and complexes are elongated meridionally, parallel to the trend of the folds in the country rocks. The individual plutons are mostly circular to elliptical in plan, with the major axes of the ellipses oriented parallel to the regional trend of the batholiths. Contacts, where observed, are very steep.

The composite nature of the batholiths is exemplified by the Berridale Batholith, which consists of about 40 separate plutons ranging up to 470 km² in outcrop area. These structurally defined units may be grouped into 20 lithologically distinct mappable units which appear as separate plutons, pairs of plutons, or groups of plutons. Screens of country rock hornfels commonly occur between adjacent plutons or between strings of plutons. Some screens may be traced in the field even though they are no more than a meter or so wide. Previous authors mapped some of these as roof pendants projecting down into the roof of a homogenous batholith, but almost invariably there are distinct differences in texture and mineralogical and chemical composition of granitoids on either side of a screen.

INTRUSIVE ENVIRONMENTS

We have previously classified granitoids of the region as *regional aureole, contact aureole* and *subvolcanic* (White and others, 1974). This has proved useful and is retained. The Cooma Granodiorite is the only regional aureole granitoid of the map region of Figure 2 east of longitude 148° E. It is surrounded by, and genetically related to, regionally metamorphosed rocks (Chappell and White, 1976). Most of the other batholiths and complexes are contact aureole types in which each pluton, except for faulted contacts, is surrounded by a narrow (1-2 km) contact aureole in which the characteristic rocks are spotted slates and hornfelses. The fairly uniform widths of the aureoles are consistent with the steep contacts of the plutons. The Gabo Granites are subvolcanic and are associated in space and time (Late Devonian) with rhyolites and ash flows of similar composition. Parts of the Young Batholith are subvolcanic

and the associated volcanic rocks are Late Silurian (Wyborn and others, 1981).

PRIMARY AND SECONDARY STRUCTURES AND INTRUSIVE MECHANISMS

Primary structures within the plutons are either inconspicuous or absent. Gneissic primary foliation and flattening of inclusions parallel to pluton contacts occur in some intrusions, notably in the Moruya Batholith; in that case the country rock hornfels is also deformed close to the granitoid margins. Although exposures are not good, the most common contact relationship is that shown for the Dalgety pluton in Figure 3. In this case, the deformation is not penetrative (graptolites are undeformed at the very contact), no mappable structures were produced by intrusion, and the pluton itself is massive. No plastic deformation is apparent, as in diapiric intrusion. Deformation is brittle and yet no evidence exists for intrusion of the Dalgety Granodiorite by stoping. Hornfels xenoliths are very rare except for some angular blocks at the immediate contact.

These data indicate that, at this level of exposure, diapirism is not the main intrusive mechanism for the batholithic rocks. Although contacts suggest brittle deformation, the absence of stoped blocks as well as arguments against contamination (see White and others, 1974) indicate that stoping is also not the mechanism of intrusion. The best interpretation of the field evidence is that the country rocks have been displaced upward as the granite moves up like a piston in a cylinder. According to this mechanism, contacts are healed cylindrical faults that were probably lubricated by water derived from dehydration reactions in the hornfels. This mechanism is also consistent with the absence of chilled margins around any of the plutons of the batholiths.

To the west, around Mt. Kosciusko, a penetrative, cataclastic foliation trends a few degrees east of north, is independent of pluton boundaries, and is therefore considered to postdate intrusion. Foliation is more intense in quartz-rich and plagioclase-poor granitoids, but even in these favorable rocks, it becomes less and less distinct to the east-southeast and it is virtually absent in most of the Berridale Batholith. A conspicuous secondary foliation is also seen to the north and northeast in the Gingera Granites and parts of the Murrumbidgee and Wyangala Batholiths.

CHEMICAL SUBDIVISION INTO SUITES

Mapping of separate distinct lithological units making up each of the batholiths and complexes studied is based on mineralogical and textural differences clearly recognizable in the field. For example, in the Berridale Batholith, the Dalgety Granodiorite is an even-grained, grey, biotite granodiorite in which the biotites characteristically occur as pseudo-hexagonal crystals. On the other hand, beyond the

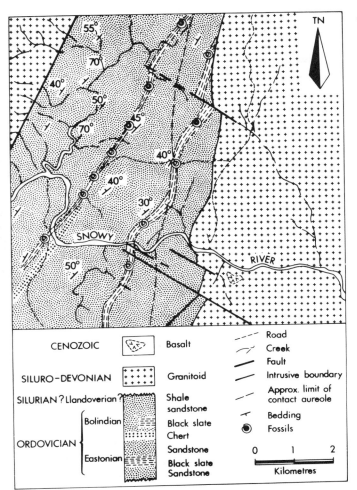

CENOZOIC — Basalt

SILURO-DEVONIAN — Granitoid

SILURIAN? Llandoverian? — Shale sandstone

ORDOVICIAN
Bolindian — Black slate Chert
Eastonian — Sandstone Black slate Sandstone

Road
Creek
Fault
Intrusive boundary
Approx. limit of contact aureole
Bedding
Fossils

0   1   2
Kilometres

Figure 3. Sketch of the western contact of the Dalgety granodiorite in the Berridale Batholith, showing the massive nature of the granitoid and the lack of structures produced by intrusion.

narrow screen on the east side of the Dalgety pluton, is the Buckleys Lake Adamellite which is a coarse-grained, pink, biotite adamellite characterized by large (up to 6 cm across) pink orthoclase crystals. Primary sphene is a common accessory mineral in the Buckleys Lake pluton, but it is absent in all Dalgety samples.

Mineralogy, and less directly texture, are functions of chemical composition and thus it is not surprising that each mappable lithological unit, which may consist of one or more structurally mappable plutons, has a distinctive chemical composition. Some units are chemically homogeneous (e.g., Wullwye Granodiorite of the Berridale Batholith; White and others, 1977), whereas others display systematic variation in chemical composition. Rarely is the variation related to precise position in the field; zoned plutons are uncommon. An example of the kind of chemical variation found within one pluton is that found within the Jindabyne Tonalite of the Kosciusko Batholith (White and others, 1977; Hine and others, 1978). This unit, which probably

consists of several small plutons, has a total area of 17 km². It consists of plagioclase (46 to 52 percent) + quartz + biotite + hornblende + K-feldspar. The opaque mineral is magnetite; allanite is also accessory, but sphene is absent. The plagioclase crystals are distinctive in that they have very calcic cores near An_{80} surrounded by sets of oscillatory zones, followed by normal zones ranging in composition from An_{35} to An_{25}. Large, well-shaped hornblende crystals are also common. The SiO_2 variation amongst the seven samples analyzed is between 62.3 percent and 65.9 percent. Other elements vary systematically when plotted against SiO_2. This is shown for Ni in Figure 4A.

Another mapped unit of the Kosciusko Batholith is the Round Flat Tonalite with an area of 7 km². It probably consists of two intrusions. Two samples from the hornblende-rich variety at the northern end contain 60.18 percent and 60.56 percent SiO_2, whereas two from the hornblende-poor southern end have SiO_2 contents of 66.07 percent and 66.71 percent. When Ni values for the Round Flat Tonalite are plotted against SiO_2, they extend the same trend determined for the Jindabyne Tonalite (Fig. 4B) and more clearly define that trend. The same relationship is found for all other elements so that the two units are assigned to the same *suite*. The Round Flat Tonalite also contains plagioclase with cores near An_{80} and has allanite but no sphene as an accessory phase. Another member of this Jindabyne Suite is the Pendergast Tonalite, which is hornblende-free and more felsic but which again has the characteristic plagioclase crystals and accessory allanite. Several other plutons exhibit these features and the Ni data for all of these lie near the same straight line (Fig. 4C). These are all members of the Jindabyne Suite.

Two other rock units forming part of the Berridale Batholith and occurring 50 km southsoutheast of the Jindabyne Suite rocks show many similar characteristics. Rocks of the Currowong Granodiorite range from hornblende-rich types with 63.76 percent SiO_2 to hornblende-poor variants containing 67.09 percent. Two specimens from the adjacent biotite-only Bimbimbie Granodiorite have SiO_2 values at 66.35 percent and 68.05 percent. All rocks from both the Currowong and Bimbimbie Granodiorites contain plagioclases with cores near An_{80} surrounded by complex oscillatory zoning. Magnetite and allanite are common accessory minerals as in the Jindabyne Suite. Both granodiorites have similar chemical characteristics to each other and are placed in the same suite, the Currowong Suite. This suite, however, has many features that are clearly distinct from the Jindabyne Suite. For example, Ni, although again low, has a distinctly different trend when plotted against SiO_2 (Fig. 4D). Both suites have similar Na-contents but the Jindabyne Suite rocks have lower K and are therefore tonalites, whereas the Currowong Suite rocks are granodiorites.

Using chemical criteria of this sort, the data for the whole

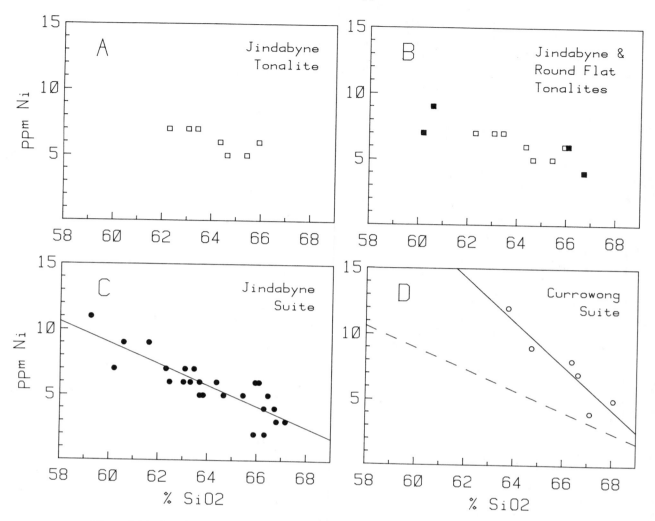

Figure 4. Harker diagrams showing Ni variation in the Jindabyne Tonalite (A), the Jindabyne and Round Flat Tonalites (B), the complete Jindabyne Suite (C), and the Currowong Suite (D).

region are used to delineate suites. Some suites consist of a single pluton, which by implication is unique, for example, the Wullwye Granodiorite of the Berridale Batholith. Other suites comprise many separate mappable units. Suites are mostly in linear belts parallel to the trend of the batholiths. An important feature is that invariably suites differ on each side of major country rock screens within the batholiths.

SOURCES OF GRANITOID MAGMAS

Suites indicate sources

Bayly (1968) said that "a suite is a group of rocks whose field relations and compositional characteristics make it appear to have a common source." Our usage of the term strictly follows this. It is clear, even from the limited data reproduced above for rocks of the Jindabyne and Currowong Suites, that one set of rocks, or suite, cannot be derived from

the other by processes such as fractional crystallization. Likewise, it must be concluded that distinct granitoid suites cannot be derived from the same source material, and in this way granitoids are imaging their source rocks (Chappell, 1979).

Variation within suites — the restite model

Curved trends on Harker diagrams are probably the result of some fractional crystallization process whereas linear trends are readily produced by mixing or unmixing. Linear trends of chemical variation for granitoid suites of southeast Australia are interpreted as resulting from varying degrees of separation or unmixing of the melt and restite components produced during the partial melting event (*restite model* of White and Chappell, 1977).

Easily recognized restite components in the Jindabyne and Currowong Suites are the calcic plagioclase cores.

These have incompletely reacted with the melt component during cooling and are preserved because they were isolated from the melt phase by the precipitation of more sodic plagioclase mantles as the magma rose. This is consistent with the negative slope of the plagioclase-out curves for dioritic melts (Eggler and Burnham, 1973). Residual zircons in granitoids are also restite. Clinopyroxene is thought to have been a common restite component of many granitoids, particularly hornblende-bearing types, but direct evidence for its previous existence is virtually confined to the presence of pyroxene cores in hornblende crystals. However, in one suite from the Bega Batholith, clinopyroxene crystals are included in plagioclase cores (Beams, 1980). Direct evidence for the existence of restite orthopyroxene is even more rare, although it has been found as small crystals in plagioclase cores. As the granite magma cools and f_{H_2O} builds up, orthopyroxene presumably reacts with the K-feldspar component in the melt to produce biotite. Confirmation of previous pyroxene restite that has undergone discontinuous reaction with melt is found in the presence of large pyroxene crystals or aggregates of crystals in volcanic rocks of equivalent composition. At this point, it can be noted that many of the granitoid suites of the Lachlan belt can be closely matched with equivalent volcanic suites (Wyborn and others, 1981).

Most inclusions in the granitoids are considered to be restite since few can be matched with the country rocks (Griffin and others, 1978). In the Currowong and Jindabyne Suites, inclusions are mafic hornblende-rich types containing complexly zoned and twinned plagioclases just as in the host granitoid. Mineralogically, inclusions match those of the host rock.

A certain chemical coherence also exists between inclusions and their hosts, but inclusion compositions commonly deviate from the compositional trends defined by the host granitoids. If inclusions represent made-over restite containing some melt, whereas the hosts are melt and some made-over restite, chemical differences are to be expected since the larger discrete inclusions probably represent original, more mafic and refractory heterogeneities in the source. Also, diffusional processes necessary to make-over the minerals of the restite inclusions to those stable at the new conditions of the cooling magma would be less efficient in large blocks. Compositions of inclusions have never been used to define suite compositions although the type of inclusion may be a good field and petrographic guide to the type of suite.

Inclusions are less abundant in the more felsic members of any particular suite. This is consistent with the restite model; more felsic magmas are those that have freed themselves of residual material. In any suite, the melt composition will have been close to that of the most felsic rock in which there is no restite. In some suites, these melts were minimum-melts with ~76 percent SiO_2; in other cases the melts were non-minimum, formed at a higher temperature, containing ~70 percent SiO_2. Variation of composition within any suite is dominantly the result of variations in the proportions of restite and melt in the magma.

SOURCE COMPOSITIONS

According to the restite model, the composition of the melt phases, the magma, and the source must lie along the same straight line. The linear trends on Harker variation diagrams — e.g., Figure 4C, defined by the composition of granitoids making up the suite — must project back through the source composition. This puts severe constraints on the nature of the source. For instance, the source rock of the Jindabyne Suite must have been very low in Ni, thus precluding a mantle source such as peridotite and even ocean-floor basalt which is also high in Ni (e.g., Sun and others, 1979). Jindabyne Suite rocks are considered to result from partial melting of rocks with an andesitic or basaltic-andesitic composition.

Using the large amount of data for each suite, estimates can be made of source rock composition for that suite. Clearly, there are many different source rocks in the Lachlan belt and, therefore, many types of granitoids resulting from separate partial melting events from a particular source. We have suggested that a first-order subdivision of the virtually infinite number of granite types can be made based on their source rock characteristics. Chappell and White (1974), recognized granitoids derived from sedimentary source rocks (S-types) and those derived from igneous sources (I-types). Loiselle and Wones (1979) proposed the term "A-type" for anorogenic granites that are usually mildly alkaline.

I-TYPE GRANITOIDS IN THE LACHLAN FOLD BELT

The geochemical characteristics of I-type granitoids are shown in Table 2. These are slightly modified from the characteristics given in 1974 because of our detailed study of more suites in recent years. The Jindabyne Suite (Hine and others, 1978) is an I-type suite derived by the partial melting of an igneous source of crustal origin. Calculations of source composition of the Jindabyne Suite (Compston and Chappell, 1979) indicate that it is one of the most mafic in the region; yet it is a basaltic andesite. Chemical characteristics (e.g., high K, low Ni) of all I-type granitoids studied in detail from the Lachlan belt show that direct derivation from mantle periodotite or ocean-floor basalt is not possible. We favor an origin from partial melting of primitive deep crust produced by the partial melting of mafic or ultramafic material at mantle depths, probably underplated beneath older crust (White, 1979).

The chemistry of I-type granitoids means that the mafic mineral assemblage is normally biotite + hornblende or

TABLE 2. GEOCHEMICAL CHARACTERISTICS OF I-TYPE GRANITOIDS

PARAMETER	CHARACTERISTIC VALUE	EXPLANATION
SiO_2	Wide range 53-76%	Relatively mafic source rocks
K_2O/Na_2O	Low	Na has not been removed by weathering
K_2O/SiO_2	Variable	Derived from source rocks of moderate and variable K-content
Ca	High in mafic rocks	High Ca in source; not removed by weathering
$\dfrac{Al_2O_3}{Na_2O+K_2O+CaO}$	Normally low	Only minimum temperature melts or fractionated I-type rocks may be peraluminous
Fe^{3+}/Fe^{2+}	Moderate	
Cr and Ni	Low	Source rocks relatively low in Cr and Ni, indicating prior fractionation
$\delta^{18}O$	Low	Primary igneous source rocks
$^{87}Sr/^{86}Sr$	Generally low	Mantle-derived igneous source rocks. Some high values for granitoids derived from old source rocks with high Rb/Sr

biotite alone. Muscovite or garnet along with biotite occurs in rare peraluminous examples, most of which appear to have been produced by fractionation. Accessory minerals of I-types normally include magnetite with or without ilmenite because of the relatively high oxidation state; ilmenite is present without magnetite in rare, more reduced types. Because of the high calcium content, sphene and/or allanite is the normal rare-earth-bearing phase; monazite occurs in certain metaluminous and in most peraluminous I-types.

Two main categories of I-type granitoid are recognized using chemical and petrographic data available on the southeastern Australian suites. For certain elements like phosphorus, the straight line on a Harker diagram projects to intersect the SiO_2 axis near 76 percent SiO_2. These are *minimum-melt* suites because members represent crystallized mixtures of a felsic melt component near 76 percent SiO_2 (minimum-melt) and restite (e.g., Moruya Suite: Griffin and others, 1978; White and Chappell, 1977). Phosphorus is an excellent element to determine the nature of the melt phase because it is virtually insoluble in pure quartz + feldspar melt. Minimum-melt suites are distinguished petrographically by the presence of restite such as calcic cores in plagioclase and clots of ferromagnesian minerals (made over restite) even in high-Si felsic members of the suite.

Nonminimum-melt suites are those in which elements such as phosphorus still have appreciable abundance in the high SiO_2 granites of that suite, particularly the member that is just restite free (e.g., Jindabyne Suite; Hine and others, 1978). The P_2O_5 variation with respect to SiO_2 has a more shallow slope than for minimum-melt suites: when the slope is zero, the abundance of P_2O_5 in the melt must be the same as that in the restite and the same as that in the source.

The residual mineralogy at the site of partial melting for nonminimum-melt I-types must be largely calcic plagioclase + clinopyroxene + orthopyroxene, so that all or most of the water is present in the melt. Given sufficient felsic components, the amount of melt produced at the site of partial melting is therefore a function of the amount of water released by breakdown of amphibole (Burnham, 1979) and biotite; if 25 percent melting is required before a magma may move (Compston & Chappell, 1979), the source rock composition will determine whether or not a magma is produced and whether it is a minimum or nonminimum-melt suite. The temperature must rise above minimum melting temperature to produce enough melt (25 percent) in the nonminimum-melt suite. Residues of minimum-melt suites are likely to contain some amphibole and biotite probably rich in F and Cl rather than OH.

S-TYPE GRANITOIDS OF THE BATHOLITHS

Granitoids of the recognized S-type suites of the region are considered to have been derived from a sedimentary source probably consisting of a mixture of greywacke and shale. It is stressed that these batholithic S-types are typical contact aureole granitoids; they occur as plutons with much the same size and shape as I-type granitoids. Volcanic rocks (ash flow tuffs and rhyolites) with equivalent compositions and containing biotite + orthopyroxene + cordierite $\pm$ garnet as common ferromagnesian assemblages are found in the Lachlan belt (Wyborn and others, 1981). These are S-type volcanic rocks.

Geochemical characteristics of S-types are shown in Table 3. High K_2O/Na_2O (Fig. 5) is explained by the fact that potassium is incorporated into clays during chemical weathering to produce sedimentary rocks, whereas sodium

TABLE 3. GEOCHEMICAL CHARACTERISTICS OF S-TYPE GRANITOIDS

PARAMETER	CHARACTERISTIC VALUE	EXPLANATION
SiO_2	Within range 65-74%	Derived from SiO_2-rich source
K_2O/Na_2O	High	K adsorbed by clays on weathering, whereas Na is removed
Ca and Sr	Low	Removed in weathering cycle
$\dfrac{Al_2O_3}{Na_2O+K_2O+CaO}$	High (>1.05) and increases as the rocks become more mafic	Weathering increases Al relative to Na+K+Ca
Fe^{3+}/Fe^{2+}	Low	Carbon common in sedimentary source rocks
Cr and Ni	High relative to I-types	Cr and Ni incorporated into clays during weathering
$\delta^{18}O$	High	Oxygen isotopes fractionate during production of clays during low temperature weathering
$^{87}Sr/^{86}Sr$	High (normally >0.708)	Rb concentrated relative to Sr during weathering and sedimentation

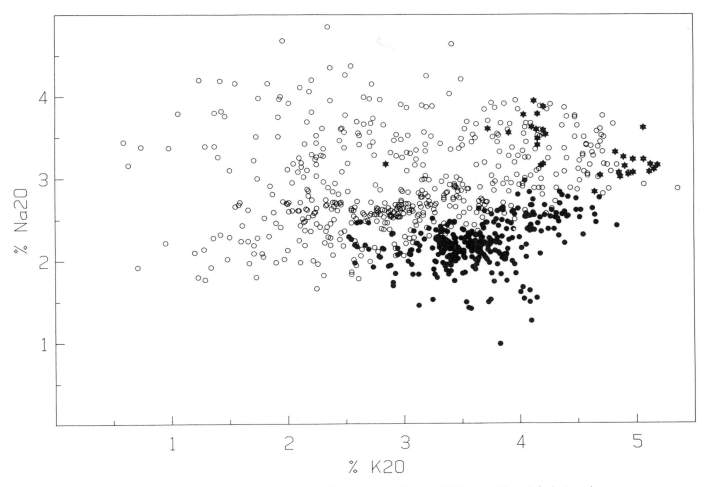

Figure 5. Plot of Na₂O versus K₂O for 532 I-type (open circles), 316 S-type (closed circles), and 31 A-type (stars) granitoids of the Lachlan Fold Belt.

is removed in solution along with Ca, Sr, and Pb. The elements removed are those present in feldspars but not in clays, so that their abundance is a function of the maturity of source sediments. It is found that abundance of these elements and hence the maturity of the source is the main factor in determining slight differences between various S-type suites. Initial $^{87}Sr/^{86}Sr$ also increases with maturity of the source.

Removal of Na and Ca relative to Al means that the source is peraluminous and thus all S-type granites are peraluminous. This feature is not, however, necessarily a diagnostic property of S-types since felsic I-types may also have this character. The low oxidation state results from the reduction of source rocks by carbon, and hence ilmenite, rather than magnetite, is the normal accessory opaque oxide phase. Transition metals are adsorbed onto clays and not removed during weathering so that at any SiO_2 value transitional elements are higher than those in I-types (Hine and others, 1978).

S-type source rocks are relatively siliceous in comparison with I-type sources. Consequently, the primary S-type restite will be relatively siliceous and quartz-bearing and will have the assemblage quartz + K-feldspar + plagioclase + biotite + cordierite + quartz + zircon + monazite at the P and T of magma consolidation. Since these are also the minerals seen in the crystallized granite, the amount of partial melting (25 percent) must be largely a function of the water content which is probably provided by the breakdown of biotite. The restite at the source of partial melting must therefore have contained hypersthene rather than biotite. The residual hypersthene reacts with the magma to produce biotite when the temperature is lowered, and f_{H_2O} increases as crystallization proceeds near the surface (Clemens and Wall, 1981). Evidence that hypersthene was a residual phase is found in the occurrence of orthopyroxene in S-type volcanic rocks (Wyborn and others, 1981).

THE I-S LINE

A clear spacial distribution of I- and S-type granitoids

exists within the southeastern Australian batholiths (Figure 6; White and others, 1976). A line separates dominantly S-type granitoids to the west from exclusively I-types in the east. This so-called I-S line passes through the center of the Berridale Batholith where it is marked by the presence of a prominent screen. The line is considered to be a major tectonic lineament. Rocks of sedimentary origin presumably occur to a greater depth in the crust in the west than in the east in spite of the similarities of the outcropping Ordovician-Silurian sediments on both sides of the line.

COOMA GRANODIORITE

The Cooma Granodiorite is a small (14 km²) body with the Al-rich mineral assemblage biotite + muscovite ± cordierite. It is a regional-aureole type of White and others (1974) and is therefore distinctly different from the batholithic granites already discussed, not only in geological setting but also in origin. However, it is similar in age to the granites of the adjacent Berridale Batholith.

Surrounding the Cooma Granodiorite is a terrane of low-pressure regional metamorphism with migmatites and K-feldspar + cordierite + andalusite + sillimanite gneisses at the highest grades, near the granodiorite contact. Progressively lower-grade rocks occur away from the contact. The regional "aureole" is widest on the western and northwestern

sides of the body where the outermost part of the biotite zone is some 10 km from the contact. The eastern side is largely covered by Cenozoic basalts, but the aureole is here much narrower. The field relationships with the migmatites and high-grade metamorphic rocks as well as the similarity in mineralogy and composition, including isotopic composition, between the Cooma Granodiorite and the country rocks (Pidgeon and Compston, 1965) led to the hypothesis that the Cooma Granodiorite was formed by partial melting of sedimentary rocks very similar or identical to those of the country rocks. it was concluded that the magma had not moved far from the site of partial melting. However, the high-grade assemblage K-feldspar + cordierite + andalusite is estimated to have crystallized at about 3 kb, and hence very high temperatures must have been attained high in the crust if the proposed origin is correct. Flood and Vernon (1978) suggested that the high-grade part of the aureole was diapirically dragged up from a deeper level with the intruding granodiorite. However, this does not alleviate the problem of exceptionally high gradients to any appreciable extent. Elsewhere in the Lachlan belt, west of longitude 148° E, Cooma-type regional-aureole granitoids are associated with larger terranes of low-pressure regional metamorphism, again indicating exceptionally high gradients.

The Cooma Granodiorite and other Cooma-type (Cooma Suite) granitoids of the Lachlan belt are characterized chemically by very low Na, Ca, and Sr contents indicating derivation from mature sediments like those of the Ordovician. The S-types of the large batholiths are richer in these elements and hence are derived from less mature sediments of a pre-Ordovician sedimentary layer not known to outcrop anywhere in the Lachlan belt (Wyborn and Chappell, 1982).

A-TYPE GRANITOIDS

Several large plutons at or near the eastern edge of the Bega Batholith (Fig. 2) and some elsewhere consist of felsic granitoids, mostly with annite-rich biotite as the main mafic mineral, and with fluorite as a common secondary (or possibly primary) phase. Some of these distinctive granitoids also contain small amounts of arfvedsonitic amphibole and/or late-stage, and possibly secondary, mattes or clusters of riebeckite needles. All of these granitoids are Late Devonian and are hence younger than the main body of granitoids. They are anorogenic and are called A-types (Loiselle and Wones, 1979).

The A-type granitoids at the eastern edge of the Bega Batholith are associated in space and time with volcanic rocks and are therefore subvolcanic according to the classification of White and others (1974). Textures such as miarolitic cavities and granophyric intergrowths are consistent with near-surface intrusion. It is this type of granitoid in a subvolcanic setting that includes the "hypersolvus" granites.

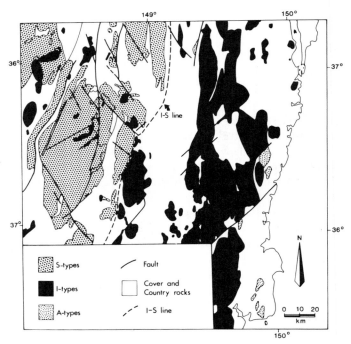

Figure 6. Map showing the distribution of S-, I-, A-type granitoids in the Lachlan Fold Belt south of Canberra. The I-S line marks the eastern limit of S-type granitoid occurrences. It separates dominantly S-type granitoids to the west from exclusively I-type granitoids to the east, and it is thought to represent the eastern limit of thick crystalline basement.

The A-type granitoids have chemical characteristics (Table 4) distinctly different from the I- and S-types discussed above. Diagnostic is the abundance of large, highly-charged cations such as Ga (Fig. 7), Nb, Sn, Zr, and REE (Collins and others, 1982). The last granitoids produced in the younger New England Batholith, northeast of the Lachlan Fold Belt, are also of this type. These are the late Permian and Triassic leucoadamellites of Shaw and Flood (1981), who describe chemical features very similar to the Lachlan A-types. Also in the New England Fold Belt, Ewart (1981) has described a suite of Late Oligocene to Early Miocene high-silica rhyolites with similar chemical characteristics.

Fractional crystallization to produce granitoids of this type from transitional to alkalic basaltic magma (Loiselle and Wones, 1979) in general fits the geochemical and textural relationships. The fact that A-type rhyolitic volcanic rocks commonly occur as flows, whereas S- and I-type volcanic rocks are either ash-flows or lava domes, indicates that the viscosity is low and this is favorable for crystal settling. On the other hand, the frequent occurrence of large

TABLE 4. GEOCHEMICAL CHARACTERISTICS OF A-TYPE GRANITOIDS

PARAMETER	CHARACTERISTIC VALUE	EXPLANATION
SiO_2	Usually high, often near 77%	Small degree of partial melting
Na_2O	High	Small degree of partial melting
CaO	Low	Small degree of partial melting, Ca not compatible with melt structure
Ga/Al	High	Ga complexed in melt, plagioclase in residue
Y and REE	High except Eu	Complexed in melt, with much Eu remaining in anorthite
Nb and Sn	High	Complexed in melt
Zr	Normally high, particularly in more mafic varieties	Complexed in melt
F and Cl	High	Source rock is a residue from an earlier melt and is rich in F and Cl

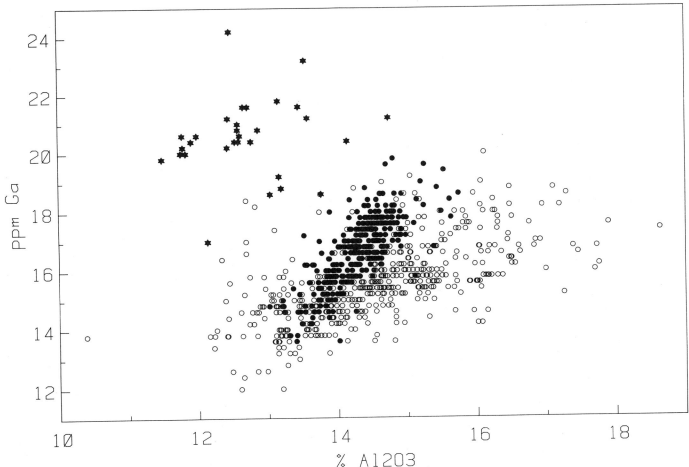

Figure 7. Plot of Ga versus Al_2O_3 for 532 I-type (open circles), 316 S-type (closed circles), and 31 A-type (stars) granitoids of the Lachlan Fold Belt.

White and Chappell

volumes of A-types, as in the Lachlan belt, without abundant associated gabbros (or basalts) and with no intermediate rocks, does not favor an origin by fractional crystallization.

Rhyolitic rocks with A-type chemical compositions are common in the western USA where Shaw and others (1976) have ascribed their unusual chemical composition to a process of thermogravitational diffusion whereby the large, highly-charged cations are concentrated at the top of I-type magma chambers. The occurrence in the Lachlan belt, of A-type plutons similar in size and shape to those of the I-types, with no detectable variation in composition over vertical distances of up to 450 meters, suggests that the process responsible for producing A-type compositions is more deep-seated than a subvolcanic environment.

Our favored hypothesis to explain the production of A-type melts involves their production from an F- and/or Cl-rich source that is essentially anhydrous. The source rocks have already been depleted in water by the extraction of minimum-melt I-type magmas. This hypothesis explains why A-types are always late in the magmatic history of a region and also why granitoids (or rhyolites) are more abundant than other igneous rocks in A-type provinces. This is similar to the origins proposed by O'Neil and others (1977) and Shaw and Flood (1981), and by Ewart (1981), for the plutonic and volcanic rocks, respectively, in the New England Fold Belt. We suggest that the unusual chemical composition of A-type granitoids results from high F and/or Cl and low H_2O activities. The highly-charged cations such as Nb, Ta, Sn, and Ga are those that readily complex with F; the melt composition is determined by its structure. A detailed account of this model for the production of A-type compositions is given by Collins and others (1982).

COMPARISON OF CHEMISTRY OF I-, S-, and A-TYPES

Average chemical analyses of I-, S-, and A-type granitoids are listed in Table 5. These are averages of those analyses for which complete data for the major and the more abundant trace elements are available. We expect complete chemical information on the granitoids of the eastern Lachlan belt to be available during 1983. This will include some 1,400 granitoid analyses.

The average I- and S-type compositions in Table 5 show many similarities, e.g., SiO_2, Al_2O_3, total Fe, MgO, and REE. However, the distinctive differences between the two types are shown. Thus, relative to the I-types, the S-types are low in CaO, Na_2O, Sr, and Fe_2O_3/FeO, and high in K_2O, Rb, Cr, and Ni.

A significant difference between the I- and S-type granitoids that is not apparent from the data in Table 5 is the much broader range in composition of the I-types. For the S-type granitoids, the complete range in SiO_2-content is

TABLE 5. AVERAGE ANALYSES OF GRANITOID TYPES FROM THE LACHLAN FOLD BELT

	I-type	S-type	A-type
No of samples	532	316	31
SiO_2	67.98	69.08	73.60
TiO_2	0.45	0.55	0.33
Al_2O_3	14.49	14.30	12.69
Fe_2O_3	1.27	0.73	0.99
FeO	2.57	3.23	1.72
MnO	0.08	0.06	0.06
MgO	1.75	1.82	0.33
CaO	3.78	2.49	1.09
Na_2O	2.95	2.20	3.34
K_2O	3.05	3.63	4.51
P_2O_5	0.11	0.13	0.09
Trace elements (ppm)			
Ba	520	480	605
Rb	132	180	199
Sr	253	139	105
Pb	16	27	29
Th	16	19	23
U	3	3	5
Zr	143	170	342
Nb	9	11	22
Y	27	32	76
La	29	31	55
Ce	63	69	134
Nd	23	25	56
Sc	15	14	14
V	74	72	10
Cr	27	46	3
Co	12	13	4
Ni	9	17	2
Cu	11	12	6
Zn	52	64	102
Ga	16	17	21

from 63.4 to 74.5 percent (only two analyses are below 65 percent), with a mean of 69.08 ± 1.90 percent (1σ). For the I-types, the range is larger at both ends. Low-SiO_2 granitoids are transitional to gabbroids, and in constructing Table 5 we have taken 53 percent SiO_2 as the lower limit, the boundary

between gabbro and gabbroic-diorite. This gives a range of SiO_2 from 53 to 77.5 percent, with a mean of 67.98 ± 5.16 percent. Again, for total iron as FeO the mean values are similar, with a greater range for the I-types (3.71 ± 1.69 percent) than the S-types (3.89 ± 0.80 percent). The more restricted range in composition at low SiO_2 for S-type granitoids results from the source rock (pelite) being relatively close to the melt in composition so that varying degrees of melt-restite separation do not generate as wide a spectrum of compositions. High-SiO_2 S-types are also less abundant; for the analyses averaged in Table 5, only five S-types have SiO_2 greater than 73.5 percent (1.6 percent of total), whereas 88 I-types (16.5 percent) are above that value. The relatively limited range of total S-type compositions is also shown by the data plotted in Figures 5 and 7.

The A-type granitoid average in Table 5 is much more felsic than the other two averages, and many of the differences are due to this factor alone. However, the high Zr, Nb, Y, REE, Zn, and Ga are distinctive. If only felsic I-and S-types are compared with A-types, the latter are seen to contain significantly higher amounts of Ba and Sc and less Al_2O_3, MgO, CaO, and Sr (Collins and others, 1982).

REFERENCES CITED

Bayly, B., 1968, Introduction to petrology: Prentice Hall, 371 p.

Beams, S. D., 1980, Magmatic evolution of the southeast Lachlan Fold Belt [Ph.D. thesis]: La Trobe University, Bundoora.

Burnham, C. W., 1979a, Magmas and hydrothermal fluids, *in* H. L. Barnes, ed., Geochemistry of hydrothermal ore deposits, 2nd ed.: New York, Wiley-Interscience, p. 71–136.

Chappell, B. W., 1979, Granites as images of their source rocks: Geological Society of America Abstracts with Programs, v. 11, p. 400.

Chappell, B. W., and White, A. J. R., 1974, Two contrasting granite types: Pacific Geology, v. 8, p. 173–174.

——,1976, Plutonic rocks of the Lachlan Mobile Zone: Excursion Guide International Geological Congress, 25th, Sydney, 41 p.

Clemens, J. D., and Wall, V. J., 1981, Origin and crystallization of some peraluminous (S-type) granitic magmas: Canadian Mineralogist, v. 19, p. 111–131.

Collins, W. J., Beams, S. D., White, A. J. R., and Chappell, B. W., 1982, Nature and origin of A-type granites with particular reference to southeastern Australia: Contributions to Mineralogy and Petrology (in press).

Compston, W., and Chappell, B. W., 1979, Sr-isotope evolution of granitoid source rocks, *in* M.W. McElhinny, ed., The Earth, its origin, structure and evolution: New York, Academic Press, p. 377–426.

Crook, K. A. W., 1980, Fore-arc evolution in the Tasman Geosyncline: the origin of the southeast Australian continental crust: Geological Society of Australia Journal, v. 27, p. 215–232.

Cull, J. P. 1979, Perturbations to the geothermal field and glacial retreat in the Snowy Mountains: Bureau of Mineral Resources, Canberra, record 79/2, p. 20–21.

Day, R. A., Nicholls, I. A., and Hunt, F. L., 1979, The Meredith ultramafic breccia pipe — Victoria's first kimberlite?: Bureau of Mineral Resources, Canberra, record 79/2, p. 22–23.

Eggler, D. H., and Burnham, C. W., 1973, Crystallization and fractionation trends in the system andesite-H_2O-CO_2-O_2 at pressures to 10 kb: Geological Society of America Bulletin, v. 84, p. 2517–2532.

Ewart, A., 1981, The mineralogy and chemistry of the anorogenic Tertiary silicic volcanics of S.E. Queensland and N.E. New South Wales, Australia: Journal of Geophysical Research, v. 86, p. 10242–10256.

Ferguson, J., Arculus, R. J., and Joyce, J., 1979, Kimberlite and kimberlitic intrusives of southeastern Australia: a review: Bureau of Mineral Resources Journal, v. 4, p. 227–241.

Ferguson, J., Ellis, D. J., and England, R. N., 1977, Unique spinel-garnet lherzolite inclusion in kimberlite from Australia: Geology, v. 5, p. 278–280.

Finlayson, D. M., Prodehl, C., and Collins, C. D. N., 1979, Explosion seismic profiles, and implications for crustal evolution in southeastern Australia: Bureau of Mineral Resources Journal, v. 4, p. 243–252.

Flood, R. H., and Vernon, R. H., 1978, The Cooma Granodiorite, Australia: an example of in-situ crustal anatexis?: Geology, v. 6, p. 81–84.

Griffin, T. J., White, A. J. R., and Chappell, B. W., 1978, The Moruya Batholith and geochemical contrasts between the Moruya and Jindabyne suites: Geological Society of Australia Journal, v. 25, p. 235–247.

Hine, R. H., Williams, I. S., Chappell, B. W., and White, A. J. R., 1978, Geochemical contrasts between I- and S-type granitoids of the Kosciusko Batholith: Geological Society of Australia Journal, v. 25, p. 219–234.

Loiselle, M. C., and Wones, D. R., 1979, Characteristics and origin of anorogenic granites: Geological Society of America Abstracts with Programs, v. 11, p. 468.

Lovering, J. F., and White, A. J. R., 1969, Granulitic and eclogitic inclusions from basic pipes at Delegate, Australia: Contributions to Mineralogy and Petrology, v. 21, p. 9–52.

O'Neil, J. R., Shaw, S. E., and Flood, R. H., 1977, Oxygen and hydrogen isotope compositions as indicators of granite genesis in the New England Batholith, Australia: Contributions to Mineralogy and Petrology, v. 62, p. 313–328.

Pidgeon, R. T., and Compston, W., 1965, The age and origin of the Cooma granite and its associated metamorphic zones, New South Wales: Journal of Petrology, v. 6, p. 193–222.

Pinchin, J., 1979, Deep crustal seismic reflections at Gundary Plains, New South Wales: Bureau of Mineral Resources, Canberra, record 79/2, p. 68–70.

Richards, D. N. G., 1980, Palaeozoic granitoids of northeastern Australia, *in* R. A. Henderson and P. J. Stephenson, eds., The geology and geophysics of northeastern Australia: Geological Society of Australia, Queensland Division, p. 229–246.

Rutland, R. W. R., 1976, Orogenic evolution of Australia: Earth Science Reviews, v. 12, p. 161–196.

Sass, J. H., and Lachenbruch, A. H., 1979, Thermal regime of the Australian continental crust, *in* M. W. McElhinny, ed., The Earth, its origin, structure and evolution: Academic Press, p. 301–351.

Scheibner, E., 1974, An outline of the tectonic development of New South Wales with special reference to mineralization, *in* N. L. Markham and H. Basden, eds., The mineral deposits of New South Wales: Geological Survey of New South Wales, p. 3–39.

Shaw, H. R., Smith, R. L., and Hildreth, W., 1976, Thermogravitational mechanisms for chemical variations in zoned magma chambers: Geological Society of America Abstracts with Programs, v. 8, p. 1102.

Shaw, S. E., and Flood, R. H., 1981, The New England Batholith, eastern Australia: geochemical variations in time and space: Journal of Geophysical Research, v. 86, p. 10530–10544.

Sun, S.-S., Nesbitt, R. W., and Sharaskin, A. Y., 1979, Geochemical characteristics of mid-ocean ridge basalts: Earth Planet: Science Letters, v. 44, p. 119–138.

Wellman, P., 1979, On the Cainozoic uplift of the southeastern Australian highland: Geological Society of Australia Journal, v. 26, p 1–19.

White, A. J. R., 1979, Sources of granite magmas: Geological Society of

America Abstracts with Programs, v. 11, p. 539.

White, A. J. R., and Chappell, B. W., 1977, Ultrametamorphism and granitoid genesis: Tectonophysics, v. 43, p. 7–22.

——,1982, Geology of the Numbla 1:100,000 Sheet 8624: Geological Survey of New South Wales.

White, A. J. R., Chappell, B. W., and Cleary, J. R., 1974, Geologic setting and emplacement of some Australian Paleozoic batholiths and implications for intrusive mechanisms: Pacific Geology, v. 8, p. 159-171.

White, A. J. R., Williams, I. S., and Chappell, B. W., 1976, The Jindabyne Thrust and its tectonic, physiographic and petrogenetic significance: Geological Society of Australia Journal, v. 23, p. 105–112.

——,1977, Geology of the Berridale 1:100,000 Sheet 8625: Geological Survey of New South Wales.

Wyborn, D., Chappell, B. W., and Johnston, R. M., 1981, Three S-type volcanic suites from the Lachlan Fold Belt, southeast Australia: Journal of Geophysical Research, v. 86, p. 10335-10348.

Wyborn, L. A. I., and Chappell, B. W., 1982, Chemistry of the Ordovician and Silurian greywackes of the Snowy Mountains, southeastern Australia: an example of chemical evolution of sediments with time: Chemical Geology (in press).

MANUSCRIPT ACCEPTED BY THE SOCIETY JULY 12, 1982

Geological Society of America
Memoir 159
1983

Multiple Mesozoic Sn-W-Sb granitoids of southeast Asia

Charles S. Hutchison
Department of Geology
University of Malaya
Kuala Lumpur 22-11
Malaysia

ABSTRACT

Southeast Asia is a complex array of granitoid belts, predominantly of Mesozoic age. An eastern belt, forming the eastern third of the Malay Peninsula and extending to the Indonesian islands of Bangka and Billiton, represents an epizonal Andean-type calc-alkaline volcano-plutonic arc which was active from the Permian to the Late Triassic. The plutonic suite ranges from gabbro, through hornblende-biotite granodiorite, to granitic, and is of mixed S- and I-type. The plutons are ringed by aureoles of andalusite-cordierite hornfels. The tin and tungsten deposits are related to the ilmenite-series members. The arc has remained isostatically stable since the Late Triassic, as reflected in the concordant Rb:Sr isotope dates.

A narrow central belt of Permian and Triassic granitoids and metamorphic complexes with local Cretaceous granites along its margins separates the eastern belt from the Main Range. The eastern margin of the Main Range is a serpentine-marked suture zone with local, strongly gneissic granites.

The Main Range batholith is the most remarkable Sn-granite in the world. It is predominantly of Late Triassic age but includes granites of Permian age. It is strictly of granitic composition and entirely of the S- or ilmenite-series. The main or eastern part is of Mesozonal emplacement into greenschist facies metasediments with no thermal overprint, and consists mainly of biotite granite which commonly contains megacrysts of maximum microcline. Muscovite results only from greisenization in hydrothermally affected areas. The Main Range belt imperceptibly grades westward through Penang, Langkawi, and peninsular Thailand to higher level plutons, which are unsheared and surrounded by andalusite and cordierite-bearing aureoles.

The north Thailand granites are predominantly Triassic and appear to be higher level than the eastern parts of the Main Range. They have some of the highest initial $^{87}Sr/^{86}Sr$ ratios of granites in the world, commonly in the range 0.722 to 0.734, and undoubtedly are entirely of upper crustal origin. Both the Main Range and northern Thai granites are devoid of contemporaneous volcanic associations, and they have been interpreted as collision-related, resulting from the closure of the central marginal basin in the Late Triassic. This represented the major orogenic event leading to the cratonization of Sundaland. Faulting, uplift, and hydrothermal circulation led to highly discordant Rb:Sr and K:Ar radiometric dates, which suggests that the region has been isostatically unstable since the Triassic; it continues to be characterized by hot springs. The granites of the western belt of Phuket and Burma are high level and characterized by good thermal aureoles, but they are not associated with obvious contemporaneous volcanic rocks. The granites and associated pegmatite emplacements were fault controlled.

The Triassic granites generally, and some of the Cretaceous granites which occur in diverse localities, are associated with tin, tungsten, and antimony deposits, representing one of the world's greatest metallogenic provinces. These elements are thought to be recycled from the continental infrastructure of Sundaland.

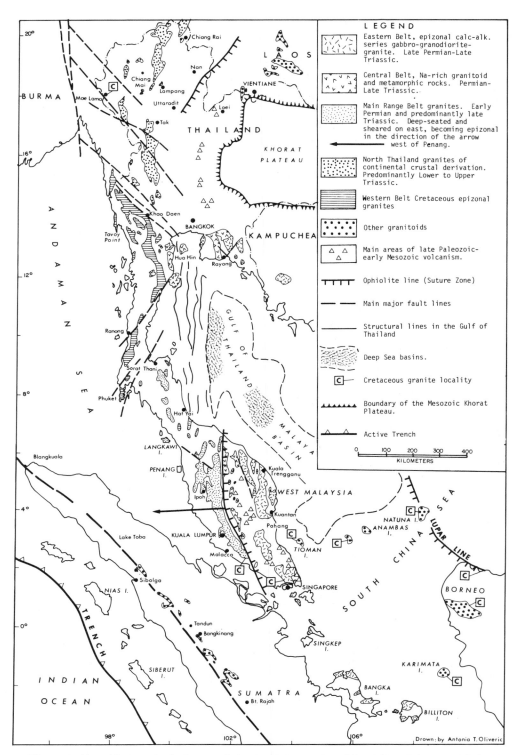

Figure 1. Simplified geological map of the Sundaland area of Thailand, Burma, Indochina, Malay Peninsula, Sumatra, and west Borneo. The various granitoid belts are taken from Hutchison (1978) and Hutchison and Taylor (1978).

INTRODUCTION

The southeast Asia granitoids form several belts of different tectonic and petrological character (Fig. 1). They are predominantly Mesozoic, but radiometric dating has encountered numerous problems, especially the discrepancies between the Rb:Sr and K:Ar methods. However, the increasing amount of careful isochron data has led to a better understanding of these plutonic belts. This understanding is fundamental to theories of the genesis and development of metallogenesis related to acid magmatism, for this region contains the greatest tin fields of the world and towards the north also produces tungsten and antimony (Hutchison and Taylor, 1978; Taylor and Hutchison, 1979).

Uplift and unroofing played an extremely important role in the formation of the giant placer cassiterite tin fields of the Malay Peninsula. The unroofed granites were deeply weathered during the Late Tertiary, and rapid erosion during the early Quaternary and Late Tertiary led to the well-developed placer tin fields of Kuala Lumpur and the Kinta Valley of Ipoh (Taylor and Hutchison, 1979).

Before the various granitoid belts are described, some of the parameters which have helped in their understanding will be briefly reviewed.

1) S- and I-Type and Ilmenite and Magnetite Series

The classification of granitoids into S- and I-type by Chappell and White (1974) and White and Chappell (1977) has greatly aided in defining models of the source of granite magmas. Their subdivision is based on chemical parameters which require whole-rock analyses for its application. The subdivision into ilmenite-series and magnetite-series, as described by Ishihara (1977), is more readily applied because only a portable magnetic susceptibility meter is needed. The classification is important for mineral exploration because Sn, W, Nb, and Ta deposits are associated with ilmenite series granitoids, and Mo, Cu, Pb, Zn, Ag, and Au deposits with the magnetite series (Ishihara, 1980). The significance of the opaque oxides in the granites was well known in southeast Asia. Aranyakanon (1961) reported that granites which accompany tin deposits contain little magnetite, but tin-barren granitoids contain an average 1.26 wt. % magnetite. Indeed, every tin mine has a pile of gangue material (amang) which is predominantly ilmenite.

The equivalence of the S-I and ilmenite-magnetite series is not simple. Takahashi et al. (1980) noted that many ilmenite-series rocks of Japan have I-type chemical characteristics, and they concluded that the two classifications are not exactly equivalent.

The S- and I-classification based on Australian samples may not be equally suitable to southeast Asian rocks, for Ishihara et al. (1980) have shown that there is a significant difference between granites of peninsular Thailand and those of Australia. This difference is shown on the normative qu-or-ab diagram. White et al. (1977) suggested that "I-types lie on the feldspar side of the quartz-feldspar liquidus phase boundary (for 500 bars water pressure) and S-types all lie in the quartz field." However, all the S-granites from south Thailand lie on the feldspar side. The inconsistency seems to depend upon the different crustal basements from which the magmas were derived. The silicic nature of the Australian granites is in keeping with the abundant quartz inclusions and metapelitic xenoliths; whereas S-type granites of the Malay Peninsula contain only metapelitic xenoliths.

2) Radiometric Dating

The problems of radiometric dating of southeast Asian granitoids have recently been highlighted by CCOP (1980) and Beckinsale et al. (1979). The most outstanding feature is the great discordance between K:Ar and Rb:Sr ages. A normal, slightly discordant pattern would be Rb:Sr whole-rock isochron age greater than the K:Ar muscovite age, which in turn is greater than the K:Ar age on biotite. This is usually interpreted as indicating the different times when each of the radiometric systems fell below the blocking temperature for diffusion of the particular radiogenic nuclide, during the protracted cooling history of the plutons. However, in southeast Asia, grossly discordant ages are a common occurrence (Beckinsale et al., 1979; Bignell and Snelling, 1977; Bignell, 1972). The simplest interpretation of these is to assume that the only true age of the igneous intrusion is given by a whole-rock Rb:Sr isochron, and to be reliable it must be demonstrated that all the samples on the isochron are from the same pluton. Paucity of exposures and field relationships makes this last requirement uncertain in southeast Asia, so that some of the existing data need to be reinvestigated. The K:Ar ages do not date the intrusion. They may represent a later thermal event or uplift of a granite which has been held at depth at a temperature higher than the blocking temperature. It is likely that the faulting allowed the convecting hydrothermal systems easy access to the granites to reset the K:Ar clocks. The K:Ar ages may therefore coincide with mineralization events.

3) Initial $^{87}Sr/^{86}Sr$ Ratios

When the granitoids have been dated by good whole-rock isochrons, the initial Sr isotope ratios offer powerful indicators of the source of the magmas (Faure, 1977). Southeast Asian granitoids show a very large range of values (Fig. 2). In general the explanation for those granitoids, which have a ratio higher than 0.710, is that the granites contain a major component of crustal strontium (Faure, 1977). One should generally expect a relationship between S- and I-type granites and the initial Sr-isotope ratios. I-type granitoids are of predominantly tonalitic or granodioritic composition and contain abundant mafic xenoliths. Their initial $^{87}Sr/^{86}Sr$ ratios of 0.704 to 0.706 suggest a mantle or lower crustal source of igneous composition. S-type granitoids contain metasedimentary xenoliths, and their initial Sr-isotope ratios are greater than 0.706.

In southeast Asia, tin and tungsten mineralization appears to be confined to the more silicic S-type granites whose whole-rock initial Sr-isotope ratios exceed 0.707, indicating that the tin is recycled from buried continental crust by polyphase anatectic events (Hutchison and Chakraborty, 1979).

4) Indicators of Level of Emplacement

Following the review by Buddington (1959), Hutchison (1977) studied the granitoids of peninsular Malaysia and found that the type of thermal aureole, the triclinicity of microcline, and the presence of muscovite were all useful indicators of the level of emplacement. More specifically, he found that high-level, or epizonal granites have been intruded into sedimentary formations or pre-existing regionally metamorphosed rocks, and that the granites are characterized by wide contact thermal aureoles (up to 3 km), which contain cordierite and andalusite. Deeper seated, or mesozonal granite batholiths were intruded into greenschist facies metasediments. The contacts are sharp, and lack cordierite and andalusite, but the dynamothermal metamorphism was sufficient to form local zones of garnet or amphibole.

High-level granites commonly contain pinkish alkali feldspars. The pink color is enhanced when the granitoid is slightly weathered on outcrop. This is attributed to the oxidation of iron oxide which was not expelled from the feldspars. Mesozonal granites of southeast Asia are always white or gray. The structural state of the alkali feldspars also gives valuable information. The epizonal granites contain alkali feldspars whose structural state varies from orthoclase, at or near the contact, to intermediate microcline deeper within the pluton. This can be determined by 2V measurements or by X-ray studies (Fig. 3), as outlined by Hutchison (1977). By contrast the deep-seated granites, such as the Main Range, contain maximum microcline and tend to be coarsely porphyritic. The attainment of the maximum microcline structural state is not just a result of the slow cooling, but is aided by the impoundment of volatiles within the granite cupolas beneath unfractured roof rocks. Thus, the hydrothermal fluids accumulate in the cupolas and the ore bodies are closely confined to the contact zones, as in the Main Range (Hosking, 1973). Epizonal granitoids are emplaced into shallowly buried country rocks, which fracture easily and allow the hydrothermal fluids to escape easily along the fault and joint systems to give exogranite ore bodies in laterally extensive hydrothermal veins within the contact metamorphic aureole (Hosking, 1977),.

Maximum microcline is found in coarsely perthitic feldspars where the polysynthetic, cross-hatched twinning extends fully across the whole crystal. The lamellae are up to 50 μm wide. By contrast, intermediate microcline never has perfectly developed cross-hatched twinning; it rarely extends fully across the crystals, and the perthitic nature is not on such a coarse scale. The lamellae are usually thinner than 5 μm.

Muscovite, where present in southeast Asian granites, is always seen to replace primary biotite, and is a product of greisenization. The high-level granitoids are less likely to be extensively greisenized because the volatiles escape from the roof zones. However, the deeper seated granites, of the Main Range, are characterized in their roof zones by a multitude of closely spaced mineralized veins, which have little vertical extent, and which are characteristically greisen-bordered (Hosking, 1977). The greisenization of the roof zones of deeper seated granite plutons, noted also by Varlamoff (1978), is likely to be accompanied by extensive tourmalinization (Hutchison and Leow, 1963).

Although the vast majority of southeast Asian granites are biotite granite, the higher level plutons of the calcalkaline series usually contain euhedral hornblendes in addition to biotite.

5) The Plutonic Suite and Orogen Type

Two contrasting granitoid series that appear to characterize different types of orogenic belts can be readily

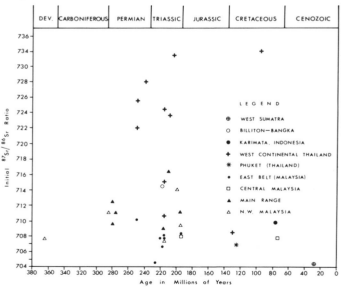

Figure 2. Summary of the dating of southeast Asian granitoids showing only those dates obtained from Rb:Sr isochrons. The age is plotted against the respective initial $^{87}Sr/^{86}Sr$ ratio. An outstanding feature is their generally extremely high initial ratios, indicating derivation from mature continental crust. Sources: Bignell and Snelling (1977), Beckinsale et al. (1979), Priem et al. (1975), Page et al. (1979).

identified. One is a compositionally expanded calc-alkaline series; the other is compositionally restricted and predominantly granitic. The difference is most clearly expressed in terms of the gabbro-diorite/tonalite-granodiorite/granite proportions which in the expanded suite are near 15/50/35, and in the restricted suite are near 2/18/80 (Pitcher, 1979).

The calc-alkaline series is characteristic of the Andinotype orogenies (Pitcher, 1979) and is taken to represent an active cordilleran plate margin, with subduction being the main orogenic mechanism. The plutonic rocks are commonly associated with volcanic rocks which range from basalt to rhyolite.

The compositionally restricted granite series was called the Hercynotype by Pitcher (1979). Hutchison (1978) and Mitchell (1979) suggested that batholiths of this kind result from collision of an island arc with a stable craton, by closure of the intervening marginal basin. If the colliding volcano-plutonic arc was characterized by a normal subduction-related calc-alkaline series, then after collision, it will end up closely parallel to the collision-related compositionally restricted granitic belt to give paired plutonic belts of contrasting character. It is considered that the Eastern and Main Range belts at Malaysia are examples of granitoids produced in that manner (Hutchison, 1978).

6) Rb/Sr Ratios and K:Rb Fractionation

The tendency for Sr concentrations to decrease and Rb to increase in a differentiated series of igneous rocks is well documented. Sr initially enters the calcic plagioclase, whereas Rb enters the alkali feldspar and micas and is thus concentrated in the later differentiates. However, the process of greisenization involves alteration of the plagioclase and should be expected to increase the Rb/Sr ratios.

The Main Range and southern Thailand granites display very high Rb/Sr ratios, whereas the Eastern Belt granitoids are characterized by low values (Hutchison, 1977). As summarized in Figure 4, the granites of the Main Range and south Thailand are more highly differentiated than those of the Eastern calc-alkaline belt. The Cretaceous granites of peninsular Malaysia and Phuket occupy intermediate positions. To some extent, these findings are paralleled by other geochemical evidence and coincide with the range of the crystallization index (Poldervaart and Parker, 1964; Hutchison, 1973a).

The fractionation of Rb and Sr with respect to potassium has been discussed by Shaw (1968, 1970). He defined an R value as = (%K × 10⁴/ppm Rb). The slight difference in ionic radius requires that Rb be enriched relative to K in residual

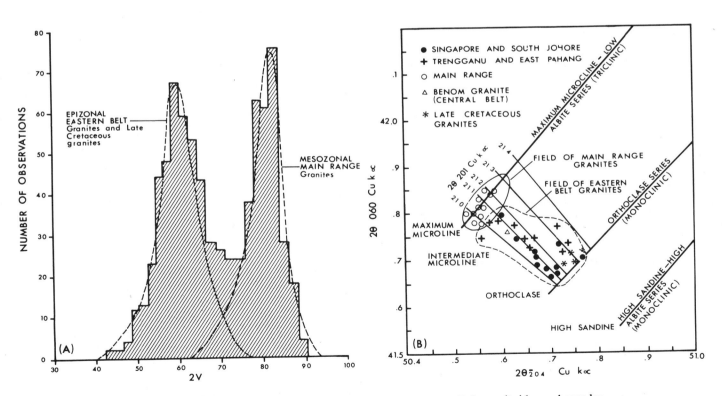

Figure 3. (A) The contrast between the Main Range and Eastern Belt granitoids, as shown by the nature of their alkali feldspars, after Hutchison (1977). The optic axial angle peaks coincide with the fields of orthoclase to intermediate microcline (left) and maximum microcline (right). (B) The X-ray data are plotted on a Wright (1968) diagram.

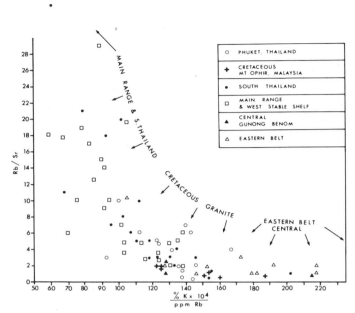

Figure 4. Rb/Sr ratio versus R = (%K × 10⁴/ppm Rb) for various granitoids of the Malay Peninsula. The more differentiated granites occupy the upper left part of the diagram. A few extreme Rb/Sr and R values have been omitted. Data from Bignell and Snelling (1977), Ishihara et al. (1980), Yeap (1980), and Garson et al. (1975).

crystallizing fluids, giving rise ultimately to pegmatites with low R values, which imply extreme fractionation. The R values are plotted against Rb/Sr ratios for a number of southeast Asian granitoids (Fig. 4) and clearly illustrate the differences between the various granitoid belts. The Eastern calc-alkaline series is the least fractionated, characterized by high R and low Rb/Sr values. The Central Belt and the Cretaceous granites of peninsular Malaysia and Phuket occupy an intermediate position. The Main Range and western stable shelf granites, including those of south peninsular Thailand, are the most differentiated, characterized by extremely high Rb/Sr ratios and low R values. Unfortunately, there are no data on the granites of northern Thailand.

GEOLOGICAL FRAMEWORK

The region may be conveniently divided by the suture lines of north Thailand (shown on Fig. 1, east of Uttaradit and Nan), northerly trending structural lines in the Gulf of Thailand, and the Bentong-Raub suture extending down the length of peninsular Malaysia to be lost under the Straits of Malacca (Hutchison, 1975). To the west of this line, the Paleozoic stratigraphy represents a miogeoclinal shallow water stable shelf (Hutchison, 1973a, and in press). The similar stratigraphy throughout the region indicates that this Shan-Thai-Main Range massif was once attached on its

western side to a major unknown craton. The recent mapping in northern Sumatra (Page et al., 1979) indicates that Sumatra was part of this massif throughout the Late Paleozoic and Triassic. The part of this western massif adjacent to the Andaman Sea has been complicated by the opening of the active marginal sea and the separation of the Sumatran Cenozoic volcanic arc from the recently extinct Burmese volcanic arc to the north. These volcanic arcs can be considered to have been immediately adjacent to the Phuket area of Thailand before the Andaman Sea opened (Mitchell, 1977), and the major transcurrent faults which outline the group E granitoids of Figure 1 played an important role in creating the present shape of the peninsula.

East of the suture lines, there is a Paleozoic-Triassic marginal basin which extends through Thailand from Loei to Bangkok and down the length of peninsular Malaysia (Fig. 1). This basin was characterized by marine clastic, volcaniclastic, and calcareous sedimentation through the Late Paleozoic, becoming shallow water in the Early Triassic. Orogenic folding and uplift were followed by continental redbed sedimentation in the Late Triassic and Jurassic (Hutchison, in press). The closing of this marginal sea in the Late Triassic was the single most important orogenic event in the creation of the impressive S-type granite batholiths of the Main Range and northern Thailand, and resulted in the cratonization of southeast Asian Sundaland.

This marginal basin has a distinct positive gravity anomaly as compared with the western Main Range and Eastern belts (Ryall, 1976). It also has mineralization characteristic of an oceanic setting, including gold and base metals, in contrast to the tin-tungsten belts which fringe either side (Hutchison and Taylor, 1978; Taylor and Hutchison, 1979).

The eastern volcano-plutonic arc in all likelihood has a continental basement. The oldest rocks exposed are iso-clinally folded Carbo-Permian metasediments, Permian limestones, and Namurian shales and sandstones. It was characterized by abundant volcanic and plutonic activity throughout the Permian and ended its active history in the Late Triassic with subaerial ignimbritic flows, which still form undisturbed geomorphological features in Johore. Vertical joints and faults have been infilled with basaltic magmas to form dikes which may have fed the Cenozoic continental alkaline basalt flows (Hutchison, 1973b). This region has therefore been tectonically and isostatically stable, and little eroded since the Late Triassic.

The central basin in Thailand is flanked on the east by the Precambrian Kontum or Indosinian Massif, which is partly overlain by the Mesozoic continental Khorat Basin (Workman, 1977). Cenozoic activity occurs only on the margins of the region, along the Barisan Mountains of Sumatra, related to the active arc-trench system along the margin of the Indian Ocean. The Lupar ophiolite suture line of Borneo marks the eastern margin of Sundaland. This margin was active in the Cretaceous, with granitoid activity in Sarawak, Natuna, and Karimata islands.

THE EASTERN CALC-ALKALINE ARC

The granitoids of the Eastern Belt form a compositionally expanded calc-alkaline series. The proportions of plutonic rock types are approximately: gabbro-diorite, 15 percent; tonalite-granodiorite, 50 percent; and granite, 35 percent of outcrop area. This suite is characteristic of what Pitcher (1979) called an Andinotype orogenic belt, and is interpreted as an ensialic volcano-plutonic island arc overlying an active Benioff Zone. The plutonic suite is associated with contemporaneous andesitic and rhyolitic volcanism.

1) Geochronology

Most of the analyzed samples from the East Coast Province appear to belong to two major intrusive episodes (Fig. 5). The granites of east Trengganu yield Rb:Sr isochrons and K:Ar data consistent in the main with Late Permian to Early Triassic intrusion at 250 ± 4 Ma. A Late Triassic episode at about 220 Ma, based on Rb:Sr isochrons

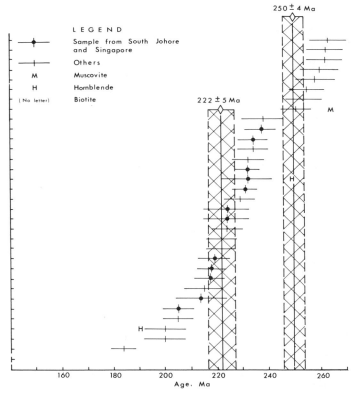

Figure 5. Eastern Belt granitoids, cumulative frequency curve of K:Ar apparent ages on micas and hornblendes. The two major intrusive events as defined by Rb:Sr isochrons are also shown. The concordance between the Rb:Sr and K:Ar dates is obvious when compared with other plutonic belts of Southeast Asia (after Bignell and Snelling, 1977).

and K:Ar data, is strongly represented in east Johore, Singapore, and northeast Kelantan. The Late Permian-Early Triassic granites yield an initial $^{87}Sr/^{86}Sr$ ratio of 0.7102, while the Late Triassic granites give values in the range 0.706 to 0.708 (Bignell and Snelling, 1977).

Although some biotites show evidence of argon loss, most mica K:Ar and Rb:Sr ages indicate little subsequent thermal disturbance (Fig. 5).

The best available age for the granitic rocks on Billiton Island, Indonesia, is about 213 ± 5 Ma based on the Rb:Sr results of Priem et al. (1975). The $^{87}Sr/^{86}Sr$ initial ratio is 0.7152. The average K:Ar age of muscovite and two cassiterite-bearing greisens is 198 ± 4 Ma, which suggests that the tin mineralization is not a simple late-stage event in the emplacement of the granitic plutons. The $^{206}Pb/^{204}Pb$ ratios of 18.5 to 18.6, $^{207}Pb/^{204}Pb$ of 15.7, and $^{208}Pb/^{204}Pb$ of 38.9 all indicate that the galena of the tin deposits is derived from underlying continental basement. The high initial Sr isotope ratio supports the deduction that the granite is derived from underlying continental crust (Jones et al., 1977). This Eastern Belt may have its northern extensions in Thailand in the Chantaburi foldbelt near Rayong (Fig. 1) and in the Kampot foldbelt of western Kampuchea around Kchol (Workman, 1977). Burton and Bignell (1969) recorded granites at Rayong with a Rb:Sr age of 272 Ma (extracted muscovite gave a K:Ar age of 72 Ma). Hornblende-biotite granodiorite from nearby Chantaburi gave mineral K:Ar dates of 196 ± 6 Ma.

All granitoids in southwest Kampuchea are probably Late Triassic (Fontaine and Workman, 1978). The largest massif at Kchol (Fig. 1) and nearby Knong-Ay were dated at 189 ± 11 Ma by Lasserre et al. (1970). Gneiss at Pailin gave an age of 228 Ma.

Possible extensions exist in Laos, north of the Khorat Plateau, where northeast-trending hornblende-biotite granodiorite has been dated by K:Ar methods at 260 ± 5 Ma (Lasserre et al., 1972).

2) Petrography

(i) Gabbros. Gabbros crop out at several places, in particular in central Singapore Island (Hutchison, 1964) and south Johore (Burton, 1973; Hutchison, 1973c). Gabbro occurs also at several localities in Trengganu and forms the northeastern and southeastern parts of Billiton Island (Aleva, 1960). Gabbro is associated with granite at Preah, southeast of Kchol, Kampuchea (Fontaine and Workman, 1978).

The suite is predominantly noritic to gabbro-norite, and its pregranite emplacement age is illustrated by its metamorphic recrystallization and assimilation by the granodiorite. Hornblende, actinolite, and biotite commonly replace the original pyroxene. The plagioclase is labradorite to bytownite. Olivine is not generally present but forms up to 8 modal percent in the south Johore occurrences. Most of the gabbro bodies are satellitic to the main granitoid masses, but poor outcrop hides their intrusive relationships.

(ii) Diorite-tonalite. The Boundary Range granitoid batholith forms the prominent mountainous country along the Trengganu-Kelantan boundary (MacDonald, 1968). The batholith is about 138 km long and 20 km wide at its maximum. Extensive assimilation of country rock xenoliths along the margin of the batholith has caused an almost continuous zone of tonalite up to 1.5 km wide. The tonalite is medium grained, granular textured, with zoned sodic plagioclase, green hornblende, dark brown biotite, and small amounts of orthoclase. The plagioclase is zoned from labradorite to oligoclase. Outside the tonalite areas, xenoliths are locally abundant. Many are of tonalite composition, but some are metasedimentary. The contaminated areas of granitoid may contain up to 80 percent hornblende and biotite (MacDonald, 1968).

Rocks of diorite composition have been described by Dawson (in MacDonald, 1968) from north Trengganu. Porphyritic diorites are fairly common. They have phenocrysts of pyroxene and intermediate plagioclase. The groundmass has a microcrystalline trachytic texture, composed of laths of plagioclase. The microdiorites are interpreted as high level and in places may grade to andesite.

Granodiorite stocks, varying from 1 to 10 km across, occur west of Veintiane between Pak Lay and Loei (Fig. 1). They are mostly in the range granodiorite-tonalite but include diorites (Page and Workman, 1968).

Diorites are found commonly throughout the Eastern Belt. Normally they contain abundant magnetite as an accessory mineral, but the diorite at Bukit Iban iron mine is free of magnetite (Taylor, 1971), and this may indicate that the diorites have hydrothermally lost their iron to form the iron-ore deposits (Hutchison and Taylor, 1978).

(iii) Granite-granodiorite. The granite-granodiorite series is usually medium to coarse, granular, and generally nonporphyritic, though weakly porphyritic varieties are not uncommon (Hutchison, 1977). Towards the south of the Malay Peninsula, the granite is even grained but locally better described as porphyritic microgranite in which 60 to 70 percent of the rock consists of even-grained phenocrysts in a fine-grained matrix. The alkali feldspar has been determined as orthoclase microperthite (Hutchison, 1964). Porphyritic microgranite has also been described from the Kuantan area (Hutchison and Snelling, 1971), where the alkali feldspar is untwinned and only slightly microperthitic. Elsewhere throughout the Eastern Belt, the alkali feldspar is untwinned, except for the patchy development of crosshatched twinning characteristic of intermediate microcline (Au, 1974). The granites may be grey or pink when weathered, and both colors may coexist in the same outcrop.

The alkali feldspar ranges from orthoclase to intermediate microcline and no maximum microcline has yet been found (Hutchison, 1977). The triclinicity depends upon the location of the collected specimen. Those samples from near a contact with the country rocks contain orthoclase. Those farther from the contact, deeper within the batholith, contain intermediate microcline. This range is to be expected from rapid cooling in an epizonal environment, with slower cooling deeper within the pluton. The alkali feldspars have a 2Va range from about 45 to 74°, with a pronounced mode around 60° (Fig. 3), in agreement with the range of structural state. Stewart (1975) gives a 2V of 60° for orthoclase and a range of 60 to 83° for intermediate microcline.

The chief ferromagnesian minerals are biotite and hornblende. The hornblende is usually euhedral, and the biotite generally subhedral. Apparently the granites that contain hornblende do not contain cassiterite (MacDonald, 1968).

The granites of Singapore and south Johore are of extremely high-level emplacement, orthoclase is the only structural state, and they are associated with distinctly rhyolitic phases (Hutchison, 1964). Burton (1973) also drew attention to the high-level nature of the intrusions and suggested their syngenetic relationship to the rhyolites and ignimbrites of south Johore.

3) The Country Rocks

The east coast granitoids are epizonally intruded into Upper Paleozoic sedimentary and volcanoclastic sediments (Gobbett, 1972). The plutonic activity was accompanied by contemporaneous volcanism which extended from Carboniferous to Late Triassic time (Hutchison, 1973b). Most of the materials are tuffs of acid to intermediate composition, occurring within the sedimentary envelope. In some regions, for example in the Kuantan-Sungei Lembing area, the country rocks are of Namurian age; in other places, they are Permian (Hutchison, in press). Most are unmetamorphosed and contain fossils. A characteristic feature of the granitoid bodies is their well-developed contact metamorphic aureole, characterized by hornfels containing andalusite, chiastolite, and cordierite (Hutchison, 1973d). Towards the southern part of the Peninsula, notably southwards from Kuantan, the country rocks consist of greenschist facies, regionally metamorphosed, isoclinally folded and steeply dipping phyllite, and fine-grained metasandstones of a flysch character. When the granitoids intruded these metasediments, which are generally held to be of Carbo-Permian age, aureoles of cordierite and andalusite hornfels were formed. Contact aureoles on unmetamorphosed and fossiliferous sedimentary rocks have been recorded by, for example, Yeap (1966), Au (1974), Goh (1973), and MacDonald (1968). The hornblende-biotite granodiorites of the area 140 km west of Vientiane are surrounded by cordierite-andalusite-bearing aureoles of 200 to 400 m width (Lasserre et al., 1972), but in the Mekong area the contact zones are narrow (Page and Workman, 1968).

In Billiton and Bangka, the country rocks comprise greenschist facies, isoclinally folded clay-slates, and fine-grained sandstones of flysch appearance, dated on Billiton by crinoids as Lower Permian (Hosking et al., 1977). On Bangka, the granites superimposed on these rocks a contact metamorphic aureole 1.5 to 2.5 km wide (Adam, 1960; Priem et al., 1975).

In several places, the granites are overlain by Mesozoic rocks with an erosional unconformity, for example, in Singapore Island by the uppermost Triassic Pasir Panjang member of the Jurong Formation and in Johore and Pahang by Upper Jurassic and Cretaceous Gagau Group molasse (Hutchison, in press).

4) Geochemistry

Geochemically, the granitoids of the Eastern Belt form an extended series, as shown by the selected analyses of Table 1 and the variation diagram of Figure 6. In common with most calc-alkaline series, the granitoid series forms rather regular variations, when plotted against either SiO_2 or

against the crystallization index of Poldervaart and Parker (1964). This has been deomonstrated by Hutchison (1973a) and Rajah et al. (1977).

Several granitoid specimens from the Eastern Belt, having been examined with a magnetic susceptibility meter (Ishihara et al., 1979), show that although a few scattered specimens belong to the magnetite series, the majority, and certainly all those around the tin mining areas of Gambang and Kuantan, belong to the ilmenite series. In this respect the Eastern Belt has similarities to the Japanese islands, where some granitoids are of the ilmenite series, and others of the magnetite series (Takahashi et al., 1980). From the bulk chemistry, the granitoids are predominantly of the S-type, but throughout the Eastern Belt there is a scatter of I-type.

In his comparison of the granitoid belts of the Malay Peninsula, Yeap (1980) compiled 49 analyses from the Eastern Belt, summarized in Table 2. Of his 51 granitoids analyzed for trace elements, the results are summarized in both Table 2 and Table 3. Of particular interest is the generally low tin content of Malaysian granites. The low niobium content is to be strongly contrasted with granites

TABLE 1. REPRESENTATIVE CHEMICAL ANALYSES (wt.%) OF THE CALC-ALKALINE PLUTONIC SUITE FROM THE EASTERN PERMO-TRIASSIC VOLCANO-PLUTONIC ARC

Oxide	1 qu. gabbro	2 eucrite	3 tonalite	4 grano-diorite	5 grano-diorite	6 granite	7 granite	8 granite	9 granite
SiO_2	53.22	48.20	56.20	66.01	63.70	75.60	74.80	76.02	74.09
Al_2O_3	15.81	19.80	14.60	14.66	15.30	13.03	12.60	12.03	12.29
Fe_2O_3	0.89	0.79	0.94	0.94	0.97	0.11	0.69	1.33	0.67
FeO	8.04	5.60	6.06	4.99	4.06	1.44	1.11	0.64	1.60
MgO	6.97	7.82	6.47	1.30	2.72	0.21	0.11	0.10	0.55
CaO	10.24	12.42	8.20	5.00	4.29	0.84	1.12	0.88	2.65
Na_2O	0.80	1.75	3.26	1.52	3.35	3.58	3.75	3.34	1.88
K_2O	0.77	0.69	1.17	3.08	2.90	4.17	4.76	4.31	4.65
H_2O-	0.30	0.27	0.22	0.21	0.24	0.13	0.09	0.05	0.10
H_2O+	1.46	2.63	1.57	1.01	1.66	0.57	0.76	0.49	0.41
TiO_2	0.57	0.30	0.77	0.62	0.57	0.16	0.10	0.24	0.41
P_2O_5	-	0.04	0.18	-	0.15	0.06	-	0.03	0.93
MnO	0.78	0.13	0.13	0.81	0.10	-	0.02	-	-
CO_2	-	0.09	0.06	-	0.07	-	0.08	0.08	-
Total	99.85	100.53	99.83	100.15	100.08	99.90	99.99	99.54	100.23

1. Djambi Hill, Billiton (Aleva, 1960).   2. Linden Hill, Johore (Hutchison, 1973a)   3. Anak Sungei Rek, Kelantan (Hutchison, 1973a)   4. Mount Blong, Billiton (Aleva, 1960).   5. Boundary Range, Gunung Gagau, Pahang (Rajah and others, 1977).   6. Singapore granite quarry (Hutchison, 1964).   7. Bukit Ubi, Kuantan (Rajah and others, 1977).   8. Ulu Sedili, Johore (Rajah and others, 1977).   9. Mount Beluru, Billiton (Aleva, 1960).

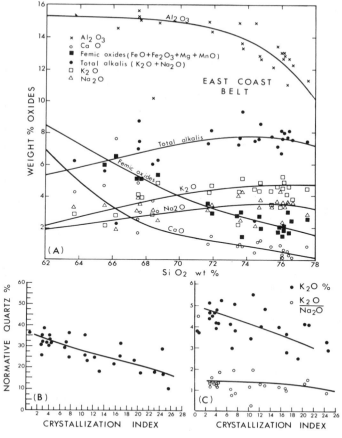

Figure 6. Eastern Belt variation diagrams for the granitoids. (A) Wt. % oxides vs. silica (after Rajah et al., 1977), (B) normative quartz vs. crystallization index of Poldervaart and Parker (1964), and (C) alkali variation with crystallization index (after Hutchison, 1973a).

which occur in anorogenic settings within Precambrian cratons. Pearce and Gale (1977) have shown that the granites of Nigeria are characterized by exceptionally high Nb contents as compared with those of Billiton and Bangka.

A characteristic feature of the Eastern Belt granitoid suite is its Rb:Sr ratio, which Table 3 shows to vary from place to place. Bignell and Snelling (1977) note that only a few East Coast granitoids have Rb:Sr ratios which exceed 4. In the plot of Rb/Sr versus R (Fig. 4), it is seen that the East Coast granitoids are not nearly as differentiated as those of the Main Range and its western extensions.

THE CENTRAL BELT

The central graben of the Malay Peninsula has been interpreted as a marginal basin, separating the continental plate west of the Bentong-Raub suture from the Eastern Volcano-plutonic arc (Hutchison, 1978). That this basin was floored by oceanic crust is clear from the gravity data (Ryall, 1976). The central basin includes a narrow line of granitoids

along the line of Benom to the Stong Complex in Kelantan, and this line includes several high-grade metamorphic complexes (Hutchison, 1973c). These metamorphic complexes may be interpreted as having resulted from high heat flow during the active spreading history of the basin.

The petrological characteristics of the Benom granite and the igneous domes to the north of it (Hutchison, 1973c) and the presence of high-grade metamorphic terranes suggest that the plutonic suite is distinct from the Main Range to the west.

1) Geochronology

The main granitic batholith of Benom has been dated both by K:Ar and Rb:Sr methods at about 200 Ma (Bignell and Snelling, 1977). Micas from the Taku Schist metamorphic mass in Kelantan indicate a minimum age of 200 Ma, and it is probable that the metamorphism took place in Permian time when the basin was actively spreading. The Kemahang Granite in Kelantan probably predates the Taku Schist; the one Rb:Sr isochron gives a mid-Paleozoic apparent age. Although the Stong migmatite complex may be contemporaneous with the Taku Schist, it has young mineral ages suggesting a Late Cretaceous thermal event. This event is also indicated from this area by other mineral age data. A muscovite granite at Batu Melintang gives a whole-rock Rb:Sr date of 76 ± 2 Ma.

Towards the south two Cretaceous granites border the marginal basin, the Gunung Ledang (Mount Ophir) on the west in the Malacca district, and the Gunung Pulai granite on the eastern margin southeast from Malacca towards Singapore (Fig. 1, C). K:Ar mineral ages and a Rb:Sr isochron suggest a Late Cretaceous age for the Mount Ophir and Gunong Pulai epizonal pink granites, with a low initial $^{87}Sr/^{86}Sr$ ratio of 0.7079.

The Benom Granite has an initial $^{87}Sr/^{86}Sr$ ratio of 0.7080, significantly lower than the granitoids of the Eastern Belt and the Main Range.

2) Petrography and Country Rocks

The plutons of the central basin are generally intrusive into Triassic country rocks (Gabbett, 1972). The Cretaceous granites on the margins of the basin are epizonal and the Mount Ophir pluton is characterized by a well-developed thermal aureole containing andalusite and cordierite (Lim, 1972). The Gunung Pulai epizonal granite is intrusive into the older granites on the margin of the Eastern Belt.

The Late Cretaceous epizonal granites are usually medium grained and granular in texture. Rarely are they porphyritic, and then only weakly. The alkali feldspar is commonly pink when slightly weathered and lacks cross-hatched twinning. The texture is similar to what Tuttle and Bowen (1958) called hypersolvus. However, not all the sodic feldspar is

TABLE 2. COMPARISON OF GRANITOIDS OF PENINSULAR MALAYSIA REGIONS IN wt.%
FOR MAJOR ELEMENTS AND ppm FOR TRACE ELEMENTS*

	Main Range and West Coast, older than L. Triassic	Main Range and West Coast, younger than L. Triassic	Central Belt	East Coast Belt	L. Cretaceous granites
	n = 36	n = 38	n = 38	n = 49	n = 7
SiO_2	71.24 (65.8 - 77.1)	74.03 (68.6 - 78.2)	73.69 (64.5 - 78.2)	72.52 (64.9 - 79.5)	75.0 (73.4 - 79.1)
Al_2O_3	14.67 (12.3 - 18.0)	13.91 (11.7 - 16.4)	14.14 (11.9 - 20.0)	14.12 (10.3 - 17.22)	13.89 (12.6 - 15.1)
FeO^{+}	2.85 (1.0 - 4.7)	1.94 (0.7 - 4.1)	2.30 (0.9 - 8.5)	2.92 (1.3 - 8.7)	1.53 (1.1 - 2.1)
MgO	0.81 (0.0 - 2.8)	0.49 (0.1 - 4.1)	0.77 (0.1 - 3.7)	0.61 (0.1 - 2.8)	0.25 (0.1 - 0.2)
CaO	1.96 (0.5 - 4.7)	1.01 (0.1 - 2.7)	0.94 (0.1 - 3.0)	1.90 (0.1 - 5.7)	1.09 (0.3 - 1.7)
Na_2O	3.45 (0.3 - 8.9)	2.89 (0.6 - 3.8)	3.21 (0.1 - 4.6)	3.32 (0.9 - 4.4)	3.26 (1.3 - 3.9)
K_2O	4.41 (0.2 - 8.9)	5.23 (3.0 - 8.4)	4.56 (1.4 - 6.5)	4.16 (2.2 - 5.6)	4.71 (3.6 - 5.3)
TiO_2	0.41 (0.1 - 1.0)	0.33 (0.1 - 0.6)	0.24 (0.0 - 0.8)	0.27 (0.1 - 0.7)	0.20 (0.1 - 0.3)
P_2O_5	0.13 (0.1 - 0.5)	0.12 (0.1 - 0.4)	0.13 (0.0 - 0.4)	0.12 (0.0 - 0.6)	0.09 (0.0 - 0.1)
MnO	0.07 (0.1 - 0.6)	0.04 (0.0 - 0.1)	0.04 (0.0 - 0.3)	0.06 (0.0 - 0.2)	0.04 (0.0 - 0.1)
	n = 50	n = 133	n = 15	n = 51	n = 7
Rb	402 (79 - 1004)	596 (188 - 1125)	330 (179 - 492)	252 (40 - 990)	228 (92 - 308)
Sr	106 (34 - 621)	33 (3 - 160)	192 (19 - 868)	132 (40 - 460)	195 (62 - 307)
Ba	611 (311 - 1379)	206 (34 - 915)	558 (246 - 1009)	729 (300 - 1727)	735 (158 - 1120)
Zr	146 (29 - 236)	81 (40 - 215)	164 (103 - 306)	132 (60 - 332)	128 (92 - 175)
Sn	6 (2 - 20)	11 (1 - 25)	6 (3 - 10)	6 (2 - 15)	5 (2 - 8)
Nb	8 (1 - 19)	8 (5 - 18)	7 (4 - 18)	6 (2 - 30)	6 (3 - 8)
W	3 (0 - 14)	6 (0 - 40)	6 (1 - 24)	2 (1 - 4)	2 (0 - 2)
$F^{§}$	1248 (745 - 1940)		835 (693 - 1012)	620 (416 - 1132	1086 (758 - 1313)
$Cl^{§}$	57 (8 - 207)		69 (53 - 81)	158 (58 - 236)	24 (18 - 30)
$Pb^{§}$	111 (46 - 154)		105 (15 - 240)	72 (17 - 167)	100 (33 - 162)
$Zn^{§}$	60 (31 - 92)		51 (41 - 60)	59 (42 - 87)	40 (36 - 42)

*After Yeap (1980).

[+]Total iron as FeO.

[§]Based on a total of 33 analyses only.

Ranges in parentheses. n = number of analyses.

within the perthite; separate plagioclase crystals are present in small amounts. The alkali feldspar is strongly perthitic, but the exsolution lamellae, blebs, and stringers are usually no thicker than 5 μm (Ng, 1974). Many alkali feldspars appear nonperthitic in thin section, but they are crypto-perthitic. The degree of development of perthitic lamellae varies from zone to zone within a crystal. The outer zones usually have more coarsely perthitic textures. The alkali feldspar is restricted to the orthoclase structural state.

The Gunong Benom has been described by Khoo (1968) as a coarse-grained, nonporphyritic biotite granite, which has a discordant contact with the country rocks. Jaafar (1979) has noted that the younger granite intrudes an older series of alkali-rich rocks that range in composition from gabbro to quartz syenite, the metamorphic nature of which was described by Hutchison (1973d). Khoo (1968) described many indications of potash metasomatism in the older

complex and has suggested that the Benom granite was the source. Whereas the Late Cretaceous granites display a good contact aureole, the Benom granite has not been described in sufficient detail to ascertain if it has a thermal aureole. The alkali feldspar of the Benom granite is inter-mediate microcline, so that the level of emplacement is not as high as for the Late Cretaceous granites (Hutchison, 1977; Yeap, 1980).

3) Geochemistry

A chemical summary of the granitoids is given in Table 2, after Yeap (1980). The variation diagrams show a rather heterogeneous scatter (Fig. 7). The most characteristic feature is that the granitoids tend to be lower in K_2O in comparison with the Main Range. As noted by Hutchison (1977), they are characterized by higher Sr contents than the

TABLE 3. SUMMARY OF Rb AND Sr ANALYSES OF GRANITES OF THE MALAY PENINSULA

	District	No. of analyses	Rb average ppm with (range)	Sr average ppm with (range)	Average Rb/Sr	Reference
Main Range and Western Shelf	Southern section (Late Paleozoic)	9	369 (256-453)	114 (73-196)	3.2	1
	Southern section (Mesozoic)	19	464 (235-627)	56 (4-143)	8.3	1
	Kuala Lumpur-Frasers Hill (Central)	23	478 (286-800)	53 (14-93)	9.0	1
	Central zone close to Kuala Lumpur	16	435 (245-767)	76 (13-169)	5.7	1
	Ipoh-Cameron Highlands area	12	453 (310-744)	62 (17-119)	7.3	1
	Bujang Melaka, Kinta Valley, Ipoh	7	389 (302-689)	74 (21-91)	5.2	1
	Kledang Range - Bintang, Dindings, Perak.	15	588 (354-1412)	70 (24-195)	8.4	1
	Northwest Malaysia - Kedah, Penang	27	370 (265-582)	84 (10-234)	4.4	1
	Peninsular Thailand (excluding Phuket)	22	435 (260-950)	87 (17-244)	5.0	2
	Penang Island, N.W. Malaysia	13	471 (396-519)	41 (3-56)	11.5	3
	Peninsular Thailand (excluding Phuket)	6	434 (295-946)	115 (101-125)	3.8	3
	Kuala Lumpur area (older than Early Triassic)	12	342 (239-612)	103 (27-158)	3.3	3
	Kuala Lumpur area (Late Triassic)	37	611 (254-1004)	54 (2-621)	11.3	3
Eastern Belt	Singapore (Late Triassic)	8	179 (147-207)	90 (32-154)	2.0	3
	Johore (Late Triassic)	8	447 (173-990)	81 (11-134)	5.5	3
	Northern section (Late Triassic)	17	224 (40-438)	161 (36-309)	1.4	3
	Northern section (Early Triassic)	16	228 (81-332)	155 (25-460)	1.4	3
	East Coast (north) 220 Ma old	9	292 (163-744)	93 (28-244)	3.1	1
	East Coast (north) 250 Ma old	*12	264 (194-337)	111 (18-232)	2.4	1
	Singapore and south Johore	6	185 (89-241)	108 (41-172)	1.7	1
	Billiton, Indonesia	7	267 (114-441)	150 (48-239)	1.8	4
	Bangka, Indonesia	4	402 (221-621)	100 (42-174)	4.0	4
	Tuju Islands, near Bangka	2	396 (324-469)	73 (23-124)	5.4	4
Central	Gunung Benom	15	330 (179-492)	192 (57-668)	1.7	3
	Gunung Benom, Late Triassic	5	308 (245-413)	99 (32-160)	3.1	1
	Late Cretaceous Mount Ophir	4	240 (175-304)	277 (128-509)	0.9	1
	Late Cretaceous Mount Ophir	4	290 (282-308)	196 (62-307)	1.4	3
	Late Cretaceous Gunung Pulai	1	173	114	1.5	1
	Late Cretaceous Gunung Pulai	3	146 (92-179)	193 (101-266)	0.8	3
Phuket	Phuket-Ranong area, Thailand	15	527 (306-1030)	80 (7-181)	6.5	3
	Phuket area	2	479 (588-370)	68 (56-80	7.0	2
	Phuket area, peninsular Thailand	12	358 (300-530)	189 (50-650)	1.9	5
Northwest Thailand	Khutan batholith, northwest Thailand	2	103 (98-108)	193 (179-208)	0.5	6
	Fang Mae Suai batholith, northwest Thailand	6	76 (48-84)	114 (53-166)	0.7	6
	Ban Hong pluton, northwest Thailand	7	86 (68-96)	99 (60-159)	0.8	6
	Li pluton, northwest Thailand	2	104 (88-121)	67 (43-91)	1.6	6
	Mae Sariang granite, west Thailand	5	97 (89-101)	162 (37-269)	0.6	6
	Gneiss northwest of Chiang Mai, northwest Thailand	8	94 (43-281)	268 (1-786)	0.4	6

1. Bignell and Snelling (1977). 2. Ishihara and others (1980). 3. Yeap (1980). 4. Priem and others (1975).
5. Garson and others (1975). 6. von Braun and others (1976).

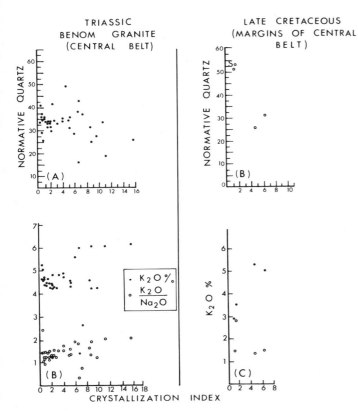

Figure 7. Normative quartz vs. crystallization index, K₂O wt % and K₂O/Na₂O vs. crystallization index for the Triassic Central Belt Benom granite and the Late Cretaceous pink granites which border the Central Belt (after Hutchison, 1973a).

TABLE 4. REPRESENTATIVE CHEMICAL ANALYSES (wt.%) OF THE COMPOSITIONALLY RESTRICTED GRANITE SERIES OF THE MAIN RANGE AND ITS WESTERN EXTENTIONS, AND EXTENSIONS INTO SOUTHERN THAILAND

Oxide	1	2	3	4	5	6	7	8	9
SiO₂	73.40	71.58	76.40	72.70	72.90	74.40	73.73	70.56	73.61
Al₂O₃	12.90	12.96	12.90	15.80	13.50	13.50	14.91	14.89	13.99
Fe₂O₃	1.03	0.61	0.42	0.22	0.62	0.86	0.52	0.60	0.44
FeO	1.58	3.12	1.02	1.24	1.22	0.81	0.61	1.65	0.54
MgO	0.41	0.35	tr.	tr.	0.15	0.21	0.11	0.87	0.16
CaO	1.70	1.78	0.55	0.32	1.34	0.82	0.41	1.80	0.69
Na₂O	3.30	3.39	2.85	1.26	3.77	3.75	3.63	3.08	3.24
K₂O	4.45	5.03	5.17	6.62	5.26	4.90	4.55	5.10	6.37
H₂O-	0.26	0.12	0.14	0.17	0.03	0.11	0.02	0.04	0.14
H₂O+	0.60	0.36	0.83	0.94	1.09	0.57	0.56	0.52	0.53
TiO₂	0.26	0.66	0.16	0.30	0.24	0.23	0.05	0.31	0.06
P₂O₅	0.04	0.11	0.05	0.11	0.04	0.10	0.28	0.10	0.02
MnO	0.08	0.15		tr.	0.02	0.08	0.06	0.06	0.02
CO₂	0.03		0.04	0.17	0.08	0.06			
Total	100.04	100.22	100.57	99.85	100.26	100.40	99.44	99.58	99.81

1. Porphyritic biotite granite; Tampin (south of Kuala Lumpur). 2. Porphyritic biotite granite, Kuala Dipang, Perak, south of Ipoh. 3. Granite, Cameron Highlands, Perak. 4. Granite-Damar Type, Baling, Kedah. 5. Biotite granite from Western Road quarry, Penang. 6. Biotite granite, Kuah, Langkawi. 7. Porphyritic tourmaline-biotite granite, Songkhla, Thailand. 8. Medium-grained, porphyritic biotite granite, Khao Kachong, S. Thailand. 9. Medium-grained, porphyritic biotite granite, Ko Samui Island, S. Thailand. (Sources: Alexander and others, 1964; Hutchison, 1973b; Ishihara and others, 1980).

Main Range (average about 195 ppm) so that their Rb/Sr ratios are generally less than 2. Some representative analyses are summarized in Table 4. In the Rb/Sr-versus-R diagram (Fig. 4), the Central Belt granitoids generally overlap with those of the Eastern Belt, and are distinctly different from the Main Range.

The Late Cretaceous granites of Mount Ophir and Gunung Pulai have lower Rb/Sr ratios than the Main Range and also a lower initial strontium ratio (0.7079). All igneous rocks in the Central Belt, including the Cretaceous granites were assigned to the magnetite series by Ishihara et al. (1979).

GNEISSIC GRANITES AND METAMORPHIC COMPLEXES ALONG THE WESTERN MARGIN OF THE CENTRAL BASIN

The plate margin, which is delineated by the Bentong-Raub and Uttaradit suture zones, is characterized by several complexes of metamorphic rocks, migmatites, and gneissic granites. The margin of the Main Range, east of Kuala Lumpur and only a few kilometers west of Bentong, is characterized by strongly gneissic granite. The shear planes dip steeply and lie parallel to the Bentong-Raub suture line. In the northern part of peninsular Malaysia, the Main Range granite appears as a complex called the Gunong Stong migmatite (Hutchison, 1973d). The granite is generally fine grained and nonporphyritic, but there are coarser and porphyritic varieties. The whole complex is migmatitic to varying degrees; the map of MacDonald et al. (1968) shows biotite gneiss and psammitic nebulite engulfing a variety of amphibolite facies schist, gneiss, and phlogopite marble, in blocks varying from a few meters to several kilometers long.

The biotite-muscovite schist of the migmatite is accompanied by venite migmatite composed partly of schlieren of sillimanite-feldspar-garnet-biotite schist (Hutchison, 1973d). Marble contains phlogopite and diopside, and psammite nebulite contains biotite and augite. The complex is only partly studied, but offers an interesting gradation from granitic to metasedimentary parts of the complex. MacDonald (1968) concluded that over a wide area granite has injected and permeated the country rock. Only two samples were analyzed by Bignell and Snelling (1977), and both have extremely low ⁸⁷Rb:⁸⁷Sr ratios so that age calculations were impossible. However, biotite from one sample gave concordant K:Ar and Rb:Sr apparent ages of 67 and 66 Ma, respectively.

These dates seem to reflect a Late Cretaceous regional resetting of the mineral clocks (Bignell and Snelling, 1977). Bignell and Snelling (1977) reported 6 analyses of rocks from the Stong region, which show an average Rb content of 492 ppm (range: 152 to 1075 ppm) and an Sr average of 246 ppm (range: 33 to 955 ppm), leading to an average Rb/Sr ratio of 2.0.

If the Bentong-Raub suture zone were extrapolated along

important horst structures evident in the basement of the Gulf of Thailand (Fig. 1), it would pass to the region west and east of Bangkok. West of Bangkok is the Hua Hin area, the geology of which has been described by Pongsapich et al. (1980) and the radiometric dating, by Beckinsale et al. (1979).

The region of Hua Hin and Pranburi consists of a regionally metamorphosed complex and associated Hua Hin, Pranburi, and Hub Kapong granite gneisses. The complex has many similarities to the Stong Complex, including metamorphic grade. In view of its position in Thailand, it may represent the eastern margin of the Shan-Thai-Main Range massif.

The complexity is due to tectonic activity along the margin of this continental plate with the marginal basin, extending northwards from Bangkok towards Loei (Fig. 1). This has also been suggested by Chantaramee (1978). Metapelites in the complex consist of quartz-biotite-sillimanite-garnet orthoclase-cordierite assemblages, and resemble the suite in the Stong Complex (Hutchison, 1973d). The suite also includes diopsidic calc-silicates and metaquartzites. The Hua Hin gneiss is a biotite-garnet or biotite gneiss. The coarse varieties are remarkably porphyroblastic. The Pranburi gneiss is a strongly sheared and mylonitic granite gneiss (Pongsapich et al., 1980).

The Hub Kapong gneiss is coarse grained and consists of quartz, microcline perthite, plagioclase, and biotite. It contains metasedimentary xenoliths. This pluton has been described by Putthapiban and Suensilpong (1978). Parts of the granite complex were originally mapped as Precambrian orthogneiss, but a good Rb:Sr isochron leaves little doubt that it is a gneissic Triassic granite, with an age of 210 ± 4 Ma and an initial $^{87}Sr/^{86}Sr$ ratio of 0.7237 (Beckinsale et al., 1979). Typically the rock has a pronounced foliation, defined by the alignment of biotite crystals and alkali feldspar phenocrysts up to 5 cm long. These phenocrysts appear to have been rotated into alignment. In this, the Hub Kapong strikingly resembles the eastern margins of the Main Range batholith, as seen on the new Karak Highway, just west of Bentong. An unfoliated granite forms the western flank of the hills. There is a spectacular discordance between the 210-Ma, Rb: Sr whole-rock isochron and the K:Ar dates on the biotites, which gave maxima at 76 and 58 Ma.

The Hub Kapong granite is highly differentiated, as shown by the chemical analyses presented by Beckinsale et al. (1979) and Pongsapich et al. (1980), plotted as a variation diagram on Figure 8 and summarized in Table 5. The analyses indicate it is of S-type.

It is interesting to compare the three gneissic granitoids of the Hua Hin-Pranburi area, using the data of Pongsapich et al. (1980) and Beckinsale et al. (1979). The Hub Kapong and Hua Hin gneisses are quite similar and both are distinctly S-type granitoids. Indeed, the Hub Kapong has a remark-

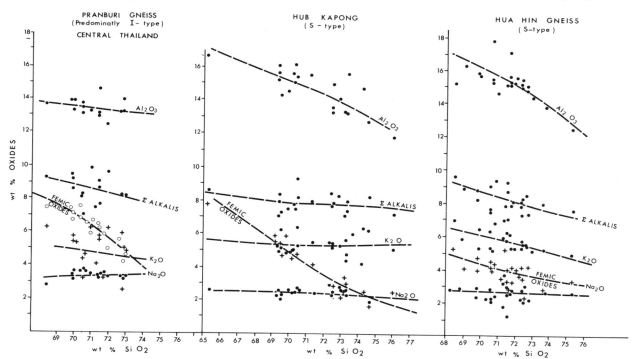

Figure 8. Variation diagrams for the Pranburi, Hub Kapong, and Hua Hin granitic gneisses of the coastal part of the peninsula southwest of Bangkok. Data are from Pongsapich et al. (1980) and Beckinsale et al. (1979). The Pranburi gneiss has higher Na₂O and lower Al₂O₃ contents than the other two and is of the I-type, whereas the other two are distinctly S-type.

TABLE 5. REPRESENTATIVE CHEMICAL ANALYSES (wt.%) OF THE COMPOSITIONALLY RESTRICTED TRIASSIC GRANITOIDS OF WEST AND NORTHWEST CONTINENTAL THAILAND*

Oxide	1	2	3	4	5	6	7	8	9
SiO_2	70.77	72.02	69.80	62.19	65.59	72.51	73.13	71.96	72.62
Al_2O_3	16.08	14.17	17.08	16.19	14.36	13.37	13.31	12.49	15.04
Fe_2O_3	0.17	0.50	0.37	2.21	3.69	1.22	0.17	0.65	0.27
FeO	1.62	1.71	1.87	4.60	1.04	0.81	1.96	0.80	0.75
MgO	1.21	1.45	1.02	1.46	0.79	0.32	0.65	0.33	0.39
CaO	1.84	1.29	1.53	4.76	2.26	1.43	1.41	3.16	0.96
Na_2O	2.59	2.60	2.66	4.49	4.34	3.48	2.41	3.33	3.07
K_2O	4.82	4.36	4.47	1.60	6.34	5.82	5.34	6.30	5.47
H_2O-	0.11	0.23	0.08	-	0.10	0.02	0.11	-	-
H_2O+	0.24	0.74	0.24	1.18	0.57	0.48	0.45	-	-
TiO_2	0.23	0.34	0.30	0.93	0.32	0.19	0.33	0.19	0.16
P_2O_5	0.14	0.24	0.19	0.01	0.19	0.05	0.11	0.04	0.04
MnO	0.04	0.05	0.05	0.18	0.08	0.03	0.05	0.03	0.03
Total	99.86	99.70	99.66	99.80	99.67	99.73	99.43	99.28	98.80

1. Biotite granite, Khuntan batholith, Lampang. 2. Biotite granite, Khuntan batholith, Lampang. 3. Biotite granite, Khuntan batholith, Lampang. 4. Quartz diorite, Tak. 5. Granodiorite, Tak. 6. Granite, Tak. 7. Hub Kapong gneiss, average of 8 analyses. 8. Pranburi gneiss, near Hub Kapong. 9. Hua Hin gneiss, near Hub Kapong.
*Based on Suensilpong and others (1977); Pongsapich and others (1980); Pongsapich and Mahawat (1977); and Beckinsale and others (1979).

ably high initial $^{87}Sr/^{86}Sr$ ratio of 0.7237 with an isochron age of 210 ± 4. By contrast, the Pranburi gneiss is I-type, lower in Al_2O_3 and higher in Na_2O than the other two (Fig. 8 and Table 5).

Across the Gulf of Thailand, southeast of Bangkok is the Cholburi-Rayong porphyritic sheared granite (Nutalya et al., 1979). Burton and Bignell (1969) suggested an Rb:Sr age of 272 and 307 Ma for the gneissic granite but reported a strongly discordant K:Ar date of 72 on its contained muscovite. The Hub Kapong granite gave a similar K:Ar date from its biotite of 63 Ma, indicating that the discordance may reflect later deformation (shearing).

The orthogneisses of the Rayong and the Hua Hin areas were formerly considered to be Precambrian (Suensilpong et al., 1978), but radiometric dating has indicated that the highly deformed gneisses are predominantly of Triassic age. There is indeed only a remote possibility of outcrops of Precambrian granites in the region. The only Precambrian granitoids of proven age in the region are clasts of trondhjemitic granite enclosed within the Paleozoic mudstone of the Singa Formation of Langkawi (Stauffer and Snelling, 1977), but they are not known to outcrop anywhere.

Farther north, immediately west of Tak (Fig. 1), a gneissic complex has been described by Chantaramee (1978). Gneisses in this area have traditionally been regarded as Precambrian, but Teggin (1975) has shown that the white granite of the gneiss complex has an Rb:Sr age of 213 ± 10 and the pink granite an age of 219 ± 12 Ma. Both have extremely high initial $^{87}Sr/^{86}Sr$ ratios of 0.7158 and 0.7104, respectively.

MAIN RANGE AND STABLE SHELF

The Main Range batholith forms the mountainous water-

shed of peninsular Malaysia. Its eastern margin is the steeply eastward-dipping, melange-like schist belt, which contains disconnected inclusions of dismembered ophiolite, referred to as the Bentong-Raub suture zone (Hutchison, 1975). This plutonic belt extends from the suture westward to include Penang, Langkawi, and the Thai Peninsula and terminates where the major transcurrent Khlong Marui fault separates from the Phuket Belt. It appears that the Bentong-Raub suture line represents a former plate margin, and that the Main Range-Thai-Shan massif was attached on its west side to a major Precambrian craton (Hutchison, in press).

Hutchison (1977) referred to the region west of the Main Range as the Western Stable Shelf because it is the region of the Paleozoic stable miogeocline, as defined by the stratigraphy of the region (Hutchison, 1973a). Recent mapping (Page et al., 1979) indicates that northern Sumatra has the same Carboniferous to Triassic stratigraphy and forms a part of this western stable shelf.

1) Geochronology

The Main Range batholith is composed of granites of Early Permian (about 280 Ma) and Late Triassic (about 230 and 200 Ma) age. K:Ar ages on micas from the southern part of the batholith show considerable argon loss. Evidence of the 280-Ma event, as defined by Rb:Sr isochrons, is completely absent from the K:Ar data, as all the K:Ar dates lie within the 80- to 200-Ma range (Fig. 9).

At least two intrusive events can be recognized in the central section of the Main Range batholith in the neighborhood of Kuala Lumpur (Fig. 9). The youngest at 199 Ma appears to be widespread (Bignell and Snelling, 1977). This event is significantly younger than the 215-Ma event of the Eastern Belt. An Early Permian event has been dated at 280 Ma, and rocks of a similar age occur in the southern extension of the batholith. Compared with the 199-Ma-old granites, these Early Permian granites have initial $^{87}Sr/^{86}Sr$ ratios of 0.710 to 0.712 (Fig. 2). A group of widely scattered samples gives Rb:Sr dates of around 237 Ma.

Bignell and Snelling (1977) suggested that the fundamental discordance between the Rb:Sr and K:Ar dates appears to be related to heat generated along the major wrench faults which traverse this part of the batholith causing some cataclasis. Hutchison (1977), however, suggested that the discordance was due to a deep-seated emplacement, with the granites remaining hotter than the blocking temperature until the region was subsequently uplifted and cooled, so that the K:Ar dates represent the time of uplift. The faulting may be related to the uplift of this deep-seated batholith and to its unroofing throughout the Cretaceous.

At least two intrusive events have been defined in the Ipoh-Cameron Highlands area of the Main Range. The best defined event is at 280 Ma with initial $^{87}Sr/^{86}Sr$ ratios of 0.7096 (Fig. 9). Younger granites with generally higher

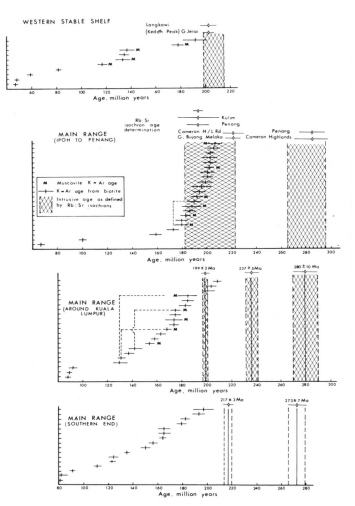

Figure 9. Summary of radiometric dating by Bignell and Snelling (1977) on the granites of the Main Range and the Stable Shelf west of it. A characteristic feature is the general discordance between the Rb:Sr and the K:Ar data, which may be explained by subsequent faulting, uplift, and hydrothermal circulation.

Rb:Sr ratios of 0.7132 lie between 210 and 230 Ma. An older event at 355 Ma is also suggested by two samples. The medium- to coarse-grained granites of the Bujang Melaka mass of Ipoh give an age of 211 Ma and a relatively high $^{87}Sr/^{86}Sr$ initial ratio of 0.7165. Granites of similar age also occur in the Kuala Lumpur area.

The Western Stable Shelf granites are represented by Penang, Kulim, and Langkawi, to the west of the Main Range. Granites from the southeast part of Penang define an isochron at 286 Ma, and the Kulim granite also appears to fit this age. There is also a good isochron dating at 197 Ma in Penang with an initial Sr ratio of 0.709. Fine- to medium-grained granites at Kulim define a 200-Ma isochron with a high initial Sr isotope ratio of 0.714. In Langkawi and Kedah Peak, the Rb:Sr data indicate a Late Triassic intrusive event, and K:Ar data are strongly discordant (Fig. 9). Radiometric dating in peninsular Thailand (Ishihara et al., 1980) indicates that most granites there are Upper Triassic, but with strongly discordant K:Ar dates. Some Cretaceous granite may be present at Yod Nam, south of Surat Thani.

2) Petrography

The rocks of the Main Range batholith are usually coarse to very coarse grained and contain large phenocrysts, up to 5 cm long (Hutchison, 1977). Finer grained varieties occur locally and include aplite and tourmaline leucogranite.

The main rock type is invariably grey to white. The megacrysts of maximum microcline show perfectly formed, coarse perthite and cross-hatched twinning. They may form a crude alignment with steep dips. None of the alkali feldspar is zoned, but it commonly displays a concentric distribution of inclusions, especially of plagioclase and quartz (mantled crystals). The alkali feldspars apparently underwent subsolidus recrystallization or continued growth (Ng, 1974). The perthite exsolution lamellae and stringers are up to 50 μm wide. Plagioclase is present as a separate phase and forms smaller phenocrysts than the alkali feldspar. Biotite forms large euhedral crystals.

The granites of the Kinta Valley area of Ipoh have been described by Ingham and Bradford (1960). They noted that no hornblende granite has been recorded and that coarsely porphyritic texture is predominant over medium-grained, nonporphyritic forms. A prominent feature of both the Kuala Lumpur and Ipoh areas of the Main Range is thick quartz veins, referred to as reefs in Malaysia.

Although the granite is usually light colored, biotite forms irregular dark schlieren. Muscovite is not a primary constituent, but it is invariably present in the mineralized areas, where it forms greisenized borders to the mineralized quartz veins (Hosking, 1977). In all cases, the muscovite is seen to replace the primary biotite. The characteristic opaque mineral is ilmenite.

Fluorite is commonly concentrated in joints and fault planes along the western margins of the Main Range batholith, especially in the mineralized areas. The tourmalinization of the granite may be shown to result from metasomatism extending outwards from the quartz veins. Tourmaline replaces the biotite and is accompanied by greisenization (Hutchison and Leow, 1963). However, tourmalinization and greisenization do not necessarily imply economic mineralization, for the alteration is much more extensive than in the hydrothermal cassiterite deposits and extends westward to the coastal exposures of Kuala Selangor and Langkawi.

The granite contains isolated, irregularly shaped, rootless pegmatitic zones containing large crystals of feldspar and tourmaline, and on Penang the pegmatite zones contain euhedral crystals growing into miarolitic cavities. Only around Kedah Peak were the pegmatitic phases able to leak away into the surrounding country rocks to form pegmatite veins. There the granite itself is cut by some spectacular zoned pegmatites containing tourmaline and some garnet (Bradford, 1972).

The eastern part of the Main Range batholith is characterized by steeply dipping shear zones, parallel to the north-trending Bentong-Raub suture zone. These are particularly well displayed on the new Karak Highway just west of Bentong. In these the porphyritic granites take on a distinctly planar foliation with parallel alignment of the phenocrysts, and in the more disturbed zones the granite becomes gneissic and mylonitic.

West of the disturbed zone the granite is strongly faulted and slickensided parallel to later faults, which characterize the area between Kuala Lumpur and Bentong.

In south Thailand, most plutons consist of coarse-grained biotite granite, containing megacrysts of alkali feldspar (Ishihara et al., 1980). They also noted the existence of north-trending aligned foliation zones in the granites.

3) Thermal metamorphism

The main envelope of the Main Range batholith is Lower Paleozoic greenschist facies phyllite and slate. Usually the contact is sharp, as seen east of Kuala Lumpur, and only local contact metamorphism is imprinted on the country rocks.

Westward from the Main Range, Foo (1964) recorded that the Langkawi granite was surrounded by a contact metamorphic aureole in which the Carboniferous Singa Formation is metamorphosed to spotted hornfels containing incipient chiastolite. Limestone in contact with the granite is characterized by a skarn zone with strong development of vesuvianite and garnet (Hutchison, 1973d).

In peninsular Thailand, the granites intrude Paleozoic and Triassic formations and produce distinct contact aureoles (Ishihara et al., 1980). The sedimentary rocks are converted to biotite hornfels and in some cases to andalusite-biotite hornfels.

4) Geochemistry

Summaries of the large number of chemical analyses of the Main Range have been given by Hutchison (1973a) and Yeap (1980), and of the southern Thailand Peninsula by Ishihara et al. (1980). Representative analyses (Table 4, Figs. 10 and 11) show that these granites are compositionally restricted and generally highly evolved (Fig. 4). In contrast to the Eastern Belt granitoids (Fig. 6), the variation diagrams (Figs. 10 and 11) show a large scatter and no well-defined trend. Chappell and White (1974) (and White and Chappell, 1977) considered this to be characteristic of S-type peraluminous granite, and thought that it reflected variation in the continental infrastructure from which the

magmas were anatectically derived. As shown by Ishihara et al. (1979), the granites of the whole region are clearly of the ilmenite series, and chemical analyses indicate that they are of the S-type. That the granites are predominantly of the ilmenite series is reflected by the fact that ilmenite is the most abundant heavy mineral discarded during the tin-ore dressing (Ingham and Bradford, 1960).

Table 2 shows that the Upper Triassic granites are more evolved than the Permian and clearly more enriched in K_2O and rubidium (Yeap, 1980). Yeap compared the granites associated with mineralization and those which are apparently barren, following the observations of Tischendorf (1977). He concluded that Malaysian granites associated with tin mineralization contain more SiO_2 and K_2O and less Al_2O_3, FeO, MgO, CaO, and Na_2O than those not associated with mineralization. Further, it must be noted that tin mineralization in the Malay Peninsula is closely associated with the Late Triassic, highly differentiated granites and not with the older ones. Bignell and Snelling (1977) deduced that the Main Range granites are more differentiated than those of the East Coast Belt and that the Late Triassic granites of the Main Range are more fractionated than the older ones (Fig. 4). The Upper Triassic granites are characterized by high Rb and low Sr contents (Table 3), giving Rb/Sr ratios which commonly may be as high as 11, whereas the Permian granites have low ratios of around 3. The granites of peninsular Thailand are similar to the lower range of the Main Range granites, with ratios as high as 5. Ishihara et al. (1980) quoted average fluorine for 22 samples to be 1683 ppm (range: 380 to 4000 ppm), and Yeap (1980) reported an average of 1248 ppm for the Main Range (Table 2).

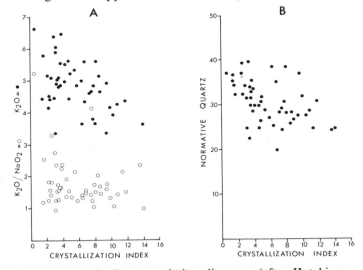

Figure 10. Main Range variation diagrams (after Hutchison, 1973a). (A) Wt. % K_2O and K_2O/Na_2O vs. crystallization index of Poldervaart and Parker (1964) for the Main Range. (B) Normative quartz vs. crystallization index. As common with most S-type granite batholiths, the points do not clearly define any variation trend. The spread probably represents the variation in the deep crust from which the granites were anatectically derived.

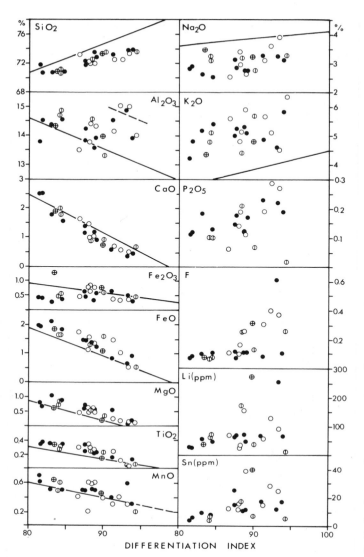

Figure 11. Southern Thailand Peninsula, variation in the granites (after Ishihara et al., 1980). The parameters are plotted against the differentiation index of Thornton and Tuttle (1960). Solid circle: eastern zone of the peninsula; open circle with bar: central zone of the peninsula; open circle: Cretaceous granites (western zone and Yod Nam); circle with cross: averages of normal granites and tin granite of Tischendorf (1977); solid line: average of the Japanese granites, taken from Aramaki et al. (1972).

The Main Range granites have an average initial Sr isotope ratio in the region of 0.710 (Bignell and Snelling, 1977). Significantly higher values are given for the Bujang Melaka granite of the Ipoh area and for the Kulim area of Kedah; both are around 0.7165. The Main Range granites show the highest fractionation of Rb, Sr, and K, as shown in Figure 4. These trace element data, when combined with the Sr isotope data, suggest that the Main Range granite magma was derived from a continental basement. This is in agreement with the metasedimentary nature of the xenoliths,

and with the observations of Ishihara et al. (1979) that specimens from the Main Range, Pehang, and Langkawi are all of the ilmenite series.

SOUTH PENINSULAR THAILAND

The granites form isolated plutons elongated generally northerly and no comagmatic volcanic rocks have been recognized (Ishihara et al., 1978, 1980). The plutonic province is separated from the granites of Phuket by the prominent Khlong Marui Fault.

Most of the plutons consist of coarse-grained biotite granite containing megacrysts of alkali feldspar, but some are equigranular. Muscovite is commonly seen replacing biotite as a result of greisenization. Many of the granites contain tourmaline. No amphibole-biotite granodiorite has been found. The plutons all plot within the granite sector of the Streckeisen (1976) diagram, and in this respect they appear to be the northward extension of the Malaysian Main Range and Western Stable Shelf. The microcline phenocrysts are generally white. Pegmatites and aplites are small but numerous. Small metasedimentary xenoliths are common. The alkali feldspar is everywhere perthitic microcline, and the biotite is generally reddish brown. Although all the granites belong to the ilmenite series with magnetic susceptibilities below 50×10^{-6} emu/g, with the exception of that at Khao Phanom Bencha on the western part of the peninsula (Ishihara et al., 1979), more samples may be assigned to the I-type using the chemical parameters of White and Chappell (1977) and Chappell and White (1974).

The granites commonly show a distinct foliation due to the parallel alignment of the microcline phenocrysts. The tin and tungsten deposits are associated with northerly trending shears and with greisen quartz veins containing muscovite, alkali feldspar, and fluorite, and disseminated in greisenized granite. Some strongly sheared granites show mylonitization textures.

The granites intrude Paleozoic and Triassic sedimentary rocks. In some cases limited aureoles of biotite and chiastolite-biotite hornfels are produced. Most of the Rb:Sr data point to a Late Triassic age for the granites, but Permian granites may be present. All K:Ar dates, however, are strongly discordant, with peaks around 190 and 55 Ma. The discordant ages may be related to the important north-trending shearing, and to hydrothermal circulation along the shear and joint planes.

Chemical variation diagrams from Ishihara et al. (1980) are plotted against the differentiation index of Thornton and Tuttle (1960) on Figure 11. The granites are distinctly different from the I-granites of Japan; in particular, they are lower in Na_2O and richer in K_2O. Of the 22 analyzed granites, the average fluorine content is 1683 ppm (range: 380 to 4000 ppm), and the average tin content is 13 ppm (range: 3 to 28 ppm). The Rb and Sr data are summarized in

Table 3. The granites contain large amounts of uranium (average 16 ppm, ranging from 5 to 57 ppm) and thorium (average 33 ppm, range: 3 to 85 ppm) (Ishihara and Mochizuki, 1980). Owing to the enrichment in uranium, Th/U ratios are low, averaging 2. This enrichment indicates a large heat-generation capacity within the continental crust. The uranium leached from the granites may have formed uranium deposits in surrounding country rocks.

NORTHWEST CONTINENTAL THAILAND

Several large granitoid batholiths were mapped by Baum et al. (1970) and their radiometric dates discussed by von Braun et al. (1976). The outstanding characteristic of these granite masses is their exceptionally high initial $^{87}Sr/^{86}Sr$ ratios (Beckinsale et al., 1979), commonly as high as the 0.728 to 0.732 range. The granites are all S-type and the high ratios clearly indicate that the magmas resulted from anatexis of highly evolved continental crust. They therefore appear to have been derived from the upper crust, as compared with the Main Range granites, whose initial ratios are high on a world scale but lower than those in northwest Thailand. It is more likely that the Main Range granites were derived from the deeper Precambrian crust. It is worthwhile to note that the granites of northwest Thailand, although associated with tin and antimony mining (Nutalya et al., 1979), do not have great tinfields like the Main Range and Phuket. It might therefore be speculated that the lower crust is the main repository of tin and that the northwest Thailand granites have not tapped the deeper crust, whereas the Main Range and Phuket granites have.

1) Khutan Batholith, Lampang

This large granitic batholith is elongated northeasterly. It separates two Cenozoic basins, that of Lampang to the east and Chiang Mai to the west. Its geology has been described by Suensilpong et al. (1977).

Using the combined results of von Braun et al. (1976) and Tegin (1975), Beckinsale et al. (1979) calculated an Rb:Sr isochron age of 212 ± 12 with an initial $^{87}Sr/^{86}Sr$ ratio of 0.7244. The age is exactly equal to the K:Ar age for extracted biotite (von Braun et al., 1976).

The batholith is intrusive into strongly folded and contorted Silurian-Devonian phyllite, schist, and quartz-feldspathic schist. The granite is partly concordant and partly discordant with the folded country rocks, which strike northeasterly, parallel with the batholith. The contact is sharp, and no aureole has been found superimposed on the phyllite and schists. This is reminiscent of the Main Range contacts of peninsular Malaysia.

The batholith is predominantly granite, with minor quartz monzonite, biotite-muscovite granite, and fine-grained tourmaline granite. The granite is strongly jointed and faulted, and quartz and pegmatite veins follow the joints (Suensilpong et al., 1977).

The granite is coarse grained, coarsely porphyritic, with microcline phenocrysts commonly 4 to 6 cm long. Mostly it is a biotite granite with part of the biotite replaced by muscovite, owing to greisenization. Tourmaline granite or aplite forms dikes.

Chemically the Khutan batholith consists entirely of S-type granitoids, as shown by the chemical analyses summarized in Table 5 and variation diagrams in Figure 12. Suensilpong et al. (1977) reported an average Sn content of 562 ppm, with a range from 40 ppm to as high as 0.15 percent. These analyses may be suspect, because such high tin values would make the granite itself an ore. The low Rb:Sr ratio of 0.5 (Table 3) is difficult to reconcile with the otherwise highly evolved character of the granite and its high initial Sr isotope ratio.

The antimony content is reported as averaging 635 ppm for the 28 analyzed samples, with a range from 100 ppm to 0.27 percent. The most important stibnite mines are at Mae Wa, west of the batholith within the Silurian-Devonian country rocks. The ore veins follow north-trending faults. Most of the tin mines are along the eastern flank of the batholith. The tin occurs as lodes or in pegmatite veins, or it is disseminated in the granite or as placer deposits derived from the erosion of the weathered batholith.

2) The Tak Batholith

This batholith outcrops over an area of 4000 km² in the vicinity of Tak. The geology has been described by Pongsapich and Mahawat (1977). It is composite and is composed of a quartz diorite-granodiorite series, and granite. Most rocks are medium grained, the more acid varieties are coarser and slightly more porphyritic. The phenocrysts of alkali feldspar in granite are both white and red or pink.

Alkali feldspar is minor or absent in the quartz diorite and increasingly abundant in granodiorite and granite. In the diorite and granodiorite, it is orthoclase and interstitial, and in the granite it forms larger subhedral crystals which range from orthoclase to intermediate microcline.

Hornblende and biotite are the major mafic minerals. Hornblende is absent from the granite but abundant in the diorite-granodiorite. The batholith formed a contact aureole in which the country rocks are normally in the albite-epidote hornfels facies. The granite series is later than the diorite-granodiorite, forms sharp contacts with it, and commonly assimilated it along the contacts.

The Tak batholith has been dated by Teggin (1975). The white granites give an Rb:Sr isochron of 213 ± 10 with an initial $^{87}Sr/^{86}Sr$ ratio of 0.7158. The pink granite gives an isochron age of 219 ± 12 and an initial ratio of 0.7104.

On the variation diagrams of Pongsapich and Mahawat (1977) (Fig. 13), the two series of diorite-granodiorite and

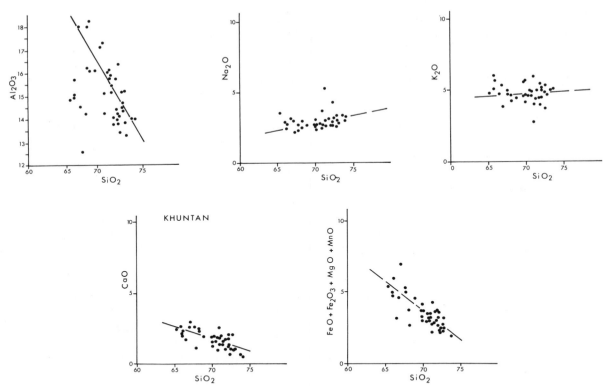

Figure 12. Variation diagrams of the Khutan batholith, which extends north-south between Chiang Mai and Lampang (data from Suensilpong et al., 1977).

granite are clearly distinguished. Although the Sr isotope ratios are high, not all of the Tak batholith is of S-type. Some analyses indicate I-type characteristics, but most suggest S-type (Table 5). The diorite-granodiorite has not been dated, but field evidence indicates it predates the granite.

Antimony mineralization is associated with the batholith, and its mixed S and I characteristics are in keeping with its association also with Pb and Cu. Barite and fluorite mineralization also occurs.

3) Other Batholiths

A north-northeast-trending series of batholiths extends into Laos from the Khutan batholith. The Fang Mae Suai batholith has been dated by von Braun et al. (1976) at 232 Ma by an Rb:Sr isochron, with an initial Sr isotope ratio of 0.728. The K:Ar mineral date is similar, at 209 Ma. It is of medium- to coarse-grained porphyritic biotite granite with alkali feldspar phenocrysts up to 10 cm long, and is undeformed.

The Ban Hong pluton is a medium-grained biotite granite containing some alkali feldspar phenocrysts, described by von Braun et al. (1976) as orthoclase. The Rb:Sr isochron gives an age of 235 Ma with an initial Sr isotope ratio of 0.725. The K:Ar mineral dates are similar, at 210 Ma.

The Li pluton is of biotite granite. On the east it is foliated, and on the west it contains feldspar phenocrysts and is unsheared. Biotite is commonly replaced by muscovite and fluorite is common in veins. It contains fresh orthoclase. The Rb:Sr isochron gives an age of 236 Ma, but it is based only on two points (von Braun et al., 1976).

West of Chiang Mai is the Mae Sariang granite chain of west Thailand. It consists of biotite granite containing some feldspar phenocrysts and some hornblende. Von Braun et al. (1976) gave its Rb:Sr-age as 190 Ma with an initial Sr isotope ratio of 0.728. K:Ar dates from mineral separates are in the range 205 Ma.

The gneiss and migmatite northwest of Chiang Mai is cut by younger, coarse-grained, slightly porphyritic biotite granite. The low Rb contents prevented useful Rb:Sr determinations on age, and the K:Ar dates on mica separates range from 201 to as low as 18 Ma. The radiometric dating is therefore inconclusive, and the previously recorded Carboniferous age (Baum et al., 1970) is now rejected (von Braun et al., 1976).

The Samoeng granite, west of Chiang Mai, has been dated by Teggin (1975) by an Rb:Sr isochron at 204 Ma with an initial Sr-isotope ratio of 0.7328.

In summary, the granites of northwest Thailand yield Early Triassic dates around 235 Ma, and all have characteristic high initial Sr-isotope ratios of about 0.725. The granites of the crystalline complex northwest of Chiang Mai are no longer thought to be Carboniferous. They are partly

Triassic, but young K:Ar dates of 194 and 204 Ma indicate Jurassic cooling, uplift, or subsequent reheating by younger granitic activity (von Braun et al., 1976). Biotite dates of 77 and 78 Ma probably are related to the Cretaceous magmatic activity which is common in the western belt of Thailand.

The northwest Thailand granites are predominantly S-type, but are unlike the Main Range of Malaysia and south Thailand in their unusually high initial Sr-isotope ratios. It is strange that such rocks have such low Rb:Sr ratios, commonly lower than one (Table 3). Beckinsale et al. (1979) suggested that they are collision related, but their level of emplacement appears to be much higher than that of the Main Range.

CRETACEOUS GRANITES OF PHUKET AND WEST THAILAND

1) Phuket Area

The granites define an Rb:Sr isochron of 124 ± 4 Ma with an $^{87}Sr/^{86}Sr$ initial ratio of 0.7073 (Garson et al., 1975), as refined by Beckinsale et al. (1979). The K:Ar dates are discordant and give ages of 55 to 65 Ma. The distribution of the granitoids is controlled by the transcurrent Khlong Marui Fault. The granitoids may be divided into coarse-grained granite, coarse-grained biotite granite, and fine- to medium-grained muscovite-biotite granite.

The granite contains pinkish-grey phenocrysts of alkali feldspar and the mafic minerals are hornblende and small amounts of clinopyroxene. The biotite granites are porphyritic and contain euhedral phenocrysts of intermediate microcline in a medium-grained groundmass. Biotite forms irregular laths. Fine- to medium-grained biotite granite and muscovite-biotite granite are leucocratic. The margins have been altered to greisen. Some foliated granites have been affected by movements along the Khlong Marui Fault and show a pronounced parallelism of the phenocrysts and biotite crystals. Adjacent to the granites and pegmatites, the Phuket Group (predominantly Carboniferous) has undergone contact metamorphism. The metamorphic aureole is up to 2 km wide in places (Garson et al., 1975). Cordierite and andalusite hornfels is typically developed.

The Phuket area is unusual in Southeast Asia for its major development of large pegmatite bodies. They are related to the northeast-trending fractures. There are two types, mica-tourmaline and lepidolite pegmatite. They can be up to 15 m wide and some are cassiterite-bearing. Variation diagrams for the Phuket granitoids are shown in Figure 14, based on data from Garson et al. (1975) and Ishihara et al. (1980), summarized in Table 6. The granites are all of the S-type and ilmenite series, except for magnetite series rocks at Khao Phanom Bencha and on the islands off the west coast (Ishihara et al., 1979).

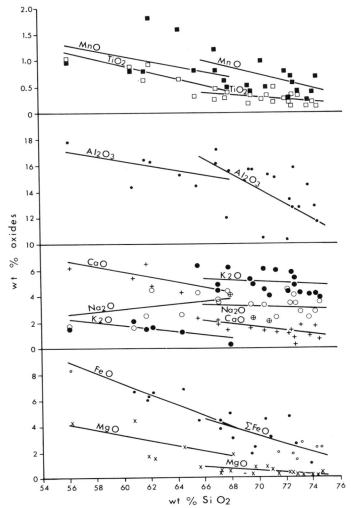

Figure 13. Variation diagrams of the Tak batholith, northwest Thailand (after Pongsapich and Mahawat, 1977).

Garson et al. (1975) have recorded the following trace element contents based on 12 samples: Sn averages 23 ppm (range: 3 to 80 ppm); Nb, 16 ppm (range: 10 to 25 ppm); Zr, 164 ppm (range: 90 to 250 ppm); and Ba, 337 ppm (range: 120 to 600 ppm). The Rb and Sr data are summarized in Table 3 and Figure 4.

2) Khao Daen Granite

The Rb:Sr isochron, which is unsatisfactory because of two groups of points at either end, indicates an intrusive age of 93 ± 4 Ma, and the separated muscovites and biotites give K:Ar ages of 72 Ma (Beckinsale et al., 1979). The granite is genetically related to the Cha-Rin tin mining area. This highly evolved granite has Rb:Sr ratios of around 20 and Rb contents of up to 1000 ppm; it seems likely that it is associated with tin mineralization.

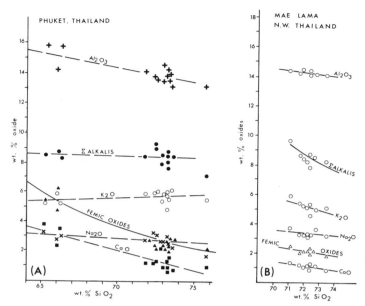

Figure 14. Variation diagram for the Cretaceous granites of (A) Phuket, and (B) Mae Lama, western Thailand, based on data from Garson et al. (1975), Ishihara et al. (1980), and Beckinsale et al. (1979).

3) *The Mae Lama Granite*

This granite forms a small stock about 4 km across, and is emplaced in Cambro-Ordovician sedimentary rocks (Pitragool and Panupaisal, 1978). It yielded a good Rb:Sr isochron age of 130 ± 4 Ma. The granite has an initial strontium ratio of 0.7086 (Beckinsale et al., 1979).

Seven mineralized vein systems in the country rocks above the granite have been worked for wolframite and cassiterite. The granite is greisenized near these veins, and von Braun et al. (1976 concluded from Rb:Sr dating of the greisen that the age is 78 Ma. The process of greisenization therefore leads to anomalously young ages.

The K:Ar ages of muscovites from the granite are concordant with those from the greisen at 72 Ma, and biotite from the granite gives average values of 53 Ma by the K:Ar method. According to Beckinsale et al. (1979), the data indicate that the granite was emplaced 130 million years ago and cooled normally within about one million years. About 70 million years ago circulation of meteoric water produced the greisenization of the granite and the mineralized quartz veins, whose filling temperatures have been estimated to be in the range 150° to 450° C (Pitragool and Punapaisal, 1978). Scattered Late Cretaceous intrusions may have provided the heat to drive the hydrothermal circulation. However, it is not necessary to invoke them, for the Main Range and north Thailand continue to be hot today, as witnessed by numerous hot springs. As in the Main Range east of Kuala Lumpur, there are prominent large veins of pure quartz, which are quarried for silica, traceable over

TABLE 6. REPRESENTATIVE CHEMICAL ANALYSES (wt.%) OF THE COMPOSITIONALLY RESTRICTED CRETACEOUS GRANITES FROM WEST THAILAND, PENINSULAR MALAYSIA, AND WESTERN BORNEO

Oxide	1	2	3	4	5	6	7	8	9
SiO_2	72.46	73.50	65.20	72.40	73.00	73.00	73.14	68.00	64.20
Al_2O_3	14.00	14.03	15.90	13.80	14.60	14.41	14.40	16.70	17.10
Fe_2O_3	0.18	0.21	2.70	1.30	1.30	1.27	1.06	1.75	1.10
FeO	1.52	1.17	1.50	0.70	0.40			1.10	3.05
MgO	0.57	0.36	0.90	0.40	0.40	0.23	0.22	0.86	1.27
CaO	1.38	0.81	3.80	1.20	0.80	1.11	1.06	2.75	4.42
Na_2O	3.34	3.18	3.30	3.30	2.60	3.62	3.59	4.90	4.45
K_2O	4.50	5.04	5.20	6.00	5.90	5.14	4.63	1.46	1.94
H_2O-	0.15	0.17	-	-	-	-	-	0.55	0.26
H_2O+	0.75	0.82	0.90	0.60	0.60	-	-	1.37	1.24
TiO_2	0.50	0.19	0.40	0.20	0.20	0.20	0.18	0.25	0.48
P_2O_5	0.16	0.19	0.10	0.10	0.10	0.04	0.06	0.15	0.26
MnO	0.04	0.03	0.10	-	0.10	0.05	0.03	0.08	-
CO_2	-	-	-	-	-	-	-	0.04	0.01
Total	99.55	99.70	100.00	100.00	100.00	99.07	98.37	99.96	99.78

1. Biotite granite, Mae Lama, West Thailand. 2. Greisenized granite, Mae Lama, West Thailand. 3. Coarse biotite-hornblende adamellite, Phuket, Thailand. 4. Porphyritic biotite granite, Phuket, Thailand. 5. Fine-grained muscovite-biotite granite, Phuket, Thailand. 6. Medium-grained biotite granite, Mount Ophir, peninsular Malaysia. 7. Medium-grained, equigranular biotite granite, Mount Ophir, peninsular Malaysia. 8. Microgranodiorite, Lundu, West Sarawak. 9. Microgranodiorite, Simanggang, West Sarawak. Sources: Beckinsale and others (1979), Garson and others (1975), Yeap (1980), and Kirk (1968).

several kilometers, and up to 0.5 kilometers wide. These were formed by the hydrothermal circulation system which is thought to have reset the K:Ar clock.

The Mae Lama granite is highly evolved, and distinctly S-type (Table 6), as shown by the chemical analyses (Beckinsale et al., 1979). The variation diagram shown in Figure 14 was constructed from their analyses.

SUMATRA

Granitoids are widely scattered throughout Sumatra, especially along the Barisan Mountains (Van Bemmelen, 1970). A compilation of their ages (Hehuwat, 1976) shows that some are Triassic, and a few may be as old as Permian, such as that at Sibolga (Fig. 1) which gave a Rb:Sr date of 257 ± 24 Ma. K:Ar dates are commonly around 215 Ma. These ages suggest that Sumatra forms part of the massif of Thailand and the Malay Peninsula. The Sibolga granite is representative of Sumatran granites, having a hypersolvus texture (characteristic of high-level granites) and pink orthoclase. The mafic minerals are biotite and euhedral hornblende.

Elsewhere in northern Sumatra, particularly near the west coast, stocks and small batholiths of mainly granitic to dioritic composition are known (Page et al., 1979). The Sikoleh batholith, which occurs near the west coast 5° N, is a multiple intrusion exhibiting a granitic core which intruded granodiorite, tonalite, and diorite. It is considered (Page et al., 1979) to be Cretaceous. The northernmost stock at Pulau Brueuh gives a Rb:Sr age of 24 Ma and a low initial Sr ratio of 0.7043. The other granitoids are similar to this and are probably related to subduction.

A large batholith at Bengkunat on the Benkulen Peninsula of south Sumatra consists of a core of coarse biotite granite, and outer parts of biotite-hornblende granodiorite (van Bemmelen, 1970). It is considered by Hamilton (1979) to be Cretaceous or Tertiary. Central Sumatra has alluvial-tin-producing centers at Tandun, Bangkinang, and Bukit Ratah, but the age of the associated granitoids is unknown.

BORNEO

Extensive plutonic activity in west Sarawak from mid- to Late Cretaceous times resulted in high-level intrusions of several large stocks of granitic rocks and smaller bodies of gabbro and dolerite (Kirk, 1968). K:Ar dates group around 77 Ma. The same intrusive event was dated on the Indonesian island of Karimata (Fig. 1) by Priem et al. (1975). The granite has a Rb:Sr isochron age of 74 ± 2 Ma and an initial Sr ratio of 0.7101. Associated amphibolite has a K:Ar age of 78 ± 5 Ma. The Karimata granites yielded the following Rb:Sr data on the basis of 8 analyses (Priem et al., 1975): Rb averages 345 ppm (range 118 to 592 ppm); Sr averages 207 ppm (range: 3 to 592 ppm); and Rb/Sr averages 1.7. West Borneo is out of the tin-tungsten granitoid belt, but west Sarawak is known for its stibnite deposits (Hutchison and Taylor, 1978). Several granitoids from the northwest extension of the Lupar line, which represents the eastern margin of Sundaland, have been dated by K:Ar methods. Their ages generally group around 85 Ma (Hehuwat, 1976).

TECTONIC CONSIDERATIONS

Only the eastern volcano-plutonic arc, in which Permian and Triassic granitoids form part of a calc-alkaline plutonic series ranging from gabbro, through diorite, to granodiorite and granite, can be classified as an Andino-type arc after the terminology of Pitcher (1979), although the initial $^{87}Sr/^{86}Sr$ ratios are considerably higher than those reported by Pitcher. The plutonic suite is associated with volcanic rocks and tuffs ranging from Carbo-Permian to Triassic age. The 220-Ma-old granites are genetically related to ignimbritic subaerial flows and the terrane has remained tectonically stable since the Late Triassic. The igneous rocks may be related to subduction processes (Fig. 15); some of the magmas are thought to have a mantle origin, and others are probably derived from the ensialic island-arc crust.

The volcanic and plutonic rocks of the central basin have lower initial Sr ratios and, although related to the Bentong-Raub suture which may be an old plate margin, their source area cannot be identified with certainty.

The Main Range and northern Thailand plutons are predominantly S-type granites of the ilmenite series. They have no direct volcanic association, and their high initial Sr ratios indicate they are derived by anatexis of continental crust. Both Mitchell (1977) and Hutchison (1978) proposed

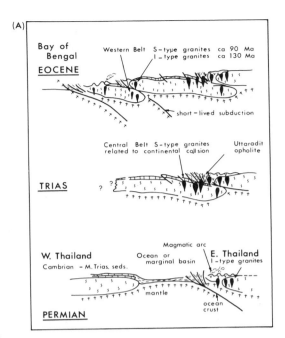

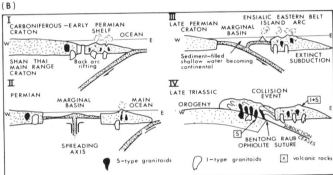

Figure 15. Two geo-cartoons showing variations on the same theme of the origin of southeast Asian granitoid belts by subduction-related and collision-related processes. (A) After Mitchell (1977) and (B) after Hutchison (1978). The Eastern Belt, with its volcanic association is an Alpino-type orogen, and the Main Range and Continental Thailand belts are collision-type related to closing of the central marginal basin. They are Hercynotype. The Western Phuket Belt and related Cretaceous granites are related to subduction from the Bay of Bengal after the Shan-Thai-Main Range massif faulted off its unknown cratonic attachment.

that these granites resulted from closure of the central margin basin. This closure was the main orogenic event which led to the cratonization of Sundaland. They have been referred to as collision-related granites (Fig. 15). In the terminology of Pitcher (1979), the Main Range and northwest Thailand granites may be classified as Hercyno-type, resulting from tectonics related to marginal basins far behind the ocean-continent margin and its arc-trench system.

The Cretaceous western belt of Phuket, western Burma, and Thailand are all S-type granites with high $^{87}Sr/^{86}Sr$

initial ratios. They may therefore have resulted from some form of crustal anatexis, possibly related to the subduction of Indian Ocean lithosphere beneath Sundaland after separation from its cratonic attachment probably in the Early Mesozoic (Fig. 15).

The granitoids of Sumatra are predominantly Cretaceous to Tertiary and are probably subduction related, but others date back to the Triassic or even the Permian.

The eastern margins of the tin-belt granites are represented by Karimata Islands, western Borneo, and Natuna where the high-level plutons are of Cretaceous age and may be related to subduction activity associated with the Lupar line suture zone, which represents the eastern margins of continental Sundaland.

Tin, tungsten, and antimony mineralization is not confined to any one age or tectonic setting of granitoid. It is associated with both epizonal and mesozonal granites. It is associated also with granites which have been interpreted as collision-related in a Hercyno-type orogenic belt. The lack of a unique setting supports the conclusion of Hutchison and Chakraborty (1979) that these elements are scavenged from the continental infrastructure by whatever orogenic process is capable of producing anatexis. Granitoids that are of the magnetite series and have initial $^{87}Sr/^{86}Sr$ isotope ratios lower than about 0.706 may have a mantle origin and therefore are devoid of any tin, tungsten, and antimony association.

ACKNOWLEDGMENTS

I am extremely grateful to Antonia T. Oliveric of the Geological Survey of Canada for drawing the illustrations from my rough sketches, and to Ms. Gemma Todd who typed the manuscript while I was on sabbatical leave at Cornell University.

REFERENCES CITED

Adam, J. W. H., 1960, On the geology of the primary tin-ore deposits in the sedimentary formation of Billiton: Geologie en Mijnbouw, 22, p. 405-426.

Aleva, G. J. J., 1960, The plutonic igneous rocks from Billiton, Indonesia: Geologie en Mijnbouw, 22, p. 427-436.

Alexander, J. B., Harral, G. M., and Flinter, B. H., 1964, Chemical analyses of Malayan rocks, commercial ores, alluvial mineral concentrates, 1903-1963: Geological Survey of Malaya Professional Paper E-64. I-C, Ipoh, Malaysia.

Aramaki, S., Hirayama, K., and Nozawa, T., 1972, Chemical composition of Japanese granites Pt. 2: variation trends and average composition of 1,200 analyses: Journal of Geological Society of Japan, 78, p. 39-49.

Aranyakanon, P., 1961, The cassiterite deposit of Haad Som Pan, Ranong province, Thailand: Royal Department of Mines, Thailand, Report of Investigation, No. 4, p. 1-182.

Au, Y. M. H., 1974, The geology of the Bukit Bintang area, Trengganu, with some aspects in geotechnics [unpubl. B.Sc. (Hons.) thesis]: University of Malaya, Kuala Lumpur, 87 p.

Baum, F., von Braun, E., Hahn, L., and others, 1970, On the geology of northern Thailand: Beihefte zum Geologischen Jahrbuch, 102, p. 1-24.

Beckinsale, R. D., Suensilpong, S., Nakapadungrat, S., and Walsh, J. N., 1979, Geochronology and geochemistry of granite magmatism in Thailand in relation to a plate tectonic model: Journal of the Geological Society of London, 136, p. 529-540.

Bignell, J. D., 1972, The geochronology of the Malayan granites [unpubl. PhD. thesis]: University of Oxford, 334 p.

Bignell, J. D., and Snelling, N. J., 1977, The geochronology of Malayan granites: Overseas geology and mineral resources, 47, 70 p.

Bradford, E. F., 1972, The geology of the Gunong Jerai area, Kedah: Geological Survey of Malaysia, Memoir 13, 242 p. + maps.

Buddington, A. F., 1959, Granite emplacement with special reference to North America: Geological Society of America Bulletin, 70, p. 671-747.

Burton, C. K., 1973, Geology and mineral resources, Johore Bahru-Kulai area, South Johore: Malaysia Geological Survey Map, Bulletin 2, 73 p.

Burton, C. K., and Bignell, J. D., 1969, Cretaceous-Tertiary events in Thailand: Geological Society of America Bulletin, 80, p. 681-688.

C.C.O.P., 1980, Studies in East Asian tectonics and resources: United Nations ESCAP, CCOP Technical Publication No. 7, U.N. Building, Bangkok, 257 p.

Chantaramee, Sompongse, 1978, Tectonic synthesis of the Lansang area and discussion of regional tectonic evolution, in Nutalaya, Prinya, ed., Proceedings of the Third Regional Conference on Geology and Mineral Resources of Southeast Asia: Asian Institute of Technology, Bangkok, p. 177-186.

Chappell, B. W., and White, A. J. R., 1974, Two contrasting granite types: Pacific Geology, 8, p. 173-174.

Dawson, J., MacDonald, S., Paton, J. R., and others, 1968, Geological map of Northwest Malaya: Geological Survey Department of West Malaysia, scale 1:250,000.

Faure, G., 1977, Principles of isotope geology: New York, Wiley-Interscience, 464 p.

Fontaine, H., and Workman, D. R., 1978, Review of the geology and mineral resources of Kampuchea, Laos and Vietnam, in Nutalaya, Prinya, ed., Proceedings of the Third Regional Conference on Geology and Mineral Resources of Southeast Asia: Asian Institute of Technology, Bangkok, p. 541-603.

Foo, K. Y., 1964, Geology of the north central region of Pulau Langkawi [unpubl. B.Sc. (Hons.) thesis]: Kuala Lumpur, University of Malaya, 62 p.

Garson, M. S., Young, B., Mitchell, A. H. G., and Tair, B. A. R., 1975, The geology of the tinbelt in Peninsular Thailand around Phuket, Phangnga and Takua Pa.: Institute of Geological Sciences, Overseas Memoir No. 1, 112 p. + maps.

Gobbett, D. J., 1972, Geological map of the Malay Peninsula: Geological Society of Malaysia, Kuala Lumpur, scale 1:1,000,000.

Goh, L. S., 1973, Geology, mineralization and geochemical studies of the Chenderong-Buloh Nipis area, Trengganu [unpubl. B.Sc. (Hons.) thesis]: Kuala Lumpur, University of Malaya, 105 p. + maps.

Hamilton, W., 1979, Tectonics of the Indonesian Region: U. S. Geological Survey Professional Paper 1078, 345 p. + map.

Hehuwat, F., 1976, Isotopic age determinations in Indonesia: the state of the art: U. N. ESCAP CCOP Technical Publication No. 3, Proceedings of seminar on isotopic dating, Bangkok, p. 135-157.

Hosking, K. F. G., 1973, Primary mineral deposits, in Gobbett, D. J., and Hutchison, C. S., eds., Geology of the Malay Peninsula: West Malaysia and Singapore: New York, Wiley-Interscience, p. 335-390.

——, 1977, Known relationships between the "hard rock" tin deposits and the granites of Southeast Asia: Geological Society of Malaysia Bulletin, 9, p. 141-157.

Hosking, K. F. G., Yancey, T. E., Strimple, H. L., and Jones, M. T., 1977,

The discovery of macrofossils at Selumar, Belitung, Indonesia: Geological Society of Malaysia Bulletin, 8, p. 113–115.

Hutchison, C. S., 1964, A gabbro-granodiorite association in Singapore island: Quarterly Journal of the Geological Society of London, 120, p. 283–297.

——,1973a, Tectonic evolution of Sundaland: a Phanerozoic synthesis: Geological Society of Malaysia Bulletin, 6, p. 61–86.

——,1973b, Volcanic activity, *in* Gobbett, D. J., and Hutchison, C. S., eds., Geology of the Malay Peninsula: West Malaysia and Singapore: New York, Wiley-Interscience, p. 177–214.

——,1973c, Plutonic activity, *in* Gobbett, D. J., and Hutchison, C. S., eds., Geology of the Malay Peninsula: West Malaysia and Singapore: New York, Wiley-Interscience, p. 215–252.

——,1973d, Metamorphism, *in* Gobbett, D. J., and Hutchison, C. S., eds., Geology of the Malay Peninsula: West Malaysia and Singapore: New York, Wiley-Interscience, p. 253–303.

——,1975, Ophiolite in Southeast Asia: Geological Society of America Bulletin, 86, p. 797–806.

——,1977, Granite emplacement and tectonic subdivisions of Peninsular Malaysia: Geological Society of Malaysia Bulletin, 9, p. 187–207.

——,1978, Southeast Asian tin granitoids of contrasting tectonic setting: Journal of Physics of the Earth, Tokyo, 26 (Suppl.) S221–S232.

——,in press, Southeast Asia (Burma, Thailand, Malaysia, Indonesia, *in* Nairn, A. E. M., and Stehli, F. G., eds., The Ocean basins and their margins — The Indian Ocean, New York, Plenum Press.

Hutchison, C. S., and Chakraborty, K. R., 1979, Tin: a mantle or crustal source?: Geological Society of Malaysia Bulletin, 11, p. 71–79.

Hutchison, C. S., and Leow, J. H., 1963, Tourmaline greisenization in Langkawi, northwest Malaya: Economic Geology, 58, p. 587–592.

Hutchison, C. S., and Snelling, N. J., 1971, Age determination on the Bukit Paloh adamellite: Geological Society of Malaysia Bulletin, 4, p. 97–100.

Hutchison, C. S., and Taylor, D., 1978, Metallogenesis in S.E. Asia: Journal of the Geological Society of London, 135, p. 407–428.

Ingham, F. T., and Bradford, E. F., 1960, The geology and mineral resources of the Kinta Valley, Perak: District Memoir 9, Federation of Malaya Geological Survey, 347 p. + maps.

Ishihara, S., 1977, The magnetite-series and ilmenite-series granitic rocks: Mining Geology, 27, p. 293–305.

——,1980, Significance of the magnetite-series and ilmenite-series of granitoids in mineral exploration: Proceedings of the Fifth Quadrennial IAGOD Symposium, Stuttgart, Schweizerbart'sche Verl, p. 309–312.

Ishihara, S., and Mochizuki, T., 1980, Uranium and thorium contents of Mesozoic granites from Peninsular Thailand: Bulletin Geological Survey of Japan, 31, p. 369–376.

Ishihara, S., Sawata, H., Arpornsuwan, S., and others, 1978, Granitic rocks in southern Thailand, *in* Prinya, Nutalaya, ed., Proceedings of the Third Regional Conference on Geology and Mineral Resources of Southeast Asia, Asian Institute of Technology, Bangkok, p. 265–267.

——,1979, The magnetite-series and ilmenite-series granitoids and their bearing on tin mineralization, particularly of the Malay Peninsula region: Geological Society of Malaysia Bulletin, 11, p. 103–110.

Ishihara, S., Sawata, H., Shibata, K., and others, 1980, Granites and Sn-W deposits of Peninsular Thailand, *in* Ishihara, S., and Takenouchi, S., eds., Granitic magmatism and related mineralization: Mining Geology Special Issue 8, p. 223–241.

Jaafar bin Ahmad, 1976, The geology and mineral resources of the Karak and Temerloh area, Pahang: Geological Survey of Malaysia District Memoir 15, 138 p.

——,1979, The petrology of the Benom Igneous complex: Geological Survey of Malaysia Special Paper 2, 141 p.

Jones, M. T., Reed, B. L., Doe, B. R., and Lanphere, M. A., 1977, Age of tin mineralization and plumbotectonics, Belitung, Indonesia: Economic Geology, 72, p. 745–752.

Khoo, T. T., 1968, A petrological study of the Sungai Ruan area, Raub, West Malaysia [unpubl. B.Sc. (Hons.) thesis]: Kuala Lumpur, University of Malaya, 134 p.

Kirk, H. J. C., 1968, The igneous rocks of Sarawak and Sabah: Geological Survey of Malaysia, Borneo Region, Bulletin 5, 210 p. + maps.

Lameyre, J., 1980, Les magmas granitique: leurs compartements, leurs associations et leurs sources: in Livre jublaire du cent cinquantenaire, 1830-1980, Societe Geologique de France, Memoire Hors Serie, Paris, No. 10, p. 51–62.

Lasserre, M., Cheymol, J., Petot, J., and Saurin, E., 1970, Geologie, chimie et geochronologie du granite de Tasal (Cambodge occidental): France, Bureau de Recherches Geologiques et Minieres, Bulletin 2e Ser. Section 4(1), p. 5–13.

Lasserre, M., Saurin, E., and Dumas, J. P., 1972, Age Permien obtenue par le methode a l'argon sur deux amphiboles extraites de granodiorites de la region de Sanakham (Laos): Societe Geologique de France, Compte Rendu Sommaire des Seances, Paris, 3, p. 65–67.

Lim, Y. K., 1972, Geology of the north-west sector of Gunung Ledang (Mt. Ophir), Johore [unpubl. B.Sc. (Hons.) thesis]: Kuala Lumpur, University of Malaya, 79 p. + map.

MacDonald, S., 1968, Geology and mineral resources of north Kelantan and north Trengganu: Geological Survey of Malaysia Memoir 10, 202 p. + map.

Mitchell, A. H. G., 1977, Tectonic settings for emplacement of Southeast Asian tin granitoids: Geological Society of Malaysia Bulletin, 9, p. 123–140.

——,1979, Rift-, subduction- and collision-related tin-belts: Geological Society of Malaysia Bulletin 11, p. 81–102.

Ng, C. N., 1974, A comparative study of some epizonal and mesozonal granites in West Malaysia [unpubl. M.Sc. thesis]: Kuala Lumpur, University of Malaya, 145 p.

Nutalaya, P., Campbell, K. V., MacDonald, A., and others, 1979, Review of the geology of Thai tin fields: Geological Society of Malaysia Bulletin, 11, p. 137–159.

Page, B. G. N., and Workman, D. R., 1969, Geological and geochemical investigations in the Mekong Valley, between Vientiane and Sayaboury and Ban Houei Sai: Report No. 9 Overseas Division Institute of Geological Sciences, London, 48 p.

Page, B. G. N., Bennett, J. D., Cameron, N. R., and others, 1979, A review of the main structural and magmatic features of northern Sumatra: Journal of the Geological Society of London, 136, p. 569–579.

Pearce, J. A., and Gale, G. H., 1977, Identification of ore-deposition environment from trace-element geochemistry of associated igneous host rocks: Volcanic processes in ore genesis, Geological Society of London Special Publication 7, p. 14–24.

Pitcher, W. S., 1979, The nature, ascent and emplacement of granitic magmas: Journal of the Geological Society of London, 136, p. 627–662.

Pitragool, S., and Panupaisal, S., 1978, Tin and tungsten mineralization of the Mae Lama mine and its vicinity, N.W. Thailand: Special Publication of Department of Geological Sciences, Chiang Mai University 2, p. 192–224.

Poldervaart, A., and Parker, A. B., 1964, The crystallization index as a parameter of igneous differentiation in binary variation diagrams: American Journal of Science, 262, p. 281–289.

Pongsapich, W., and Mahawat, C., 1977, Some aspects of Tak granites, northern Thailand: Geological Society of Malaysia Bulletin, 9, p. 175–186.

Pongsapich, W., Vedchakanchana, S., and Pongprayoon, P., 1980, Petrology of the Pranburi-Hua Hin metamorphic complex and geochemistry of gneisses in it: Geological Society of Malaysia Bulletin, 12, p. 55–74.

Priem, H. N. A., Boelrijk, N. A. I. M., Bon, E. H., and others, 1975, Isotope geochronology in the Indonesian tin belt: Geologie en Mijnbouw, 54, p. 61–70.

Putthapiban, P., and Suensilpong, S., 1978, The igneous geology of the granitic rocks of the Hub Kapong-Hua Hin area: Journal of the Geological Society of Thailand, 3, M1, p. 1–22.

Rajah, S. S., Chand, F., and Singh, D. S., 1977, The granitoids and mineralization of the eastern belt of Peninsular Malaysia: Geological Society of Malaysia Bulletin, 9, p. 209–232.

Ryall, P. J., 1976, Gravity traverse of Peninsular Malaysia — preliminary results: Abstract of Geological Society of Malaysia Ipoh Meeting, December 1976.

Shaw, D. M., 1968, A review of the K-Rb fractionation trends by covariance analysis: Geochimica et Cosmochimica Acta, 32, p. 573–601.

——, 1970, Trace element fractionation during anatexis: Geochimica et Cosmochimica Acta, 34, p. 237–243.

Stauffer, P. H., and Snelling, N. J., 1977, A Precambrian trondhjemite boulder in Palaeozoic mudstone of N.W. Malaya: Geological Magazine, 114 (6), p. 479–482.

Stewart, D. B., 1975, Optical properties of alkali feldspars, in Ribbe, P. H., ed., Feldspar mineralogy: Mineralogical Society of America Short Course Notes, 2, St23–St30.

Streckeisen, A., 1976, To each plutonic rock its proper name: Earth Science Review, 12, p. 1–33.

Suensilpong, S., Meesook, A., Nakapadungrat, S., and Putthapiban, P., 1977, The granitic rocks and mineralization at the Khuntan batholith, Lampang: Geological Society of Malaysia Bulletin, 9, p. 159–173.

Suensilpong, S., Burton, C. K., Mantajit, N., and Workman, D. R., 1978, Geological evolution and igneous activity of Thailand and adjacent areas: Episodes, No. 3, 1978, p. 12–18.

Takahashi, M., Aramaki, S., and Ishihara, S., 1980, Magnetite-series/ ilmenite-series vs. I-type/S-type granitoids, in Ishihara, S., and Takenouchi, S., eds., Granitic magmatism and related mineralization, Mining Geology, Special Issue No. 8, p. 13–28.

Taylor, D., 1971, An outline of the geology of the Bukit Ibam orebody, Rompin, Pahang: Geological Society of Malaysia Bulletin, 4, p. 71–89.

Taylor, D., and Hutchison, C. S., 1979, Patterns of mineralization in Southeast Asia, their relationships to broad-scale geological features and the relevance of plate-tectonic concepts to their understanding, in Jones, M. J., ed., Proceedings of the 11th Commonwealth Mining and Metallurgy Congress, Hong Kong, Institute of Mining and Metallurgy, London, p. 93–107.

Teggin, D. E., 1975, The granites of northern Thailand [unpubl. Ph.D. thesis]: University of Manchester.

Thornton, C. P., and Tuttle, O. F., 1960, Chemistry of igneous rocks, Part 1, Differentiation index: American Journal of Science, 258, p. 664–684.

Tischendorf, G., 1977, Geochemical and petrographic characteristics of silicic magmatic rocks associated with rare-element mineralization, in Stemprok, M., Burnol, L., and Tischendorf, G., eds., Metallization associated with acid magmatism: Geological Survey of Praha, 2, p. 41–96.

Tuttle, O. F., and Bowen, N. L, 1958, Origin of granite in the light of experimental studies in the system $NaAlSi_3O_8$-$KalSi_3O_8$-SiO_2-H_2O: Geological Society of America Memoir 74, 153 p.

Van Bemmelen, R. W., 1970, The geology of Indonesia: Vol. 1A, General geology of Indonesia and adjacent archipelagoes, Hague, Martinus Nijhoff, 732 p.

Varlamoff, N., 1978, Classification and spatio-temporal distribution of tin and associated mineral deposits, in Stemprok, M., Burnol, L., and Tischendorf, G., eds., Metallization associated with acid magmatism, 3, Prague Geological Survey, p. 139–159.

Von Braun, E., Besang, C., Eberle, W., and others, 1976, Radiometric age determinations of granites in northern Thailand: Geologisches Jahrbuch, Reihe B, 21, p. 171–204.

White, A. J. R., and Chappell, B. W., 1977, Ultrametamorphism and granitoid genesis: Tectonophysics, 43, p. 7–22.

White, A. J. R., Williams, I. S., and Chappell, B. W., 1977, Geology of Berridale 1:100,000, sheet 8625: Geological Survey of New South Wales, 138 p.

Workman, D. R., 1977, Geology of Laos, Cambodia, South Vietnam and the eastern parts of Thailand: Overseas Geology and Mineral Resources, No. 50, 33 p.

Wright, T. L., 1968, X-ray and optical study of alkali feldspars: II. An X-ray method for determining the composition and structural state from measurements of 20 values for three reflections: The American Mineralogist 53, p. 88–104.

Yeap, C. H., 1966, Geology of the Sungei Lembing area, Pahang, West Malaysia [unpubl. B.Sc. (Hons.) thesis]: Kuala Lumpur, University of Malaya, 109 p. + maps.

——, 1980, A comparative study of Peninsular Malaysian granites with special reference to tin mineralization [unpubl. Ph.D. thesis]: Kuala Lumpur, University of Malaya, 395 p. + maps.

MANUSCRIPT ACCEPTED BY THE SOCIETY JULY 12, 1982

Geological Society of America
Memoir 159
1983

Granitoids of the Tertiary continent — island arc collision zone, Papua New Guinea

T. J. Griffin*
Geological Survey of Papua New Guinea
P.O. Box 778
Port Moresby
Papua New Guinea

ABSTRACT

The Tertiary I-type calcalkaline granitoids of Papua New Guinea occur within three structural-tectonic zones; island arc, continental orogenic, and cratonic. Preliminary data indicate that the granitoids of one zone are not distinct from those of any other, but that within any one zone there are large variations in textures and composition, a predominance of basic and intermediate rock types, examples of comagmatic volcanics, and porphyry Cu-Au mineralization. The similarities among the three zones indicate a common origin. Underplating with mantle-derived magmas during a subduction event provides a common source material; however, partial melting at the base of the crust to produce the granitoids could result from subduction, crustal thickening, or crustal buckling.

INTRODUCTION

Papua New Guinea is in a complex tectonic setting between the leading edge of the cratonic Australian plate and the oceanic Pacific plate. The geology can be summarized in terms of three structural-tectonic areas; Papuan Basin, Central Orogenic Belt, and the Bismarck-Solomon region (Fig. 1).

The Papuan Basin covers relatively low country drained by the Fly, Strickland, and Purari Rivers. It is characterized by Mesozoic and Cenozoic sedimentary rocks 3000 m to 10,000 m thick. The basement is believed to be mainly Paleozoic granite and metamorphic rocks. Shelf sediments in the southwestern part of the region give way to trough sediments and volcanics in the east and northeast, with a corresponding increase in the amount of deformation of the rocks from flat lying in the southwest to tightly folded and thrust faulted in the north, northeast, and east.

The Central Orogenic Belt forms a high mountain spine to mainland Papua New Guinea and is characterized by mainly crystalline rocks and by widespread faulting. The rocks are almost entirely Mesozoic and Cenozoic in age with the exception of small areas of Permian and Mesozoic granites and metamorphics. Remnants of oceanic crust consisting of ultramafic, gabbroic, and basaltic rocks outcrop in several areas. Collision with island arcs had a major influence on the tectonics in this region.

The Bismarck-Solomon region is made up of Cenozoic rocks which are predominantly volcanic and associated intrusives, volcanic sediment, and limestone. These formed in an island arc environment.

Tertiary tectonics has given rise to volcanics and significant volumes of granitoids in all three areas. Figure 2 illustrates the relationships of the granitoid intrusions to the stratigraphy in each of the areas of plutonism throughout Papua New Guinea discussed below.

Papua New Guinea is very rugged and densely vegetated except for the highest peaks where the vegetation is restricted by low temperatures and occasional snow. Despite the difficult access, many intrusions were located by exploration companies searching for porphyry copper mineralization. Systematic mapping by the Australian Bureau of Mineral Resources and the Geological Survey of Papua New Guinea at 1:250,000 scale and more recently at 1:100,000 scale forms the basis of this review. In addition detailed studies, most commonly in areas with porphyry mineralization, have

---

*Present Address: Geological Survey of Western Australia, 66 Adelaide Terrace, PERTH, Australia

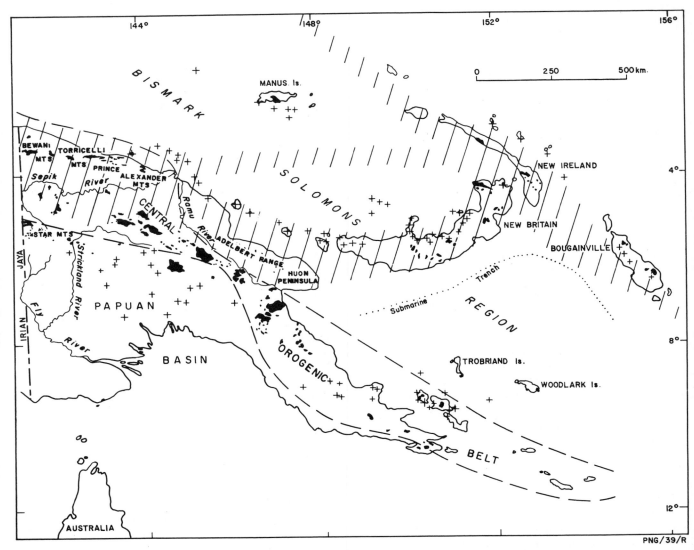

Figure 1. Papua New Guinea showing the distribution of granitoids in relation to the three major structural zones. Also shown are late Cenozoic volcanic centres (crosses) (Johnson, 1976), and zones of seismic activity (hachured).

provided valuable material; see the compilations of Knight (1975) and Gustafson and Titley (1978).

GRANITOIDS OF THE ISLAND ARCS

The Cenozoic island arcs of Papua New Guinea lie in the Bismarck-Solomon region (Fig. 1) and include the islands of Bougainville, New Ireland, Manus, New Britain, Woodlark, and Trobriands. The Adelbert Range-Huon Peninsula area is also included in the island arc province because of its obvious geological similarity to the other island arcs (Jaques and Robinson, 1977).

The island arcs are located in the most seismically active areas and are associated with active volcanoes (Fig. 1). New Britain has the classic island arc profile containing a submarine trench off the south coast and a northward-

dipping seismic zone beneath the island (Johnson, 1976). The other major islands are believed to have formed in a similar arc environment; however, the trenches and seismic zones are no longer obvious. In addition, the active volcanoes in all but the New Britain arc cannot be related to a simple subduction process and some modification is proposed by Johnson and others (1976) and Johnson and Smith (1974). This deviation from the classic system is not surprising when it is considered that there are at least two minor plates between the Australian plate and the Pacific plate in this region.

Bougainville

The intrusive rocks on Bougainville (Fig. 3) (Blake, 1967; Blake and Miezitis, 1967) occur within both? Oligocene

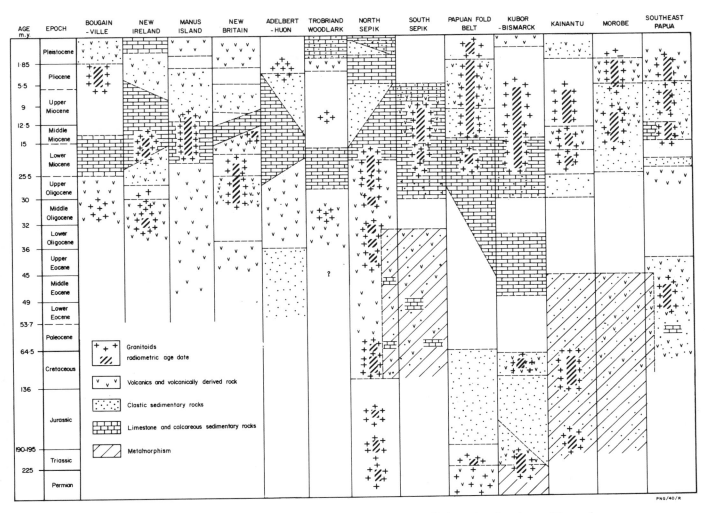

Figure 2. A time-space plot for various regions of Papua New Guinea showing the positions of the granitoids. Data sources: Bougainville, Blake (1967), Page and McDougall (1972b); New Ireland, Hohnen (1978); Manus Island, Jaques and Webb (1975), Jaques (1976); New Britain, Page and Ryburn (1977); Adelbert-Huon, Jaques and Robinson (1977); Trobriand-Woodlark, McGee (1978); North Sepik, Hutchison (1975), Hutchison and Norvick (1978); South Sepik, Dow and others (1972), Page and McDougall (1972a); Papuan Fold Belt, Davies and Norvick (1974), Page and McDougall (1972a), Arnold and others (1979); Kubor-Bismarck, Bain and others (1975), Page (1976), Pigram and others (in preparation); Kainantu, Tingey and Grainger (1976), Page (1976); Morobe, Page and McDougall (1972a), Dow and others (1974); Southeast Papua, Smith and Davies (1976).

Kieta Volcanics and the Pleistocene Bougainville Group. Ages of intrusion are thought to range from Oligocene to Pleistocene.

The largest body is an irregular mass 12 km across; smaller plugs and dikes are more common consisting predominantly of porphyritic microdiorite. Rock types include diorite, microdiorite, granodiorite, monzonite, syenite, and granophyre. Some intrusions are surrounded by hornfels zones.

The oldest intrusions, which outcrop in southern Bougainville, are apparently comagmatic with the mainly andesite and basalt flows and pyroclastics of the Kieta Volcanics.

The Bubia Formation, which has a similar age to the Kieta Volcanics, is also andesitic but predominantly a sedimentary unit.

The massive Lower Miocene Keriaka Limestone was deposited unconformably on the Kieta Volcanics prior to the resumption of volcanism in the Pleistocene. These are the Bougainville Group volcanics and occupy most of the island; they include rocks of several readily identifiable volcanoes. They are predominantly andesitic and dacitic and range to Recent in age.

Sulphide mineralization is associated with most intrusions and copper and gold with a few, e.g., Panguna Mine. The

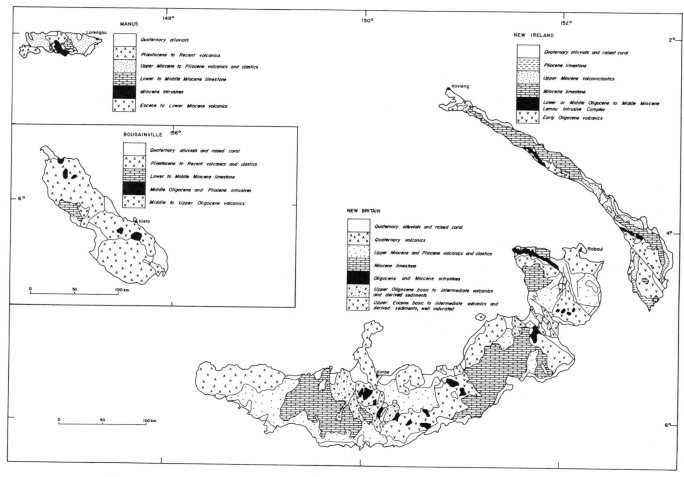

Figure 3. Simplified geology of the island arc region; Manus Island, New Ireland, New Britain, and Bougainville Island.

intrusive rocks of this mine differ from elsewhere as they contain andesitic and dioritic xenoliths (Blake, 1967). K-Ar ages from the Panguna area show that the premineralized phase is 4 to 5 m.y. old and the mineralized phase is 3.4 m.y.; a younger dike has been dated at 1.6 m.y. (Page and McDougall, 1972b).

New Ireland

The Lemau Intrusive Complex, consisting of gabbro, norite, diorite, tonalite, trondhjemite, granodiorite, and leucocratic dikes intrude the Lower to Middle Oligocene andesitic Jaulu Volcanics. The intrusive rocks are mostly overlain by the Miocene Lelet Limestone which forms large plateaus throughout New Ireland. Some of the younger intrusive phases intrude the base of the limestone (Hohnen, 1978).

Igneous activity postdating the Miocene limestone is restricted to the andesitic and dacitic volcanics in the Upper Miocene Rataman Formation. These rocks were deposited in a small graben which separated the northern and southern parts of the island in the Late Miocene. A volcanic island chain parallels the northeast coast and contains strongly alkaline rocks ranging from basalt to phonolite; these are pre-Middle Miocene to Pleistocene in age (Johnson and others, 1976).

K-Ar ages of the Lemau Intrusive Complex range from 31.8 m.y. to 13.8 m.y. and at least part of the complex may be comagmatic with the Jaulu Volcanics (Hohnen, 1978). Exposure of the granitoids is restricted to small outcrops over most of the southern part of New Ireland and to an 8-km-wide zone in the northeast of the island (Fig. 3). Granodiorite forms large irregular stocks with continuous outcrop of up to 20 km, whereas gabbro, norite, diorite, trondhjemite, and rhyodacite form small plug-like bodies and dikes. The intrusions have narrow metamorphic contact aureoles.

The mafic rocks contain a relic mineralogy believed to be the remnants of volcanic xenoliths (Hohnen, 1978). However, these textures could be equally due to assimilation of cogenetic mafix xenoliths. Alteration involving chlorite, fibrous amphibole, and biotite are present. Porphyry copper mineralization associated with the alteration has been investigated.

Manus Island

The geology of the Admiralty Islands has been studied in detail by Jaques (1976). The oldest rocks are Middle Eocene to earliest Miocene Tinniwi Volcanics which outcrop in the centre of Manus Island. These are unconformably overlain by the Miocene Mundrau Limestone and intruded by the multiphase Yirri Complex. The Mundrau Limestone is overlain by the coarse volcanic Tingau Conglomerate which passes laterally away from the centre of the island into the thick calcareous and tuffaceous siltstone and sandstone of the Lauis Formation of Late Miocene to Pliocene age.

Widespread Miocene to Pliocene basalts in the southeast postdate and are partly contemporaneous with the clastic sequence. In the west the clastics are overlain by Pliocene andesitic agglomerate.

Adjacent islands are composed of Quaternary volcanics, and some coral islands are considered to have volcanic submarine basement.

The Yirri Complex intrudes the Tinniwi Volcanics as a discordant batholith which outcrops over 200 km² in the centre of Manus Island. The complex is Miocene in age, 10 to 18 m.y. (Jaques and Webb, 1975). Fresh, medium-grained quartz monzodiorite comprises the bulk of the complex; fine- to medium-grained porphyritic phases are confined largely to the upper levels and margins. Rock types range from quartz diorite to quartz monzonite.

Jaques (1976) interprets the Yirri Complex as the upper part of a batholith containing volcanic, subvolcanic, and plutonic rocks. The relationships between these have been studied at Mount Kren, a porphyry copper prospect. Highly altered and brecciated andesitic volcanics and micromonzonite porphyries (12 m.y.) overlie moderately altered micromonzonite and andesite porphyries which pass at depth to weakly altered porphyritic quartz monzonite (18 m.y.)

Mineralization, involving widespread pyrite associated with extensive brecciation and alteration of the high-level porphyries and volcanics at Mount Kren, postdate the major intrusives and Jaques (1976) suggested the mineralization is the "result of late-stage resurgent boiling of the batholith."

New Britain

The granitoids in New Britain (Fig. 3) intrude the indurated Upper Eocene Baining Volcanics and the Upper Oligocene basic to intermediate volcanics and derived sediments (Macnab, 1970; Mackenzie, 1971; Davies, 1973a; Ryburn, 1974, 1975, 1976; Page and Ryburn, 1977). During the Early, Middle, and possibly Late Miocene, the thick Yalam Limestone was deposited unconformably on the older volcanics and intrusives. Volcanism resumed in the Late Miocene or Early Pliocene giving rise to lavas, volcaniclastics, and marine and terrestrial sediments in varying proportions throughout the island. Uplift, faulting, and volcanism followed in the Late Pliocene or Early Pleistocene (Page and Ryburn, 1977). The still active volcanism has lavas which include high-Al basalts, andesite, dacite, and minor rhyolite (Johnson and others, 1973).

The larger plutons, up to 15 km across, are mostly complex bodies of tonalite and diorite, and less commonly gabbro, granodiorite, monzonite, and adamellite; small cognate mafix xenoliths are abundant in the mafic intrusions (Hine and Mason, 1978). The intrusives are generally rich in quartz, hornblende, and augite and poor in biotite and potash feldspar. Accessories include magnetite, apatite, zircon, and sphene (Macnab, 1970; Mackenzie, 1971; Hine and Mason, 1978). The Baining Volcanics are hornfelsed adjacent to the larger plutons; related porphyries and fine-grained equigranular rocks are common as small dikes and stocks. K-Ar ages (Page and Ryburn, 1977) range from 30 m.y. to 22 m.y., and one rock was dated at 14 m.y. These results are from ten separate medium to coarse plutons throughout the island and indicate a period of major intrusive activity from mid-Oligocene to earliest Miocene.

The porphyritic phases, which are often associated with copper mineralization, intrude the equigranular granitoids at Plesyumi (Titley and Bell, 1971). They are believed to have slightly younger ages than those rocks dated by Page and Ryburn (1977) but still predate the Yalam Limestone.

Close genetic relationship between the plutonism and equivalent age volcanism has been recognized (Page and Ryburn, 1977; Titley, 1978). Many porphyry copper prospects have been investigated in detail. Hine and others (1978a), Hine and Mason (1978), and Titley (1978) discussed the extensive brecciation and alteration in these mineralized areas.

Adelbert Range-Huon Peninsula Region

Jaques and Robinson (1977) interpret the Adelbert Range-Finisterre Range-Huon Peninsula region (Fig. 1) as a Paleogene island arc which collided with the Australian plate, initially during the Early Miocene. However, sedimentation in this region continued into the Pliocene. The former plate boundary is now defined by a transcurrent fault located in the Ramu-Markham Valley (Fig. 6).

Several small intrusions less than 5 km across intrude the island arc succession which is composed of the prevolcanic Eocene Gusap Argillite overlying oceanic crust, Oligocene basic and andesitic Finisterre Volcanics, and postvolcanic Late Oligocene to Pliocene Gowop Limestone and clastic sediments. The intrusives consisting of gabbro, diorite, clinopyroxene diorite, hypersthene gabbro, dolerite, micro-diorite, quartz gabbro, tonalite, microtonalite andesite porphyry, and granophyric differentiates are thought to be Pliocene in age (Robinson and others, 1976). Minor metasomatic mineralization involving pyrite and chalcopyrite occur along contacts with limestone.

Trobriand-Woodlark Region

The Trobriand, Marshall Bennett, Woodlark, and Laughlan island chain are coral reef islands built on Tertiary volcanic basement. The largest exposure of the basement is on Woodlark Island, and the geology has been described by Trail (1967) and McGee (1978). In summary, it consists of Lower Tertiary andesitic and basic volcanics overlain by Upper Oligocene to Lower Miocene Nassai Limestone. Volcanism resumed in the Late Pliocene (Egum Volcanics) and was followed by reef-lagoon deposits in the Pleistocene. Intrusive rocks are basic to intermediate and are possibly cogenetic with the lower volcanics. They are described as predominantly basic feldspar porphyry and porphyritic monzonite (McGee, 1978). Alteration has given rise to sericite and biotite which defines a foliation; K-Ar age of 11 m.y. was obtained on the biotite (McGee, 1978). Over 200,000 ounces of gold was recovered from Woodlark Island before the First World War.

GRANITOIDS OF THE CENTRAL OROGENIC BELT AND THE AUSTRALIAN CRATON

The granitoids in the Central Orogenic Belt and the Papuan Basin region are located within or on the edge of the late Paleozoic Australian craton with the possible exception of those in southeast Papua. The simple picture of Tertiary plutons, which make up the bulk of Papua New Guinea granitoids, intruding the continental edge is complicated by several Mesozoic granitoids associated with metamorphics in the North Sepik region (Hutchison, 1975) and the Kainantu region (Page, 1976); the significance of these granitoids is not fully understood and requires further work.

Apart from the Kubor Complex which is treated later, outcrops of basement rocks are restricted. Pink granite with chlorite, calcite and sericite alteration, leucocratic biotite granite, and porphyritic hornblende diorite in the Strickland River (Jenkins and White, 1970; Davies and Norvick, 1974) have been dated at 222 m.y. A similar age of 236 m.y. has been obtained from biotite granite in a petroleum exploration well (Aramia No. 1), 270 km to the south (Page, 1976).

These ages together with the 244 m.y. age for the Kubor Complex (Page, 1976) suggest a widespread phase of Late Permian and Early Triassic igneous intrusion. Further evidence of basement granitoids is from clasts in the pre-Middle Jurassic Bol Arkose, which lies unconformably on the basement. It includes pink, coarse-grained granite as the predominant clast type containing chloritized biotite as the only mafic phase (Arnold and others, 1979). Basement granite and acid volcanics have been encountered in petroleum exploration wells scattered throughout southeast Papua.

The Carboniferous Badu Granite (295 m.y.; Willmott, 1972) is exposed on the southern coast of western Papua. It consists of leucocratic biotite granite, porphyritic biotite granite and adamellite, and some hornblende-biotite adamellite and granodiorite. Mafic xenoliths are present at some localities. The Badu Granite intrudes Carboniferous acid volcanics.

North Sepik Region

The dominant feature in the North Sepik region (Fig. 4) is the basement complex which forms the Bewani and Torricelli Mountains (Hutchison, 1975; Hutchison and Norvick, 1978; Norvick and Hutchison, in press). Upper Cretaceous to Eocene Ambunti Metamorphics outcrop south of the range. The northern sequence consists of ?Paleocene to earliest Miocene marine Bliri Volcanics and comagmatic intrusives of the Torricelli Intrusive Complex in the west, and the igneous and metamorphic rocks of the Prince Alexander and Turu Complexes in the eastern ranges. Faulting within this basement sequence is intense.

Unmetamorphosed mid-Permian (249 m.y.; Hutchison, 1975) intrusives occur as float near the Irian Jaya border 70 km south of the Bewani Mountain range and are associated with metamorphosed intrusions (Amanab Metadiorite) and probable Upper Cretaceous to Eocene greenschist to amphibolite grade metamorphics.

The volcanics and intrusives in the main range are unconformably overlain by a polymict conglomerate or bioclastic limestone, both of Miocene age. The youngest sediments are Pleistocene terriginous gravels and Pleistocene to Holocene reef limestone.

The Torricelli Intrusive Complex is predominantly medium-grained, nonporphyritic gabbro, massive dolerite and diorite, subordinate monzonite, granodiorite and adamellite, and rare ultramafics. The complex is invaded in places by dolerite dikes; fine-grained porphyries and late-stage pegmatites are also present. Faulting obscures intrusive relationships and produces crush breccias and serpentinization of gabbros. Alteration affects most rocks but is most intense in the intermediate to acid types. The most common alteration minerals are chlorite, epidote, sericite, and kaolinite. Calcite, uralite, and clinozoisite are present in some basic rocks. Recrystallization involving quartz is common in sheared rocks. K-Ar age data (Hutchison, 1975) include some Late Cretaceous, and a majority of Late Eocene to Early Miocene ages; some unreliable Early Devonian to Late Jurassic ages were also obtained.

The Prince Alexander Complex is highly deformed and

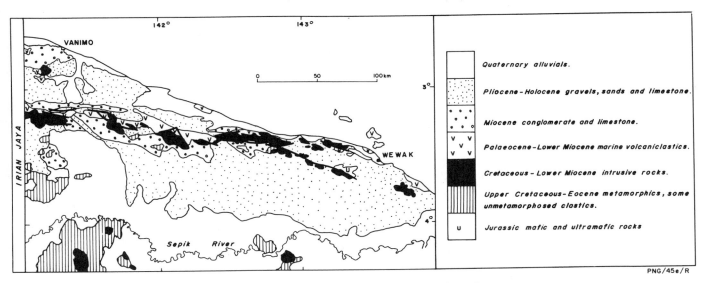

Figure 4. Simplified geology of the North Sepik region.

contains a variety of rocks. Igneous rocks are sheared granodiorite, diorite, and dolerite; less deformed andesitic dikes, biotite adamellite, and granodiorite are subordinate. The metamorphics are high-grade amphibolite and ortho-gneiss, and subordinate mica schist. Pyrite, disseminated and in veins, is common throughout the complex. The metamorphics including the sheared granodiorite give an Early Cretaceous K-Ar age (114 to 106 m.y.). Andesite porphyry dikes, a pegmatite, and a biotite-bearing intrusive range from Late Oligocene to Early Miocene (25 to 20 m.y., Hutchison, 1975).

The Mount Turu Complex is predominantly mafic and ultramafic rock and minor diorite, with rare dolerite, biotite adamellite, and granodiorite. Ultramafic rock types include clinopyroxenite, websterite, wehrlite, and lherzolite with textures ranging from granular to granoblastic. Mafic rocks have ophitic to subophitic textures and range from troctolite through olivine gabbro to hornblende gabbro; orthopyroxene is absent as a primary phase. Olivine, clinopyroxene and hornblende diorites make up the intermediate rocks. Alteration within the complex is common especially in shear zones and typically involves serpentinization and saussuritization. Secondary amphibole defines a foliation in some rocks. One K-Ar age (188±50 m.y.) on gabbro suggests the bulk of the complex is Jurassic (Hutchison and Norvick, 1978); a biotite adamellite sample was dated as 18 m.y. old.

Alluvial gold has been recovered from the North Sepik region, and some hydrothermal sulphide mineralization has been investigated.

South Sepik Region

The granitoids in the South Sepik region are widely distributed as batholithic complexes and small stocks south of the Sepik River and north of the Lagiap Fault Zone and the main dividing range (Fig. 5). The granitoids, on and south of the divide, are in the Papuan Fold Belt region. The Lagiap Fault Zone just north of the divide marks a break in sedimentation; shelf sediments were deposited to the south, and deeper water marine facies dominate the Late Cretaceous to Eocene rocks to the north.

Most South Sepik granitoids intrude the Upper Cretaceous to Lower Tertiary Salumei Formation in the south and the Ambunti Metamorphics in the north. The Salumei Formation contains marine clastic sediments which have a basic volcanic component in some areas and variable degrees of metamorphism ranging to high-P/T glaucophane-bearing assemblages and eclogite.

The Ambunti Metamorphics to the north range up to amphibolite and are possible high-T/P equivalents of the Salumei Formation (Dow and others, 1972). Following a break in sedimentation in the Early Tertiary, a sequence of volcanically derived sediments was deposited in the Miocene. Tectonic activity during this break in sedimentation is evidenced by volcanism, intrusion of granitoids, emplacement of ultramafic bodies, and metamorphism of parts of the Salumei Formation. In contrast, Miocene sediments south of the Lagiap Fault Zone lack volcanics and consist of limestone and fine calcareous sediments, and intrusions are mainly Late Miocene to Pliocene in age.

In the east, the Miocene Yuat Batholith and some small stocks are exposed, in part, within Triassic volcanics and younger sediments. These are separated from the Salumei Formation by the Maramuni Fault Zone which was possibly the locus of considerable movement prior to the Miocene intrusions (Dow and others, 1972).

The Frieda volcanic-intrusive complex is an important copper prospect (Hall and Simpson, 1975; Asami and Britten, 1979); the volcanics are interbedded with Middle Miocene sediments and overlain by Middle Miocene lime-

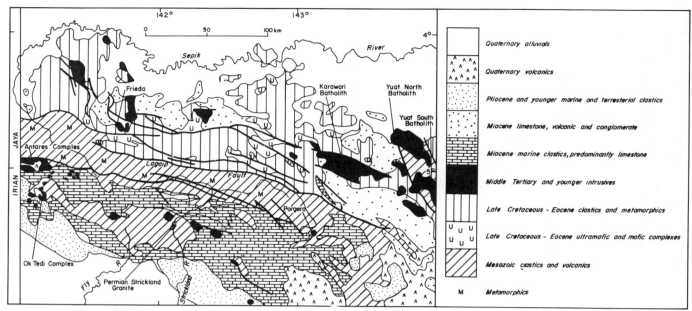

Figure 5. Simplified geology of the South Sepik region, and the western part of the Papuan Fold Belt.

stone. The intrusive rocks are predominantly hornblende plagioclase diorite porphyry (Mason, 1975). The larger pluton north of the Frieda Complex consists mainly of fine-grained diorite and minor granodiorite, monzonite, and gabbro. Middle Miocene K-Ar ages (17 to 13 m.y.) for the Frieda Complex are reported by Page and McDougall (1972a).

The Karawari Batholith is the largest intrusive body in the South Sepik region (650 km²). Both geological and radiometric age constraints support a Middle Miocene age (Dow and others, 1972; Dekker and Faulks, 1964; Page, 1976). This batholith also has cogenetic volcanics. Intrusive rocks include hornblende clinopyroxene quartz diorite and hornblende biotite tonalite, and minor gabbro and diorite porphyries; mafic xenoliths are common (Mason, 1975).

Major phases in the Yuat North Batholith are diorite and granodiorite containing mafic xenoliths; gabbro and adamellite are also present. The Yuat South Batholith is described as predominantly hornblende biotite granodiorite containing gabbroic and dioritic margins, some of which are porphyritic (Mason, 1975; Mason and McDonald, 1978).

Some of the smaller stocks south of the Karawari and Yuat Batholiths are apparently more basic and are similar to the Oipo intrusives farther east in the Central Highlands region. These stocks are heterogeneous, containing ultramafic and mafic cumulates, gabbro, diorite, and granodiorite. Pyroxenites and hornblendites have been described, and in at least one stock, a late felsic phase is recognized (Mason, 1975). Page (1976) concludes from 27 K-Ar age determinations in the Karawari and Yuat Batholiths and

adjacent stocks that these rocks intruded over the period 15 to 10 m.y. ago.

The remaining intrusive bodies in the South Sepik region have been collectively mapped as diorite, granodiorite and gabbro, and some felsic porphyry (Dow and others, 1972).

A crude foliation has been observed in some intrusions adjacent to the high-T/P metamorphics. Such a diorite dated by Page (1976) is Early Miocene and is about 5 m.y. younger than the metamorphism in nearby rocks.

Papuan Fold Belt Region

The granitoids of the Papuan Fold Belt include igneous complexes and stocks extending from the Star Mountains in the west (Figs. 1 and 5) to Mount Michael, 40 km south of Goroka (Fig. 6). They intrude a thick sequence (over 4500 m in the Ok Tedi area; Arnold and others, 1979) of Jurassic to Middle Miocene platform sediments which overlie the Paleozoic basement of the Australian craton. The platform sediments were not affected by the Eocene to Miocene tectonism in the central Orogenic Belt to the north. Deformation and uplift accompanied the Plio-Pleistocene plutonism, resulting in a foreland thrust-fold belt (Arnold and Griffin, 1978). This deformation affected the outer margin of the present structural basin and involved basement in some of the structures.

The granitoids are interpreted as mainly high-level intrusions, in many cases only recently unroofed. Coarse, equigranular phases are present in the larger stocks and complexes. Recent mapping by Arnold and others (1979)

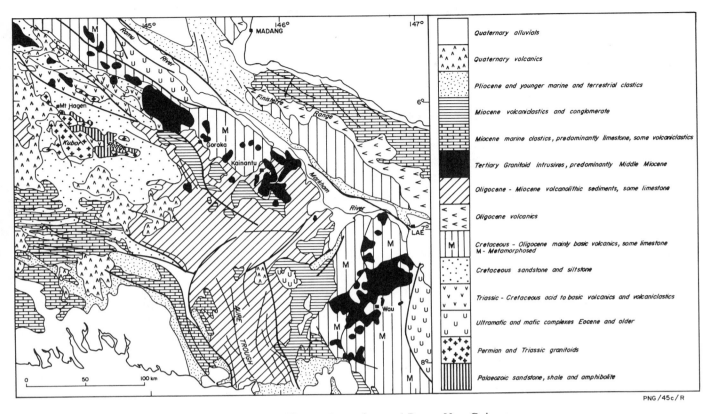

Figure 6. Simplified geology of central Papua New Guinea.

has outlined a large area of volcanics in the eastern end of the Antares Complex that are interpreted as part of a Pliocene volcanic pile into which the granitoids intruded. Associated volcanism may have been minor or absent at other centres of igneous activity. Doming of the surrounding sediments has been observed in two widely separate areas; the southern stocks in the Star Mountains, and Mount Michael. There is a concentration of generally small intrusions associated with the larger Antares Complex (over 300 km², half of which is in Irian Jaya) in the Star Mountains (Arnold and others, 1979). The Antares Complex comprises mainly equigranular monzodiorite, diorite, and microdiorite, and an altered volcanic sequence of andesite, volcanolithic siltstone agglomerate, and tuff. These volcanics are possibly part of an extensive volcanic pile that has since been eroded and would be a convenient source for the Miocene to Pliocene volcaniclastic sediments of the Birim Formation to the south. The granitoids were possibly intruded between 4 m.y. and 7 m.y. ago (Late Miocene to Pliocene), and the volcanics which predate the intrusives are probably also Late Miocene in age (8 m.y.) (Page and McDougall, 1972a; Arnold and others, 1979). The smaller stocks are irregularly distributed west and south of the Antares Complex. Extensive diking is a common feature of some of these intrusions. Rock types are commonly monzodiorite, quartz diorite, and diorite, but some gabbro, tonalite, and quartz monzonite

are present. Most are porphyries with phenocryst plagioclase and hornblende, sometimes clinopyroxene, and rarely quartz and biotite. Radiometric ages for these stocks range from 12 m.y. to 1 m.y.; one 23 m.y. age has also been obtained. A significant porphyry copper-gold prospect exists at Ok Tedi, and porphyry and skarn mineralization is present in other stocks in the Star Mountains region (Arnold and Griffin, 1978).

Nine other areas of granitoid intrusives have been mapped in the Papuan Fold Belt between the Star Mountains and Mount Michael. Rock types include monzonite, micromonzonite, diorite, microdiorite, and porphyritic varieties. Volcanics are possibly associated with some of these intrusions. Ages ranging between 1.5 m.y. to 5 m.y. are reported by Hutchison (1977). The porphyry gold mineralization at Porgera (Cotton, 1975) is associated with diorite intrusions in the Papuan Fold Belt region.

The Mount Michael Diorite is a 7.3 m.y. old, high-level intrusion of microporphyritic hornblende diorite and hornblende quartz diorite (Mason, 1975). Bain and others (1975) recognized hydrothermal alteration within this stock. Between the Mount Michael Diorite and the large batholith of the Bismarck Complex, sills, dikes and stocks of commonly porphyritic gabbro, quartz monzodiorite, diorite, and granodiorite intrude the Lower to Middle Miocene marine clastics of the Movi Beds.

Kubor-Bismarck Region

The Late Permian (244 m.y.; Page, 1976) Kubor Complex intrudes the low-grade greenschists and amphibolites of the Paleozoic Omung Metamorphics in the Kubor Range in the south of the Kubor-Bismarck region (Fig. 6). These rocks are exposed in a large, anticlinal structure (Bain and others, 1975). Similar equigranular biotite granodiorite has been mapped and dated (231 m.y.; Pigram and others, in preparation) from the Jimi-Wahgi divide 20 km north of the Kubor Complex. Unconformably overlying the Permian-Triassic granitoids are small patches of Upper Permian-Triassic limestone and arkose (Kuta Formation), and the extensive Triassic Kana Volcanics containing basic to dacitic flows and clastics. Recent mapping indicates a complex intrusive-extrusive relationship between the granitoids and volcanics north of the Kubor Range (Pigram and others, in preparation). A thick sequence (about 7000 m) of Upper Jurassic to Upper Cretaceous clastics and volcanics overly the Kana Volcanics. The Tertiary rocks are predominantly shelf siltstone and limestone.

Middle Miocene granitoids intrude the sediments north of the Kubor Range (18 m.y. to 15 m.y.; Page, 1976). Except for the Bismarck Intrusive Complex, they are mostly small, isolated stocks. Several large, predominantly basaltic Quaternary stratovolcanoes (Mackenzie, 1976) are present to the west and south of the Kubor Range.

The Kubor Complex consists of large plutons of mainly coarse-grained granodiorite and tonalite, and small stocks and dikes consisting of gabbro, diorite, and adamellite. The gabbro commonly occurs on the margins of the granodiorite and tonalite. Contact hornfels zones are generally narrow. Xenoliths are rare and late-stage aplite and pegmatite dikes have been dated as young as 90 m.y. (Page, 1976).

The biotite and hornblende in the granodiorite and tonalite are often altered, ranging from partly chloritized tonalite to complete replacement by chlorite, epidote, sphene, and opaques in some granodiorite. Actinolite and chlorite partly replace the hornblende in gabbro. Accessory phases include magnetite, ilmenite, apatite, and little sphene; garnet and zircon are rare. The aplites contain accessory sphene, monazite, pink garnet, and muscovite (Bain and others, 1975).

The Bismarck Intrusive Complex is a large batholith which intrudes at the boundary between the metamorphics of the Goroka Formation in the east and the Mesozoic sequence to the west. Extensive K-Ar and Rb-Sr dating (Page and McDougall, 1972a; Page, 1976) indicates Middle to Late Miocene intrusion around 12.5 m.y. ago. The complex consists of gabbro, diorite, quartz monzodiorite, and late-stage muscovite pegmatites (8 to 9 m.y.) in the western and northeastern parts; and granodiorite tonalite and rare granite elsewhere. Ultramafic rocks, including pyroxenite, hornblendite, peridotite, dunite, and anorthosite

are present (McMillan and Malone, 1960). A large number of rock types have been described from the Yandera porphyry copper prospect (Grant and Nielson, 1975; Titley and others, 1978) which lies inside the northeast margin of the batholith; these include quartz diorite porphyry, microdiorite, dacite porphyry, aplitic quartz monzonite porphyry, and both intrusion and intrusive breccias.

West of the Bismarck Complex, many Miocene stocks intrude the Mesozoic sequence. The northern-most group is extremely variable ranging from pyroxenites to granodiorite. Textures range from accumulate to pegmatitic, and felsic stockworks and complex veining are common (Dow and Dekker, 1964). Page (1976) reported K-Ar ages ranging from 18 to 15 m.y. The intrusions to the south in the Jimi-Wahgi divide have variable textures and include diorite, gabbro, tonalite, granodiorite, and andesite porphyry, dolerite, basalt, and minor trachyandesite dikes. Hydrothermal alteration and low-grade burial metamorphism have caused the development of chlorite, actinolite, epidote, sericite, biotite, sphene, laumontite, prehnite, and pumpellyite (Pigram and others, in preparation). Both Miocene and Triassic ages have been obtained from this group of intrusives.

Mount Pugent, 50 km north-northwest of Mount Hagen is an unusual pyroxene biotite microsyenite (Mason, 1975) that intrudes the Cretaceous Kondaku Tuff (Dow and Dekker, 1964).

Kainantu Region

Jurassic-Cretaceous and Miocene granitoids in the Kainantu region (Fig. 6) intrude metamorphics of the Bena Bena Formation (schist, gneiss, phyllite, metasandstone) and the Goroka Formation (schist, slate, phyllite, metasandstone, minor recrystallized limestone, and calcareous siltstone) (Tingey and Grainger, 1976). The Bena Bena Formation is more metamorphosed and possibly older than the Jurassic to Cretaceous Goroka Formation.

The Urabagga Intrusives, 15 km northwest of Goroka, are small, mainly granodiorite and diorite bodies up to 2 km across which intrude the Goroka Formation and are overlain by Oligocene limestone (McMillan and Malone, 1960). Rock types range to chloritized leucogranite and are intruded by dolerite and aplite dikes. The Urbagga Intrusives are believed to be Jurassic in age (190 m.y.; Page, 1976).

The Karmantina Gneissic Granite also has a Jurassic age (172 m.y.; Page, 1976) and has intruded the Bena Bena Formation 20 km northwest of Kainantu. This gneissic, biotite-muscovite granite is possibly older than the measured age having been reset by late-stage metamorphism at 25 to 20 m.y. (Tingey and Grainger, 1976).

The Mount Victor Granodiorite is a small, mid- to Upper Cretaceous (95 to 77 m.y.; Page, 1976) intrusion of biotite hornblende granodiorite 20 km southeast of Kainantu. It is intruded by the Upper Miocene Elendora Porphyry in the south and west, and overlain by Upper Tertiary sediments in the north and east.

Elendora Porphyry is the youngest intrusive body in this region (12 to 7 m.y.; Page, 1976) consisting of hornblende andesite porphyry and subordinate porphyritic microdiorite and minor serpentinite. The Elendora Porphyry contains minor gold and copper mineralization and has propylitic alteration in some parts.

The Middle Miocene Akuna Intrusive Complex consists of a large, irregular body east of Kainantu and many small stocks to the northwest. It intrudes both the Bena Bena and Goroka Formations and also the Lower Miocene Omaura Greywacke (Page, 1976). K-Ar ages (Page, 1976) indicate complex intrusion in the Early to Middle Miocene (17 to 13 m.y.). Rock types include olivine gabbro, hornblende gabbro, porphyritic dolerite, diorite, granodiorite, minor peridotite, and serpentinite (Tingey and Grainger, 1976).

Wau-Bulolo Region

The granitoids of the Wau-Bulolo region are dominated by the Middle Miocene Morobe Granodiorite, consisting of a 1200 km² batholith and isolated stocks mainly to the north and south. These intrude the predominantly low-grade metamorphics of the Owen Stanley Metamorphics (Fig. 6). The intrusive rocks are generally unaltered granodiorite and adamellite; subordinate monzonite, diorite, hornblendite, and pegmatites are developed at the margins (Dow and others, 1974; Rebek, 1975). Ten rocks dated by whole rock Rb-Sr, and K-Ar on biotite and hornblende give ages ranging from 15 to 11 m.y., Middle Miocene (Page, 1976).

The Pliocene (3.8 m.y.; Page and McDougall, 1972a) Edie Porphyry intrudes the central part of the Morobe Granodiorite and is possibly associated with adjacent volcanic activity of a similar age. Biotite andesite, hornblende andesite, dacite, and quartz biotite porphyry stocks and dikes make up the Edie Porphyry.

The porphyry bodies in upper Edie Creek and the latest dikes in the Wau area are extremely sericitized and locally silicified and pyritized, and were apparently the main gold mineralizers for the important Morobe Goldfield, (Dow and others, 1974).

Southeast Papua

Most of the mainland peninsula of southeast Papua (Fig. 7) consists of Upper Cretaceous to Middle Eocene submarine basalts with minor lenses of Cretaceous limestone and chert (Kutu Volcanics). Some of these rocks are metamorphosed (Davies, 1978). In the northwest, the peridotite and gabbro of the Papuan ultramafic belt are associated with Cretaceous? basalt and represent oceanic crust which was thrust over the Owen Stanley Metamorphics in the Eocene.

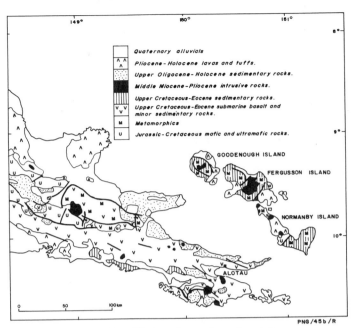

Figure 7. Simplified geology of southeast Papua.

Small stocks and dikes of Middle Miocene age intrude the Kutu Volcanics. The Upper Oligocene to Middle Miocene tuffs and marine sediments, which overlie the Cretaceous and Eocene rocks in places, are possibly related to the stocks and dikes. Middle Miocene and younger sedimentary rocks and Upper Miocene and younger volcanics are widespread. Intrusion and extrusion in the late Cenozoic are characterized by calcalkaline and shoshonitic magmas (Smith and Davies, 1976).

The Miocene to Pliocene (K-Ar ages range from 16.5 to 4.4 m.y.; Davies and Smith, 1974; Smith, 1972) granitoids consist of rock types ranging from gabbro to granite. The calcalkaline rocks in the Suckling-Dayman massif area (Davies, 1978) include granite, adamellite, granodiorite, monzonite, diorite, and hornblendite; one intrusive body is porphyritic. All rocks contain hornblende and biotite, and the hornblendite is a possible xenolith or mafic accumulate in which pyroxene has been replaced by hornblende and magnetite.

The shoshonitic intrusives consist of gabbro, monzonite syenite, and biotite pyroxenite and have been divided into near-saturated and undersaturated groups (Smith, 1972). The basic rocks of the near-saturated group contain more K-feldspar and biotite (ranges up to 20 percent) than the equivalent rocks of the undersaturated group.

Monzonite of the undersaturated group commonly contains small phenocrysts of melanite and zoned aegerine-augite or aegerine; the syenites of the near-saturated group contain aegerine-augite, pale green hornblende, or rarely riebekite. Undersaturated syenites contain melanite and biotite in addition to pyroxene and amphibole. The fine-grained rocks associated with these intrusions are also shoshonitic and are characterized by biotite phenocrysts; sanidine and melanite phenocrysts are present in one porphyritic phase. Biotite is a rare phenocryst phase in fine-grained porphyritic calcalkaline rocks. Alteration is generally minor involving epidote, calcite, zeolite, chlorite, and secondary amphibole.

K-Ar results (Davies, 1973b) indicate a possible Late Pliocene age for at least some of the granitoid intrusions in the large offshore islands. They are predominantly grano-diorite but range from tonalite to granite; some intrusions contain xenoliths of metamorphic and ultramafic country rocks. The granitoids are regarded as probable subvolcanic plutons related to the peralkaline acid volcanics which cover part of southwest and southeast Fergusson Island. They intruded an eroded dome of metamorphosed? Cretaceous rocks which was overthrust by oceanic crust possibly in Eocene times. A similar interpretation is given by Davies (1978) for the Suckling-Dayman massif, 150 km to the southwest on the Papua New Guinea mainland.

CHEMISTRY

Many of the granitoids in Papua New Guinea have not been systematically mapped and sampled. Some representative analyses of unaltered rocks containing around 60 percent SiO_2 are listed in Table 1. The whole rock chemical data available are very patchy and often related to porphyry copper systems, e.g., Ok Tedi (Ayres and Bamford, 1977; Mason and McDonald, 1978). Figure 8 illustrates the extreme chemical variation in the rocks from the Ok Tedi Complex when compared to fresh rocks from adjacent Star Mountain intrusive suites (Arnold and others, 1979). Data from the Yandera prospect (Watmuff, 1978) also may reflect alteration similar to that of the Ok Tedi prospect.

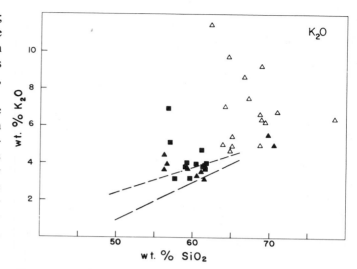

Figure 8. K_2O vs SiO_2 plot of the Ok Tedi Complex; △ — mineralized Fubilan Quartz Monzonite Porphyry, ■, ▲ — unmineralized rocks from other phases of the Ok Tedi Complex (Ayres and Bamford, 1977; Mason, 1975). The two regression lines are from unmineralized intrusive rocks of the Antarex Complex (lower) (Arnold and others, 1979; Hutchison, 1977) and the Mount Ian Complex (upper) (Arnold and others, 1979; Ayres and Bamford, 1977).

Mason and McDonald (1978) conclude from a review of data from the different tectonic settings that no major variation in the chemistry of calcalkaline intrusive rocks occurs as a function of tectonic setting in Papua New Guinea. They did, however, highlight some geochemical features that could be tested with further work. Unfortunately, the additional data now available do not include minor elements and are insufficient to test subtle variations.

At this stage, the most striking feature of whole rock

TABLE 1.   SELECTED ANALYSES OF GRANITOIDS IN PAPUA NEW GUINEA

	1	2	3	4	5	6	7	8	9	10	11	12	13	14	15	16	17
SiO_2	59.70	60.41	62.4	60.8	59.1	62.9	62.2	63.0	59.62	64.1	63.0	62.07	60.6	60.18	59.38	61.61	62.0
TiO_2	0.28	0.52	0.61	0.55	0.83	0.68	0.37	0.51	0.69	0.48	0.65	0.71	0.47	1.11	0.90	0.51	0.76
Al_2O_3	19.70	15.22	16.1	18.11	17.5	16.0	14.4	17.3	17.74	17.3	15.3	16.33	15.9	16.29	17.33	17.12	15.8
Fe_2O_3	1.24	4.88#	2.7	nd	2.50	2.31	1.78	1.92	3.45	2.40	2.10	2.34	3.05	3.56	2.72	5.39#	3.3
FeO	0.78	nd	2.55	4.27*	4.40	2.93	1.66	2.03	2.86	2.03	3.40	3.99	3.95	3.63	3.02	nd	2.75
MnO	0.06	0.09	0.08	0.10	0.15	0.10	0.0	0.03	0.15	0.12	0.09	0.10	0.12	0.08	0.07	0.09	0.14
MgO	0.29	3.38	1.7	1.06	2.65	2.13	1.35	2.35	2.99	1.80	3.15	2.79	2.90	2.45	2.51	1.91	2.3
CaO	2.20	4.93	4.1	5.96	5.60	4.78	4.33	3.93	6.89	5.99	6.10	2.79	6.50	6.44	5.36	5.24	5.5
Na_2O	2.50	4.25	4.1	3.96	3.35	3.71	4.15	4.94	3.68	4.54	3.40	3.55	3.25	3.77	4.56	4.20	3.5
K_2O	10.90	5.76	4.6	3.61	2.80	3.06	3.08	1.84	1.20	0.90	1.40	1.16	0.55	0.44	2.73	1.91	2.3
P_2O_5	0.04	0.61	0.34	0.21	0.21	0.22	0.19	0.26	0.29	0.22	0.11	0.11	0.09	0.28	0.35	0.19	0.17
S	nd	nd	nd	0.01	nd	0.07	1.33	0.18	0.19	0.06	nd	0.01	0.03	0.05	0.00	nd	nd
H_2O^+	1.34	nd	0.58	0.98	0.37	0.61	0.89	1.50+	1.12	0.48	0.77	0.82	1.51	0.98	0.71	nd	nd
H_2O^-	0.13	nd	0.32	0.31	0.43	0.20	0.83	nd	0.70	0.03	0.23	0.09	0.61	0.17	0.21	nd	1.13
CO_2	0.05	nd	0.05	0.24	0.07	0.19	2.93	nd	0.22	0.10	0.09	0.00	0.10	0.24	0.10	nd	0.17
Total	99.24	100.05	100.23	101.15	99.96	99.89	98.82△	99.70△	101.79	100.55△	99.79	100.24	99.53	99.67	99.95	98.17	99.87

*Total iron as FeO; # Total iron as Fe_2O_3; + Loss of ignition; △ Total adjusted S≡0; References as for Figure 9.

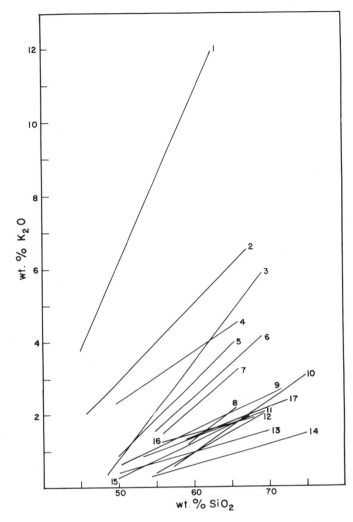

geochemistry is the similarity of the granitoids from the three major tectonic settings and the variation within any one setting.

Figure 9 shows the range in K_2O throughout Papua New Guinea. The rocks plotted are post-Oligocene intrusives and exhibit a wide range in composition although maintaining a calcalkaline character (Fig. 10); the very high-K rocks from southeastern Papua are regarded as shoshonitic by Smith and Davies (1976). The data from New Britain for the South and Central Baining Mountains (Macnab, 1970) and the Kulu Complex (Hine and Mason, 1978) show the extent of the range in K_2O in an island arc setting.

Many major granitoid complexes are not represented on Figures 9 and 10, and Table 1; for example, the Morobe Granodiorite, the Akuna Complex, and most of the Bismarck Complex. No chemical data are available for these or for the pre-Oligocene granitoids.

Additional data from Plesyumi, New Britain (Titley, 1978) do not define a trend but range over the spectrum defined by other suites in New Britain. The data from Mount Kren, Manus Island have a limited SiO_2 range but are in the high-K group with about 2.5 percent K_2O at 60 percent SiO_2 (Mason, 1975). Gabbros, gabbroic diorites, and other mafic intrusives associated with the calcalkaline granitoids are not included in this discussion. They are poorly studied, apparently form a separate chemical group (Arnold and others, 1979), and possibly have a different origin to associated granitoids (Hine and Mason, 1978; Griffin and others, 1978).

Mount Michael, the only granitoid other than the Star Mountain Intrusives in the Papuan Fold Belt region for which there is chemical data, has about 1.5 percent K_2O at

Figure 9. K_2O vs SiO_2 for granitoid suites in Papua New Guinea after Mason and McDonald (1978). 1 = Southeast Papua shoshonites, undersaturated (Smith and Davies, 1976); 2 = Southeast Papua shoshonites, near-saturated (Smith and Davies, 1976); 3 = Central and South Baining Mountains, New Britain (Macnab, 1970); 4 = Mount Ian Complex, Papuan Fold Belt region (Arnold and others, 1979; Ayres and Bamford, 1977); 5 = Antares Complex, Papuan Fold Belt region (Arnold and others, 1979; Hutchison, 1977); 6 = Yuat North batholith, South Sepik region (Mason and McDonald, 1978); 7 = Panguna high-K suite, Bougainville (Mason and McDonald, 1978; Ford, 1978); 8 = Frieda Complex, South Sepik region (Mason and McDonald, 1978); 9 = Yuat South batholith, South Sepik region (Mason and McDonald, 1978); 10 = Karawari batholith, South Sepik region (Mason and McDonald, 1978); 11 = Lemau Complex, New Ireland (Mason and McDonald, 1978; Hohnen, 1978); 12 = Esis-Sai Complex, New Britain (Hine and Mason, 1978); 13 = Torricelli Complex, North Sepik region (Hutchison, 1975); 14 = Kulu Complex, New Britain (Hine and Mason, 1978); 15 = Plesyumi Complex, New Britain (Hine and Mason, 1978; Mason and McDonald, 1978); 16 = Panguna low-K suite, Bougainville (Mason and McDonald, 1978; Ford, 1978); 17 = North Baining Mountains, New Britain (Macnab, 1970).

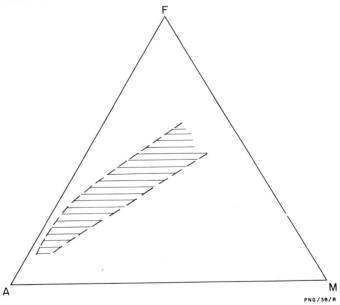

PNG/38/R

Figure 10. Field for post-Oligocene granitoids of Papua New Guinea on an AFM plot.

60 percent SiO_2 (Mason, 1975) and shows that the high-K nature of the Star mountains is not characteristic of the essentially stable craton.

DISCUSSION

Attempts by Mason and McDonald (1978) and Titley and Hendrick (1978) to find criteria which characterize granitoids in the three major structural zones (Bismarck-Solomon, Central Orogenic Belt, and Papuan Basin) have been unsuccessful. The work of Titley and Hendrick (1978) re-emphasizes the existence of the structural provinces about which there is little argument save defining the exact positions of the boundaries. Certainly the intrusion of granitoids in differing structural regimes will produce certain differences such as level of intrusion and the sympathetic response to major fault styles. However, do the different structural-tectonic zones produce granitoids with unique petrographic and geochemical features characteristic of a particular zone?

This paper demonstrates that at the present state of knowledge, granitoids in three quite different structural settings appear indistinguishable from one another. The exception is the shoshonitic granitoids in southeastern Papua (Smith and Davies, 1976).

In summary, the similarities are:

a) The wide range in rock types, texturally, mineralogically, and chemically (within the constraints of the calcalkaline suite).
b) The predominance of basic to intermediate types (see Mason and McDonald, 1978, Fig. 2).
c) The presence of gabbros with variable mineralogy and chemistry. Calcalkaline suites in eastern Australia also have mafic rocks with variable chemistry (Griffin and others, 1978; Hine and others, 1978b; Hine and Mason, 1978).
d) An established or inferred volcanic-intrusive relationship in many areas in all three provinces.
e) The presence of porphyry-style Cu-Au mineralization, alteration, and brecciation.

The Carboniferous granites in the platform are likely to include S-type granites and are excluded from the above observations. The Permian and Mesozoic granitoids are not completely understood within current tectonic models for Papua New Guinea but probably are I-type granites of Chappel and White (1974), as are the Tertiary granitoids.

If the granitoids of the three tectonic zones are similar, then it follows that the granitoids most probably were generated by the same process. This may have been dipping seismic zones beneath all the areas containing intrusive rock as various authors have proposed. However, it should be stressed that despite the presence of large Quaternary stratovolcanoes and very young hydrothermal activity (e.g., Ok Tedi Complex) in the Papuan Basin, there is no convincing evidence for a recent southward-dipping sub-duction zone beneath Papua New Guinea (Mackenzie, 1976). Also the presence of volcanics or intrusives, even on an active continental margin, need not automatically imply subduction. Alternatively, if an underplating hypothesis (Chappell, 1978) is accepted, then this process becomes the genetic feature common to all structural zones and the granitoids are produced by partly melting this material at the base of the crust.

The amount of underplating may be quite variable but is likely to be related to crustal thickness and the ease with which a magma has access through the crust. In young island arcs, there may be little underplating because of the relatively thin crust; however, the material in the island arc itself is similar in composition to the underplating material. The magmas produced during subduction will find it increasingly difficult to reach the surface as the island arc crust thickens. It is suggested that these magmas underplate the crust to an increasing amount with a resultant increase in the geothermal gradient and partial melting giving rise to granitoids. Old island arcs and continental areas are likely to contain underplated material from a previous orogenic event. This is the source material for I-type granitoids in these zones if partial melting occurs.

The mechanism that initiates partial melting and intrusion of granitoids in Papua New Guinea could involve one or all of three tectonic processes related to the collision of the Australian and Pacific plates: 1) subduction, 2) crustal thickening by island arc accumulation as suggested by Hine and Mason (1978) for New Britain, and 3) crustal buckling related to collision as speculated by Arnold and Griffin (1978) for the Star Mountain Intrusives.

ACKNOWLEDGMENTS

In this synthesis, I have extensively used the work of geologists who have been active in Papua New Guinea and I trust I have not misrepresented them. I thank H. L. Davies for his criticisms of the draft manuscript. The paper is published with the permission of the Secretary of the Department of Minerals and Energy, and the Chief Government Geologist, Port Moresby.

REFERENCES CITED

Arnold, G. O., and Griffin, T. J., 1978, Intrusions and porphyry copper prospects of the Star Mountains, Papua New Guinea: Economic Geology, 73, 785–795.

Arnold, G. O., Griffin, T. J., and Hodge, C. C., 1979, Geology of the Ok Tedi and southern Atbalmin 1:100,000 sheets: Geological Survey of Papua New Guinea Report, Parts I, II & III.

Asami, M., and Britten, R. M., 1980, The porphyry copper deposits at the Frieda River prospect, Papua New Guinea: Mining Geology, v. 30, p. 101–102 (in Japanese).

Ayres, D. E., and Bamford, R. W., 1977, The Mt. Fubilan (Ok Tedi) porphyry copper deposit: geology, geochemistry and origin (unpublished manuscript).

Bain, J. H. C., MacKenzie, D. E., and Ryburn, R. J., 1975, Geology of the Kubor Anticline, central highlands of Papua New Guinea: Bureau of Mineral Resources, Australia, Bulletin, 155.

Blake, D. H., 1967, Bougainville Island North and South, Papua New Guinea, 1:250,000 gological series: Bureau of Mineral Resources, Australia, Explanatory Notes, SB/56-8 and SB/56-12.

Blake, D. H., and Miezitis, Y., 1967, Geology of Bougainville and Buka Islands: Bureau of Mineral Resources, Australia, Bulletin, 93 (PNG 1).

Chappell, B. W., 1978, Granitoids from the Moonbi District, New England Batholith Eastern Australia: Journal of the Geological Society of Australia, 25, 267–283.

Chappell, B. W., and White, A. J. R., 1974, Two contrasting granite types: Pacific Geology, 8, 173–174.

Cotton, R. E., 1975, Porgera gold deposits, *in* Knight, C. L., ed., Economic geology of Australia and Papua New Guinea, 1. Metals: Australasian Institute Mining and Metallurgy, Monograph, 7, 872–874.

Davies, H. L., 1973a, Gazelle Peninsula, Papua New Guinea, 1:250,000 geological series: Bureau Mineral Resources, Australia, Explanatary Notes, SB/56-2.

——, 1973b, Fergusson Island, Papua New Guinea, 1:250,000 geological series: Bureau Mineral Resources, Australia, Explanatory Notes, SC/56-5.

——, 1978, Folded thrust fault and associated metamorphics in the Suckling-Dayman Massif, Papua New Guinea: Geological Survey of Papua New Guinea Report, 78/16.

Davies, H. L., and Norvick, M., 1974, Blucher Range, Papua, 1:250,000 geological series: Bureau Mineral Resources, Australia, Explanatory Notes, SB/54-7.

Davies, H. L., and Smith, I. E., 1974, Tufi-Cape Nelson, Papua New Guinea, 1:250,000 geological series: Bureau Mineral Resources, Australia, Explanatory Notes, SC/55-8 and SC/55-4.

Dekker, F. E., and Faulks, I. G., 1964, The geology of the Wabag area, New Guinea: Bureau Mineral Resources, Australia, Record, 1964/137 (unpubl.).

Dow, D. B., and Dekker, F. E., 1964, Geology of the Bismarck Mountains, New Guinea: Bureau of Mineral Resources, Australia, Report, 76.

Dow, D. B., Smit, J. A. J., Bain, J. H. C., and Ryburn, R. J., 1972, The geology of the South Sepik region, New Guinea: Bureau Mineral Resources, Australia, Bulletin, 133.

Dow, D. B., Smit, J. A. J., and Page, R. W., 1974, Wau, Papua New Guinea, 1:250,000 geological series: Bureau of Mineral Resources, Australia, Explanatory Notes, SB/55-14.

Ford, J. H., 1978, A chemical study of alteration at the Panguna porphyry copper deposit, Bougainville, Papua New Guinea: Economic Geology, 73, 703–720.

Grant, J. N., and Nielson, R. L., 1975, Geology and geochronology of the Yandera porphyry copper deposit, Papua New Guinea: Economic Geology, 70, 1157–1174.

Griffin, T. J., White, A. J. R., and Chappell, B. W., 1978, The Moruya batholith and geochemical contrasts between the Moruya and Jindabyne Suites: Journal of the Geological Society of Australia, 25, 234–247.

Gustafson, L. B., and Titley, S. R., eds., 1978, Porphyry copper deposits of the southwestern Pacific islands and Australia: Economic Geology, 73, p. 597–985.

Hall, R. J., and Simpson, P. G., 1975, Frieda porphyry copper prospect, Papua New Guinea, *in* Knight, C. L., ed., Economic geology of Australia and Papua New Guinea, 1. Metals: Australasian Institute of Mining Metallurgy Monograph, 7, 836–845.

Hine, R., Bye, S. M., Cook, F. W., Leckie, J. F., and Torr, G. L., 1978, The Esis porphyry copper deposit, East New Britain, Papua New Guinea: Economic Geology, 73, 761–765.

Hine, R., and Mason, D. R., 1978, Intrusive rocks associated with porphyry copper mineralisation, New Britain, Papua New Guinea: Economic Geology 73, 749–760.

Hine, R., Williams, I. S., Chappell, B. W., and White, A. J. R., 1978, Contrasts between I- and S-type Granitoids of the Kosciusko Batholith: Journal of the Geological Society of Australia, 25, 219–234.

Hohnen, P. D., 1978, Geology of New Ireland, Papua New Guinea: Bureau Mineral Resources, Australia, Bulletin, 194 (PNG 12).

Hutchison, D. S., 1977, Small stocks in the west Sepik survey area: Geological Survey of Papua New Guinea Report, 77/16.

——, 1975, Basement geology of the north Sepik region, Papua New Guinea: Bureau of Mineral Resources, Australia, Record, 1975/162 (unpubl.)

Hutchison, D. S., and Norvick, M., 1978, Wewak, Papua New Guinea, 1:250,000 geological series: Bureau Mineral Resources, Australia, Explanatory Notes, SA/54-16.

Jaques, A. L., 1976, Explanatory Notes on the Admiralty Islands geological map: Geological Survey of Papua New Guinea Report, 76/15.

Jaques, A. L., and Robinson, G. P., 1977, The continent/island-arc collision in northern Papua New Guinea: Bureau of Mineral Resources, Journal, 2, 289–304.

Jaques, A. L., and Webb, A. W., 1975, Geochronology of "porphyry copper" intrusives from Manus Island, Papua New Guinea: Geological Survey of Papua New Guinea Report, 75/5.

Jenkins, D. A. L., and White, M. F., 1970, Report on the Strickland River survey, Permit 46, Papua: BP Petroleum Development Australia Pty. Ltd., Company Report (unpubl.)

Johnson, R. W., 1976, Late Cainozoic volcanism and plate tectonics at the southern margin of the Bismarck Sea, Papua New Guinea, *in* Johnson, R. W., ed., Volcanism in Australasia: Amsterdam, Elsevier, 101–116.

Johnson, R. W., MacKenzie, D. E., Smith, I. E., and Taylor, G. A. M., 1973, Distribution and chemistry of Quaternary volcanoes in Papua New Guinea, *in* Coleman, P. J., ed., The Western Pacific; island arcs, marginal seas, geochemistry. Perth University of Western Australia Press, 523–534.

Johnson, R. W., and Smith, I. E., 1974, Volcanoes and rocks of St. Andrew Strait, Papua New Guinea: Journal of the Geological Society of Australia, 21, 333–351.

Johnson, R. W., Wallace, D. A., and Ellis, D. J., 1976, Feldspathoid bearing potassic rocks and associated types from volcanic islands off the coast of New Ireland, Papua New Guinea: a preliminary account of geology and petrology, *in* Johnson, R. W., ed., Volcanism in Australasia: Amsterdam, Elsevier, 297–336.

Knight, C. L., ed., 1975, Economic geology of Australia and Papua New Guinea, 1. Metals: Australasian Institute of Mining and Metallurgy Monograph, 7.

Mackenzie, D. E., 1976, Nature and origin of late Cainozoic volcanoes in Western Papua New Guinea, *in* Johnson, R. W., ed., Volcanism in Australasia: Amsterdam, Elsevier, 221–238.

——, 1971, Intrusive rocks of New Britain: Bureau Mineral Resources, Australia, Record, 1971/70 (unpubl.)

MacNab, R. P., 1970, Geology of the Gazelle Peninsula, Papua New Guinea: Bureau of Mineral Resources, Australia, Record, 1970/63 (unpubl.).

Mason, D. R., 1975, Geochemistry of intrusive rock suites and related porphyry copper mineralisation in the Papua New Guinea-Solomon Island region [unpubl. Ph.D. thesis]: Canberra: Australia National University.

Mason, D. R., and McDonald, J. A., 1978, Intrusive rocks and porphyry copper occurrences of the Papua New Guinea-Solomon Islands region: A reconnaissance study: Economic Geology, 73, 857–877.

McGee, W. A., 1978, Contributions to the geology of Woodlark Island:

Geological Survey Papua New Guinea Report, 78/10.

McMillan, N. J., and Malone, E. J., 1960, The geology of the eastern central highlands of New Guinea: Bureau of Mineral Resources, Australia, Report, 48.

Norvick, M., and Hutchison, D. S., in press, Aitape-Vanimo, Papua New Guinea, 1:250,000 geological series: Bureau of Mineral Resources, Australia, Explanatory Notes.

Page, R. W., 1976, Geochronology of igneous and metamorphic rocks in the New Guinea Highlands: Bureau of Mineral Resources, Australia, Bulletin, 162.

Page, R. W., and McDougall, I., 1972a, Ages of mineralisation of gold and porphyry copper deposits in the New Guinea highlands: Economic Geology, 67, 1034–1048.

——, 1972b, Geochronology of Panguna porphyry copper deposit, Bougainville Island, New Guinea: Economic Geology, 67, 1065–1074.

Page, R. W., and Ryburn, R. J., 1977, K-Ar ages and geological relations of intrusive rocks in New Britain: Pacific Geology, 12, 99–105.

Pigram, C. J., Arnold, G. O., and Griffin, T. J., in prep., The geology of the Minj 1:100,000 sheet: Geological Survey of Papua New Guinea Report.

Rebek, R. J., 1975, Edie Creek and Wau gold lodes, *in* Knight, C. L., ed., Economic geology of Australia and Papua New Guinea, 1. Metals: Australasian Institute of Mining and Metallurgy Monograph, 7, 867–872.

Robinson, G. P., Jaques, A. L., and Brown, C. M., 1976, Madang, Papua New Guinea, 1:250,000 geological series: Bureau of Mineral Resources, Australia, Explanatory Notes, SB/55-6.

Ryburn, R. J., 1974, Pomio, Papua New Guinea 1:250,000 geological series: Bureau of Mineral Resources, Australia, Explanatory Notes, SB/56-6.

——, 1975, Talasea-Gasmata, Papua New Guinea 1:250,000 geological series: Bureau of Mineral Resources, Australia, Explanatory Notes, SB/57-5 and SB/56-9.

——, 1976, Cape Raoult-Arawa, Papua New Guinea 1:250,000 geological series: Bureau of Mineral Resources, Australia, Explanatory Notes, SB/55-8 and SB/55-12.

Smith, I. E., 1972, High-potassium intrusives from southeastern Papua: Contributions to Mineralogy Petrology, 34, 167–176.

Smith, I. E., and Davies, H. L., 1973, Abau, Papua New Guinea 1:250,000 geological series: Bureau of Mineral Resources, Australia, Explanatory Notes, SB/55-12.

——, 1976, Geology of the southeast Papuan mainland: Bureau of Mineral Resources, Australia, Bulletin, 165.

Tingey, R. J., and Grainger, D. J., 1976, Markham, Papua New Guinea 1:250,000 geological series: Bureau of Mineral Resources, Australia, Explanatory Notes, SB/55-10.

Titley, S. R., 1978, Geologic history, hypogene features, and processes of secondary sulphide enrichment at the Plesyumi copper prospect, New Britain, Papua New Guinea: Economic Geology, 73, 768–784.

Titley, S. R., Fleming, A. W., and Neale, T. I., 1978, Tectonic evolution of the porphyry copper system of Yandera, Papua New Guinea: Economic Geology, 73, 810–828.

Titley, S. R., and Bell, E. B., 1971, The porphyry copper prospect of Plesyumi, New Britain, PNG, *in* porphyry copper deposits of the Southwest Pacific; Specialist Group in the Genesis of Ore Deposits (Geological Society of Australia, Inc.), Sydney, p. 45.

Titley, S. R., and Hendrick, T. L., 1978, Intrusion and fracture styles of some mineralized porphyry systems of the southwestern pacific and their relationship to plate interactions: Economic Geology, 73, 891–903.

Trail, D. W., 1967, Geology of Woodlark Island, Papua: Bureau of Mineral Resources, Australia, Report, 115.

Watmuff, G., 1978, Geology and alteration-mineralization zoning in the central portion of the Yandera porphyry copper prospect, Papua New Guinea: Economic Geology, 73, 829–856.

Willmott, W. E., 1972, Daru-Maer, Papua and Queensland, 1:250,000 geological series: Bureau of Mineral Resources, Australia, Explanatory Notes, SC/54-8 and SC/55-5.

MANUSCRIPT ACCEPTED BY THE SOCIETY JULY 12, 1982

Geological Society of America
Memoir 159
1983

Some aspects of tin granite and its relationship to tectonic setting[1]

S. Suensilpong
P. Putthapiban[2]
N. Mantajit
Geological Survey Division
Department of Mineral Resources
Bangkok, Thailand

ABSTRACT

Within the belt of tin granites, the tectonic setting is more complicated than might appear at first glance. Formation of different phases of granite was confined to certain periods in the tectonic history of the belt.

Reconstruction of the paleotectonic scene for periods as old as Permian has been attempted, and the formation of each phase of granite has been related to a particular tectonic setting. It is clear that both "S-type" and "I-type" granites are present throughout the country and that probably the younger, Upper Cretaceous-Tertiary, "S-type" granites are the most important for tin mineralization. Their origin appears to be connected with the remelting of crustal material during the subduction that resulted from the collision of the Indian and Eurasian plates in Late Cretaceous-Tertiary time.

INTRODUCTION

The granitic belt of Southeast Asia has attracted considerable geological study because of its tin and tungsten mineralization. The annual tin production of the three major producing countries (Malaysia, Thailand, and Indonesia) amounted to between 58 percent and 62 percent of the total world production between 1976 and 1980 (Table 1). The same granitic belt extends into Burma, but insufficient data are available for production from there. The relationship between tin and granite is not fully understood, but clearly an acceptable interpretation must account for the field evidence. The tectonic activity in Southeast Asia seems to be complicated and characterized by radical changes over time spans as short as 20 to 30 million years. Younger tectonism to a greater or lesser extent destroyed the imprint of older events. Areas of overlapping isotopic ages are extensive, and isotopically undisturbed areas are rare. This paper deals mainly with the geology of the tin granites in

Thailand, that is, the central part of the Southeast Asia tin granite belt. It is derived partly from data generated by the current Granite Project.

GRANITOID ROCKS

Among the igneous rocks found in Thailand, those of acidic compositions such as granite, granodiorite, and

TABLE 1. PRODUCTION OF TIN CONCENTRATES OF MALAYSIA, INDONESIA, AND THAILAND DURING 1976-1980
(unit in tonnes)

	1976	1977	1978	1979	1980*
Malaysia	63,401	58,703	62,650	62,500	61,229
Indonesia	23,418	25,921	27,410	28,800	32,527
Thailand	20,453	24,205	30,186	31,800	33,685
Total	107,272	108,829	120,246	123,100	127,441
World's production[†]	179,800	188,500	197,800	198,700	201,200
% as compared to the world's production	59.66	57.73	60.79	61.95	63.34

*Source from ITC Statistics No. 149, issued on march 12, 1981.
[†]Excluding productions of Albania, People's Republic of China, The German Democratic Republic, Mongolia, North Korea, U.S.S.R., and Vietnam.

[1]Publication authorized by Director-General, Department of Mineral Resources, Bangkok.
[2]Present address: Department of Geology, LaTrobe University, Bandoora, Victoria, Australia.

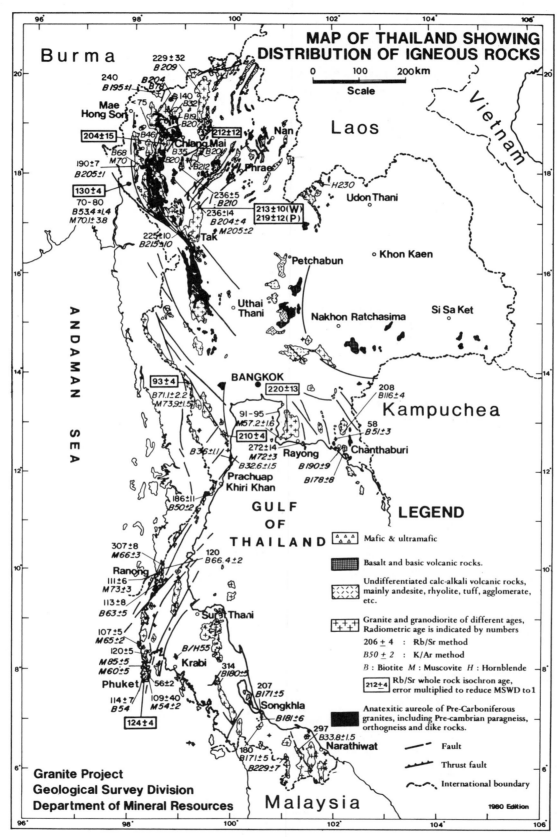

Figure 1. Distribution of igneous rocks in Thailand.

rhyolite are the most abundant (Fig. 1). The more basic igneous rocks such as diorite, gabbro, and basalt, are relatively minor. The granitic belt extends northerly across the western part of the country for a distance of about 1700 kilometres. It can be traced farther to the north, and west into Burma. It extends southward also into Malaysia, Singapore, and the Bangka and Belitung Islands of Indonesia.

Tin mineralization is confined mostly to the highly silicic parts of the S-type granitoids of Chappell and White (1974), whereas tungsten, molybdenum, and porphyry copper mineralization is associated mainly with I-type granitoids. Not all S-type granites, however, are associated with tin mineralization.

The geochemistry and mineralogy show that S-type granites predominate in Thailand. Many granites of Triassic-Jurassic age (on the basis of Rb/Sr whole-rock isochrons) are of the S-type, for example, Mae Sariang granite (190 ± 7 Ma, K-Ar date of 205 ± 1 Ma), Samoeng granite (204 ± 15 Ma), Hub Kapong granite (210 ± 4 Ma), and Khuntan granite 212 ± 12 Ma). High initial $^{87}Sr/^{86}Sr$ ratios of the Permian-Triassic granites (Braun and others, 1976) indicate that they also may be of the S-type.

Cretaceous granites can be divided into both S- and I-types. For instance, in the Mae Lama granite, a small stock underlying about 15 km², S-type granite has been injected into I-type granite. On Phuket Island about four phases of granites have been recognized, including both I- and S-types. The I-type granites on Phuket have been dated as Cretaceous, 124 ± 4 Ma, according to Snelling and others (1970), but some uncertainty exists as to the age of the S-type granites.

PALEOTECTONIC MODELS OF THAILAND

On the basis of geochronological data together with the distribution of the igneous rocks (Fig. 1), paleotectonic models for periods as far back as Late Paleozoic can be constructed. Evidence for orogenic or tectonic movements in the Precambrian is sparse or lacking. Construction of a tectonic model for the Late Paleozoic can be done, although with more difficulty than for later periods because of destruction or dislocation of older structures by later events.

Permian

For the Permian, the model is based on stratigraphic sequences represented by flysch-like and turbidite basins together with the distribution of granites in northern Thailand and of ophiolites in Uttaradit (Baum and others, 1970; Bunopas and Vella, 1978). A west-dipping subduction zone possibly existed in the central part of northern Thailand and is thought to be marked by the undated ophiolite suite in Uttaradit. Subduction there may be correlated with the formation of granite dated at 235 to 240 Ma (Braun and others, 1976), and also with some acidic to intermediate volcanic rocks found in northern Thailand. Geological data are not sufficient to permit correlation to the south with the Permian subduction in Malaysia proposed by Hutchison (1973).

Triassic-Jurassic

A tectonic model for the Triassic-Jurassic can be constructed in more detail as more data are available. The distribution of the Mesozoic plutonic and volcanic rocks can be interpreted as the products of a west-dipping subduction zone passing east of Luang Prabang in Laos (Workman, 1977). The basic and ultrabasic rocks marking the southern and northern margins of the present Khorat Plateau may be interpreted as an ophiolite suite. At the southern margin strongly sheared complex ultrabasic rocks, striking northwesterly, are found at Khao Tham Makok, about 115 km directly north of Chanthaburi, and at Khao Sam Sib, Ban Bo Nang Ching, which lies about 15 km northeast of Khao Tham Makok (Salyapongse, 1979, personal communication). Certain parts of that ultrabasic belt are concealed under the thick alluvial plain derived from the younger clastic rocks of the Khorat Series but can be detected aeromagnetically, especially east of Kabin Buri. Farther north gabbroic rocks are found in the Petchabun area, intruding Karnian sedimentary rocks. The northern margin is marked by ophiolite-like rocks northeast of Loei, which may extend into Laos. Field evidence shows that those basic and ultrabasic rocks are older than the clastic sequences of the Khorat Series. Thus the major part of the ophiolite belt is thought to be concealed under the Jurassic-Cretaceous sediments of the present Khorat Plateau. The suture recording the line of former subduction apparently lies east of the calc-alkali volcanic belt (Fig. 2), and strikes more or less northerly. Both the calc-alkali volcanic belt and the suture zone can be extended to southern Thailand and Malaysia, and ultimately into the South China Sea, east of the Malay Peninsula (Hutchison, 1973).

As shown in Table 2, Mesozoic plutonic rocks are widely distributed throughout the country. No isotopic data are yet available, but field evidence suggests that the calc-alkali volcanic rocks are younger than the granites in northern Thailand. For example, the Khuntan granite, which has been isotopically dated as $212 \pm$ Ma by Rb/Sr, is intruded by andesite dikes. That age seems to set the maximum age for the calc-alkali rocks. On the other hand, calc-alkali rock is not known to intrude the clastic strata of the Khorat Series which is generally accepted as being Upper Jurassic. If the volcanic rocks fall in the Upper Triassic-Lower Jurassic interval of the possible range, they may belong to the same tectonic episode as the Upper Triassic granites.

There is little to suggest that tin deposits are genetically

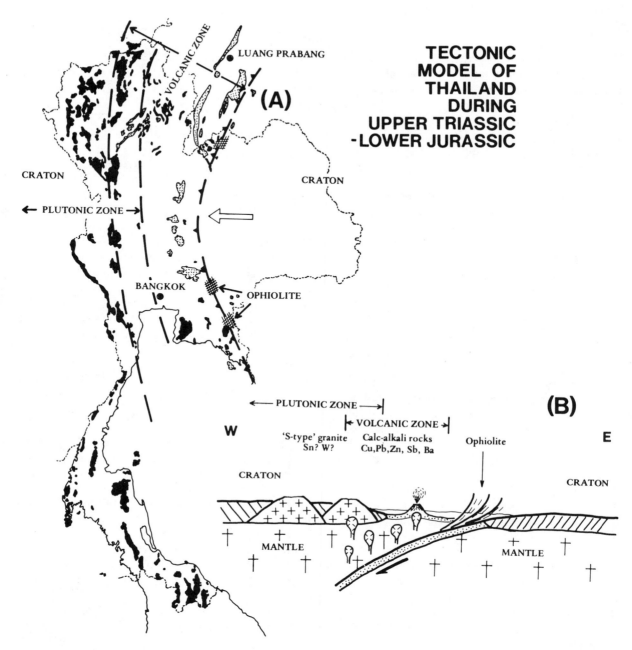

Figure 2. Sketch showing (A) the distribution of plutonic and volcanic rocks and the trench of subduction, (B) tectonic model of northern and central Thailand illustrating the formation of plutonic and volcanic rocks.

associated with the Triassic S-type granite. Most of the tin deposits in the vicinity of Triassic granite are related to a younger granite intruded into the Triassic granite. The younger granite has been dated in places, such as Samoeng in northern Thailand, as Tertiary. The main part of the Triassic batholith consists of porphyritic granite, originally dated at 195 ± 5 Ma (Teggin, 1975), but was subsequently corrected to 204 ± 15 Ma (Beckinsale and others, 1979). In the same area a fine- to medium-grained leucocratic granite

associated with tin deposits yielded a pattern of trace elements that differs from the Triassic granite, and K-Ar ages of 43 ± 1 Ma and 49 ± 1 Ma. If the tin mineralization is related to the younger intrusive, it is clearly Tertiary. Braun and others (1976) discovered yet another phase of granite (also intrusive into the Triassic granite) which yielded a whole-rock Rb/Sr age of 60 ± 5 Ma. This fine-grained granite forms microgranite veins within the Triassic batholith northwest and west of Mae Taeng, about 45 km north of

TABLE 2. RADIOMETRIC AGES OF SOME THAI GRANITES

Location	Rb/Sr age Ma	K/Ar age Ma	Sr^{87}/Sr^{86} ratio	References
Samoeng (leucogranite)	-	43 ± 1	?	Teggin (1975)
Phuket Island	56 ± 2	-	?	Garison and others (1975)
Khao Daen	93 ± 4	-	0.7338 ± 7	Beckinsale and others (1979)
Mae Lama	70 - 80	69, 60	?	Braun, v. and others (1976) Beckinsale and others (1979)
Phuket Island	108 ± 5	-	0.7293 ± 5	Beckinsale, R.D. (1979)
Phuket Island	124 ± 4	-	0.7073 ± 13	Snelling and others (1970) Beckinsale and others (1979)
Huai Yang	186 ± 11	Biot. 50 ± 2	0.823 ± 3	Burton and Bignell (1969)
Mae Sariang	190 ± 7	205 ± 1	0.7280	Braun, v. and others (1976)
Samoeng	204 ± 15	-	0.7328 ± 21	Teggin (1975) Beckinsale and others (1979)
Hub Kapong	210 ± 4	-	0.7237 ± 6	Beckinsale and others (1979)
Khuntan	212 ± 12	-	0.7244 ± 20	Teggin (1975) Braun, v. and others (1976) Beckinsale and others (1979)
Tak (white	213 ± 10	-	0.7158 ± 13	Teggin (1975)
Tak (pink)	219 ± 12	-	0.710 ± 19	Teggin (1975)
Rayong-Bang Lamung	220 ± 13	-	0.7265 ± 13	Beckinsale, R. D. (1979)
Fang-Mae Suei	240 ± 64	-	0.7280 ± 66	Braun, v. and others (1976) Beckinsale and others (1979)
Li	244 ± 28	-	0.7220 ± 44	Braun, v. and others (1976) Beckinsale and others (1979)
Ban Hong	342 ± 9	-	0.7253 ± 15	Braun, v. and others (1976)

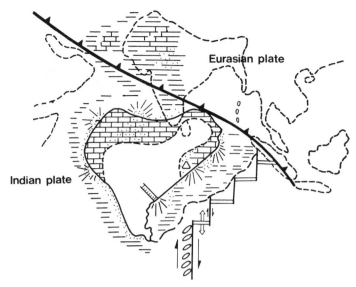

Figure 3. Sketch showing the approximate position of the collision zone between the Indian and Eurasian plates, and the alignment of the trench (after Curray and Moore, 1974, except for the time of collision which is taken from Curray, et al., 1978).

Chiang Mai (not shown on the map) in northern Thailand. These findings are in general accord with views expressed by Hosking (1973), 1974) who stated that tin mineralizaton at Salangor and in the Kinta Valley in Malaysia is related to the underlying cusps of Upper Cretaceous or Tertiary granite rathern than to the Triassic granite.

In all of Southeast Asia, the only primary tin deposit known to be related to Triassic granite is found at Belitung in Indonesia. That granite was dated by Rb/Sr method on whole rock as 213 ± 5 Ma, with an initial $^{87}Sr/^{86}Sr$ ratio of 0.7139. An average K-Ar age of 198 ± 4 Ma was also obtained by Jones and others (1977). One must conclude that the S-type Triassic granite in Southeast Asia, although not commonly mineralized, may contain tin deposits.

Cretaceous

In the Cretaceous, a subduction zone formed as a consequence of the collision of the Indian oceanic plate with the Eurasian continental plate (Fig. 3). The process is believed to have begun early in the Cretaceous (Curray and others, 1978) and subsequently accounted for the formation of I-type granite on the continental side (Fig. 4A). The isochron ages of the known I-type granites, such as those at

Mae Lama and Phuket Island, are about 130 Ma (Beckinsale and others, 1979) and coincide with the beginning of Cretaceous subduction.

The northerly movement of the Indian plate continued after the first collision (Curray and Moore, 1974), and the marginal basin became narrow as stress increased in the crust on the continental side. Heat production increased enormously (Lubinova, 1979, and Fig. 4C), and partial melting began in the crust. The fusion probably involved pre-existing granitic and other crustal material, which may account for the discordance between K-Ar and Rb/Sr ages. These new S-type granites give isotopic ages in the range of 50 to 80 Ma (Garson and Mitchell, 1975; Teggin, 1975; Braun and others, 1976) and are often found intruded into the older granites. The partial melting in the crust apparently caused enrichment (Taylor, 1974; Suensilpong, 1977) in tin which became concentrated in the higher parts of the melt. Some such hypothesis may explain how the younger, fine-grained, S-type granite became associated with tin deposits.

Stress resulting from subduction produced a foliation in most of the older granites (both Triassic and Lower Cretaceous) in the direction perpendicular to compression, that is, northwesterly. The intensity of the foliation varies from place to place and may be absent as in the Triassic Mae Sariang granite. However, the area underlain by foliated granite is extensive and almost 700 km wide. The deformation is useful in distinguishing most of the older granites from the younger granites, which are more apt to be tin-related. In certain areas the direction of foliation appears to have been reset by younger movement, especially in southern Thailand.

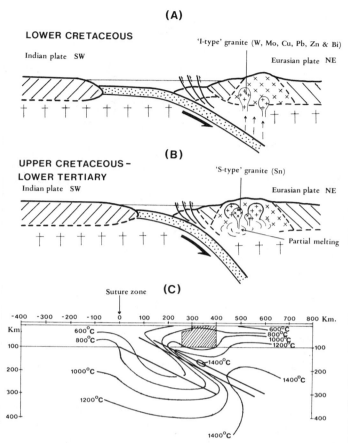

Figure 4. Sketch showing subduction during the collision of the Indian plate with the Eurasian plate. (A) Formation of I-type granite during Early Cretaceous time. (B) Formation of S-type granite in Late Cretaceous-Early Tertiary time. (C) Example of temperature gradient in the subduction zone, after the proposal of Lubimova, and others (1979) for Mexico.

TIN GRANITES

In the most promising tin areas, Mae Lama, Hub Kapong, and Phuket Island, intensive studies of the granites have been conducted. Being representative of tin granites in both northern and southern Thailand, they are thought to be particularly significant. Their chemistry and modes are summarized in Table 3.

Mae Lama

As stated earlier, at Mae Lama two generations of granite are present: the I-type, which has been dated as 130 ± 4 Ma (Beckinsale and others, 1979), and the fine-grained, aplitic S-type, which gives ages between 70 and 80 Ma (Braun and others, 1976). Both ages were determined by the Rb/Sr methods. The I-type granite is probably responsible for the major tungsten-bearing quartz vein at Mae Lama Mine

which also contains molybdenite, sphalerite, chalcopyrite, and galena. The S-type granite is probably responsible for the tin mineralization which is found in the lower parts of Mae Lama, Huai Luang, and Mae Sariang mines.

The latter type, or tin granite, is commonly light gray to white and medium- to coarse-grained. In composition, it varies from muscovite granite to muscovite-tourmaline granite. It is found in places around the margins of the Mae Lama granitic stock in contact with both the country rock and the I-type granite. It also forms veins and dikes. In most places, however, the contact between the older and younger granites cannot be clearly defined. This may be due to the fact that in the marginal zone the I-type granite has differentiated to become a fine-grained rock exhibiting textures similar to those of the younger S-type granite. The main mass of younger granite is thought to underlie the I-type intrusion.

Microscopic investigation shows that the tin granite has a hypidiomorphic granular texture. Microcline and microcline-perthite form anhedral to subhedral crystals, whereas plagioclase forms smaller crystals. The composition of the plagioclase varies from albite to albitic oligoclase. Zoning is rare in the plagioclase. Quartz appears as an interstitial mineral in which plagioclase, muscovite, and apatite form inclusions. Biotite is present but only in very small amounts. Most of it is chloritized. Muscovite is thought to be a primary constituent. Accessory minerals are apatite, tourmaline, garnet, zircon, and ilmenite.

Hub Kapong

At Hub Kapong, about 200 km by road southwest of Bangkok, at least three phases of granite (Fig. 5) are present in the batholith (Putthapiban and Suensilpong, 1978). The tin granite is characteristically a nonfoliated, fine- to medium-grained rock with a hypidiomorphic granular texture. In mineral composition, it ranges from biotite-muscovite granite to a muscovite-tourmaline granite. No isotopic age is available for this granite, but it intrudes a very coarse-grained, porphyritic gneissose granite that has been dated (Rb/Sr method) as 210 ± 4 Ma (Beckinsale and others, 1979). Being the only nonfoliated rock among strongly foliated granites, the tin granite may be Upper Cretaceous or Tertiary, similar to the tin granite in Phuket. The chemical composition of this granite (Table 3) suggests that it is an S-type, as is the major phase in this area. The tin granite intrudes the northern and western parts of the Hub Kapong batholith, and the tin deposits, of course, are found in the same area.

Phuket Island

On Phuket Island the dominant granite is an I-type, which has been dated at 124 ± 4 Ma (Snelling and others, 1970;

TABLE 3. CHEMICAL ANALYSES, NORMS, AND MODES OF SOME TIN-RELATED GRANITES

	M98	M99	M136	M144	M150	HK14	HK15	HK18	HK50	HK52	KP1	KP5	KP13	KP52	KP63
SiO_2	72.69	72.84	71.26	71.75	71.67	73.57	68.55	67.84	72.57	72.93	71.38	69.91	69.60	69.61	68.33
TiO_2	0.04	0.04	0.07	0.07	0.07	0.26	0.79	0.81	0.01	0.23	0.27	0.24	0.42	0.35	0.23
Al_2O_3	15.54	15.22	16.81	16.55	15.19	13.66	13.96	15.19	16.00	13.95	13.94	14.45	14.66	14.62	14.51
Fe_2O_3	0.73	0.76	0.00	0.08	0.17	0.63	0.67	0.40	0.50	0.21	0.27	0.48	0.11	0.23	0.79
FeO	0.38	0.37	1.04	1.43	1.59	1.98	3.82	3.49	0.92	2.84	3.98	4.19	3.69	2.41	3.87
MnO	0.01	0.02	0.03	0.01	0.01	0.04	0.03	0.03	0.02	0.03	0.02	0.05	0.04	0.04	0.06
MgO	0.73	0.24	0.18	0.13	0.12	0.07	0.24	0.26	0.17	0.45	0.13	0.13	0.56	0.33	0.55
CaO	0.62	0.73	0.72	0.75	1.09	1.02	2.05	2.36	0.51	1.40	1.42	1.52	1.37	1.45	1.50
Na_2O	3.59	4.79	5.16	4.58	4.30	3.01	2.32	2.31	3.62	2.58	2.36	2.53	2.19	2.43	2.72
K_2O	4.61	3.45	3.26	3.60	3.74	4.39	7.02	5.70	4.49	4.77	5.93	5.96	5.77	5.95	5.66
P_2O_5	0.23	0.29	0.35	0.19	0.69	0.01	0.12	0.30	0.07	0.12	0.07	0.07	0.66	0.38	0.22
H_2O^+	0.17	0.17	n.d.	n.d.	n.d.	0.23	0.32	0.38	0.38	nil	0.01	0.01	0.07	0.72	1.01
H_2O^-	nil	0.17	0.09	0.02	0.02	0.08	0.08	0.03	n.d.	n.d.	0.22	0.31	0.29	0.17	0.14
Total	99.34	99.09	98.97	99.16	98.66	98.95	99.97	99.09	99.18	99.51	100.00	99.85	99.43	98.70	99.87

Norms

	M98	M99	M136	M144	M150	HK14	HK15	HK18	HK50	HK52	KP1	KP5	KP13	KP52	KP63
Q	32.40	30.66	27.09	28.95	30.60	36.78	21.72	24.96	32.76	33.72	28.98	25.92	28.56	30.53	24.63
Or	27.24	20.39	19.26	21.27	21.10	26.13	41.70	33.92	26.69	28.36	35.03	35.03	33.92	35.16	33.45
Ab	30.38	40.53	43.66	38.75	36.38	25.68	19.39	19.39	29.87	22.01	19.91	22.22	18.34	20.56	23.02
An	1.57	1.73	1.28	2.48	0.90	4.73	6.95	11.18	2.78	6.95	6.83	7.51	6.55	0.00	6.00
Co	4.07	2.97	4.32	4.21	3.74	3.16	0.00	0.92	4.28	1.94	1.12	1.12	2.45	4.18	1.71
Hy	1.82	0.62	2.29	2.78	2.89	3.10	5.75	5.32	1.72	5.72	6.90	5.55	7.47	4.52	7.53
Hen	1.82	0.60	0.45	0.32	0.30	·	·	·	·	·	0.00	0.00	0.00	0.82	1.37
Ffs	0.00	0.02	1.84	2.45	2.59	·	·	·	·	·	0.00	0.00	0.00	3.72	6.16
Wo	1.14	1.10	0.00	0.12	0.25	0.00	0.00	0.00	0.00	0.00	0.00	0.00	0.00	0.00	0.00
Mt	0.08	0.08	0.13	0.13	0.23	1.02	1.02	0.75	0.70	0.23	0.46	0.70	0.23	0.33	1.14
Il	6.94	0.00	0.00	0.00	0.00	0.46	1.52	1.52	0.00	0.46	0.46	0.46	0.76	0.66	0.44
Ap	0.53	0.67	0.81	0.44	1.60	0.06	0.00	0.05	0.01	0.00	0.00	0.00	n.d.	n.d.	n.d.

Modes

	M98	M99	M136	M144	M150	HK14	HK15	HK18	HK50	HK52	KP1	KP5	KP13	KP52	KP63
Q	32.41	32.98	28.63	28.13	31.49	30.10	28.99	33.00	26.50	33.33	29.13	30.69	30.92	30.18	29.45
Plag.	32.24	28.33	24.83	26.13	22.41	27.50	27.23	12.50	25.69	19.23	27.84	31.20	23.39	18.54	17.48
K-felds.	34.04	34.04	40.76	41.64	38.52	36.00	35.63	46.50	37.10	35.26	32.86	26.09	39.51	41.09	40.18
Biot.	n.d.	n.d.	n.d.	n.d.	n.d.	2.00	6.93	9.00	n.d.	12.18	7.28	6.39	5.12	8.09	12.88
Musc.	1.10	1.07	5.78	9.44	7.59	3.00	1.22	n.d.	10.71	n.d.	3.88	5.63	1.07	2.82	n.d.
Tour.	0.81	1.75	n.d.	n.d.	n.d.	1.40	n.d.	n.d.	n.d.	n.d.	n.d.	n.d.	n.d.	n.d.	n.d.

Note : M = Mae Lama granites, Lat 17°40 Long 97°50
HK = Hub Kapong granites, Lat 12°40   Long 99°50
KP = Phuket Island granites, Lat 7°50   Long 98°20

See also Figure 1 for location.

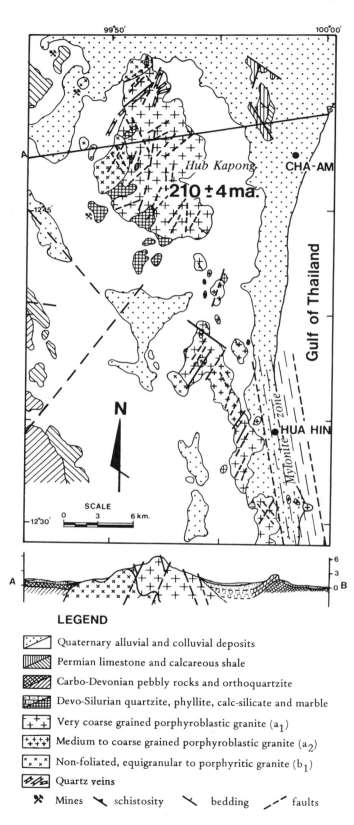

LEGEND

```
[.:.]  Quaternary alluvial and colluvial deposits
[▨]    Permian limestone and calcareous shale
[▩]    Carbo-Devonian pebbly rocks and orthoquartzite
[▦]    Devo-Silurian quartzite, phyllite, calc-silicate and marble
[+++]  Very coarse grained porphyroblastic granite (a₁)
[++++] Medium to coarse grained porphyroblastic granite (a₂)
[×××]  Non-foliated, equigranular to porphyritic granite (b₁)
[///]  Quartz veins
 ✖ Mines   ╲ schistosity   ╲ bedding   ⟋ faults
```

Figure 5. Geological map of Hub Kaporig Batholith.

Beckinsale and others, 1979). It contains sphene, hornblende, and allanite, but no tin deposits. In the areas covered by the tin-mining leases, S-type granite is exposed and was dated (Rb/Sr method) as 56 ± 2 Ma (Garson and Mitchell, 1975). The age, however, was not regarded as significant before the difference between I- and S-type granite had been recognized. The S-type granite on Phuket Island is characterized by muscovite, tourmaline, and garnet, in addition to K-feldspar, plagioclase, and some biotite. It is generally fine- to medium-grained and equigranular, but in places it is greisenized and albitized. Garson and Mitchell (1975) referred to it as a two-mica granite, and Aranyakanon (1961) called it "Haad Som Pan" type granite. The accessory minerals include garnet, zircon, rutile, and opaque minerals. Cassiterite is found in a few places disseminated in greisenized muscovite granite. No sphene has been detected in this rock type.

This granite forms random intrusions within older granite, especially in the Lower Cretaceous I-type granite, as can be seen at Patong Beach and Ban Nakha Le. The relationship between this granite and the country rock is best seen at Khao Sapam, Khao Phan Thu Rat, Khao To Sae, and Ka Thu mining district. The granite is closely associated with pegmatitic dikes, some of which carry cassiterite of economic significance.

CONCLUSIONS

The Southeast Asia granite belt comprises more than one phase of granite and each phase belongs to a particular tectonic regime. The paleotectonic models for both Permian and Late Triassic-Early Jurassic include west-dipping subduction zones. For the Early Cretaceous and Late Cretaceous-Early Tertiary, the tectonic scene changed and involved east-dipping subduction. Subduction in the latter two intervals is thought to be associated with the collision between the Indian oceanic plate and the Eurasian plate.

Tungsten, molybdenum, and copper mineralization are related to I-type granite of Early Cretaceous age, whereas tin is related to the S-type granite. Not all granites of the S-type, however, are of equal economic significance with respect to tin; S-type granites of Late Cretaceous-Early Tertiary age are the most important. The S-type granite that is important for tin mineralization commonly does not show any foliation. The tin granite is usually a fine-grained rock with muscovite, tourmaline, and garnet.

ACKNOWLEDGMENTS

The authors are greatly indebted to Prabhas Chakkaphak, the Director-General of Mineral Resources, for permission to publish this paper. Many thanks are due to Manus Veeraburus, the Director of the Geological Survey Division for his full support and encouragement.

They wish to thank also their colleagues in the analytical laboratory for their work on the chemical analyses. Many thanks are also due to A. Khamchu, V. Bupphasiri, and Miss N. Tanareongsakulthai for their help in drawing the diagrams and typing the manuscript. Lastly, they wish to thank E. J. Cobbing, Institute of Geological Sciences for reading the manuscript and for some useful comments.

REFERENCES CITED

Aranyakanon, P. 1961, The cassiterite deposit of Haad Som Pan, Ranong Province, Thailand: Royal Department of Mines, Report of Investigation, no. 4, 182 p.

Baum, F., Braun, E. von, Hahn, L., Hess, A., Koch, K. E., Kruse, G., Quarch, H., and Seibenhuner, M., 1970, On the geology of northern Thailand: Beihefte zum Geologischen Jahrbuch, v. 102, 24 p.

Beckinsale, R. D., 1979, Granite magmatism in the tin belt of Southeast Asia: *in* Atherton, M. P., and Tarney, J., eds., Origin of granite batholiths, geochemical evidence: Shiva Publishing Ltd., Kent, U.K., p. 34–44.

Beckinsale, R. D., Suensilpong, S., Nakapdungrar, S., and Walsh, J. N., 1979, Geochronology and geochemistry of granite magmatism in Thailand in relation to a plate tectonic model: Journal of the Geological Society of London, v. 136, p. 529–540.

Braun, E. von, Besang, C., Eberle, W., Harre, W., Kreuzer, H., Lenz, H., Muller, P., and Wendt, I., 1976, Radiometric age determinations of granites in northern Thailand: Geologisches Jahrbuch, v. 21, p. 171–204.

Bunopas, S., and Vella, P., 1978, Late Paleozoic and Mesozoic structural evolution of northern Thailand; a plate tectonic model: *in* Nutalaya, P., ed., Proceedings of the Third Regional Conference on the geology and mineral resources of Southeast Asia: Geological Survey of Thailand, p. 133–140.

Burton, C. K., and Bignell, J. D., 1969, Cretaceous-Tertiary events in Southeast Asia: Geological Society of America Bulletin, v. 80, p. 681–688.

Chappell, B. W., and White, A. J. R., 1974, Two contrasting granite types: Pacific Geology, no. 8, p. 173–174.

Curray, J. R., and Moore, D. G., 1974, Sedimentary and tectonic processes in the Bengal deep-sea fan and geosyncline: *in* Burk, C. A., and Drake, C. L., eds., The geology of continental margins: Springer-Verlag, p. 617–627.

Curray, J. R., Moore, D. G., Lawver, L. A., Emmel, F. J., Raitt, R. W., Henry, M., and Kieckhefer, R., 1978, Tectonic of the Andaman Sea and Burma: American Association of Petroleum Geologists, Memoir 29, p. 189–198.

Garson, M. S., Young, B., Mitchell, A. H. G., and Tair, B. A. R., 1975, The geology of the tin belt in peninsular Thailand, Phuket, Phangnga and Takua Pa: Institute of Geological Sciences, Overseas memoir 1, 112 p.

Hosking, K. F. G., 1973, Primary mineral deposits: *in* Gobber, D. J., and Hutchison, C. S., eds., Geology of the Malay Peninsula, West Malaysia and Singapore: New York, Wiley Interscience, p. 335–390.

——,1974, The search for deposits from which tin can be profitably recovered now and in the foreseeable future: Fourth World Conference on Tin, Kuala Lumpur 1974, Paper 1.1, 55 p.

Hutchison, C. S., 1973, Tectonic evolution of Sundaland: Geological Society of Malaysia Bulletin, no. 6, p. 61–86.

Jones, M. T., Reed, B. L., Doe, B. R., and Lanphere, M. A., 1977, Age of tin mineralization and Plumbotectonics, Belitung, Indonesia: Economic Geology, v. 72, p. 745–752.

Lubimova, E. A., Luboshits, V. M., and Nikitino, V. N., 1979, Analysis of geothermal models of seismo-volcanic Mexican-type belt: Section B IV, XIV Pacific Science Congress, Khabarovsk, U.S.S.R.

Putthapiban, P., and Suensilpong, S., 1978, The igneous geology of the granitic rocks of Hub Kapong-Hua Hin area: Journal of the Geological Society of Thailand, v. 3, p. M1.1–M1.22.

Snelling, N. J., Hart, R., and Harding, R. R., 1970, Age determinations on samples from the Phuket region of Thailand: Institute of Geological Sciences, Report no. IGU 70.19 (unpublished).

Suensilpong, S., Meesook, A., Nakapadungrat, S., and Putthapiban, P., 1977, The granitic rocks and mineralization at the Khuntan Batholith, Lampang: Geological Society of Malaysia Bulletin, no. 9, p. 159–173.

Taylor, D., 1974, The liberation of minor elements from rocks during plutonic igneous cycles and subsequent concentration to form workable ores with particular reference to copper and tin: Geological Society of Malaysia Bulletin, no. 7, p. 1–16.

Teggin, D. E., 1975, The granites of northern Thailand: unpub. Ph.D. thesis, Manchester University, U.K.

Workman, D. R., 1977, Geology of Laos, South Vietnam and the eastern parts of Thailand: Institute of Geological Sciences, Overseas Geology and Mineral Resources, no. 50, 33 p.

MANUSCRIPT ACCEPTED BY THE SOCIETY JULY 12, 1982

PRINTED IN U.S.A.

Geological Society of America
Memoir 159
1983

Summary of igneous activity in South Korea

Ok Joon Kim
Dai Sung Lee
Department of Geology
Yonsei University
Seoul, Korea

ABSTRACT

The Korean peninsula, unlike nearby Japan, is underlain mainly by Precambrian terrane. Superimposed on these ancient rocks in South Korea is a northeasterly trending geosynclinal assemblage which forms the Ogcheon Zone. It is about 60 kilometres wide, and consists of both metamorphic and unmetamorphosed strata ranging in age from Precambrian to Cretaceous. The Precambrian crystalline terrane that emerges north of the geosynclinal belt is known as the Gyonggi Massif and that to the south is the Ryongnam Massif. Farther southeast is the Cretaceous Gyongsang sedimentary basin. It occupies most of the southeastern part of the Korean peninsula, but the southeastern shore is overlapped by the margin of a Tertiary basin of sedimentary and volcanic rocks.

Mesozoic plutonic rocks are widespread, forming major intrusions in all of the main geologic provinces of South Korea. The batholithic rocks are divided into two main groups, the Daebo granites of Early Jurassic to Early Cretaceous age, the Bulgugsa granites of Late Cretaceous to Early Tertiary age. Most of the older granites are aligned in the regionally dominant, northeasterly direction known as the Sinian trend. The younger granites show no clear pattern and seem to be related to later block movements. Most of both the Jurassic and Cretaceous plutonic rocks are granodiorite and granite, but small plutons of basic to intermediate composition exist in both groups. As in Japan, the ages of the granitic rocks are generally younger to the southeast, but in both countries there are numerous complications. In the Ogcheon Zone is a major thrust fault which separates the older and metamorphosed part of the Ogchean assemblage from the younger, unmetamorphosed strata. Aligned along the fault is a sequence of basic rocks (amphibolite, gabbro, basalt, andesite, and chlorite schist) that may indicate the fault is an ancient geosuture. Base metal deposits are associated with both the Daebo and Bulgugsa granites but not, as far as is known, with any of the volcanic rocks.

INTRODUCTION

A minor craton in eastern Asia, known as the "Korean-Chinese Heterogen," extends from north China through Manchuria to Korea, where it includes most of the peninsula. In the southern part of the Korean peninsula is a series of diagonal lineaments that form the 60-km-wide Ogcheon Zone. Its northeastern trend is parallel with the "Sinian" direction of the regional structure (Pumpelly, 1866), which extends from Fukien province in southern China to Sikhota Alin in northeastern U.S.S.R. According to Lin (1971), the

Korean region of eastern Asia is situated at the intersection of the Tethys and the circum-Pacific fracture and fold belts (Fig. 1). Tectonically, the Sinian direction is parallel with the circum-Pacific belt, and the most remarkable part of this belt appears in South Korea as the folded and fractured geosynclinal belt, known as the Ogcheon Zone.

More than half of the Korean peninsula, mainly the northern and middle parts, is underlain by Precambrian basement which is divided broadly into three massifs, which

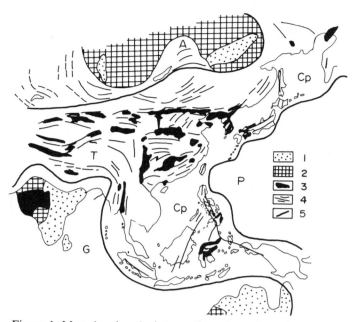

Figure 1. Map showing the intersection of two fracture and fold-belts in eastern Asia (modified from Lin, 1971). 1: Platform cover, 2: Old platform basement, 3: Basement of Precambrian fold-belt, 4: Caledonian to Alpian fold-belts, 5: Boundary, P: Pacific block, G: Gondwana block, A: Angara block, T: Tethys fracture and fold-belt, Cp: Circum-Pacific fracture and fold-belt.

from north to south are Pyongbug-Gyema, Gyonggi, and Ryongnam.[1] These massifs can be correlated, respectively, with the Chiao Liao, Chi-Lu-Chiao Liao, and Yangchin massifs of China (Sato, 1975). Only the Gyonggi and Ryongnam massifs are in South Korea, where they formed the basement on which numerous episodes of sedimentation and igneous activity took place during the Phanerozoic.

The tectonics of South Korea were outlined by Kobayashi (1953, 1957), later revised by Kim (1970), and then modified by S. M. Lee (1974) and D. S. Lee (1977b). Kim put the geological boundary between southern and northern Korea along the Chugaryong rift-valley, which trends north-northeast across the middle of Korea, and divided South Korea into five provinces based on tectonic units and geologic ages, namely: (1) Gyonggi Massif, composed mainly of Precambrian metamorphic rocks and Jurassic granites; (2) Ryongnam Massif, mainly Precambrian metamorphic rocks and Jurassic to Cretaceous igneous rocks; (3) Ogcheon geosynclinal zone, mainly Precambrian to Cretaceous metamorphic and sedimentary rocks, and Jurassic to Cretaceous granitic rocks; (4) Gyongsang sedimentary basin, consisting of Cretaceous sedimentary and igneous rocks; and (5) a Tertiary sedimentary and volcanic basin. The Plio-Pleistocene volcanic rock of the area was regarded as a separate province by D. S. Lee (1977b), as seen in Figure

[1]Alternative spellings of Korean place-names are listed at the end of this paper.

2. It forms the western border zone of the "circum-Japan Sea alkali-rock province" (Aoki, 1958).

The Gyonggi and Ryongnam massifs flank both sides of the Ogcheon Zone (Fig. 2) and are closely related to each other tectonically, lithologically, and in geologic age. These massifs may, in fact, form a single unit, the Gyonggi-Ryongnam Massif, although an ophiolite suite has been recognized in the area between them, i.e., in the Ogcheon

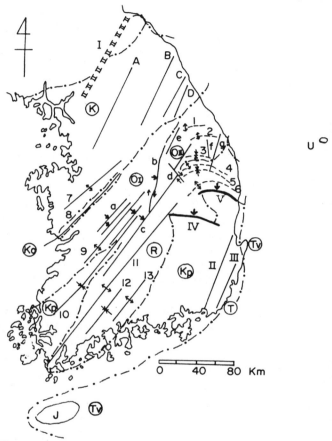

Figure 2. Geologic provinces and tectonic map of South Korea (after O. J. Kim, 1970; Tv is added by D. S. Lee).
Provinces (circled letters): K, Gyonggi land; R, Ryongnam land; OI, Ogcheon paleogeosynclinal zone (Lee's Pibanryong type zone); OII, Ogcheon neogeosynclinal zone (Lee's Hambaeg-san type zone); Kp, Gyongsang basin proper; Ko, Gyongsang trough in Ogcheon zone; T, Tertiary basins; Tv, Neogene alkali-rock province.
Pre-Triassic faults: A, Pyunggang fault; B, Inje fault; C, Hyonri fault; D, Changchon fault.
Triassic deformation: 1, Jongson syncline; 2, Jungbongsan anticline; 3, Hambaek syncline; 4, Sobaegsan anticline; 5, Yulri syncline; 6, Andong anticline.
Jurassic deformation: 7, Charyong anticline; 8, Gongju syncline; 9, Ogcheon anticline; 10, Yongdong syncline; 11, Dugyusan anticline; 12, Gure syncline; 13, Jirisan anticline; a, Ogcheon thrust; b, Bonghwajae thrust; c, Jeomchon thrust; d, Danyang fault; e, Pyongchang fault; f, Samchog fault; g, Osipchon fault.
Cretaceous-Tertiary deformation: I, Chugaryong rift; II, Yangsan fault; III, Tongrae fault, IV, Andong thrust; V, Ilwolsan thrust; J: Jeju volcanic island; U: Ulnung volcanic island.

Zone (O. J. Kim and others, 1976b; M. S. Lee and others, 1980; D. S. Lee, 1980).

STRATIGRAPHY

The stratigraphy in South Korea is relatively well understood except in the Precambrian massifs and parts of the Ogcheon Zone. C. H. Cheong summarized the geologic sequences of Korea in 1956 from the data then available. His outline was the accepted standard until the Precambrian stratigraphy was established (Kim 1968, 1970, 1973) and the complete geologic sequence of South Korea was compiled (Table 1). The stratigraphy, structure, and geologic age of the Ogcheon Group, which underlies the middle of the Ogcheon geosynclinal zone, are the subject of different opinions, as shown in Table 2. The Ogcheon Zone was previously divided into a northeast subzone of unmetamorphosed Paleozoic and Mesozoic strata and a southwest subzone of metamorphosed but presumed equivalent rocks (Kobayashi, 1953; Son, 1969b). Kim (1973, 1970), however, recognized a thrust fault marked by a discontinuity in lithology and structure between the subzones, and stated

that the Ogcheon Group in the metamorphosed zone must be Precambrian. Accordingly, he designated the metamorphosed subzone as the "paleogeosynclinal zone" and the unmetamorphosed subzone as the "neogeosynclinal zone." The upper member of the metamorphosed subzone, the Hwanggangri Formation, contains a pebble-bearing phyllite that is thought to be a tillite (Kim, 1970) and has been correlated with Precambrian tillite in the Yangtze Valley of central China (Reedman and Fletcher, 1976).

An index fossil, Archaeocyatha, was discovered in a dolomitic limestone bed in the Daehyangsan Formation, a lower member of the Ogcheon Group, and suggests that the formation is Lower to Middle Cambrian rather than Precambrian (D. S. Lee and others, 1972b). Thus, D. S. Lee (1977b) proposed that the name of the metamorphosed zone be revised to "Pibanryong-type Ogcheon subzone" and the unmetamorphosed zone to "Hambaegsan-type Ogcheon subzone." These two subzones are coeval, but the depositional environments were different. However, O.J. Kim suggested that the Archaeocyatha-bearing limestone, which was regarded as a lower member of the Ogcheon Group, may be a tectonic remnant of the Cambrian limestone that outcrops in the northeast because the fossiliferous limestone is everywhere discontinuous.

REGIONAL TECTONICS AND RELATED IGNEOUS ACTIVITY

In the Gyonggi Massif two great unconformities have been recognized (Kim, 1973), and in the Ryongnam Massif three unconformities have been identified (Kim and others, 1963), as shown in Table 1. The metamorphic rocks separated by these unconformities show different degrees of metamorphism and deformation. Two or three periods of deformation and metamorphism affected both massifs, but the timing is uncertain, and events in the two terranes cannot be correlated with confidence. For this reason and because later structural breaks in Korea are definitely known to be of Late Triassic (Songrim Disturbance), Jurassic (Daebo Orogeny), Late Cretaceous (Bulgugsa Disturbance), and mid-Tertiary age (Yonil Disturbance), the structures in both Precambrian terranes are assigned simply to the pre-Triassic.

It must be noted that a great hiatus extending from Late Ordovician to Early Carboniferous time exists for which practically no stratigraphic record is known in South Korea. Recently, however, H. Y. Lee (1980) concluded from his conodont study that a Silurian formation exists in the northeastern part of the Ogcheon Zone.

Pre-Triassic Deformation

In the Gyonggi Massif, the trend of foliation in the metamorphic complex is very diverse but north-northwest trends prevail in the western part and north-northeast trends

TABLE 1. THE GEOLOGICAL SEQUENCE OF SOUTH KOREA

Age		System	Series					
Cenozoic		Quaternary	basalts					
		Tertiary	Yeonil					
			Janggi					
			～～ granites(Bulgugsa gr.) volcanics ～～					
Mesozoic	Cretaceous	Gyongsang	Silla					
			Nagdong — Granites(Daebo granites)～～					
	Jurassic	Daedong	Chungnam, Dansan, Bansong					
	Triassic		Nogam					
Paleozoic	Permian	Pyongan	Gobangsan					
	Carbonif.		Sadong					
			Hongjeom					
	Devonian		absent					
	Silurian							
	Ordovician	Chosun	Great Limestone					
	Cambrian		Yangdeong					
Late Precambrian		Sangwon system (in N. Korea)	Ogcheon group*	Gunjasan				
				Hwanggangri				
				Changri Munjuri Hyangsanri				
Mid. to late Precambrian		Chunchon system	～ granite gn. ～	Chunseong group	Gemyongsan	?	～ granite gn.～ Taebaegsan	
			Jangragsan group				～granite gn.～ Goseonri Gaghwasa	Yulri system
Early-mid. Precambrian	Yonchon system	～ granite gn.～ Yangpyong c.† Sihung c. Puchon c.				～ granite gn.～ Wonnam Gisong Pyonghae	Ryongnam system

*The stratigraphy and geologic age of the group after Kim (1968, 1970).
†c: complex.

TABLE 2.   CORRELATION CHART OF THE OGCHEON SYSTEM

	M. S. Lee et. al., (1965)	O. J. Kim (1968, 1970)	D. S. Lee (1971	Reedman et. al., (1973)
Mid-Ordov. - Carbo.	Munjuri Hwanggangri Myongori Buknori Daehyangsan Hyangsanri Gyemyeongsan		Triassic: Hwanggangri	
Camb.-Ord.	— Unconformity —	Great Limestone Series	Ogcheon group: Majeonri / Changri* / Munjuri / Daehyangsan†	Chungju group: Great Limestone series / Gyemyeongsan / Daehyangsan
Late Precambrian		— fault — Gunjasan Hwanggangri Changri Munjuri Daehyangsan Gyemyeongsan (Ogcheon System)	? Gyemyeongsan	? Ogcheon Group: Munjuri / Hwanggangri / Myongori / Buknori / Seochangri / Gounri

*Contains conodont fragments.
†Contains "Archaeocyatha."

in the central and eastern parts of the massif. Four major faults and a synclinorium trend north-northeast but are truncated by the Daebo Granite at the southeastern end of the massif (Fig. 2). The foliation of the complex was not affected by the intrusion of the Daebo Granite. Metamorphic facies in the basement complex indicate low pressures and intermediate temperatures (Na, 1978).

In the Ryongnam Massif, the trend of foliation is also very diverse. The northeasterly trend in the southwestern part is cut at a low angle by the Jeomchon thrust which is related to the Daebo Orogeny (Fig. 2). During this deformation, granite, granite gneisses, and some ultrabasic and intermediate igneous rocks, such as anorthosite, gabbro-syenite, and diorite were emplaced into the Gyonggi-Ryongnam terrane. The ages of anorthosite and syenite plutons have not been determined although their stratigraphic relationship indicates that they are similar in age.

Triassic Deformation (Late Triassic, Songrim Disturbance)

In the northeastern part of the Ogcheon Zone, fold axes in

the Paleozoic and Triassic sedimentary formations trend west-northwest. This deformation is thought to be the result of the Songrim Disturbance at the end of the Triassic, as the Jurassic Daedong Formation was not affected by the deformation. The western end of those folds, however, were bent in the Jurassic to conform with the Sinian direction.

In the northeastern part of the Ryongnam Massif are several anticlines (Sobaegsan, Yulri, and Andong), trending generally west-northwest, which may be Precambrian structures modified by Songrim Deformation (Fig. 2). Related to this deformation are some andesitic, basaltic, and liparitic lavas and tuff layers in the Bansong area, and some granite plutons in North Korea. Hornblendite, monzonite, and syenite are thought to be associated with this activity.

Jurassic Deformation (Daebo Orogeny)

The Daebo Orogeny extended from Early Jurassic into the Early Cretaceous, as determined by dating the Daebo Granite. This orogeny severely folded and faulted some sedimentary strata in the Ogcheon, Joseon, Pyongan, and

Daedong Groups, and formed their northeasterly trend (Fig. 2). The Daebo granitic batholiths lie parallel with this trend, as does the elongation of the Ogcheon synclinal zone and adjacent Precambrian terranes (Fig. 3). Four anticlinoria and three intervening synclinoria extend from the southern border of the Gyonggi Massif through the Ogcheon Zone, to the Ryongnam Massif (Fig. 2). The main Ogcheon thrusts (within the Ogcheon Zone) are the Bonghwajae, which separates the two subzones of the Ogcheon Zone, and the Jeomchon thrust, which separates the Ogcheon Zone from the Ryongnam Massif on the southwest. The anticlinoria form major mountain ranges. Younger Jurassic and Cretaceous sediments, which may be molasse, are found in several isolated places in the synclinoria. Three phases of Daebo igneous activity have been recognized in the middle of the Ogcheon Zone; the two early phases are syntectonic and the last phase is late tectonic. The first phase is transitional from pretectonic to syntectonic and yielded some intermediate to mafic volcanic rocks. The second phase is marked by metasomatic plutonism that formed bodies of schistose and/or porphyroblastic adamellite and granodiorite, migmatite, and granitic gneiss along the southeast margin of the Ogcheon Zone. The third, and most active, phase led to the emplacement of large granodiorite batholiths, accompanied by small dioritic and gabbroic stocks parallel with the Sinian direction along the core of the Ogcheon folded belt (D. S. Lee, 1971, 1977a). The granite plutons of this phase also intruded the Gyonggi and Ryongnam massifs.

Cretaceous Deformation (Bulgugsa Disturbance)

Only minor folds are present in the Cretaceous Gyongsang sedimentary basin, although the formations show a homoclinal structure dipping generally to the southeast. At the end of this disturbance, Andong and Ilwolsan thrusts were formed, marking the northern limit of the Cretaceous sediments. In the Late Tertiary, movement along the Chugaryong rift-valley began, accompanied by extrusions of small amounts of olivine-free basalt and abundant andesitic pyroclastics.

Several other fault systems, trending north-northwest in the main part of Gyongsang Basin and north-northeast in the southeast corner, originated at the end of this period (Fig. 2). They are thought by Min and others (1982) to be related to Tertiary plate movements. Won (1968) divided the igneous cycle in Gyongsang Basin into three sequential phases: volcanism, plutonic intrusion, and dike intrusion. A similar cycle has been recognized also in the Ogcheon Zone (Fig. 4; D. S. Lee, 1971; Lee and others, 1977a).

The volcanism began during deposition of the Silla conglomerate, a basal member of the mid-Cretaceous sequence. It comprises in ascending order the Hagbong lava flows of andesitic porphyry and andesitic basalt, tuffaceous

sediments, the Chaeyagsan lava flows of andesitic porphyry and andesitic basalt, and the Jayangsan lava flow of andesite porphyry and andesitic tuff.

Plutonic masses intruded the middle and eastern parts of Gyongsang Basin, the middle and the southwest end of the Ogcheon Zone, and the middle of the Ryongnam Massif (Fig. 3). Emplacement of these plutons was controlled by the structure of the region, which is marked by faults trending N60W, N20E, and N30W.

Dikes related to the last stage are distributed evenly throughout South Korea, mainly in shear zones, especially in Gyongsang Basin, and the middle part of the Ogcheon Zone. The dikes are commonly granophyre, felsite, pegmatite, porphyry, and lamprophyre; north, northeast, and northwest trends dominate. Three granite masses, outcropping on the east coast near Ulsan city, were recently found to be Early Tertiary by K-Ar dating which yielded a range of 58 to 63 Ma (Lee and Ueda, 1976).

Mid-Tertiary Deformation (Yonil Disturbance)

An angular unconformity separates the Lower and Upper Miocene formations. The lower formation is folded, whereas the upper ones exhibit no deformation. The unconformity indicates a geologic disturbance, which Son (1969b) named the Yonil Disturbance. In the Eoil Basin, near Pohang city, basalt flows and tuff layers in the lower part of the Yonil Formation represent the igneous activity of this period.

Plio-Pleistocene Volcanism

Plio-Pleistocene volcanic rocks on Jeju and Ulnung islands in the east and south seas, and on the tip of Janggi peninsula (Fig. 2) are included in the western border zone of the circum-Japan sea alkali-rock province. The province extends northward to the Baegdu volcanic mountain range in North Korea, and south and southeast to the northern coast of Kyushu and the inner coastal zone of the Japan island arc. The basaltic extrusion in the Chugaryong rift-valley is also related to this period (Fig. 3).

On Jeju Island, five eruptive cycles were recognized by Won (1976). Each cycle began with basalt flows and ended with pyroclastic explosions followed by trachyte or trachyandesite eruptions. On Ulnung Island, Harumoto (1970) divided the volcanic sequence into five events: (1) extrusion of vast olivine-basalt flows, (2) emplacement of trachyte and phonolitic dikes, (3) extrusion of trachyte and phonolite lava flows from central vents, (4) intra-caldera eruptions of leucite-bearing trachyandesite, and (5) parasitic eruptions of trachyandesite. In Chugaryong rift-valley, three volcanic events have been recognized (D. S. Lee, 1977b). The activity there began in the Late Cretaceous with olivine-free basalt flows and andesitic to rhyolitic pyroclastic explosions, and concluded in the Quaternary with olivine-basalt flows which covered valley floors. The basalt sheets that, in part, intrude the Tertiary Gampo Formation, and the olivine basalt and pyroxene basalt flows on Janggi peninsula have been long recognized as products of Pleistocene volcanism (Tateiwa, 1924).

PLUTONIC ROCKS

Plutonic rocks, which underlie about 43 per cent of South Korea (Fig. 3), form two different types of bodies. To one type belong the elongate batholiths which trend parallel with the Sinian direction and occur mainly in the Gyonggi Massif, Ogcheon Zone, and Ryongnam Massif; to the other type belong stocks or dike-like plutons which are irregularly distributed in the Gyongsang Basin. The batholithic rocks are related to the Daebo Orogeny (Jurassic mainly) and are synorogenic or late orogenic whereas the latter type are related to block movements of the Bulgugsa Disturbance (Late Cretaceous to Early Tertiary). Most of the large plutons are granitic in composition but some of the stocks and dikes consist of intermediate to basic rocks.

Plutons in Gyonggi-Ryongnam Massic

Although most of the batholithic masses in the Gyonggi-Ryongnam Massif are Jurassic, several smaller bodies of basic to intermediate composition are pre-Jurassic. These include the gabbro-monzonite complexes, four of which are known in South Korea. One such stock is located east of Seoul and consists of hornblende gabbro, diorite, and porphyritic monzonite, thought to be a comagmatic series (Lee and Kim, 1974).

Also pre-Jurassic is a northerly elongated pluton of anorthosite in the Ryongnam Massif (Son and Cheong, 1972; oral communication, 1979). The anorthosite exhibits many complexities. Its contacts with the Precambrian metamorphic rocks are gradational and metasomatic. The anorthosite itself grades from a leucocratic phase in the central area, through lineated melanocratic rock to gabbro along the margin. The pluton is cut by diorite, schistose granite, and syenite, all of unknown age. Various replacement features and the paragenetic sequence in the anorthosite suggest that it is not only a product of simple fractional crystallization of basic magma, but also a product of metasomatism of the metamorphic and plutonic rocks into which the anorthositic melt (of unknown source) soaked.

The dominant Jurassic batholiths in the Gyonggi-Ryongnam Massif include both I-type and S-type granites (Chappell and White, 1974). The I-type is characterized by sharp contacts against the country rocks, whereas the S-type, found in the Ryongnam Massif, forms schistose hybrid complexes which exhibit gradational contacts. The Jurassic granitic masses consist chiefly of granodiorite but also include adamellite, granite, and tonalite. Chemical analyses

from 18 granitic and 54 intermediate and basic rocks show an average Peacock Index of 66 (Table 3), which falls in his calcic series.

Plutons in the Ogcheon Zone

The widely distributed plutonic bodies in the Ogcheon Zone consist of schistose and migmatitic granites along the southeast marginal zone, batholithic granite masses along the northwest margin, and irregular-shaped granitic stocks in the central part of the zone. Also, acidic hypabyssal intrusives, acidic to intermediate paleovolcanics, and intermediate to ultrabasic plutonic bodies are exposed in the central part. Diorite and gabbro plutons are also present, and peridotite-nodule-bearing basalt intrudes metasediments of the Ogcheon Group.

The plutons in the Ogcheon Zone are grouped into three subzones (Fig. 4). The SE-subzone comprises those along southeast margin; the NW-subzone, those along the northwest margin; and the C-subzone, those of the central area (D. S. Lee, 1971, 1977a). This subdivision represents not only a geographic grouping but also reflects a grouping of tectonic features.

SE-subzone. The plutons in this subzone consist mostly of migmatite and schistose granitic rock which are closely related to the Precambrian metamorphic rocks. Their margins commonly grade into porphyroblastic gneiss or biotite gneiss, and also into mylonite. All samples of the plutonic rock contain microcline, and in composition, most are adamellite to granodiorite (Fig. 5).

Major mineral components of the migmatite and schistose granite are plagioclase, microcline and quartz with minor biotite and epidote, and rare accessory sphene and tourmaline. Myrmekitic texture is fairly abundant between microcline and plagioclase, and also between other minerals. Modal and chemical compositions are much more varied than those of granitic rocks in other subzones.

The major bodies in this subzone yielded isotopic ages ranging from 148 to 181 Ma (Fig. 3). More recent data (Yun, 1978), however, indicate that in the northeast migmatization may have taken place in Carboniferous time.

NW-subzone. This subzone includes batholithic rocks ranging from gabbro, through tonalite and granodiorite, to adamellite (Fig. 6). Crosscutting relationships indicate that the order listed is also an age sequence, which might be accounted for by magmatic differentiation. The complex of plutons is elongate and parallel with the general trend of the Ogcheon Zone but partly cuts across the folded metasediments of the C-subzone. This relationship indicates that the plutons of the batholithic complex were emplaced by forceful injection in the later stages of the Daebo Orogeny.

The gabbro forms a small stock intruding metasediments midway along the southeastern margin of this subzone. Major mineral constituents are cummingtonite and plagio-

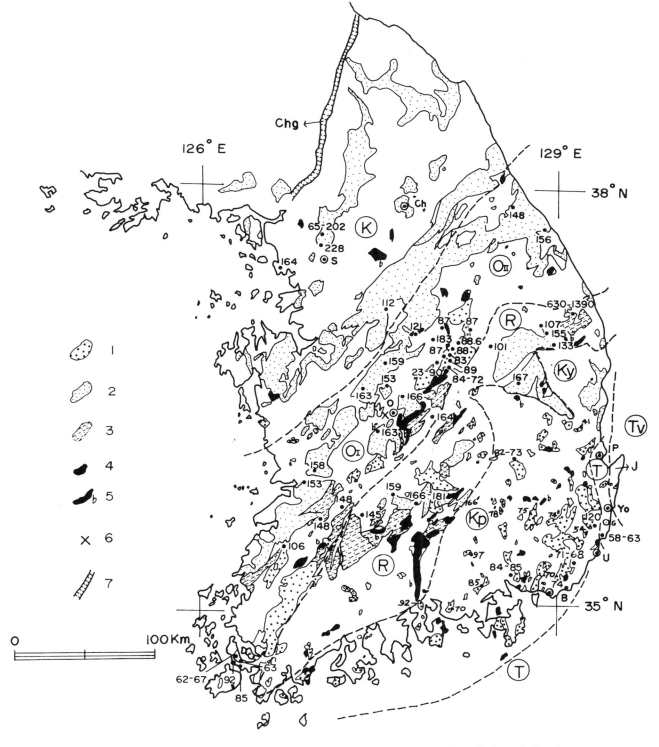

Figure 3. Distribution of plutonic rocks within tectonic provinces and isotopic dates in South Korea. Circled letters are the same as those of Fig. 2. Chg: Chugaryong rift-valley, J: Janggi peninsula, S: Seoul, Ch: Chuncheon, O: Ogcheon, P: Pohang, Yo: Yonil, G: Gampo, U: Ulsan, B: Busan. Numbers in granite bodies indicate isotopic age. 1: Late Cret.-Paleogene granite, 2: Jurassic-Early Cretaceous granite, 3: Jurassic schistose granite, 4: Intermediate rock, 5: Mafic and meta-mafic rocks, 6: Peridotite, 7: Rift valley. (After Geological Survey of Korea, 1972-1973.)

TABLE 3.   PETROCHEMICAL PROPERTIES OF IGNEOUS ROCK ASSEMBLAGES IN S. KOREA*

Province	Alkali-lime index(% SiO_2)	Type of differentia-tion trend	Degree of contam. with pelitic rocks	K_2O : Na_2O	
Gyonggi-Ryongnam	66	World calc alkali rock series	very high	$K_2O < Na_2O$	gr.[†] 18 oth.[§] 54
Ogcheon	62	World calc alkali rock series	very high	$K_2O \fallingdotseq Na_2O$	gr. 77 oth. 55
Gyongsang	58	Karroo dolerite	high	$K_2O < Na_2O$	gr. 58 oth. 20
Alk. rock ⌈Jeju	48	Mull volcano	rather high	$K_2O < Na_2O$	oth. 31
Ulnung	50	Mull volcano	very low	$K_2O < Na_2O$	oth. 25
⌊Chuga-ryong	50 ?	Alkali olivine basalt plateau	low	$K_2O < Na_2O$	oth. 31

*D. S. Lee, 1977b.
[†]gr. : granitic rock.
[§]oth. : other igneous rocks.

clase accompanied by magnetite and apatite. Euhedral plagioclase crystals are commonly included in large, euhedral to subhedral cummingtonite crystals.

Exposures of the tonalite masses are mostly along the southeastern margin of this subzone. Essential mineral constituents of the rock are plagioclase, quartz, hornblende, and biotite, accompanied by apatite, zircon, magnetite, and interstitial calcite.

In spite of its huge mass, the composition of the granodi-

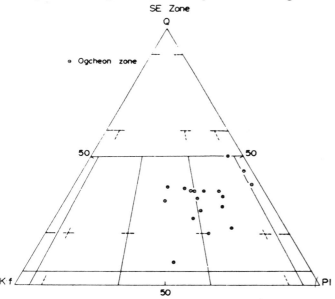

Figure 5. Modal composition of granitic rocks in SE-subzone, Ogcheon zone (D. S. Lee, 1971). Dashed lines indicate the boundaries of IUGS classification.

orite-adamellite complex is not particularly variable. It is mainly granodiorite to adamellite, although along the southeast margin it grades into muscovite granodiorite, two-mica adamellite, and two-mica granite. All samples from this subzone contain microcline rather than orthoclase or perthite, and myrmekite is rare. The mineral assemblage distinguishes it from the younger granites of the C-subzone. Some samples from the middle of this subzone contain corroded garnet grains, and others have remarkable amounts of sphene. The muscovite granite commonly contains tourmaline crystals.

The triangular plots from modal analyses of the granites from this subzone and C-subzone (Fig. 6) suggest a serial variation in mineral compositions (D. S. Lee, 1971, 1977a).

The isotope ages of 10 samples from this subzone range from 164 to 112 Ma, indicating a Middle Jurassic to Early Cretaceous age.

C-subzone. Outcrops of granitic rock are rare in the northeastern part of the C-zone where much of the region is covered by a thick, Paleozoic limestone sequence. The granitic rocks share the southwestern part of the subzone with extensive Cretaceous volcanic rocks.

The plutons in the C-subzone can be classified on the basis of isotopic age into an older group of Jurassic granodiorite-adamellite bodies and a younger group of Cretaceous to Early Tertiary adamellite-granite masses. The former have ages ranging from 166 to 163 Ma and the latter, from 109 to 62 Ma (Kim and Kim, 1978b).

The rocks of the older group are dominantly granodiorite and subordinately adamellite. Their mineral constituents are quartz, microcline and/or microperthite, plagioclase

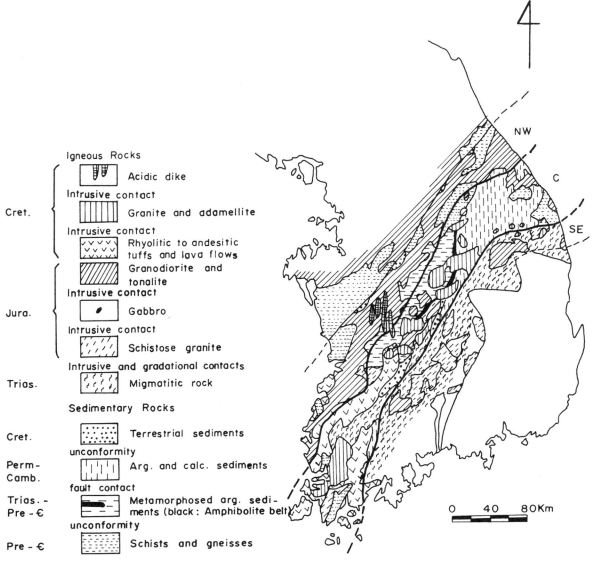

Figure 4. Zoning of Ogcheon Orogenic Zone (after S. A. Lee, 1977a).

and commonly biotite, with minor epidote, chlorite and sericite.

The rocks of the younger group are generally massive, fine- to coarse-grained, pink to gray, and miarolitic in places. Most are more leucocratic than those in other subzones. The potash feldspar is chiefly perthite (cryptoperthite or microperthite); microcline is, with few exceptions, relatively rare. Usually the amount of myrmekite is less than in other rock types, and some masses contain rather abundant muscovite. In the C-subzone, orthoclase is rare throughout and microcline or microperthite predominates. The older group contains sphene and epidote as in the SE- and NW-subzones, while the younger group lacks these accessories. The plutons of the older group resemble those of the NW-subzone in geologic age and lithology. The plutons of the younger

group in the C-subzone are completely different from those in other subzones in age and lithology.

An elongate exposure of so-called "green rock" in the central part of the Ogcheon Zone has been known since 1927 when Shimamura identified it as hornblendite. According to Lee and Woo (1970), the green rock is a complex of several varieties of amphibolite and chlorite schist which were derived from calcareous and ferruginous sediments, and basic intrusives by regional metamorphism. Further work in the area revealed a sequence of hornblende-gabbro amphibolite, basalt and andesite (in part metamorphosed), and green schist, that is overlain exclusively by cherty pelagic sediments. The basic rocks are aligned along a great thrust fault which separates the paleogeosynclinal zone from the neogeosynclinal zone (Kim and Kim, 1976). Comparing the

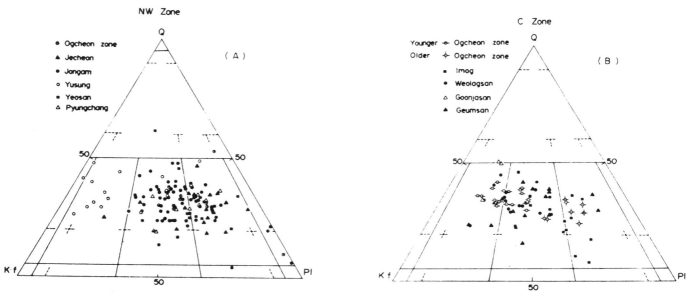

Figure 6. Modal compositions of granites in Ogcheon Zone. (A): NW-subzone, (B): C-subzone. (D. S. Lee, 1971; Park et al., 1977; Cho, 1977; Kim et al., 1978b; D. S. Lee et al., 1978). Dashed lines indicate the boundaries of IUGS classification.

Ogcheon basic rock suite to the ophiolite suites in other parts of the world, Kim and Kim (1976) suggested that the basic rock suite may be an ophiolite marking the site of an ancient geosuture.

Plutons in Gyongsang Province

The Gyongsang province is a Cretaceous basin of sedimentary and igneous rocks in the southeastern part of the Korean peninsula. The basin originated along the southeastern periphery of the Ryongnam Massif in Early Cretaceous time after the Jurassic Daebo Orogeny.

In the basin, thick fluvial and lacustrine sediments were deposited during several vertical crustal movements that affected the basin probably until the early Paleogene. Neogene marine and terrestrial sedimentary formations were laid down upon the Cretaceous sequences only in isolated, small Tertiary basins along the east coast, and tuff and andesitic basalt flows covered the Neogene sediments during Plio-Pleistocene time.

The Gyongsang Basin covers approximately 24,000 km², and is about 200 km from north to south and about 120 km from east to west. It is subdivided into three minor basins by two major tectonic lineaments, the Cheongsong Ridge and the Palgongsan Uplift zone (Won and others, 1978). The minor basins are, from north to south, Yeongyang, Uiseong, and Yoocheon (Fig. 7a).

Son (1969a, 1969b), Won (1968), and Won and others (1978) dealt mainly with the igneous activity in the basin as a whole, and Cha (1976) and others studied petrographically the igneous rocks in the southern part of Yoocheon minor

basin. Won proposed that the igneous activity produced a sequence of phases derived from a single comagmatic source. But the relationship between the geologic sequences of the Gyongsang system and igneous activity differs from place to place as his summary shows (Table 4).

In Yeongyang minor basin, only the first stage, volcanism, is well represented; in the Yoocheon minor basin, all four stages of igneous activity are represented; and in Uiseong minor basin, dikes and late volcanism are prominent.

The investigators revealed again that plutonism, especially granitic emplacement, was spatially and temporally related to major faulting and warping caused by block movements during the Bulgugsa Disturbance. The granitic plutons are aligned along major fault systems trending N60W and northerly (Fig. 7-b), but the block movements continued after magma consolidation. The sequence of the two fault systems has not been determined, but some geologists believe that the N60W system is the older (Won and others, 1978).

The granite forms stocks and small batholiths, but their mechanism of emplacement varies considerably depending on tectonic position and size. Forceful injection or doming are predominant for the stocks located along tectonic lines in the north and east border areas of the basin. They are characterized by a monolithic composition. In contrast, geometric forms, attributable to magmatic stoping, are common in the batholiths located in the central part of the basin; they are characterized by polylithic phases. The intrusions are estimated to have penetrated upward to within 1 to 3 km of the surface (Cha, 1976). The plutons consist dominantly of granite, adamellite, and quartz monzonite, but of diorite and gabbro in places. Generally,

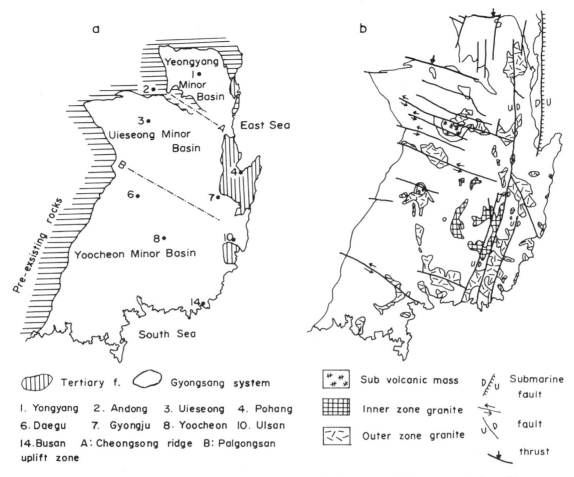

Figure 7. Subdivision, tectonics, and igneous plutons in Gyongsang Basin (modified from Won and others, 1978).

quartz monzonite and diorite form marginal facies of granitic or granodioritic bodies.

The granitic plutons in Yoocheon minor basin are distributed roughly in inner and outer concentric zones. The eastern border of the zonal arrangement is cut by the north-trending fault system (Fig. 7-b). Although the inner zone is polylithic and the outer zone monolithic, the mineralogy of both is similar, but generally more inequigranular and variable than the granites in the other minor basins. The marginal phase of single masses varies remarkably. For instance, a body in the inner zone consists of biotite hornblende granite but its margin contains monzonite, adamellite, monzonite porphyry, masanite (a kind of miarolitic alkali granite), hornblende granite and hornblende granite porphyry, biotite granite and biotite granite porphyry, micrographic porphyry, feldspar porphyry, and quartz porphyry. They grade into each other and locally into subvolcanic masses (Won, 1968). Granite plutons in the northern and eastern parts of the Gyongsang Basin, however, are relatively equigranular and homogeneous in composition. Modal compositions of some granitic masses in the

basin and the petrochemical properties based on the chemical analyses are shown in Figure 8 and Table 3, respectively. The isotope ages of the granites in the basin range from 41 to 120 Ma.

MINERALIZATION

After distinguishing Jurassic granites (Daebo) from Late Cretaceous granites (Bulgugsa), both of which were previously known as younger granite and thought to be Late Cretaceous, Kim (1971a, b) proposed four metallogenic epochs for South Korea, that is, Precambrian, Paleozoic, Jurassic to Early Cretaceous, and Late Cretaceous to Early Tertiary. The classification of the first two are independent of their relationship (if any) to plutonism and/or volcanism. The most important metallogenic epochs, however, are the last two, and both are associated with granites forming syntectonic plutons.

Belonging to the Jurassic-Early Cretaceous epoch are ten zones of gold and silver, six of lead and zinc, seven of tungsten and molybdenum, three of fluorite, and five of

TABLE 4.   RELATION BETWEEN MAJOR ROCK UNITS AND IGNEOUS ACTIVITY IN GYEONGSANG BASIN*

Rock Unit			Igneous Activity				Age
		Igneous Cycle		Yeongyang Minor Basin	Uiseong Minor Basin	Yoocheon Minor Basin	
GYEONGSANG SUPERGROUP	Silla Group	Acidic and intermed. dikes	Minor intrusions		few	common	some	Late Cretaceous
		Bulgugsa Granite	Plutonic intrusions		few	some	abundant	
		Acidic volcanics	Volcanism and sub-volcanic intrusion	4th stage	absent	few	common	
		Jayangsan Fm.		3rd stage (Jayangsan)	absent	few ?	abundant	
		Jain Formation		2nd stage (Chaeyagsan)	absent	some ?	common	
		Banyaewol Fm.						
		Haman Formation						
		Silla Conglomerate		1st stage (Hagbong)	abundant	few ?	some	
	Nagdong Group	Chilgog Fm.						Early Cretaceous
		Jinju Fm.						
		Hasandong Fm.						
		Nagdong Fm.						

*After Won and others, 1978.
Question mark indicates uncertain identification.

magnetite (Fig. 9a). Nearly all of the zones are aligned along the Sinian trend of the Daebo Granite.

Late Cretaceous-Early Tertiary metallogenic epochs are represented by various metallogenic zones, mostly restricted to the southern parts of the Korean peninsula. They comprise a gold zone, a manganese zone, four lead and zinc zones, two copper zones, two tungsten and molybdenum zones, four pyrophyllite zones, and two magnetite zones (Fig. 9b).

A few exceptions to the above generalization exist since the ages of some plutons that seem to be related to the ore genesis had not been determined when the 1971 classification was published. For example, the age of mineralization of the Shinyemi lead-zinc deposits (Kim and Kim, 1978a) is now known to be Late Cretaceous although they were previously grouped with the Jurassic-Early Cretaceous epoch.

Mineral deposits associated with volcanism in Korea are very scarce, and no stratabound sulfide nor Kuroko-type deposits are known. J. H. Lee (1968) described a native-copper deposit in the Yeongyang minor basin in southeastern Korea, which is associated with Cretaceous basalt flows. He thought that the copper mineralization was derived from the basalt itself. Kim and Kim (1974) described the Red Hill deposit of the Dongjeom copper mine, which was the first discovery of porphyry copper in Korea. Subsequent to this, and because certain pyrophyllite deposits were suspected as possibly overlying porphyry copper deposits, they were drilled in a few places in the southern parts of Korea but without success.

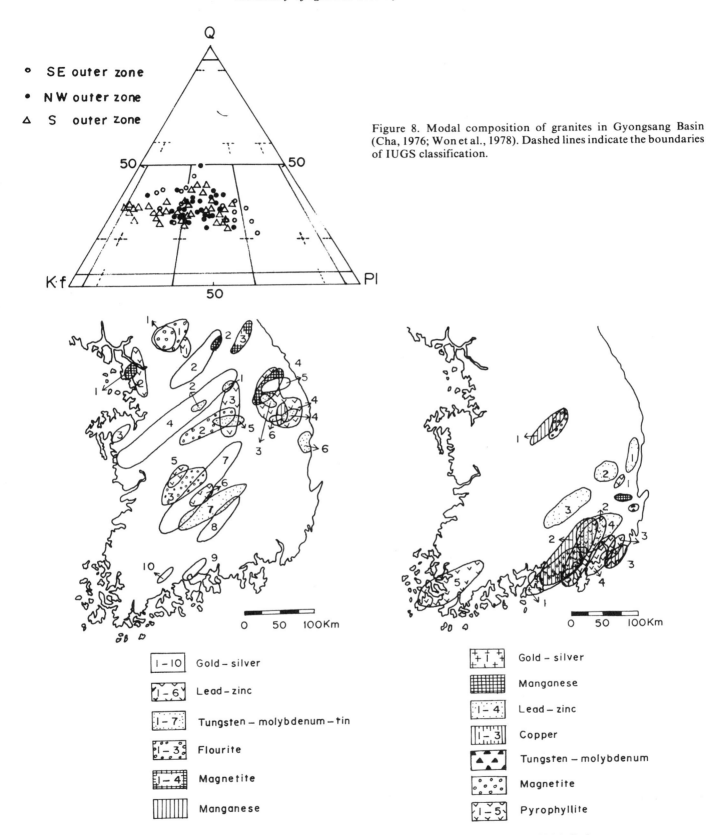

Figure 8. Modal composition of granites in Gyongsang Basin (Cha, 1976; Won et al., 1978). Dashed lines indicate the boundaries of IUGS classification.

Figure 9. Metallogenic provinces: A. Jurassic to Early Cretaceous (after Kim, 1976a). B. Late Cretaceous to Early Tertiary (after Kim, 1976a).

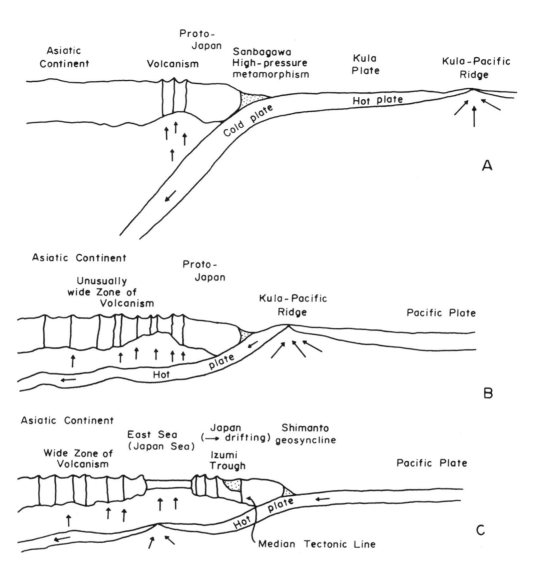

Figure 10. Schematic diagrams showing the supposed sequence of events related to the subduction of the Kula-Pacific Ridge (not drawn to scale) (after Uyeda and Miyashiro, 1974).
A. 120 m.y. ago. The ridge is approaching the Asiatic continent. The rapid underthrusting of the cold plate causes the Sanbagawa high-pressure metamorphism in the subduction zone with the Ryoke metamorphism on the continental side.
B. 90 m.y. ago. The ridge is so close to the continent that high-pressure metamorphism is no longer taking place. The light, hot Kula plate is underthrust with a very small dip. Its thermal effect reduces the thickness of the continental plate above and causes extensive volcanism.
C. 70 m.y. ago. The ridge is submerged beneath the continental plate. Its thermal effect further reduces the thickness of a part of the continental plate, which eventually is broken by tensional force. The oceanic-side fragment of the continental plate drifts away to form the Japanese Islands, leaving the newly opened Japan Sea behind. The same system of tensional force produces the Izumi Trough in Japan, where sandstone formations as thick as 10 km are deposited.

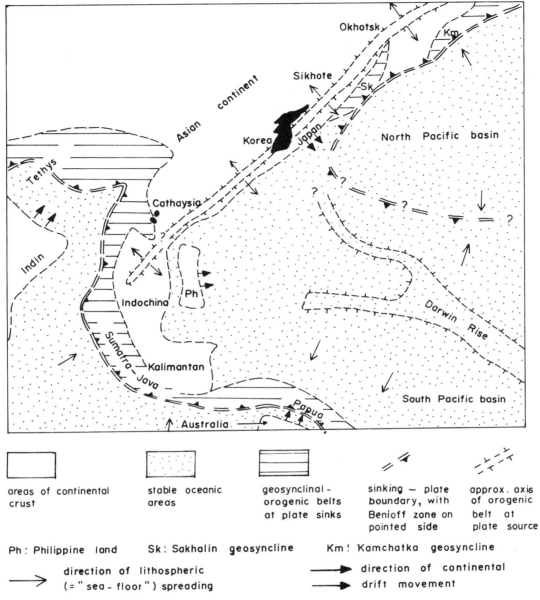

Figure 11. Possible distribution of lithospheric plates and related orogenic belt in the northwest Pacific region during Cretaceous time (modified from Workman, 1972).

IGNEOUS ACTIVITY RELATED TO PLATE TECTONICS

The principal geologic features of the Korean peninsula may be relevant to plate tectonics. The Daebo granites are distributed along the Sinian direction in the central part of the Korean peninsula, whereas the Bulgugsa granites have a random distribution at its southern end. Kim (1971) and Lee (1971) pointed out that the ages of granites in Korea are progressively younger toward the south. Triassic and older granites are known in a few places in North Korea; Jurassic to Early Cretaceous granites, in the central part of the

peninsula; and Late Cretaceous to Early Tertiary granites, in the south.

In the Ogcheon mobile zone, basic rocks such as gabbro, amphibolite, basalt, andesite, green schist, and peridotite-nodule-bearing basalt outcrop along the Bonghwajae thrust fault, which may represent a paleo-geosuture (Kim and Kim, 1976b). The associated sedimentary rocks that overlie the basic rocks comprise chert and other flysch sediments. Kim and others (1964) described basic rocks north of Mt. Jiri, but they lie in the Ryongnam Massif and are not connected with those in the Ogcheon Zone.

In the Cretaceous Gyongsang sedimentary basin, andesitic volcanic rocks and pyroclastics are widely distributed and some minor basalt flows are intercalated with Cretaceous sediments.

Late Tertiary to Pleistocene alkali basalts are distributed along the periphery of the peninsula and on the islands off the coast.

On the basis of the main geological features, one school postulates that during the Mesozoic a subduction zone was near or at the southern end of the peninsula, which is regarded as a continental margin. Uyeda and Miyashiro (1974) stated that the Kula plate was subducted beneath the Asiatic continent during the Early Cretaceous (120 m.y. ago, Fig. 10-a); the mid-oceanic Kula-Pacific ridge underthrust the Asiatic continent (90 m.y. ago) at a low angle causing a thermal rise which reduced the thickness of the continental plate above and caused extensive volcanism (Fig. 10-b); tensional forces on the overlying continental plate caused it to break, forming the Japanese island chain which drifted southward (about 70 m.y. ago) leaving the newly opened Sea of Japan behind (Fig. 10-c). Uyeda and Miyashiro estimated that the subduction of the Kula-Pacific ridge began approximately 80 to 90 m.y. ago and is generally thought to have continued until the Early Paleogene, based on the isotopic age determinations of Paleogene granites in Korea by Lee and Ueda (1976). Their interpretation might be extended by postulating that the Triassic to Early Cretaceous granites originated during the Kula-plate subduction, and Late Cretaceous to Paleogene granites originated during the Kula-Pacific ridge subduction.

S. M. Lee (1974) proposed on the bases of metamorphism, heat flow, and magmatism that the subduction zone migrated southward but was not specific as to time and place. Park and Kim (1971) and Park and So (1972) also discussed very briefly a metamorphic facies that they thought could be interpreted by plate subduction near the Korean peninsula.

Contrary to the idea of the plate subduction, Workman (1972) proposed that the Mesozoic granites of Korea may lie along a Mesozoic thermal rise and perhaps a lithospheric plate source. As shown in Figure 11, the Korean peninsula was located along the Okhotsk-Cathaysia belt which was regarded by Workman as an orogenic belt marking a spreading ridge, once connected with the Darwin Rise; but later it separated and formed a T-shaped or three-way junction. In order to explain the ages of granites in Korea which are progressively younger towards the south, Workman postulated the migration of the intrusive activity from the spreading ridge towards the Pacific basin through Mesozoic and Early Tertiary time.

ROMANIZATION OF LOCALITY NAMES

Busan = Pusan
Chosun = Josun = Joseon
Chugaryong = Chugaryeong = Choogaryong
Gemyongsan = Kemyongsan
Gisong = Kisong
Gosonri = Goseonri = Kosonri = Koseonri
Gyonggi = Gyeonggi = Kyonggi = Kyeonggi
Gyongsang = Gyeongsang = Kyongsang = Kyeongsang
Hambaek = Hambaeg
Jomchon = Jumcheon
Ogcheon = Okcheon = Okchon
Pyongan = Pyeongan
Ryongnam = Ryeongnam = Yeongnam
Sobaegsan = Sobaeksan
Taebaegsan = Taebaeksan
Yangdeog = Yangdeok
Yoncheon = Yeoncheon = Yonchon

REFERENCES CITED

Aoki, K., 1958, Petrology of alkali rocks from Iki Island and Higashimatsuura District (in Japanese with English abstract): Kazan, II, v. 3, p. 1–16.

Cha, M. S., 1976, Petrological study on the Bulgugsa acidic igneous rocks in Busan area (in Korean with English abstract): Journal of the Korean Institute of Mining Geology, v. 9, p. 93–97.

Chappell, B. W., and White, A. J. R., 1974, Two contrasting granite types: Pacific Geology, v. 8, p. 173–174.

Cheong, C. H., 1956, Outline of geology of Korea (in Korean): Bulletin of the Geological Survey of Korea, no. 1, p. 2–31.

Cho, S. H., 1977, Petrological study on the Jongam granites in the Ogcheon geosynclinal zone: Journal of the Korean Institute of Mining Geology, v. 10, p. 185–198.

Geological Survey of Korea, 1972, Isotope ages and geologic map of Korea: scale 1:2,000,000.

——, 1973, Geologic map of South Korea: scale 1:250,000.

Harumoto, A., 1970, Volcanic rocks and associated rocks of Utsuryoto Island (Japan Sea): Harumoto Memorial Edition, p. 1–29

Kim, O. J., 1968, Stratigraphy and tectonics of Ogcheon system in the area between Chungju and Munkyong (in Korean with English abstract): Journal of the Korean Institute of Mining Geology, v. 1, p. 35–46.

——, 1970, Geology and tectonics of the mid-central region of South Korea: Journal of the Korean Institute of Mining Geology, v. 2, p. 73–90.

——, 1971a, Study on the intrusion epochs of younger granites and their bearing on orogenesis in South Korea (in Korean with English abstract): Journal of the Korean Institute of Mining Geology, v. 4, p. 1–9.

——, 1971b, Metallogenic epochs and provinces of South Korea: Journal of the Geological Society of Korea, v. 7, p. 37–59.

——, 1973, The stratigraphy and geologic structure of the metamorphic complex in the northwestern area of the Gyeonggi Massif: Journal of the Korean Institute of Mining Geology, v. 6, p. 201–218.

——, 1976a, Mineral resources of Korea: American Association of Petroleum Geologists, Memoir 25, p. 440–447.

Kim, O. J., and Kim, K. H., 1974, A study on Red Hill copper deposits of the Dongjeom mine (in Korean with English abstract): Journal of the

Korean Institute of Mining Geology, v. 7, p. 157–174.

——,1976b, Petrochemical study on the basic rocks in the Okcheon Zone (in Korean with English abstract): Journal of the Korean Institute of Mining Geology, v. 9, p. 13–26.

——,1978b, On the genesis of the ore deposits of Yemi District in the Taebacksan metallogenic province: Journal of the National Scientific Research Institute , v. 2, p. 71–94.

Kim, O. J., Clark, A. H., and Farrar, E., 1978a, Age of the Sangdong tungsten deposit, Republic of Korea, and its bearing on the metallogeny of the southern Korean peninsula: Economic Geology, v. 73, p. 547–552.

Kim, O. J., Hong, M. S., and Park, H. I., 1964, Geologic atlas of Unbong quadrangle sheet: Geological Survey of Korea.

Kim, O. J., Hong, M. S., Park, H. I., and Kim, K. T., 1963, Geologic atlas of Sangeunri quadrangle sheet: Geological Survey of Korea.

Kobayashi, T., 1953, Geology of South Korea: Tokyo Univ. Japan, p. 155–167.

——,1957, Geology of South Korea: Geological and Mining Research, Far East, v. 1, p. 25–134.

Lee, D. S., Chang, K. H., and Lee, H. Y., 1972b, Discovery of Archaeocyatha from Hyangsanri dolomite formation of the Ogcheon system and its significance (in Korean with English abstract): Journal of the Geological Society of Korea, v. 8, p. 191–197.

Lee, D. S., and Kang, J. N., 1978, Study on the contact metamorphism of Weolagsan granite (in Korean with English abstract): Journal of the Korean Institute of Mining Geology, v. 11, p. 169–182.

Lee, D. S., and Kim, Y. J., 1974, Petrology and petrochemistry of the Yangpyeong igneous complex (in Korean with English Abstract): Journal of the Korean Institute of Mining Geology, v. 7, p. 123–152.

Lee, D. S., and Lee, H. Y., 1976, Geologic and geochemical study on the rock sequences containing oily materials in southwestern coast area of Korea (in Korean with English abstract): Journal of the Korean Institute of Mining Geology, v. 9, p. 45–74.

Lee, D. S., and Oh, M. S., 1972a, Petrological study on the volcanic rocks in the Unjangsan area (in Korean with English abstract): Journal of the Geological Society of Korea, v. 8, p. 129–155.

Lee, D. S., and Woo, K. Y., 1970, A study of basic metamorphic rocks in the Chungsan-Okcheon area (in Korean with English abstract): Journal of the Geological Society of Korea, v. 6, p. 29–52.

Lee, H. Y., 1980, Discovery of Silurian conodont fauna from South Korea (in Korean with English abstract): Journal of the Geological Society of Korea, v. 16, p. 114–123.

Lee, M. S., and Park, B. S., 1965, Geologic atlas of Hwanggangri quadrangle sheet: Geological Survey of Korea.

Lee, S. A., 1971, Study on the igneous activity in the middle Ogcheon geosynclinal zone, Korea: Journal of the Geological Society of Korea, v. 7, p. 153–216.

——,1977a, Granitic plutons and ore deposits in the Okcheon zone, Korea: Geological Society of Malaysia Bulletin 9, p. 99–116.

——,1977b, Chemical compositions of petrographic assemblages of igneous and related rocks in South Korea: in Nozawa, T., and Yamada, N. (eds.), Plutonism in relation to volcanism and metamorphism: International Geological Correlation Program, Circum-Pacific Plutonism Project, Toyama, Japan, p. 101–126.

——,1980, Igneous activity and geotectonic interpretation in the Ogcheon geosynclinal zone, Korea (in Korean with English abstract): Yonsei Nonchong, no. 17, p. 109–137.

Lee, S. M., 1974, The tectonic setting of Korea, with relation to plate tectonics: Journal of the Geological Society of Korea, v. 10, p. 25–36.

Lee, Y. J., 1980, Granitic rocks from the southern Gyeongsang Basin, southeastern Korea (Japanese with English abstract): Journal of the Japanese Association of Mining Petrologists and Economic Geologists, v. 75, p. 105–116.

Lee, Y. J., and Ueda, Y., 1976, K-Ar dating on granitic rocks from the Eonyang and the northwestern part of Ulsan-quadrangle, Gyeongsang-Namdo Korea (in Korean with English abstract): Journal of the Korean Institute of Mining Geology, v. 7, p. 123–152.

Lin, C., 1971, Outline of geotectonic problems in China mainland (translated into Japanese by K. Motojima): Bulletin of Oil Geology, Taiwan, no. 9, p. 227–238.

Min, K. D., et al., 1982, Applicability of plate tectonics to the post-Late Cretaceous igneous activities and mineralization in the southern part of South Korea (in Korean with English abstract): Journal of the Korean Institute of Mining Geology, v. 14, (in press).

Na, G. C., 1978, Regional metamorphism in the Gyeonggi Massif with comparative studies on the Yeoncheon and Ogcheon metamorphic belt (1): Journal of the Geological Society of Korea, v. 14, p. 195–211.

Park, B. K., and Kim, S. W., 1971, Recent tectonics in the Korean peninsula and sea floor spreading: Journal of the Korean Institute of Mining Geology, v. 4, p. 39–43.

Park, B. K., and So, C. S., 1972, The Ogcheon system in the central part of southern Korean peninsula as an ancient island arc: Journal of the Geological Society of Korea, v. 8, p. 198–210.

Park, H. I., Lee, J. D., and Cheong, J. G., 1977, Geologic atlas of Yuseong quadrangle sheet: Geological Survey of Korea.

Pumpelly, 1866, Geological Research In China, Mongolia And Japan: Smithsonian Contributions To Knowledge.

Reedman, A. J., and Flectcher, C. J. N., 1976, Tillites of the Ogcheon Group and their stratigraphic significance: Journal of the Geological Society of Korea, v. 12, p. 107–112.

Reedman, A. J., et. al., 1973, The geology of the Hwanggangri mining district, Republic of Korea: Anglo-Korean Mineral Exploration Group.

Sato, S., 1975, Recent studies on the geologic development of the Chinese continent: Tsikyukakaku, v. 29, p. 297–299.

Shimamura, S., 1927, Geological atlas of Chosen: Eito and Seizan sheets: Geological Survey of Korea.

Son, C. M., 1969a, Geology of Korea (in Korean): Gwangjin, v. 3, p. 13–14.

——,1969b, On the crustal movements in Korea (in Korean with English abstract): Journal of the Geological Society of Korea, v. 5, p. 167–210.

Son, C. M., and Cheong, J. G., 1972, On the origin of anorthosite in the area of Hadong, Sancheong, Gyeongsang-Namdo, Korea (in Korean with English abstract): Journal of the Korean Institute of Mining Geology, v. 5, p. 1–20.

Tateiwa, I., 1924, Geological atlas of Chosen: Yonil, Guryongpo and Choyang sheets: Geological Survey of Korea.

Uyeda, S., and Miyashiro, A., 1974, Plate tectonics and the Japanese islands: a synthesis: Geological Society of American Bulletin, v. 85, p. 1159–1170.

Won, C. K., 1968, The study on the Cretaceous igneous activities in the Gyeongsang Basin (1) (in Korean with English abstract): Journal of the Geological Society of Korea, v. 4, p. 215–236.

——,1976, Study of petrochemistry of volcanic rocks in Jeju Island (in Korean with English abstract): Journal of the Geological Society of Korea, v. 12, p. 207–226.

Won, C. K., Kang, P. C., and Lee, S. H., 1978, Study on the tectonic interpretation and igneous pluton in the Gyongsang Basin (in Korean with English abstract): Journal of the Geological Society of Korea, v. 3, p. 79–92.

Workman, D. R., 1972, The tectonic setting of the Mesozoic granites of Korea: Journal of the Geological Society of Korea, v. 8, p. 67–76.

Yun, S., 1978, Block tectonics of the Taebaegsan Basin and en echelon sedimentary wedges of the Yeonhwa-Ulchin district, mideastern South Korea: Journal of the Korean Institute of Mining Geology, v. 11, p. 127–141.

MANUSCRIPT ACCEPTED BY THE SOCIETY JULY 12, 1982

Geological Society of America
Memoir 159
1983

Felsic plutonism in Japan

Tamotsu Nozawa
CPPP National Committee of Japan
Geological Survey of Japan
1-1-3 Higashi, Yatabe, Ibaraki, 305
Japan

ABSTRACT

The Japanese granitic rocks cover 12 percent of the surface and consist mainly of granite and granodiorite. The plutonism that began in Silurian time and closed in the Miocene is subdivided into four episodes: Paleozoic, Triassic-Jurassic, Cretaceous-Paleogene, and Miocene. The location of the plutonism is inferred to have migrated oceanward in zones nearly parallel with the Honshu arc. This migration is in harmony with the migration of associated geosynclines.

Some of the granitic rocks are closely related to regional metamorphic rocks and others to volcanic rocks commonly represented by a vast volume of welded tuffs. The Cretaceous-Paleogene and Miocene granites are especially notable for their associated volcanic activity.

The major element chemistry of the granitic rocks is characterized by high SiO_2 content and relatively low Fe_2O_3, Na_2O, and K_2O. The initial $^{87}Sr/^{86}Sr$ ratio varies between 0.704 and 0.710 but averages about 0.705 or slightly higher. Biotites of Cretaceous-Paleogene granites yield $Mg/Mg+Fe^{2+}+Fe^{3+}+Mn$ values of between 0.2 and 0.6 and Si of around 5.5 on the basis of O,OH=24. Amphiboles of the same granites are mostly magnesio- or ferro-hornblende.

The Japanese granitic rocks can be grouped into magnetite series and ilmenite series. The two series seem to correspond roughly to areal variations in chemistry and mineralogy and also to the distribution of associated ore deposits.

INTRODUCTION

A vast volume of granitic rocks is exposed in the Japanese Islands. Granitic rocks including subordinate mafic plutonic rocks underlie about 12 percent of the surface of Japan (Fig. 1).

The plutonic rocks range from quartz diorite to granite, with granite and granodiorite* being dominant. They are characterized by high silica content compared with other granitic rocks in the circum-Pacific belt. Alkali rocks are rare.

Plutonism is inferred to have started in the Silurian and became most vigorous in Cretaceous-Paleogene time. The youngest plutonism is Miocene. Diverse in occurrence, the plutons form both concordant intrusions closely related to regional metamorphism and discordant intrusions commonly accompanied by volcanism.

Their radiometric ages and mineralogical and geochemical characteristics form zones which are in harmony with the geotectonic framework of the Japanese Islands.

The Japanese granitic rocks are considered to have played an important role in circum-Pacific granite plutonism. The predominant group, Cretaceous-Paleogene granites, is thought to be related to the late Yanshanian Orogeny on the Asiatic mainland and to the Nevadan-Laramide Orogeny in North America, though not strictly coeval.

---

* Classification of plutonic rocks is according to IUGS (Streckeisen, 1973) recommendations.

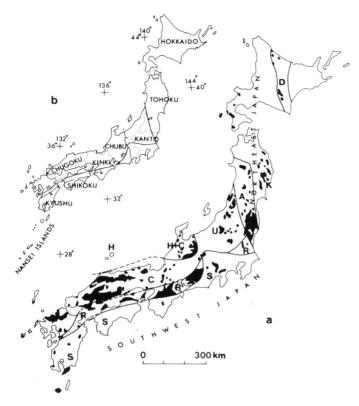

Figure 1a. Distribution and division of granitic rocks in Japan (Nozawa, 1977a). Abbreviations: D, Hidaka granitic province; K, Kitakami granitic province; A, Abukuma granitic province; U, Uetsu granitic province; H, Hida granitic province; C, Chugoku granitic province; R, Ryoke granitic province; S, Shimanto granitic province. The provinces of Green Tuff and Northwest Kyushu are excluded from the map. They are indicated on Fig. 3. b. Major tectonic lines and geographic division. Abbreviations: t, Tanakura Tectonic Line; i, Itoigawa-Shizuoka Tectonic Line; m, Median Tectonic Line; n, Dai (Nagato Structural Belt); k, Kurosegawa Tectonic Line.

CHRONOLOGICAL AND SPATIAL DEVELOPMENT

(i) Chronological and Spatial Division

The oldest radiometric ages from Japanese granitic rocks are Silurian (Fig. 2), but older plutonic rocks must have existed. Although not now exposed, pre-Silurian plutonic rocks are indicated by granite pebbles in the Silurian conglomerate which forms the lowest horizon of the Honshu geosyncline. Since then, plutonism repeatedly waxed and waned until the Miocene. The granitic rocks fall into four episodes, each of which has its own climax of plutonism (Table 1):

1. Paleozoic granites   430-250 m.y.
2. Triassic-Jurassic granites   240-180 m.y.
3. Cretaceous-Paleogene granites   130-35 m.y.
4. Miocene granites   30-5 m.y.

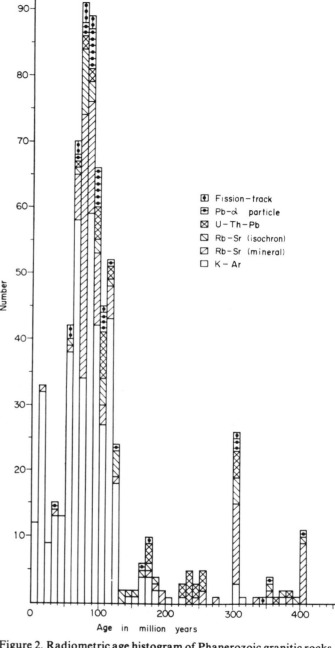

Figure 2. Radiometric age histogram of Phanerozoic granitic rocks in Japan (Nozawa, 1977a).

Spatially, the Japanese granitic rocks form six divisions, within each of which the plutons have similar age, petrography, and field relations (Fig. 3).

1. Hida granite division   Triassic-Jurassic
2. Honshu granite division   Cretaceous-Paleogene
3. Shimanto granite division   Miocene
4. Hidaka granite division   Miocene
5. Green Tuff granite division   Miocene

TABLE 1. GEOHISTORY OF MAIN GEOSYNCLINAL MOVEMENT AND FELSIC PLUTONISM IN JAPAN*

Province / Age	Southwest Japan — Inner Side		Southwest Japan — Outer Side			Northeast Japan (W ———— E)		Hokkaido	
	Hida	N ——— S	N	Middle	S	S. Kitakami and Abukuma	N. Kitakami — SW. Kitakami	Central	East
Cenozoic Quaternary									
Tertiary — Neogene		Green Tuff ++ ++			+ +++++ +	Green Tuff	+ ++	+ ++	Green Tuff
Tertiary — Paleogene		+ ++++++++ +++++++ +++			Shimanto			Yezo	Nemuro ?
Mesozoic Cretaceous						++++++++ ++++++++	+++++++ +++++++		
Jurassic		++++			?		Taro	Hidaka	
Triassic				Sambosan			Iwaizumi ?	?	
Paleozoic Permian				?		Honshu	?	?	
Carboniferous		Honshu				South Kitakami +	?		
Devonian		+ ++							
Silurian									
Pre-Silurian	++ Hida		Kuro-se-gawa			Abu-kuma			

.·.·. Geosyncline, ++++ Plutonism, ∼∼ Unconformity, /// Break in sequence, Blank area: Shallow sea or non-marine facies

*Geosynclinal movements are mainly after T. Yoshida, personal communication.

---

6. Northwest Kyushu granite
   division   Miocene

For convenience of description, a more detailed areal division is commonly made, namely, Hidaka, Kitakami, Abukuma, Uetsu, Hida, Chugoku, Ryoke, Shimanto, Green Tuff, and Northwest Kyushu (Fig. 1).

Among the four age groups, the Triassic-Jurassic, Cretaceous-Paleogene, and some of the Miocene granites have exclusive territory, whereas the Paleozoic granites have no definite territory and the Green Tuff and Northwest Kyushu granites, both Miocene, overlap other granite divisions.

The Paleozoic granites are small in total area, forming less than 1 percent of Japanese granitic rocks. With a few exceptions, the individual plutons are also small. Most are lenticular and are exposed along structural belts, such as

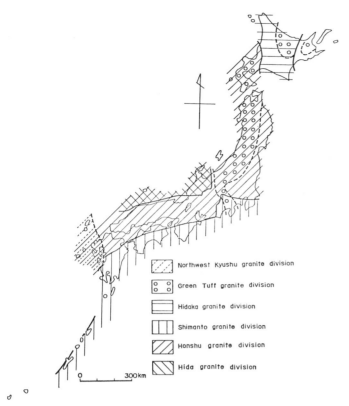

Figure 3. Major (petrographic division of granitic rocks in relation to major tectonic division of the Japanese Islands (Nozawa, 1977b).

Dai, Maizuru, Hikami, and Kurosegawa. Their radiometric ages indicate two climaxes of plutonism, at about 400 m.y. and 300 m.y. ago.

The Triassic-Jurassic granites are restricted to the Hida granite division and are called Funatsu Granitic Rocks. They form about 3 percent of the Japanese granitic rocks and constitute the oldest granitic mass of batholithic dimensions in the Japanese Islands. In most places, they are closely associated with the Hida Metamorphic Rocks. Their radiometric age indicates a definite climax at 180 m.y. ago.

The Cretaceous-Paleogene granites are the largest group, forming about 90 percent of Japanese granitic rocks, and are exposed in the Honshu granite division. In the Hida province, they overlap the Triassic-Jurassic granite division. Most form batholiths of various sizes. Some are closely related to low-pressure-type metamorphic rocks, and some are associated with vast volumes of felsic volcanic rocks. Their radiometric ages indicate climaxes of activity at three peaks, about 120, 90, and 60 m.y. ago.

The Miocene granites are exposed in four granite divisions, Shimanto, Hidaka, Green Tuff, and Northwest Kyushu. They are small in total, comprising about 7 percent of Japanese granitic rocks. They commonly form small masses and are scattered over wide areas. In the Shimanto division, they consist mostly of small stocks — rarely batholiths — and show a definite climax of activity at 13 m.y. ago. In the Hidaka division, most of them form discordant intrusions of Miocene age, but some of the Oligocene intrusions are partly concordant. All are intruded into the Hidaka Metamorphic Rocks. In the Green Tuff division, they are mostly small stocks widely disseminated over the Green Tuff terrane and have a close relationship with the volcanic rocks of the Green Tuff Formation. The radiometric ages of the granitic rocks of the Green Tuff division range widely from 18 to 5 m.y. In the Northwest Kyushu division, a few granitic masses are exposed, mostly small stocks cutting Tertiary strata. One yielded a radiometric age of 12 m.y.

(ii) Large-scale Oceanward Migration of Plutonism

If we delete from consideration some minor granite masses, that is, the Paleozoic granites and Miocene granites of the Northwest Kyushu and Green Tuff divisions, the zonal arrangement of ages of granitic rocks in the Japanese Islands is nearly parallel with the geotectonic zonal framework. The age zones have a distinct tendency to become younger toward the Pacific Ocean (Table 1).

In northern Japan, the older granites of the Honshu division, 130 to 60 m.y. old, are on the continental side; younger granites of the Hidaka division, 30 to 16 m.y. old, are on the Pacific side. Also in western Japan, the oldest granites of the Hida division, 240 to 180 m.y. old, are on the continental side; the youngest, those of the Shimanto division, 20 to 13 m.y. old, are on the Pacific side. Between these two, granites of the Honshu division of intermediate age, 120 to 35 m.y., occupy a large area.

The oceanward migration is in harmony with the migration direction of orogenesis in the Japanese Islands, which has evolved from continental side to Pacific side, namely, Hida Belt, Honshu Belt, Shimanto Belt, and Hidaka Belt.

(iii) Continent-ward Migration of Plutonism Inside the Honshu Granite Division

The largest division, Honshu, displays some internal variations that oppose the large-scale oceanward migration. In Southwest Japan, ages of granitic rocks in the southern half of the Chugoku region are 100 to 80 m.y. and decrease northward to 60 to 35 m.y. A similar tendency is also recognized in Northeast Japan in the Kitakami and Abukuma provinces where they decrease from 120 to 90 m.y. Though the reason for the continentward migration is not yet clear, the zoning of age is roughly in harmony with geochemical and mineralogical variations which will be stated in a later section.

(iv) Relation Between Plutonism and Orogenic Cycles

In the Hida division, plutonism took place mainly in the Early Jurassic, far later than deposition of the original sediments of the Hida Metamorphic Rocks, which is presumably pre-Permian. However, it corresponds chronologically to the latest phase of the Honshu geosynclinal sedimentation (Table 1).

In the Honshu division, the Honshu geosynclinal movement started in the late Paleozoic and continued until middle Mesozoic time. There, in Abukuma province, plutonism occurred several times in the Paleozoic and Mesozoic. Some of the Paleozoic bodies have a radiometric age of about 400 m.y. Paleozoic plutonism is nearly coeval with country rocks of Middle Paleozoic age and probably with the younger metamorphism of the Gosaisho-Takanuki Metamorphic Rocks (believed to have suffered polymetamorphism). Other granitic rocks of the Honshu division were emplaced in Early Cretaceous time, a little later than the end of the Honshu geosynclinal sedimentation, and are nearly coeval with sedimentation of the main part of the Shimanto Geosyncline. In the Ryoke province, the Honshu geosynclinal sediments were changed regionally to low-pressure-type metamorphic rocks. The metamorphism was accompanied by felsic plutonism.

In the Shimanto division, the radiometric age of granites, around 14 m.y., means that the plutonism took place near the end of the Shimanto geosynclinal sedimentation and the subsequent low-grade metamorphism.

In the Hidaka division, even the older part of the plutonism took place several tens of millions of years later than the end of the Hidaka geosynclinal sedimentation.

In the Green Tuff and Northwest Kyushu divisions, granitic rocks were emplaced during or not much later than the volcanism and sedimentation of Miocene age.

Emplacement of the Japanese granitic rocks is not always associated chronologically with the geosynclinal deposits in which they occur. However, in Southwest Japan, plutonism seems to have followed the migration of the geosynclines; that is, during plutonism in the Hida division, most of the adjacent Honshu Geosyncline was sinking, and when the plutonism in the Honshu division was active, the adjacent Shimanto Geosyncline was sinking.

PETROGRAPHY OF GRANITE GROUPS

(i) Pre-Silurian Granitic Rocks

As stated previously, the oldest granite mass actually exposed in Japan is of Silurian age. However, there is much evidence that pre-Silurian granitic mountains were once exposed (Minato and others, 1968). For instance, the Silurian sediments, the oldest fossiliferous formation in Japan, commonly contain granitic pebbles and arkosic sandstone. A Silurian-Devonian conglomerate exposed in the southern Kitakami region contains granitic pebbles, some of which are nearly 10 m across. They are irregular in shape and ill-sorted, and consist mostly of light-colored granitic rocks. The Silurian formation in the Kurosegawa Structural Belt contains arkosic sandstone which includes abundant fragments of potassium-feldspar megacrysts. The Devonian formations of the Abukuma Mountains and the Hida Marginal Belt also contain granitic pebbles.

The existence of a granitic layer under the main geosynclinal sediments is supported also by seismological investigations. Therefore, some granitic plutonism must have taken place prior to the Honshu geosynclinal movement, even though its products may be neither ubiquitous nor voluminous. At present, however, there is no direct evidence on whether the plutonism belongs to Precambrian or Caledonian orogeny.

(ii) Paleozoic granites

Most of the Paleozoic granites are distributed along tectonic lines, such as Kurosegawa, Dai, and others. They are usually small and lenticular except for the Hikami Granite and some older intrusives in Abukuma province. They are mostly granodiorite associated with tonalite, quartz diorite, diorite, and gabbro. They are also commonly associated with ultramafic rocks, metamorphic rocks, and Silurian-Devonian formations. Direct geological evidence of age is poor because the relationship with Paleozoic formations is rarely known. However, granitic rocks from Kurosegawa and Dai yield Silurian radiometric ages, and those from Hikami in the Kitakami Mountains yield Carboniferous ages.

They are commonly tectonically deformed, being partly mylonitized, and most are fault-bounded. Chemically, they are characterized by relatively high Na_2O and low K_2O content.

The Hikami Granite is an isolated mass in the Paleozoic terrane and is composed mainly of granite and granodiorite, mylonitized to some extent. It yields an Rb-Sr whole-rock isochron age of about 340 m.y.

In Abukuma province, two groups of granitic rocks are exposed. The older intrusives comprise mainly quartz diorite, tonalite, granodiorite, and a small amount of granite, and are commonly associated with gabbro and ultramafic rocks. They are intruded into the Gosaisho-Takanuki Metamorphic Rocks and have an Rb-Sr whole-rock isochron age of about 400 m.y. (Maruyama, 1978). The younger intrusives are discussed later under Cretaceous-Paleogene granites.

(iii) Triassic-Jurassic Granites

Triassic-Jurassic granites, called Funatsu Granitic Rocks collectively, are exposed in the Hida division (Nozawa, 1977c). They form large, concordant intrusive bodies that extend along the southern and northern rim of the main mass of the metamorphic rocks, partly enclosing them. They are closely associated with the Hida Metamorphic Rocks, having the same K-Ar age and close spatial distribution. In addition, small related bodies are distributed also inside the metamorphic terrane.

The Funatsu Granitic Rocks are roughly classified into two types: Shimonomoto type — tonalite and quartz diorite, and Funatsu type — granodiorite and granite. The two types occur in a close spatial relationship. The Shimonomoto type is not homogeneous. Although mainly tonalite, it becomes in a few places granodiorite as a result of permeation of microcline from the Funatsu type. It usually has weak foliation, commonly disrupted by agmatite which in a few places exhibits orbicular rocks. Small dioritic masses are often found in both types, especially in the Shimonomoto type.

Being intruded into the Shimonomoto type, the Funatsu type is younger and consists mainly of porphyritic granodiorite with megacrysts of microcline. By metasomatic processes, it commonly converted the Hida Metamorphic Rocks and also the Shimonomoto type to intermediate granitic rock. The Funatsu type is mylonitic to some extent, especially near the border where porphyritic granodiorite takes on the appearance of augen gneiss. Most of it is strongly altered and most of the biotite is chloritized.

The Funatsu Granitic Rocks commonly sharply cut or show strongly deformed contacts with the Carboniferous and probably Permian rocks. Contact effects are slight in most places. The Funatsu Granitic Rocks are covered by Jurassic conglomerate unconformably.

Their K-Ar and Rb-Sr ages are mainly concentrated near 180 m.y. However, K-Ar determinations on hornblende indicate 200 to 230 m.y. or even older ages.

Many synplutonic dikes of mafic, intermediate, and in a few cases, felsic nature are observed in the Funatsu Granitic Rocks as well as in the Hida Metamorphic Rocks. They include dikes, metamorphosed to various degrees, from basalt or andesite of effusive appearance, through low-grade amphibole schist, to amphibolite and hornblende gneiss.

(iv) Cretaceous-Paleogene Granites

A large quantity of felsic to intermediate plutonic and volcanic rocks, mostly of Cretaceous age, are distributed widely in the Honshu division, from southwest Hokkaido, through Honshu and northern Shikoku, to northern Kyushu. In Southwest Japan, they are developed abundantly to the west of Fossa Magna, but some of them extend to the east of Fossa Magna. In Northeast Japan, they are exposed largely in the southern part of the Kitakami and Abukuma provinces and sporadically in the northern parts of both provinces.

a. Kitakami province. In the Kitakami province, granitic rocks are exposed mainly in the Kitakami Mountains and commonly form masses of batholithic dimensions (Katada and others, 1974). They range from gabbro to granite but are mostly tonalite and granodiorite. Some are associated with Lower Cretaceous volcanic formations and are thought, therefore, to be of relatively high-level emplacement. Contact aureoles are widely developed and are characterized by sillimanite. Chemically, most of the granitic rocks are calcic in the Peacock's calc-alkali index, but alkali-gabbro and quartz monzonite occur in a few places.

b. Abukuma province. The younger intrusives of the Abukuma province are exposed mostly in the Abukuma Mountains. They form two facies. The earlier facies comprises mainly quartz diorite, tonalite, and granodiorite, intruding the Abukuma Metamorphic Rocks concordantly. The later facies is composed mainly of granite and granodiorite, forming discordant intrusions. Both facies have nearly the same K-Ar age of mid-Cretaceous, but Rb-Sr whole-rock isochron ages indicate Jurassic or older for the earlier facies and Early Cretaceous for the later facies.

To the north of the Abukuma Mountains, several masses of granitic rocks of the younger group are found, some associated with metamorphic rocks of unknown age.

c. Uetsu province. A large amount of granitic rock, mainly granite and granodiorite, is exposed in the Uetsu province. Some bodies are strongly mylonitic and others are gneissose. Some are closely related to thermally metamorphosed rhyolitic-andesitic volcanic rocks. Their K-Ar ages are mostly Late Cretaceous but rarely extend into the Paleogene.

In the northern Kanto region, several large masses of granitic rocks are exposed. Upper Cretaceous rhyolite divides the granitic rocks chronologically into an older granodiorite and quartz diorite, about 120 m.y. old, and a younger granite and granodiorite, about 70 to 60 m.y. old.

d. Chugoku province. To the north of Ryoke province, a series of high-level granitic rocks and felsic volcanic rocks is widely exposed.

In the Chubu region, granitic rocks are divided into two parts. The granitic rocks of the southern part are made up mainly of granite, 70 to 60 m.y. old, and are intruded into rhyolite. On the other hand, those of the northern part, also composed mainly of granite and some mafic intrusives, are 60 to 40 m.y. old and are closely associated with rhyolitic and andesitic volcanic rocks (Yamada, 1977). Similar granite-rhyolite association is found in the northern Kinki region on a relatively small scale.

In the Chugoku region, felsic intrusive and extrusive rocks are widely exposed. The granitic rocks there are also divided into two zones (Murakami, 1974). The southern zone contains a large batholith, mostly granite 90 to 75 m.y. old, associated with Upper Cretaceous volcanic rocks. Small bodies of quartz syenite are found in this batholith and are considered to have been formed by alkali-metasomatism in a pneumatolytic-hydrothermal stage. The northern zone contains granitic masses of batholithic dimensions and sporadic, minor, high-level intrusive masses. The batholithic masses are composed mainly of granite and granodiorite, 65 to 40 m.y. old. The minor intrusives, however, are highly variable, porphyritic granodiorite, quartz diorite, granophyre, porphyrite, etc. of both pre- and post- batholithic age (the latter about 40 m.y. old). Most of the post-batholithic intrusives are closely related to andesite-rhyolite volcanic rocks surrounded by large ring fractures suggesting cauldron subsidence.

In northern Kyushu, three groups are recognized. The first, schistose granodiorite and tonalite, is intruded into the Sangun Metamorphic Rocks concordantly. The second, massive granodiorite and tonalite, is intruded into Lower Cretaceous formations discordantly. The last and youngest is made up mainly of massive granite, which is intruded into the other two groups and into Paleozoic formations, and is unconformably overlain by Paleogene formations. All three groups have similar K-Ar ages of 95 to 80 m.y.

e. Ryoke province.

In the Ryoke province, various granitic rocks form many intrusive masses, ranging from quartz diorite to granite, accompanied by subordinate gabbro and cortlandite. They are divided into Older Granitic Rocks and Younger Granitic Rocks, demarcated by the Ryoke metamorphism (Hayama and Yamada, 1977). The Older Granitic Rocks are intruded concordantly in the higher grade part of the Ryoke metamorphic terrane, are mostly foliated, and comprise granodiorite, tonalite, quartz diorite, and granite. On the other hand, the Younger Granitic Rocks are intruded discordantly in the lower grade part of the metamorphic terrane and are mostly free from foliation. They comprise granodiorite and granite. Some of the Younger Granitic Rocks are intruded into the Cretaceous volcanic rocks, mainly rhyolitic welded tuffs, and probably are of high-level emplacement.

Their K-Ar ages are mostly Late Cretaceous but some Rb-Sr ages are Early Cretaceous.

(v) Miocene Granites

Miocene granites are exposed in four divisions. Two of them, Green Tuff and Northwest Kyushu, are, in general, high-level plutons, associated with volcanic rocks. Another one, Hidaka division, is related to Mesozoic and Tertiary orogeny and is accompanied by violent tectonic movement and metamorphism. The last, the Shimanto division, is mostly restricted to the Outer Zone of Southwest Japan. The Green Tuff and Northwest Kyushu divisions overlap

the older divisions. The Hidaka and Shimanto divisions have their own territory.

a. Hidaka division. In the Hidaka Metamorphic Belt of Hokkaido, there are exposed granitic rocks associated with the Hidaka orogeny (Minato and others, 1968). The western part of the Hidaka Belt contains considerable gabbro and the eastern part comprises mostly granodiorite and granite. Some are cordierite-bearing tonalites, enveloped by gneissose rocks. They are migmatitic and are considered to be products of anatexis. The gabbros, probably the earlier facies, are usually foliated. The later facies form granodiorite- and granite- discordant intrusions.

(b) Green Tuff division. Various sized granitic bodies occur sporadically throughout the Green Tuff terrane, in Hokkaido, in the western part of northern Honshu, in the northern part of Southwest Japan, and in the Fossa Magna region. In most areas they are quartz diorite, tonalite, and granodiorite, but in some places, they include granite. In the northern Fossa Magna region, small masses of diorite are found, while in the southern Fossa Magna, relatively large masses of tonalite, granodiorite, and granite are exposed.

They are not only closely associated with the Green Tuff volcanic rocks but also, in places, grade into granodiorite porphyry, quartz prophyry, and even into rhyolite toward their margins. Some are much richer in Na_2O than in K_2O.

(c) Shimanto division. In the Outer Zone of Southwest Japan, various sized discordant granitic masses of the Shimanto division are exposed (Oba, 1977). They are intruded into rocks of the Outer Zone of Southwest Japan, namely, the Shimanto and Chichibu Groups and Sambagawa Metamorphic Rocks. They comprise mostly granite and granodiorite, and in a few cases, include a tonalitic facies. Some are closely related to felsic and intermediate volcanic rocks, but none is related to regional metamorphism. Some have hypabyssal or effusive facies near their top or margin. Gabbro and diorite are also distributed in this division but are not everywhere accompanied by felsic intrusives.

Several large masses contain roof pendants of the host rocks and also aplitic or pegmatitic facies at their tops. Thus, it is inferred that the tops of the masses are exposed at the present ground surface. Chemically, they are characterized by abundant K_2O. Their K-Ar ages are concentrated at 13 m.y. throughout the Shimanto division. The narrow range of age and similarity of petrography and chemistry are remarkable features of the granitic rocks of the Shimanto division.

On the Nansei and Ryukyu Islands, granitic rocks ranging from granite to quartz diorite are exposed sporadically. Some have similar K-Ar ages with those in the Shimanto division, but others have older ages (Eocene). It is not certain whether they belong to the Shimanto division, although they have similar petrographic characters.

(d) Northwest Kyushu division. In the Northwest Kyushu region, including nearby offshore islands scattered small masses of granite and granodiorite commonly have margins that grade into granite porphyry or quartz porphyry. Rarely, small bodies of diorite and gabbro are also found. These granitic rocks are intruded into Tertiary rocks whose upper part is rich in volcanic materials. One K-Ar age is 12 m.y..

VOLCANISM RELATED TO GRANITE PLUTONISM

Direct relationship between granite plutonism and volcanism is demonstrated mainly by the Cretaceous-Paleogene and Miocene granites. There is little evidence that the other two granite groups, Paleozoic and Triassic-Jurassic granites formed high-level intrusions. Their association with felsic volcanic rocks is supposed only indirectly. In the case of the Paleozoic granites, Silurian formations into which the Silurian granites are intruded are characteristically made up partly of felsic volcanic rocks, including some welded tuffs. As for the Triassic-Jurassic granites, the Jurassic conglomerate covering the Triassic-Jurassic granites contains abundant pebbles of felsic volcanic rocks thought to be derived from the upper part of the Triassic-Jurassic granites, although their origin is not definitely known.

The largest granite group, the Cretaceous-Paleogene granites, is accompanied by huge amounts of felsic volcanic rocks evident in most exposures from Hokkaido to Kyushu. The greater part of the volcano-plutonic association is exposed in the western part of the Honshu division. There, plutonism and volcanism took place in one continuous period of igneous activity (Ichikawa and others, 1968). In the western Honshu division, it started with eruption of andesitic rocks in the late Early Cretaceous in western Chugoku and northern Kyushu, and was accompanied by intrusion of porphyrite, diorite, and granodiorite on a small scale. In the early Late Cretaceous, eruption of andesite to rhyolite and minor intrusions of gabbro to granite took place in Chugoku. Then followed large-scale eruptions of pyroclastic flows, such as the Nohi Rhyolite (Yamada and others, 1971), which spread over Chugoku, Chubu, and even some parts of northern Japan. This stage also included major intrusions of granite and granite porphyry. Subsequently, in the middle Late Cretaceous, granite to granodiorite batholiths were intruded into the northern Kyushu, Chugoku, Kinki, and Chubu regions. Then small-scale eruptions of either andesite or rhyolite occurred at numerous localities. In the Paleogene, there followed a large-scale intrusion of granite. In the final stage, Oligocene granitic rocks were emplaced crosscutting those of Cretaceous to Paleogene age, mainly in northern Chugoku. Thus, through the entire history of volcano-plutonic activity, the main phase of plutonism followed the main phase of volcanism.

The volcanic activity is thought to have taken place mostly on land and differs from that associated with geosynclinal movement. It produced mostly felsic rocks with a subordinate amount of andesite and often formed large pyroclastic piles, as for example, the Nohi Ryolite, which has a volume of more than 16,000 km³.

Most of the Miocene granites are high level and are associated with volcanic extrusions, except those of the Hidaka division. The Miocene volcanic rocks of the Green Tuff Formation occupy a vast area from Hokkaido to north Chugoku. They include basalt, andesite, dacite, and rhyolite, which are on average more mafic than the volcanic rocks associated with the Cretaceous-Paleogene granites. The granitic rocks of the Green Tuff division yield radiometric ages from 20 to 5 m.y. and intrude the volcanic rocks. The Miocene formations in the Green Tuff region have been subjected to extensive zeolite-facies metamorphism and in some areas, steep geothermal gradients resulted in prehnite-pumpellyite to greenschist-facies metamorphism. In these more highly metamorphosed areas, the granitic rocks range from gabbro to granite, among which quartz diorite, tonalite, and granodiorite are predominant, and are on average, more mafic than the Cretaceous-Paleogene granites, as are the associated volcanic rocks.

The Miocene granites of the Shimanto division are also high-level intrusions and are mainly granodiorite and granite. They are commonly associated with felsic volcanic rocks (Aramaki and others, 1977). The volcanic rocks form large bodies in some places, as in west Shikoku and central Kyushu. The volcanic rocks are of andesite-dacite-rhyolite affinity, and some contain large volumes of welded tuff. Other granite masses in this division are associated with small amounts of volcanic rock, mainly felsic and hypabyssal. Although a few batholithic granite masses are not associated with volcanic effusives, most of the granites and felsic volcanic rocks in the Shimanto division are inferred to have resulted from large-scale, volcano-plutonic activity (rhyolite-granite) in the Miocene, within a relatively narrow time range centered around 13 m.y. ago. The large-scale, Miocene felsic volcanism in the Shimanto division is inferred to have taken place mostly on dry land.

In the Northwest Kyushu division also, the Miocene granites are associated with felsic volcanic rocks. They are granite porphyry, quartz porphyry, and granophyre.

Overall, in Japan two large volcano-plutonic episodes of the rhyolite-granite type are recorded, one of Cretaceous-Paleogene age in the Honshu division, and the other of Miocene age in the Shimanto division. The older episode in the Honshu division produced a much larger volume of rock and spanned a much longer time than that in the Shimanto division. The older episode formed one of the important members of the granite-rhyolite series which characteristically fringes the Asiatic continental margin, and so, it is correlated with the igneous activity of the Yanshanian Orogeny, Bulgugsa granite plutonism, and Okhotsk volcanic activity.

CHEMISTRY AND MINERALOGY

Studies of the geochemistry and mineralogy of individual plutons have been in progress for many years, but recent studies cover broader areas of the granitic terranes in various regions. Now it is possible to draw general conclusions on the regional variation of chemical and mineralogical characters of the granites.

(i) Major Element Chemistry

A statistical study of about 1,500 analyses of Japanese granitic rocks (Table 2, and Aramaki and Nozawa, 1978) shows that regardless of their age, mode of occurrence, petrography, and mineralogy, most Japanese granitic rocks fall within a relatively narrow range of chemical variation. When plotted on Thornton-Tuttle's variation diagrams, a remarkable straight-line regression results for the major elements (Fig. 4). Projections on the Q-Or-Ab-An system show good concentration of points within well-defined provinces (Fig. 5).

In comparison with Washington's 5,000 igneous rocks of the world, Japanese granitic rocks are generally slightly higher in SiO_2 and lower in Fe_2O_3, Na_2O, and K_2O contents. Most of the variation trends on the Thornton-Tuttle's diagram are similar to other circum-Pacific granitic rocks, except for the relatively high SiO_2 content. When plotted on the Q-Or-Ab-An tetrahedron, the Japanese granitic rocks cluster near the center of the basal plane Q-Or-Ab, extending towards the middle of the Ab-An edge as An content increases. The center of concentration of points for the An-poor granitic rocks of Japan is significantly lower in Or than that for similar granitic rocks from the Washington table.

The areal and chronological variation of major element chemistry of the Japanese granitic rocks is small. However, minor variations in trend appear among large granite groups. For instance, in Southwest Japan major granite divisions have a zonal arrangement from the continental side to the Pacific side, formed by Hida division of Triassic-Jurassic age, Chugoku and Ryoke province in Honshu division of Cretaceous-Paleogene age, and Shimanto division of Miocene age. The following elements show a parallel gradual change, from the continental side to the Pacific side. Fe_2O_3 displays an oceanward decrease, as shown by the average values 1.30, 1.11, 0.70, 0.75 percent with a little irregularity in the Ryoke division. FeO increases oceanward 2.13, 2.28, 2.34, 2.89 percent, as does MgO, 1.21, 1.30, 0.90, 1.43 percent, with a little irregularity in the Ryoke division. CaO decreases oceanward 3.33, 3.11, 3.19, 2.75 percent, also with a little irregularity in the Ryoke division. K_2O increases oceanward 2.84, 2.85, 3.04, 3.56 percent, while Na_2O decreases 3.97, 3.66, 3.30, 3.23 percent. The reason for these variation trends is not understood.

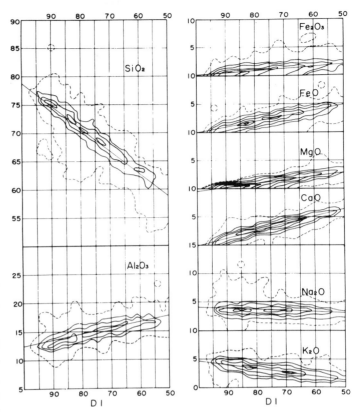

Figure 4. Contour diagrams showing distribution of points for the Japanese granitic rocks in the variation diagrams of major chemical elements (Aramaki and others, 1977b). Rectangular counting area 0.02 percent wide is used. All analyses fall within the outermost contours (dashed line). In all the diagrams, the second contours (the outermost solid lines) indicate the area containing points more than 0.5 percent of the total in a 0.02 percent area. Similarly, the third contour delineates the area with more than 1 percent density, the fourth more than 2 percent, the fifth more than 3 percent, and the sixth more than 4 percent. In the silica diagram, the innermost solid line indicates a density of 1.5 percent (instead of 2 percent) in a 0.02 percent area. Straight lines indicate linear regression by least squares.

(ii) Isotopic Chemistry

The initial $^{87}Sr/^{86}Sr$ ratio of most Japanese granitic rocks is relatively low and ranges from 0.704 to 0.710. A diversity of Sr-initial ratios is apparent in each region and seems to be independent of the diversity of rock type. The ratio in the Hida division is about 0.706; in the Kitakami province of the Honshu division, about 0.704; in the Abukuma province, 0.704 to 0.706; and in other provinces of the Honshu division, 0.705 to 0.710. Within the Honshu division, it is remarkably high in the Ryoke province, 0.707 to 0.710, but low, about 0.705, in northern Chugoku and Kyushu in the Chugoku province. The Miocene granites of the Shimanto and Hidaka divisions are relatively high (0.706 to 0.707). In places this areal variation is not in good harmony with the areal variation in Cenozoic volcanic rocks.

T. Nozawa

TABLE 2. AVERAGE CHEMICAL COMPOSITIONS OF JAPANESE GRANITIC ROCKS*

					Average, hydrous						
DI	100	95	90	85	80	75	70	65	60	55	50
SiO_2	78.75	76.76	74.77	72.78	70.78	68.79	66.80	64.81	62.82	60.83	58.83
Al_2O_3	12.50	13.02	13.53	14.05	14.57	15.09	15.61	16.12	16.64	17.16	17.68
TiO_2	(0.00)	0.06	0.14	0.23	0.31	0.39	0.47	0.56	0.64	0.72	0.80
Fe_2O_3	0.25	0.42	0.59	0.76	0.93	1.10	1.27	1.44	1.61	1.78	1.95
FeO	(0.00)	0.44	0.92	1.41	1.89	2.38	2.86	3.35	3.83	4.32	4.80
MnO	0.03	0.03	0.04	0.05	0.06	0.07	0.07	0.08	0.09	0.10	0.11
MgO	(0.00)	(0.00)	0.20	0.54	0.89	1.23	1.57	1.91	2.26	2.60	2.94
CaO	(0.00)	0.51	1.17	1.82	2.47	3.13	3.78	4.43	5.09	5.74	6.39
Na_2O	4.07	3.97	3.87	3.78	3.68	3.58	3.49	3.39	3.29	3.19	3.10
K_2O	4.50	4.18	3.87	3.55	3.24	2.92	2.61	2.29	1.98	1.66	1.35
Total	(100.03)	(99.38)	99.09	98.95	98.81	98.66	98.52	98.38	98.23	98.09	97.95
					CIPW norm						
Q	37.91	36.40	34.19	31.77	29.36	26.94	24.52	22.10	19.77	17.45	15.13
C	0.94	1.03	0.86	0.69	0.53	0.36	0.19	0.02	-	-	-
or	26.55	24.69	22.83	20.97	19.11	17.25	15.39	13.53	11.67	9.81	7.95
ab	34.41	33.59	32.77	31.94	31.12	30.30	29.47	28.65	27.83	27.01	26.18
an	(0.00)	2.54	5.78	9.02	12.26	15.50	18.74	21.98	24.81	27.60	30.37
wo	-	-	-	-	-	-	-	-	0.17	0.36	0.55
en	(0.00)	(0.00)	0.50	1.35	2.20	3.06	3.91	4.77	5.62	6.47	7.33
fs	-	0.42	1.05	1.68	2.30	2.94	3.56	4.19	4.82	5.45	6.08
mt	0.08	0.60	0.85	1.10	1.34	1.59	1.84	2.08	2.33	2.58	2.83
il	(0.00)	0.11	0.27	0.43	0.58	0.74	0.90	1.06	1.21	1.37	1.53
hm	0.19	-	-	-	-	-	-	-	-	-	-
					Average, anhydrous						
SiO_2	78.69	77.24	75.45	73.55	71.64	69.72	67.80	65.88	63.95	62.01	60.06
Al_2O_3	12.49	13.10	13.66	14.20	14.75	15.29	15.84	16.39	16.94	17.50	18.05
TiO_2	(0.00)	0.06	0.14	0.23	0.31	0.40	0.48	0.57	0.65	0.74	0.82
Fe_2O_3	0.25	0.42	0.59	0.76	0.94	1.11	1.29	1.46	1.64	1.81	1.99
FeO	(0.00)	0.44	0.93	1.42	1.91	2.41	2.91	3.40	3.90	4.40	4.90
MnO	0.03	0.03	0.04	0.05	0.06	0.07	0.08	0.08	0.09	0.10	0.11
MgO	(0.00)	(0.00)	0.20	0.55	0.90	1.25	1.60	1.95	2.30	2.65	3.01
CaO	(0.00)	0.52	1.18	1.84	2.50	3.17	3.84	4.50	5.18	5.85	6.53
Na_2O	4.07	4.00	3.91	3.82	3.72	3.63	3.54	3.44	3.35	3.26	3.16
K_2O	4.49	4.21	3.90	3.59	3.27	2.96	2.64	2.33	2.01	1.69	1.37
Total	100.00	100.00	100.00	100.00	100.00	100.00	100.00	100.00	100.00	100.00	100.00
					CIPW norm						
Q	37.88	36.62	34.51	32.11	29.72	27.30	24.89	22.46	20.13	17.79	15.45
C	0.94	1.04	0.87	0.70	0.53	0.36	0.19	0.02	-	-	-
or	26.53	24.85	23.04	21.20	19.34	17.49	15.62	13.75	11.88	10.00	8.11
ab	34.38	33.80	33.06	32.28	31.49	30.71	29.91	29.13	28.32	27.53	26.73
an	(0.00)	2.55	5.83	9.12	12.41	15.71	19.02	22.34	25.26	28.13	31.01
wo	-	-	-	-	-	-	-	-	0.17	0.37	0.56
en	(0.00)	(0.00)	0.50	1.36	2.23	3.10	3.97	4.84	5.72	6.60	7.48
fs	-	0.42	1.06	1.70	2.33	2.98	3.62	4.26	4.91	5.56	6.21
mt	0.08	0.61	0.86	1.11	1.36	1.61	1.86	2.12	2.37	2.63	2.89
il	(0.00)	0.11	0.27	0.43	0.59	0.75	0.91	1.07	1.23	1.40	1.56
hm	0.19	-	-	-	-	-	-	-	-	-	-

*Aramaki and others, 1972.

In case the computed values become negative, they are indicated as (0.00). The second set of values is recalculated from the first to make the total of 10 oxides equal 100 percent. This may be considered to be roughly equal to anhydrous average compositions of the Japanese granitic rocks.

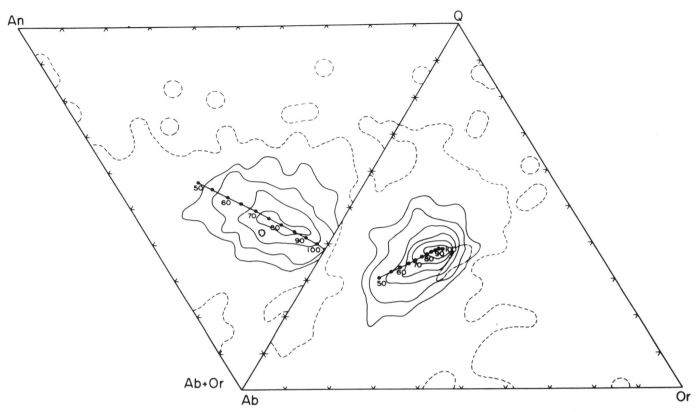

Figure 5. Contour diagram illustrating distribution of points for the Japanese granitic rocks in the Q-Or-Ab-An tetrahedron (Aramaki and others, 1978). All the points fall within the outermost contour (dashed lines). Solid contours more than 0.5, 1, 2, 3, 4 and 5 percent density in a 0.25 percent counter. Dash-dot curve is reproduced from the 6 to 7 percent contour in Fig. 42 of Tuttle and Bowen's paper (1958). Average compositions (Table 1) are shown as points connected by solid lines. Numbers indicate Differentiation Index.

On the basis of oxygen isotopes, the Upper Cretaceous to Paleogene granitic rocks are divided into two types. One has a $\sigma^{18}O$ range of 8.5 to 10.0 o/oo for quartz and the other, 11.5 to 14 o/oo. Genetic implications of this variation are discussed by Honma (in this volume).

Recent hydrogen isotopic study based on D/H and Fe/(Fe+Mg) distribution between coexisting biotite and hornblende also reveals two types of the Japanese granitic rocks, an equilibrium type and a disequilibrium type, as dealt with by Kuroda (in this volume).

The reasons for these areal variations are in dispute, and are usually attributed to either assimilation of crustal materials or heterogeneity of original materials in the mantle or crust.

(iii) Chemistry of Biotite and Hornblende

The chemistry of biotite and hornblende from the Japanese granitic rocks varies areally, roughly in accordance with their division into a magnetite series and an ilmenite series (Ishihara, 1977), and will be discussed later.

In biotites of the Cretaceous-Paleogene granites, mg-value ($= Mg/Mg+Fe^{2+}Fe3^{+}+Mn$) is between 0.2 and 0.6, and Si is around 5.5 on the basis of O,Oh=24. Biotites of magnetite series rocks from the Kitakami province and the northern Chugoku region are rich in Mg, poor in Al^{IV}, and rich in phlogopite molecule; those from the Ryoke division, however, are rich in Fe^{2+}, Al^{IV}, and also in siderophyllite molecule. Biotites from the Abukuma province are intermediate between the above-mentioned two. Biotites from the southern Chugoku region are rich in Fe and divergent in Al. Biotites from some parts of the Kitakami and Abukuma provinces and northern Chugoku region coexist with potassium feldspar and magnetite, and are inferred from their Fe^{3+}:Fe^{2+}:Mg ratio to have been formed under oxygen fugacity near the Ni-NiO buffer (Wones and Eugster, 1965) or a little higher. Biotites from the Ryoke division are thought to have been formed under oxygen fugacity near the Ni-NiO buffer or a little lower. Uncertainty exists because of the absence of coexisting magnetite, but low oxygen fugacity is suggested by the paucity of Mg and abundance of Fe^{2+}.

Most of the hornblendes belong to the magnesio- or ferro-hornblendes of Leake (1968). Like the biotites, hornblendes of the Kitakami province and northern Chugoku region are magnesio-hornblende, poor in Fe and Al^{IV}, whereas those of the Ryoke division are ferro-hornblende or ferro-tscher-makitic hornblende. Hornblendes of the Ryoke division are rich in both Al^{IV} and Al^{VI} and have about 1.5 Al^{IV} and an Mg-value of 0.18 to 0.46. In contrast, those of the Kitakami province and the northern Chugoku region are often devoid of Al^{IV} and have about 1.0 Al^{IV} replacing Si, and mg-value more than 0.43. Those of the Abukuma division are intermediate between the above-mentioned two.

The chemistry of biotite and hornblende from granitic rocks of the magnetite series of the Abukuma province and northern Chugoku region is similar to that from Mesozoic granitic rocks of the Sierra Nevada; on the other hand, that of the ilmenite series of the Ryoke division is similar not to Sierra Nevadan but to Adirondack granites.

MAGNETITE SERIES AND ILMENITE SERIES OF JAPANESE GRANITIC ROCKS

The Japanese granitic rocks can be grouped into the magnetite-series and ilmenite-series, which may be separated by the Ni-NiO buffer. Although the two series of granitic rocks imply relatively oxidized or reduced status of the solidifying magmas, the fO_2 differs two to three orders of magnitude (Czamanske and others, 1981) (Table 3).

TABLE 3. SELECTED CHEMICAL DATA (AVERAGE) OF MASSIVE GRANITOIDS OF THE MAGNETITE-SERIES AND ILMENITE-SERIES TERRANES*

Belt	DI(%)	F(ppm)	F/Cl	Sn(ppm)	$\sigma^{18}O(\%)$ At $SiO_2$74%	$^{87}/^{86}Sr_0$
Magnetite-series belt						
Kitakami belt	63(n=31)	271	1.5	1.2	8.5	0.7045
Sanin belt	81(n=66)	360	1.7	1.8	7.2	0.7062
Ilmenite-series belt						
Sanyo belt	82(n=50)	825	5.3	4.2	9.5	0.7084
Outer belt of southwest Japan	80(n=53)	679	4.3	4.0	10.7	0.7070

*Ishihara 1978.

The great variation in magnetite content of granitic rocks was first realized in studies of the Chugoku batholith (Ishihara, 1971). During the following years, Ishihara's group produced data on available assemblage and chemistry of the opaque oxides (Tsusue and Ishihara, 1974, 1975), and on the regional variation in magnetic susceptibility of granitic rocks (Kanaya and Ishihara, 1973), which is related to magnetite content. The magnetite-series of granitic rocks, in which magnetite (0.1 to 2.0 percent) can be observed by routine microscopic study, is characterized also by the presence of ilmenite, hematite, pyrite, chalcopyrite, sphene, and epidote, whereas the ilmenite-series is practically free of opaque oxides but may contain a very small amount of ilmenite (less than 0.1 vol. %) and rare pyrrhotite.

These two series do not seem to correlate well with the "I-type" and "S-type" of Chapell and White (1974). The rather rare occurrence of magnetite and sphene in the I-type (O'Neil and others, 1977) and the great overlap in the

Fe_2O_3/FeO ratio of I- and S-types (White and others, 1977) indicate that a large part of the I-type belongs to the ilmenite-series. Using chemical criteria for these two types, Takahashi and others (1980) revealed that almost all of Japanese granitic rocks fall into the I-type, but the criteria for magnetite-series and ilmenite-series divide Japanese granitic rocks into two groups, roughly equal in size (Fig. 6).

Distribution of the magnetite-series and ilmenite-series granitic rocks is related not to their age but more or less to the island arc-trench system in Japan. Overall, the magnetite-series predominates in the marginal-sea side, whereas the ilmenite-series appears to dominate in the oceanic side. This tendency is most prominent in the Miocene granitic belts, that is, granites of the Green Tuff terrane are of the magnetite-series whereas those of the Hidaka and Outer

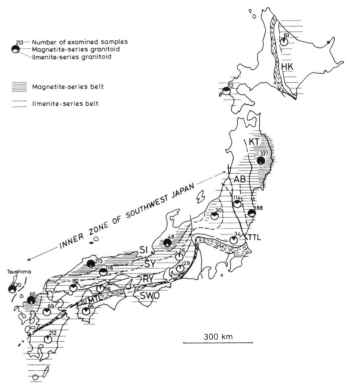

Figure 6. Distribution of the magnetite-series and ilmenite-series granitic rocks in Japan (Ishihara, 1978). Ratios of the two series in one tectonic unit or one area are shown in circles. The Inner Side of Southwest Japan is subdivided into northern Kyushu, Chugoku-Kinki, Chubu and Niigata-Kanto districts. Abbreviations: HK, Hidaka Belt (Tertiary); KT, Kitakami Belt (Early Cretaceous); AB, Abukuma Belt (Cretaceous and minor older rocks); RY, Ryoke Belt (Cretaceous and minor older rocks); SY, San-yo Belt (Cretaceous-Paleogene); SL, San-in Belt (Cretaceous-Paleogene); SWO, Outer Zone of southwest Japan (Miocene); TTL, Tanakura Tectonic Line; MTL, Median Tectonic Line; SM, Sambagawa Metamorphic Belt; KM, Kamuikotan Metamorphic Belt. Triassic-Jurassic Funatsu Granitic Rocks and Miocene granites of the Green Tuff terrane are not shown. Both consist of magnetite-series rocks. Granitic rocks of Tsushima are of Miocene and belong probably to an independent belt, Northwest Kyushu division.

Zone of Southwest Japan are mostly of the ilmenite-series. In Cretaceous granitic terranes, the ilmenite-series occurs mainly in the low-pressure-type metamorphic region of the Ryoke Belt. Thus the genesis of the two series of granitic rocks must also be connected with that of metamorphism and probably with faulting and other regional scale tectonic activity.

The two series of granitic rocks have similar assemblages of ferro-magnesian silicates, but muscovite-biotite granitic rocks with or without garnet are characteristic of the ilmenite-series. Since the total iron content of the two series is more or less the same, presence or absence of the opaque oxides reflects the Mg/Fe and Fe^{3+}/Fe^{2+} ratios of hornblende and biotite, which are summarized by S. Kanisawa (in this volume).

In the Neogene plutonic rocks, the magnetite series-rocks are more mafic than those of the ilmenite-series, which are characterized by higher SiO_2, K_2O, and other sialic components; but this difference is not obvious in the Late Cretaceous-Paleogene rocks. The same is true for their minor-element content; those known to accumulate with silica and potassium are generally more abundant in the ilmenite-series. Most significant in their chemistry is the

different variation patterns of the minor elements versus silica (or D.I.) (Fig. 7). Fluorine, Rb, Li, Pb, Sn, and Be, for example, do not increase with increasing silica in the magnetite-series but do increase with silica in the ilmenite-series. This difference was explained, in the case of tin, by the "Goldschmidt rule" which proposes differing valence status for this element in magmas, namely, Sn^4 and Sn^2 in the magnetite-series and ilmenite-series, respectively (Ishihara and Terashima, 1977; Ishihara, 1978). There is, however, also a possibility of volatile transport, because the ilmenite-series rocks contain a large amount of fluorine.

Interaction of granitic magma with sedimentary or metamorphic basement is commonly seen in the ilmenite-series terranes but not in the magnetite-series granitic rocks. The interaction was best shown by oxygen isotope studies in the Ryoke Metamorphic Belt (see Honma's summary in this volume). The ilmenite-series granitic rocks in nonmetamorphic terranes also have high $\sigma^{18}O$ values relative to those of the magnetite-series. Thus the contribution of crustal materials is obvious for the ilmenite-series, including organic carbon which acts as a reducing agent in the magma. However, a problem remains as to how much the crustal materials contributed relative to the original magma.

Distinct chemical difference between the two series (Table 3) and the paired occurrence in their regional tectonic setting indicate that the two types of magma are intrinsically different.

In their rock sulfur isotope composition, the magnetite-series has positive $\sigma^{34}S$ values from +1 to +9 o/oo, while the ilmenite-series is dominated by mainly negative values between −11 and +1 o/oo (Fig. 8) (Sasaki and Ishihara, 1979). Thus the ilmenite-series magma must have received biogenic sedimentary sulfur rich in ^{32}S. The sulfur-isotope

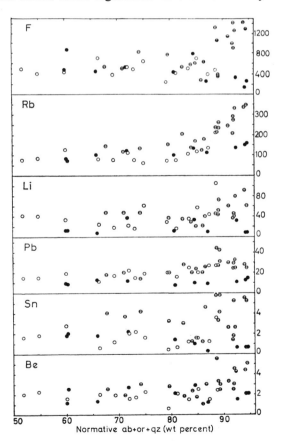

Figure 7. Minor elements (in ppm) plotted against D.I. of the Cretaceous-Paleogene granites across the Chubu transection (Ishihara and Terashima, 1977).

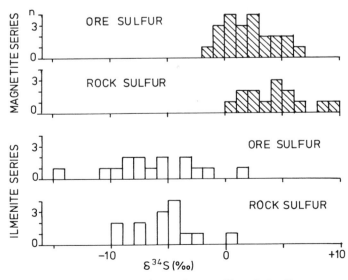

Figure 8. $\sigma^{34}S$ of rock sulfur and ore sulfur of the Cretaceous-Paleogene and Miocene granites and mineral deposits (Sasaki and Ishihara, 1980).

data on the magnetite-series granitic rocks can be grouped into one having $\sigma^{34}S$ about +1 to +2 o/oo and the other near +5 o/oo. The latter, heavy group is difficult to explain if the sulfur is simply derived from the upper mantle or lowest crust material. Hence introduction of certain heavy sulfur from seawater has necessarily combined with the magma. This heavy sulfur-isotope magma may have originated in altered oceanic crust in a subduction zone. This type of the magnetite-series granitic rocks is characteristic of Neogene plutonic belt. (Sasaki and Ishihara, 1979).

MINERALIZATION RELATED TO GRANITE PLUTONISM

Mineral deposits related to felsic magmatism occur in the volcano-plutonic terranes, but no important deposits are found in granitic belts where no comagmatic volcanism is known, such as the Hidaka Belt. Pegmatites, Mo, Sn, and W vein and skarn deposits, and base-metal skarn deposits are the best known examples of ore deposits of granitic affinity. It has been recently proposed, however, that many base-metal, Mn, and Au-Ag veins, and even the Kuroko deposits are genetically connected with the volcanic equivalents of hidden granitic bodies (Ishihara and Sasaki, 1978).

In the Cretaceous-Paleogene granitic belts, asymmetric ore-metal zoning is apparent, as is illustrated in Figure 9. In Southwest Japan, for example, the barren province which corresponds to the Ryoke Metamorphic Belt, is followed by the W-Sn-Cu province of the San-yo belt and by the Mo-Pb-Zn province in the San-in belt toward the marginal-sea side. Related granitic rocks in the W-Sn-Cu province are of the ilmenite series, whereas those of the Mo-Pb-Zn province belong to the magnetite series. Thus, the genesis of these metallogenic provinces is closely connected with two different source magmas of the granitic rocks (Fig. 9).

Hundreds of tungsten and base-metal deposits occur in the W-Sn-Cu province. The tungsten deposits are of quartz-vein type in noncalcareous host rocks and the skarn type where the host rocks are limestone. The base-metal deposits are skarn and vein types, and large deposits tend to have high Cu/Pb+Zn ratios. Pyrrhotite and not pyrite is abundant in the copper deposits.

The main ore mineral in the tungsten-quartz veins is wolframite, with scheelite commonly occurring as an accessory mineral; but scheelite is the main ore mineral at the Otani and Kaneuchi mines in the Kinki region. Some veins contain workable amounts of cassiterite and chalcopyrite. Molybdenite, Bi-minerals, beryl and iron sulphides occur in negligible amounts. The veins are surrounded by so-called greisen envelopes which consist of muscovite, quartz, and a little topaz. Li-micas, fluorite, and green biotite are accessories in the altered envelope. There are scattered fluorite deposits in the northern marginal zone of the W-Sn-Cu province.

Veins of the subvolcanic type in the Ikuno-Akenobe area are characterized by:

1. a polymetallic nature, with economic amounts of tin, tungsten, copper, lead, and zinc, and even gold and silver.
2. a multiple supply of the ore fluids.
3. apparently a reversed vertical zoning showing tin-tungsten zones at higher levels than base-metal zones. This has been explained by a hypothetical model of upheaval and withdrawal of a concealed granitic cupola.

In the Mo-Pb-Zn province, the Kamioka mine in the Hida province is famous. It is a lead-zinc mine but also contains workable amounts of tungsten, molybdenum, gold, and silver. Galena and sphalerite occur in the Mokuji skarn which consists of hedenbergite and some garnet and actinolite, and the Shiroji replacement ores, consisting mainly of quartz and calcite within calcareous horizons of the strongly folded Hida Metamorphic Rocks. Late Cretaceous porphyritic, aplitic monzogranite dikes are located in the center of the mining area. They contain Mo-bearing minerals.

Similar lead-zinc skarn deposits are known at Nakatatsu, and minor ones occur at other localities in the Hida province. Lead-zinc deposits in the rest of the Mo-Pb-Zn province are few, consisting of small vein and skarn types in the central part, but there are none in the western part. Thus, concentrations of lead and zinc in the province are limited to the area of the Hida Metamorphic Belt. Molybdenum deposits are widely scattered in the province, but the mineral assemblages are slightly different from one area to another. In the largest mining area of the central part, most molybdenite vein deposits consist solely of molybdenite and a little pyrite. No greisen minerals are present, but andalusite, potassium-feldspar, and sericite occur in the altered envelopes of the high-temperature types. Molybdenite deposits in this province are accompanied by accessory amounts of fluorite, and tungsten, tin and bismuth minerals, which are also common minerals in the tungsten deposits of the W-Sn-Cu province.

In northern Kyushu, small molybdenite veins contain approximately equal amounts of molybdenite and chalcopyrite.

A reverse zoning and absence of tin deposits are characteristics of Northeast Japan, herein defined as that part of Japan east of the Tanakura Tectonic Line (Fig. 9). In the W-Cu province, tungsten veins do not contain wolframite, cassiterite, or fluorite, but they contain scheelite and gold in a quartz matrix. Some veins have high gold/scheelite ratios. Magnetite, pyrrhotite, pyrite, chalcopyrite, and scheelite are the main ore minerals in the skarn-type deposits. The skarn minerals consist mainly of garnet, hedenbergite, diopside, and epidote. Wollastonite is characteristically absent from these skarns.

Along the easternmost zone of northeast Honshu, the

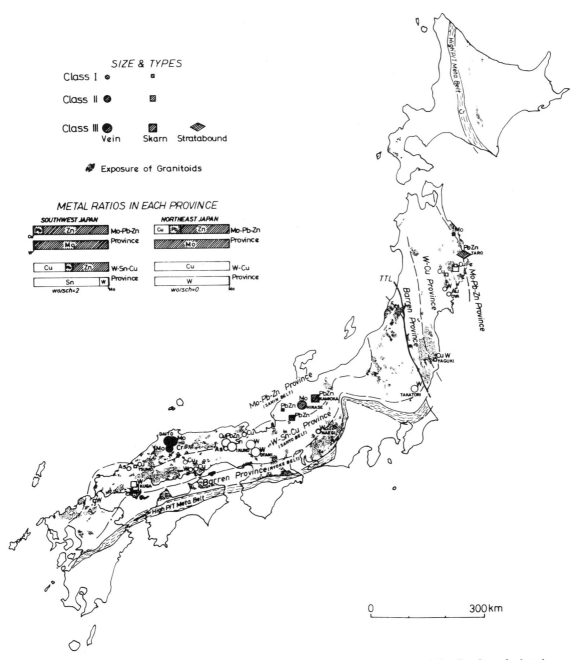

Figure 9. Geotectonic division of the Cretaceous-Paleogene granites and distribution of related ore deposits (Ishihara, 1978).
Size class I, II, and III: Mo, 0.1–0.5, 0.5–1, 1–3×10³ tons; W, 0.1–0.5, 0.5–3, 3–10×10³ tons; Sn 0.1–1, 1–20, 20–100×10³ tons; Cu and Pb+Zn, 10–100, 100–300, more than 300×10³ tons, respectively. Only major ones for As (1–10×10³ tons). For Au, see Fig. 11. Cr, major chromite deposits area. Abbreviations: wo, wolframite; sch, scheelite.

granitic rocks are associated with abundant volcanic rocks of nearly the same age, which may imply a higher level of erosion in the Mo-Pb-Zn province than in the W-Cu province. Molybdenite occurs in veins and skarns and is very locally disseminated in leucocratic monzogranite. Kuroko-type deposits are associated with the volcanic rocks

at the Taro mine. There, more than ten strata-bound deposits, consisting of chalcopyrite, galena, sphalerite, and pyrite, occur at restricted stratigraphic horizons in the Upper Jurassic to Lower Cretaceous Rikuchu Group. The host rocks are shale and sandstone alternating with andesitic to rhyolitic lavas and pyroclastics, all weakly metamor-

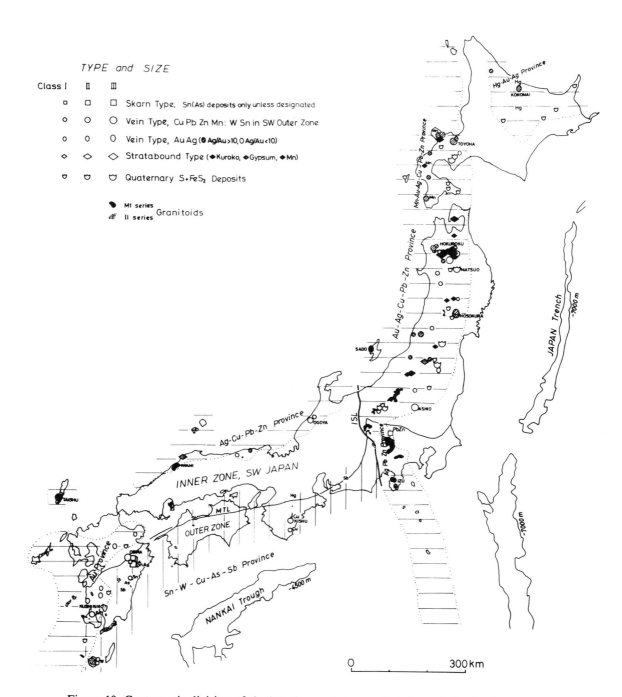

Figure 10. Geotectonic division of the late Cenozoic magmatic belts and distribution of the related ore deposits and Miocene granites (Ishihara, 1978).
Size class I, II, and III: Cu and Pb+Zn, 30–100, 100–500, 500–2000×10³ tons; Au, 10–20, 20–50, 50–100 tons; Mn, 30–300, 300–1000×10³ tons; S, 30–300, 300–1000, 1000–6000×10³ tons, respectively. For W and Sn, see Fig. 10. For the base- and precious-metal deposits, lead-zinc-rich ones (Cu/Pb+Zn<2) and silver-rich ones (Ag/Au>10) are shaded. Major antimony and mercury deposits are shown by chemical symbols.

phosed. Chalcopyrite-pyrite ore bodies tend to occur in the lower stratigraphic units, with galena-sphalerite-pyrite deposits in the upper stratigraphic units.

Two contrasting metallogenic provinces are distinct in the Miocene granitic belts, each of which corresponds to either the ilmenite-series or magnetite-series granitic belt. In the Outer Side of Southwest Japan, in the Shimanto division for example, the granitic rocks are of the ilmenite-series and the ore deposits are characterized by Sn, W, and Sb skarn and veins. The Green Tuff division of the magnetite-series, on the other hand, is famed for abundant base-metal ores, especially of Pb and Zn, and for Au-Ag and Mn mineralization (Fig. 10). Compared with the Cretaceous-Paleogene mineralization, the Neogene is rich in so-called low-temperature metals, such as Au-Ag, Mn, Sb, and Hg. However, consistent correlation between the magma series and the kind of ore metals observed in the Cenozoic-Mesozoic granitic terranes is distinct also in the Neogene. This correlation would rule out the recently common view on the origin of ore metals, which maintains that they were brought from shallow materials through a hydrothermal plumbing system of meteoric or sea water.

PLATE TECTONIC HYPOTHESIS AND FELSIC PLUTONISM IN JAPAN

The Japanese Islands are located on the triple junction of three plates, Asian, Pacific, and Philippine, but they are largely separated from the Asian plate by a back-arc basin, the Japan Sea.

Igneous activity in Japan has been studied with the plate tectonic concept in mind since the early 1970s. As previously mentioned, zonal variation in the character of the felsic plutonic rocks is roughly parallel with the arc-trench system. The zonal variations pertain to petrography, chemistry, mineralogy, oxygen fugacity, and oxygen and sulfur isotopes in the Cretaceous-Paleogene granites, and especially to the distribution of magnetite-series and ilmenite-series rocks. Recent study revealed that the Shimanto granites also have a zonation parallel with the arc-trench system in Japan. It is represented by an oceanward increase of SiO_2, K_2O, and normative corundum, and a decrease of Fe_2O_3/total FeO and CaO/total FeO.

Among several problems about which plate tectonic hypothesis has not yet given any satisfactory explanation is the vast volume of Cretaceous-Paleogene plutonic rocks of felsic nature with only rare mafic representatives, not only in Japan but also over a wide area of the Asiatic continental margin from Japan to mainland China. Also anomalous is the oceanward increase of K_2O in the Shimanto granites, and less clearly in the Cretaceous-Paleogene granites in Southwest Japan, which is disharmonious with the oceanward decrease of K_2O in Quaternary basaltic rocks.

One of the reasons that the petrographic and chemical characteristics of Japanese granitic rocks do not show a clearer directional trend is the presence of paired magnetite-series/ilmenite-series granitic rocks (Ishihara, 1977). Since the ilmenite-series rocks on the oceanic side appear to have thoroughly interacted with continental crust materials, the initial character of mafic magma is considered to have changed greatly.

The wide distribution of Mesozoic plutonic rocks was explained by the proposal of Uyeda and Miyashiro (1974) that the proto-Pacific plate was emplaced beneath these terranes by a hot and shallow-dipping subduction zone.

As for origin of the Japan Sea, the plate tectonic concept involves the opening of the Japan Sea probably in late Mesozoic to early Tertiary time. Some schools of Japan and USSR, however, maintain that the Japan Sea was originally underlain by continental crust but was changed later wholly or partly to oceanic crust, and if opening was a factor it was confined to a restricted area. The continuation of geology of Southwest Japan to that of the Korean peninsula has been another aspect of this problem and the entire matter remains hotly disputed. Although geohistory of the granitic and metamorphic rocks in Korea shows an oceanward migration of activities from Precambrian to Tertiary time as in Southwest Japan, geologic correlation between these two regions has not yet been established satisfactorily.

Contrary to plate tectonic hypothesis, some schools emphasize vertical movements as the fundamental mechanism for all geologic phenomena in the crust, and another school developed an earth-expanding hypothesis (Gorai, 1979). Although the plate tectonic hypothesis accounts for many geological phenomena in or near the Japanese Islands, numerous problems remain and await further research.

REFERENCES CITED

Aramaki, S., Hirayama, K., and Nozawa, T., 1972, Chemical composition of Japanese granites, Part 2, Variation trends and average composition of 1200 analyses: Journal of the Geological Society of Japan, v. 78, p. 39–49.

Aramaki, S., Takahashi, M., and Nozawa, T., 1977, Kumano acidic rocks and Okueyama Complex: Preprints Issue, 7th Circum-Pacific Plutonism Project Meeting, Toyama, Japan, p. 127–147.

Aramaki, S., and Nozawa, T., 1978, A reference book of chemical data for Japanese granites: Contribution from Geodynamic Project of Japan, 78–1, 88 p.

Chappell, B. S., and White, A. J. R., 1974, Two contrasting granite types: Pacific Geology, v. 8, p. 173–174.

Czamanske, G. K., Ishihara, S., and Atkin, S. A., 1981, Chemistry of rock-forming minerals of the Cretaceous-Paleogene batholith in southwestern Japan and implications for magma genesis: Journal of Geophysical Research, v. 86, B11, p. 10431–10469.

Gorai, M., 1978, Evolution of the earth; Expanding earth (in Japanese): Otsuki-Shoten, Tokyo, 224 p.

Hayama, Y., and Yamada, T., 1977, Ryoke metamorphic belt in the Komagane-Kashio district: in Yamada, N., ed., Mesozoic felsic igneous activity and related metamorphism in central Japan; Guidebook for Excursion 4, Geological Survey of Japan, p. 7–32.

Ichikawa, K., Murakame, N., Hase, A., and Wakatsume, K., 1968, Late Mesozoic igneous activity in the Inner Side of Southwest Japan: Pacific Geology, v. 1, p. 97–118.

Ishihara, S., 1971, Modal and chemical composition of the granitic rocks related to the major molybdenum and tungsten deposits in the Inner Side of Southwest Japan: Journal of the Geological Society of Japan, v. 77, p. 441–452.

——, 1977, The magnetite series and ilmenite series granitic rocks: Mining Geology, v. 27, p. 293–305.

Ishihara, S., and Terashima, S., 1977, Chemical variation of the Cretaceous granitoids across southwestern Japan; Shirakawa-Toki-Okazaki transection: Journal of the Geological Society of Japan, v. 83, p. 1–18.

——, 1978, Metallogenesis within the Japanese island-arc system: Journal of the Geological Society of London, v. 135, p. 389–406.

Ishihara, S., and Sasaki, A., 1978, Sulfur of Kuroko deposits; a deep seated origin: Mining Geology, v. 26, p. 361–367.

Kanaya, H., and Ishihara, S., 1973, Regional variation of magnetic susceptibility of the granitic rocks in Japan (in Japanese): Journal of the Japanese Association of Mining, Petroleum and Economic Geologists, v. 68, p. 211–224.

Katada, M., 1974, Granitic rocks in the southern Kitakami Mountains and zonal arrangement of granitic rocks in the entire Kitakami Mountains: in Cretaceous granitic rocks in the Kitakami Mountains: Geological Survey of Japan, Report no. 251, p. 121–133 (in Japanese with English abstract).

Leake, B. E., 1968, A catalog of analysed calciferous and subcalciferous amphiboles together with their nomenclature and associated minerals: Geological Society of America Special Paper, no. 98, 210 p.

Maruyama, T., 1978, Geochronological studies on granitic rocks distributed in the Gosaisho-Takanuki district, southern Abukuma Plateau, Japan: Journal of the Mining College, Akita University, ser. A, v. 5, no. 3, p. 53–102.

Minato, M., Gorai, M., and Funahashi, M., eds., 1968, The geologic development of the Japanese Islands: Tzukiji-Shoten, Tokyo, p. 224–232.

Murakami, N., 1974, Some problems concerning late Mesozoic to Early Tertiary igneous activity on the Inner Side of Southwest Japan: Pacific Geology, v. 8, p. 139–151.

Nakada, S., and Takahashi, M., 1979, Regional variation in the chemistry of the Miocene intermediate to felsic magmas in the Outer Zone and the Setouchi Province of Southwest Japan: Journal of the Geological Society of Japan, v. 85, p. 571–582.

Nozawa, T., 1977a, A review of the radiometric ages of the Japanese granitic rocks: Geological Society of Malaysia Bulletin, v. 9, p. 91–98.

——, 1977b, Plutonic rocks: in Tanaka, K., and Nozawa, T., eds., Geology and mineral resources of Japan, 3rd Edition, v. 1: Geological Survey of Japan, p. 282–289.

——, 1977c, Hida Belt and Hida Marginal Belt in central Hida Mountains:

in Yamada, N., ed., Mesozoic felsic igneous activity and related metamorphism in central Japan; Guidebook for Excursion 4: Geological Survey of Japan, p. 61–84.

Oba, N., 1977, Emplacement of granitic rocks in the Outer Zone of Southwest Japan and geological significance: Journal of Geology, v. 85, p. 383–393.

O'Neil, J. R., Shaw, S. E., and Flood, R. H., 1977, Oxygen and hydrogen isotopic compositions as indicators of granite genesis in the New England Batholith, Australia: Contributions to Mineralogy and Petrology, v. 62, p. 313–328.

Sasaki, A., and Ishihara, S., 1979, Sulfur isotope composition of the magnetite-series and ilmenite-series granitoids in Japan: Contributions to Mineralogy and Petrology, v. 68, p. 107–115.

——, 1980, Sulfur isotope characteristics of granitoids and related mineral deposits in Japan: Proceedings of the Fifth Quadrennial IAGOD Symposium, Volume 1, 1978, Stuttgart, E. Schweizerbart'sche Verlagsbuchhandlung, p. 325–335.

Takahashi, M., Aramaki, S., and Ishihara, S., 1980, Magnetite-type/ Ilmenite-type, I-type/S-type, granitoids: Mining Geology Special Issue, no. 8, p. 12–28.

Tsusue, A., and Ishihara, S., 1974, The iron-titanium oxides in the granitic rocks of Southwest Japan (in Japanese): Mining Geology, v. 24, p. 13–30.

——, 1975, Residual iron-sand deposits of Southwest Japan: Economic Geology, v. 70, p. 706–716.

Tuttle, O. F., and Bowen, N. L., 1958, Origin of granite in the light of experimental studies in the system $NaAlSi_3O_8$-$KAlSi_3O_8$-SiO_2-H_2O: Geological Society of America Memoir, v. 74, 153 p.

Uyeda, S., and Miyashiro, A., 1974, Plate tectonics and the Japanese Islands: A synthesis: Geological Society of America Bulletin, v. 85, p. 1159–1170.

White, A. J. R., Beams, S. D., and Cramer, J. J., 1977, Granitoid types and mineralization with special reference to tin: Preprints Issue, 7th Circum-Pacific Plutonism Project Meeting, Toyama, Japan, p. 89–100.

Wones, D. R., and Eugster, H. P., 1965, Stability of biotite; Experiment, theory, and application: American Journal of Science, v. 50, p. 1228–1272.

Yamada, N., 1977, Nohi Rhyolite and associated granitic rocks: in Yamada, N., ed., Mesozoic felsic igneous activity and related metamorphism in central Japan, Guidebook for Excursion 4: Geological Survey of Japan, p. 33–60.

Yamada, N., Kawada, K., and Morohashi, T., 1971, The Nohi Rhyolite and pyroclastic flow deposits (in Japanese): Chikyu Kagaku, v. 25, p. 25–88.

MANUSCRIPT ACCEPTED BY THE SOCIETY JULY 12, 1982

Geological Society of America
Memoir 159
1983

Hydrogen isotope study of Japanese granitic rocks

Yoshimasu Kuroda
Department of Geology, Shinshu University, Matsumoto, 390 Japan

Tetsuro Suzuoki
Marine Department, Meteorological Agency, Otemachi, Chiyoda-ku, Tokyo, 100 Japan

Sadao Matsuo
Department of Chemistry, Tokyo Institute of Technology, O-okayama, Meguro-ku, Tokyo, 152 Japan

ABSTRACT

Based on the D/H and Fe/(Fe + Mg) distribution between coexisting biotite and hornblende, Japanese granitic rocks can be divided into equilibrium and disequilibrium types (Kuroda and others, 1977a). In the former rocks, the relationship between biotite and hornblende in the σD-X_{Fe} plot follows the empirical equilibrium relationship found by Suzuoki and Epstein (1976), whereas in the latter rocks, it does not.

The origin of the two types can be interpreted as follows. The magma of the equilibrium type is oversaturated with water; i.e., it contains free hydrous fluid as well as dissolved water, and later, the free hydrous fluid interacts with the hydrous silicates as shown by the Suzuoki-Epstein experiment. In contrast, disequilibrium magma is undersaturated, and hydrous fluid does not separate but remains as dissolved water and is used up consecutively by the hydrous silicates (Kuroda et al., 1977a, 1978).

Most of the Japanese granitic rocks belong to the equilibrium type, and only in the Ryoke belt is the disequilibrium type found.

D/H FRACTIONATION AND Fe^{2+} IN COEXISTING BIOTITE AND HORNBLENDE

The authors were greatly surprised to find that the inclination of the tie-lines in the σD-X_{Fe}* plot for coexisting biotite and hornblende from the granites of the Kitakami mountainous district is equal to or nearly equal to –64, as shown in Figure 1. This inclination is exactly the same as that derived from Suzuoki and Epstein's experimental results (1976) discussed in the next section.

During our study, we encountered a quite different type of granite in which the inclination of the tie-lines is far from –64, being, in fact, nearly vertical. Most Japanese Cretaceous granites seem to belong to the former type, and many granites in the Ryoke belt belong to the latter type, as shown in Fig. 2.

We tentatively assign those having tie-lines near –64 to the "equilibrium" type and those having nearly vertical tie-lines to the "disequilibrium" type.

No individual granitic mass was found to contain a mixture of equilibrium and disequilibrium types. There are, however, modified equilibrium types, characterized by tie-lines which are only gently inclined (Fig. 3).

SIGNIFICANCE

According to Suzuoki and Epstein (1976), the fractionation factor of D/H between a hydrous silicate (mica or amphibole) and water can be expressed in the following form:

$$10^3 \ln\alpha_{m-w} = -22.4 \times 10^6 T^{-2} + (2X_{Al} - 4X_{Mg} - 68X_{Fe}) \quad (1).$$

where α_{m-w} is equal to $(D/H)_{mineral}/(D/H)_{water}$, T ($^\circ$K) is the equilibrium temperature, and X is the atomic fraction of

---

* σD denotes $\{([D/H]$ sample $/ [D/H]$ standard sea water$) - 1\} \cdot 10^3$, and X_{Fe} stands for the atomic fraction of Fe^{II} in the octahedral site of the minerals.

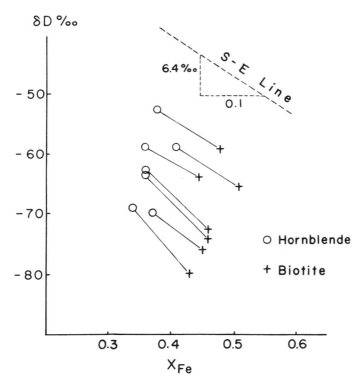

Figure 1. σD-X_{Fe} relationship between coexisting biotite and hornblende from the Cretaceous granitic rocks in the southern Kitakami mountainous district (after Kuroda and others, 1974).

cations in the six-coordinated position. In an equilibrium condition, biotite and hornblende can be regarded for this purpose as the same mineral.

Eq. (1) was obtained by the partial exchange technique using a hydrothermal bomb and is valid for the temperature range 400 to 850°C. When biotite and hornblende are in isotopic exchange equilibrium with coexisting water, equilibrium is also established between the biotite and hornblende.

The fractionation factor of D/H ratio in equilibrium between biotite and hornblende is presented in the following way using Eq. (1):

$$10^3 \ln\alpha_{bi-ho} = (2X_{Al} - 4X_{Mg} - 68X_{Fe})_{bi} - (2X_{Al} - 4X_{Mg} - 68X_{Fe})_{ho} \quad (2).$$

where α_{bi-ho} denotes $(D/H)_{biotite}/(D/H)_{hornblende}$. In most cases, X_{Al} is negligible for both biotite and hornblende so that X_{Mg} can be put equal to $1-X_{Fe}$. Therefore, Eq. (2) can be approximated by

$$\sigma D_{bi} - \sigma D_{ho} \simeq 64[(X_{Fe})_{ho} - (X_{Fe})_{bi}] \quad (3).$$

From Eq. (3) it is seen that the slope of the tie-line between biotite and hornblende in the σD-X_{Fe} plot is -64 when the isotopic exchange equilibrium is established at a definite temperature. The slope of the line given by Eq. (3) is shown by a broken line in Figure 1.

When the slope of the tie-line of a pair of biotite and

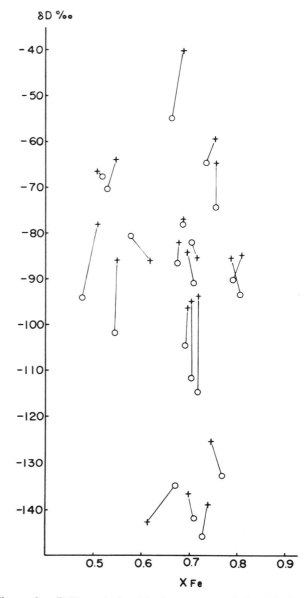

Figure 2. σD-X_{Fe} relationship between coexisting biotite and hornblende from the disequilibrium-type rocks in the Ryoke belt (after Kuroda and others, 1974, 1975, and unpublished data).

hornblende is equal to -64 in the σD-X_{Fe} plot, the pair is regarded to be in hydrogen isotope exchange equilibrium. In such a case, we can estimate the σD value of the mineral pair at $X_{Fe} = 0(\sigma D°_{mineral})$. Then using Eq. (1),

$$\sigma D°_{mineral} - \sigma D°_{water} \simeq -22.4 \times 10^6 T^{-2} + 24.2 \quad (4).$$

As seen in Eq. (4), when the biotite and hornblende pair is in equilibrium, we can estimate the temperature of equilibrium if σD value of the coexisting water is known, or we can estimate the σD value of the coexisting water if the equilibrium temperature is known by other means. How-

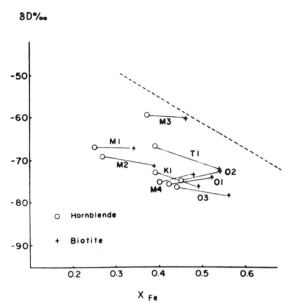

Figure 3. σD-X_{Fe} relationship between coexisting biotite and hornblende from somewhat modified equilibrium-type rocks in the northern Kitakami mountainous district (after Kuroda and others, 1975).

ever, the above conclusion is valid only when water vapor coexists with the hydrous silicates.

If a granitic magma is oversaturated with water, the "aqueous fluid" separates; this is equivalent to the water vapor in Suzuoki and Epstein's experiment (1976). In such a case, the experimentally determined fractionation factor, α_1 (Fig. 4a), is applicable to the system, aqueous fluid, and hydrous silicate. However, if a magma is undersaturated with water, thus containing only dissolved water (Fig. 4c), α_2 holds for the system of dissolved water and hydrous silicate and is not yet known. If a magma is just saturated with water (Fig. 4b), i.e., a minute amount of free aqueous fluid is present together with the dissolved water, we presume that both biotite and hornblende are in isotopic exchange equilibrium with both the free aqueous fluid and the dissolved water. With the knowledge of σD_{bi} and σD_{ho} and the equilibrium temperature, we may estimate the σD value of the aqueous fluid. In this case, however, we can not estimate the σD value of the whole magmatic water since the ratio of the dissolved water and free aqueous fluid is not known.

In a special case, we can estimate the σD values of the magmatic water. When a magma is undersaturated with water to the extent that consecutive (or simultaneous) crystallization of hornblende and biotite uses up the dissolved water, the σD value of the magmatic water is simply the mass ratio of hornblende and biotite.

APPLICATION TO JAPANESE GRANITIC ROCKS

The equilibrium type rocks in Japan are found in the

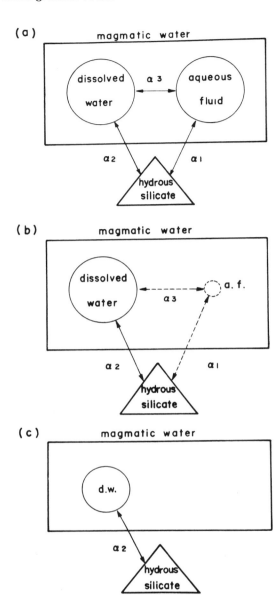

Figure 4. Schematic relationship of D/H fractionation factors between the aqueous fluid, dissolved water, and hydrous silicate.

following districts: the Kitakami mountainous district (Kuroda and others, 1974, 1975), the San-in and Sanyo belts in southwest Japan (Kuroda and others, 1977b), and the eastern part of the Kofu basin. The first two are typical Japanese Cretaceous granites, and the last may be a Tertiary (Miocene) granite mass. In the Kitakami and Kofu rocks, the σD value of water for free aqueous fluid could be estimated at about –30 o/oo. In the case of the San-in and Sanyo belts, it is somewhat lower and rather scattered.

One mass in the Abukuma plateau, the Tabito complex, is also composed of equilibrium-type rocks and σD value of the aqueous fluid was estimated also to be about –30 o/oo for the younger facies considered to be part of the Cretaceous granites (Kuroda and others, 1976, Fig. 3). However, many granitic masses in the Abukuma plateau have not yet

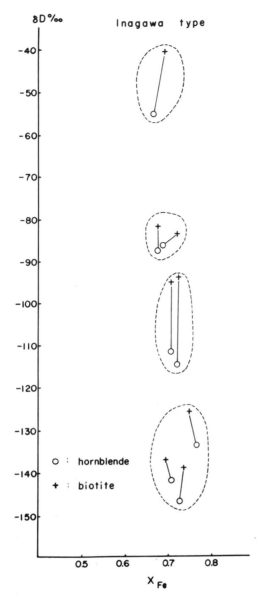

Figure 5. σD-X_{Fe} plots of the biotite-hornblende pairs from the Inagawa-type granite in the Ryoke belt. (Enclosed areas by broken lines correspond to the respective bodies.)

been investigated. Because of the complicated history of the plateau, it can not be said that all of the granitic masses in the plateau belong to the equilibrium type.

The Tertiary granitic masses in the Outer Zone of Southwest Japan seem to be classified as the equilibrium type, but the supporting data are very few.

The most typical disequilibrium type rocks of granitic compositions were found in the rapakivi granite of Finland (Kuroda and others, 1978). They are characterized by the great difference between σD values of coexisting biotite and hornblende, which generally exceeds 50 o/oo. In addition, the difference of X_{Fe} between the two minerals is very small, X_{Fe}(biotite)/X_{Fe}(hornblende) being nearly unity.

In Japan the disequilibrium-type rocks are found only in the Ryoke belt. In the belt there are several groups of granitic rocks, some early intrusions — for example, the Hiji type — and some later intrusions — for example, the Inagawa type — but almost all produce data characteristic of disequilibrium-type rocks. However, the difference between σD values of biotite and hornblende (10 to 20 o/oo) is not as large as that of the rapakivi granite (Kuroda and others, 1978). The difference between X_{Fe} values of the two minerals is also very small. In the Ryoke belt there is a great variation in σD values, even within rock masses which are classified as the same type. The variation can be seen, for example, in several rock masses of the Inagawa type (Fig. 5). It is difficult to account for this in terms of hydrogen isotope fractionation.

Furthermore, the equilibrium-type rocks have recently been found among some granitic masses in the Ryoke belt. They belong to the Katsuragi type, the Shinshiro type, etc., which are massive and discordant to the regional structure. We cannot, however, find any petrographic difference that would distinguish them from the normal Ryoke granitic masses.

Lastly, the average water content of hornblendes from disequilibrium-type rocks is lower than that from equilibrium-type rocks, possibly reflecting the water content of the magma. However, no difference between the water contents of biotites from the two types has been recognized.

ACKNOWLEDGMENTS

We are very much obliged to our colleagues and students who offered the specimens and helped us in the laboratory and field. Because of their large number, however, their names are not listed. We also appreciate the financial support provided by the Ministry of Education, Science and Culture of Japan.

REFERENCES CITED

Kuroda, Y., Suzuoki, T., Matsuo, S., and Kanisawa, S., 1974, D/H fractionation of coexisting biotite and hornblende in some granitic masses: Journal of the Japanese Association of Mining, Petroleum and Economic Geologists, v. 69, p. 95–102.
——, 1975, D/H fractionation of coexisting biotite and hornblende in some granitic masses; A supplement: Journal of the Japanese Association of Mining, Petroleum and Economic Geologists, v. 70, p. 352–362.
Kuroda, Y., Suzuoki, T., Matsuo, S., and Tanaka, H., 1976, D/H fractionation of coexisting biotite and hornblende in Tabito composite mass, Abukuma Plateau, Japan: Journal of the Japanese Association of Mining, Petroleum and Economic Geologists, v. 71, p. 1–5.
Kuroda, Y., Suzuoki, T., and Matsuo, S., 1977a, Equilibrium and disequilibrium type of the Japanese granitic rocks derived from hydrogen isotope study: *in* Plutonism in relation to volcanism and metamorphism, Preprint Issue, 7th Circum-Pacific Plutonism Project Meeting, Toyama, Japan, p. 184–193.
Kuroda, Y., Suzuoki, T., Matsuo, S., and others, 1977b, D/H ratios of biotites and hornblendes from some granitic rocks in the Chugoku

district, Southwest Japan: Journal of the Geological Society of Japan, v. 83, p. 719–724.

Kuroda, Y., Suzuoki, T., and Matsuo, S., 1978, Hydrogen isotope fractionation in granitic magma genesis: Abstracts of papers, International Geodynamics Conference on the Western Pacific and Magma Genesis, Tokyo, p. 282–283.

Suzuoki, T., and Epstein, S., 1976, Hydrogen fractionation between OH-bearing silicate minerals and water: Geochimica et Cosmochimica Acta, v. 40, p. 1229–1240.

MANUSCRIPT ACCEPTED BY THE SOCIETY JULY 12, 1982

Geological Society of America
Memoir 159
1983

Chemical characteristics of biotites and hornblendes of Late Mesozoic to Early Tertiary granitic rocks in Japan

Satoshi Kanisawa
Department of Earth Sciences, College of General Education
Tohoku University, Sendai 980, Japan

ABSTRACT

Granites carrying magnetite, such as those in the Kitakami, Abukuma, and San-in—Shirakawa zones, have Mg-rich and Al-poor biotite and hornblende, whereas granites carrying ilmenite instead of magnetite, such as those of the Ryoke, Sanyo—Naegi zones, have Fe- and Al-rich ones. These differences of chemistry are conspicuous and reflect the oxygen fugacities during crystallization. Trace elements and fluorine content in the Kitakami and the Ryoke granites are also presented. In the Kitakami, Abukuma, and San-in granites, fluorine is distinctly concentrated into biotite and hornblende in the late stages of crystallization, but in some Ryoke granites, the relation between fluorine content and differentiation is indistinct. The granites in the former districts may have been closed with respect to fluorine, while those in the latter may have been open, and fluorine in the granitic magma is thought to have reacted to some degree with the surrounding rocks.

INTRODUCTION

Mesozoic acid plutonism in Japan took place during the Early to Late Cretaceous and the Early Tertiary, and the rocks related to those activities are widely distributed throughout the Japanese Islands. Acid plutonism of this age is also common elsewhere in the circum-Pacific region.

Chemical composition of the Mesozoic granitic rocks in Japan shows a remarkable straight-line regression when plotted in Thornton and Tuttle's diagrams for oxides (Aramaki and others, 1972), and thus, these granitic rocks may belong to a large, single rock province. However, division into smaller rock provinces is practical on the bases of chemical composition, magnetic susceptibility, types of ore deposits related to the granites, K-Ar ages, and other isotopic data. From the petrographic and geochemical characteristics mentioned above, the granitic rocks in Japan show remarkable zonal arrangements (Kawano and Ueda, 1967; Kanaya and Ishihara, 1973).

GRANITIC ROCKS IN VARIOUS DISTRICTS IN JAPAN

The pre-Miocene geology of the Japanese Islands is divided into the southwest and northeast by the Tanakura Tectonic Zone (Isomi and Kawada, 1968). Granitic rocks of the northeast division are distributed through the Kitakami Mountains and the Abukuma Plateau and are characterized by magnetite (Kanisawa, 1972, 1974; Kanisawa and others, 1975; Tanaka, 1975, 1977). On the other hand, granitic rocks of the southwest division are subdivided further into those of the Outer Zone, the Ryoke Zone, the Sanyo-Naegi Zone, and the San-in—Shirakawa Zone of the Inner Zone, and contain ilmenite instead of magnetite except for those in the San-in—Shirakawa Zone (Kanaya and Ishihara, 1973; Tsusue and Ishihara, 1974; Kanisawa, 1976). Hornblende and biotite in the San-in—Shirakawa granites, however, are associated with magnetite (Tsusue and Ishihara, 1974) and have similar chemical features to those of the Kitakami granites (Kanisawa, 1976). These differences are reflected in the Fe^3/Fe^2 ratio or oxidation ratio of the rocks; that is, granitic rocks in the Kitakami Mountains and the San-in—Shirakawa Zone have high ratios, while those in the Ryoke Zone, Sanyo—Naegi Zone, and the Outer Zone of southwest Japan have low ratios.

CHEMISTRY OF BIOTITE AND HORNBLENDE

Major element chemistry of biotite and hornblende in the Early Cretaceous to Early Tertiary granitic rocks in Japan have been reported by many authors. Data from the Kitakami Mountains were obtained by Onuki and Tiba (1964), Kanisawa (1969, 1972, 1974), Kato (1972, 1974), and Kato and Tanaka (1973); those from Abukuma Plateau by Tanaka (1975, 1977); those from the Ryoke Zone by Honma (1974; Kanisawa, 1975); those from the Sanyo Zone by Murakami (1969); those from the San-in Zone by Murakami (1969; Kanisawa, 1976); and those from the Okueyama granites of Miocene in the Outer Zone by Oba and others (1977). Further, Shibata and others (1966) have reported many chemical data of biotites and hornblendes from various granites in Japan.

Biotite composition of the Japanese granites and that from various districts in the world are illustrated in annite-siderophyllite-phlogopite-eastonite diagrams (Fig. 1). They show that $Mg/(Mg+Fe^{2+}+Fe^{3+}+Mn)$ ratios are concentrated between 0.2 and 0.6, and Si is about 5.5. Each province, however, has a characteristic biotite chemistry; that is, biotites from the Kitakami and San-in—Shirakawa granites are rich in Mg and poor in Al^{IV}, while those in the Ryoke granites are rich in Fe^2 and Al^{IV}. Thus, biotites of the former

districts are rich in the phlogopite molecule, and those of the latter are rich in siderophyllite molecule. Biotites of the Tabito mass in the Abukuma Plateau show intermediate characteristics between the Ryoke and the Kitakami biotites. Biotites of the Sanyo Zone are rich in Fe, but show a wide range of Al^{IV}.

Generally, Fe-rich biotites tend to be rich also in Al. This relationship is also observed in biotites from the central and northern Sierra Nevada batholiths, Coast and Transverse Range, northern Portugal, and the Ben Nevis complex in Scotland. Biotites from shallow intrusives such as the Hirota, Orikabe, Tabashine, and Taro masses and those from the Ichinohe alkali gabbro complex in the Kitakami Mountains are associated with clinopyroxene and are poor in Al, and thus, deficient in tetrahedral $Si+Al^{IV}$ sites. These biotites, having crystallized at higher temperatures, are generally rich in Ti, and Ti may take the place of Al^{IV} in the tetrahedral sites. These Ti-rich and Al-deficient biotites were reported from Ben Nevis (Haslam, 1968). Generally, the biotites from Kitakami are rich in Ti, whereas those from Ryoke are poor in Ti and rich in Al.

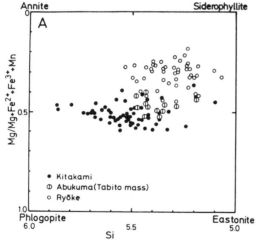

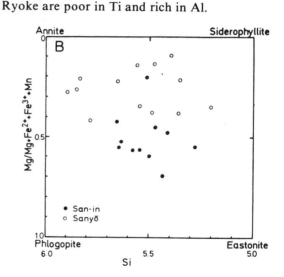

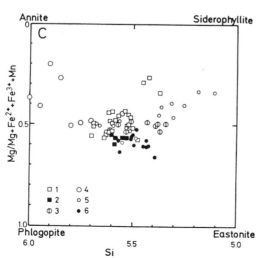

Figure 1. Composition of biotites of granitic rocks from various districts. A: Kitakami, Abukuma (Tabito mass), and Ryoke. B: San-in and Sanyo. C: 1. Central Sierra Nevada. 2. Northern Sierra Nevada. 3. Coast and Transverse Range. 4. Southern California. 5. Northern Portugal. 6. Ben Nevis Complex.

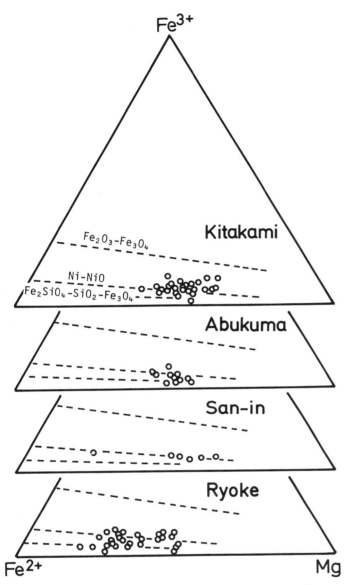

Figure 2. Fe^{3+}-Fe^{2+}-Mg diagrams for biotites in granitic rocks from various districts, Japan.

Most calciferous amphiboles in Japanese granites are magnesio- or ferro-hornblende according to Leake's (1968) classification. The Ryoke hornblendes are rich in Al^{IV} and Fe and thus belong to ferro-hornblende or ferro-tschermakitic hornblende. On the other hand, the Kitakami and the San-in granites are poor in Al^{IV} and Fe and thus are magnesio-hornblendes (Fig. 3). Hornblendes of the Ryoke granites have large amounts of both Al^{IV} and Al^{VI}, but those of the Kitakami Mountains and San-in Zone are commonly free of Al^{VI}.

Hornblendes in shallow intrusives closely associated with volcanic rocks are especially poor in both Al^{IV} and Al^{VI}. Al^{IV} replacing Si amounts to nearly 1.5 per formula unit in the Ryoke hornblendes but is approximately 1 in the Kitakami and San-in hornblendes when calculating the structural formulae on the basis of 24(O,OH). In the Al^{IV}-(Na+K) and Al^{IV}-(Al^{VI}+Fe^{3+}+Ti) relations (Deer and others, 1963), it is evident that hornblendes of the Ryoke granites are richer in Na+K, Al^{IV}+Fe^{3+}+Ti than those of the Kitakami and San-in—Shirakawa Zone. Niggli's mg value of the Ryoke hornblendes ranges from 0.18 to 0.46, whereas that of the Kitakami and San-in hornblendes is larger than 0.43.

Elevated pressure is favorable to some extent for increasing the aluminum content in hornblendes, especially the maximum amount of Al^{VI} (Leake, 1965). Thus, the Ryoke hornblendes may have crystallized at deeper levels than those of Kitakami. However, the amount of Al^{VI} in hornblende is not directly proportional to pressure during crystallization because of the importance of magma composition. The Al_2O_3 content of the Ryoke granites is higher than that of the Kitakami Mountains. Thus, the host rock chemistry may also be an important factor bearing on the Al content of hornblendes. Tabito hornblendes in the Abukuma Plateau show characteristics intermediate between the Ryoke and the Kitakami hornblendes. Both biotites and hornblendes from the Okueyama granodiorite mass in the Outer Zone of southwest Japan are poor in Al and have low Mg/Fe^2 ratios. They are similar in composition to those of volcano-plutonic complexes of other districts (Oba and others, 1977). Granitic hornblendes associated with biotites in Japan are always richer in Fe^3 than the coexisting biotites.

TRACE ELEMENTS CHEMISTRY OF GRANITES, BIOTITES, AND HORNBLENDES

Some trace elements in the Ryoke and the Kitakami granites and their biotites and hornblendes are known (Goto and others, 1977), and the fluorine content in various granites in Japan has been measured (Kanisawa, 1979). In the Ryoke granites, Zn, Co, and V decrease with increasing differentiation index, whereas Li, Cr, and Ni tend to be variable and Pb and Cu are almost invariable. The relations between Mg and V and between Fe and V show positive correlation, but that between Cr and Mg is not so distinct. In

As shown in the Fe^3-Fe^2-Mg diagram (Fig. 2), biotites in Japanese granites are commonly defined by oxygen fugacities near the Ni-NiO buffer estimated by Wones and Eugster (1965). Most biotites in the Kitakami, Abukuma, and San-in granites are associated with K-feldspar and magnetite, and thus, the oxygen fugacities of the granitic magmas of these districts were nearly the same. Biotites of the Ryoke granites fall near the Ni-NiO or fayalite-quartz-magnetite buffers; however, as these biotites do not coexist with magnetite, estimation of oxygen fugacities is uncertain. Nevertheless, magnetite-free and ilmenite-bearing granites in the Ryoke and the Sanyo granites crystallized at lower oxygen fugacities than magnetite-bearing granites (Tsusue and Ishihara, 1974; Kanisawa, 1975; Shimazaki, 1976).

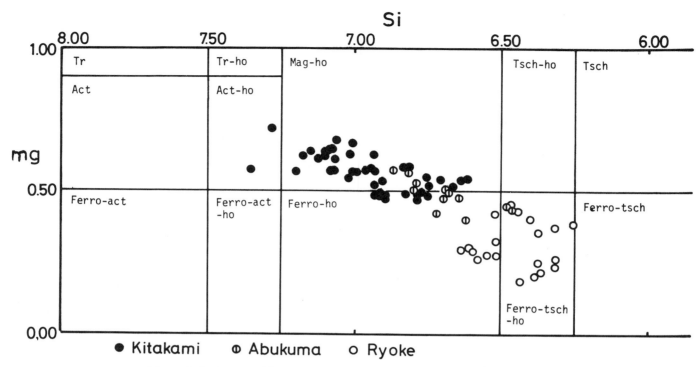

Figure 3. Leake's classification of hornblendes in some Japanese granitic rocks.

biotites, Ni, Co, Cr, and V decrease and Zn increases with decreasing mg-value. A remarkable concentration of Zn and V appears in the biotites and hornblendes, and Pb, Cu, Ni, and Cr are limited in these minerals. Li concentration is conspicuous in the biotites. Although the Ryoke granites are divided into several stages on the bases of their succession of intrusion and rock facies, their major and trace elements show nearly the same differentiation trend. However, the Inagawa granites, belonging to the Younger Ryoke granites, have less Ni and Cr than any other granites, and their biotites show slightly different variations in Zn, V, and Li than do the latter. The contents of some trace elements, especially Cr, Ni, and Li are different for each rock mass in the Kitakami Mountains. Generally, Co, Cu, Zn, V, and Li decrease with increasing differentiation index, whereas Pb is nearly constant and is less than one-half that of the Ryoke granites. Co, Cr, Ni, and V decrease and Zn increases in biotites with advancing differentiation of the host rocks. Pb in biotites is very limited.

In hornblendes, Ni, Cr, and V decrease and Zn increases with advancing differentiation of the host rocks. Li, Co, and Cu are nearly constant. Zn is lower in the Fe-poor Kitakami biotites and hornblendes, except those in the Miyako granites, than in the Fe-rich biotites in the Ryoke granites. The reason is that Zn enters the Fe^2 position and the Zn^2/Fe^2 ratio increases during differentiation. Trends of major and trace elements of biotites and hornblendes in the Miyako granites of the Kitakami Mountains differ from those of

other Kitakami granites in that the Fe/Mg ratio, Ni, Cr, and V in these minerals decrease and Zn increases with advancing differentiation because of increasing oxygen fugacities during crystallization.

Behavior of fluorine in the crystallization of granitic magma is interesting. In the Kitakami Mountains, fluorine is most abundant in alkali-rich rocks and steeply decreases with advancing differentiation. It is comparatively abundant in shallow intrusive granites. In other granites of the Kitakami Mountains, the fluorine content is moderate and remains nearly constant throughout differentiation. In some granites, fluorine is distinctly concentrated into biotite and hornblende during the later stages of crystallization (Fig. 4) (Kanisawa, 1979).

Fluorine content in granitic rocks from the Abukuma Plateau decreases with advancing differentiation but is concentrated into biotite and hornblende during later stages of crystallization. In contrast, the relation between fluorine content and differentiation index in some Ryoke granites, i.e., the Tenryukyo and the Inagawa granites, is rather inconsistent (Fig. 5). In granites of the San-in Zone, fluorine is also concentrated into biotite and hornblende during late stages. Fluorine content of biotite is nearly twice as much as that in coexisting hornblende in all granites. It is considered that the fluorine in Kitakami, Abukuma, and San-in granites may be magmatic in origin and that neither the surrounding rocks nor the ground water participated in the granite chemistry. However, in some Ryoke granites, the

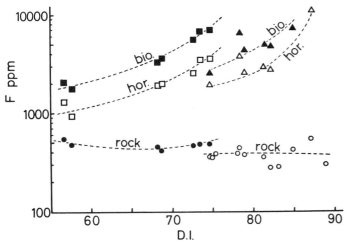

Figure 4. Behavior of fluorine in some granitic rocks in the Kitakami Mountains. Closed circle and square: Miyako granites. Open circle and triangle: Granites in the Taro Belt.

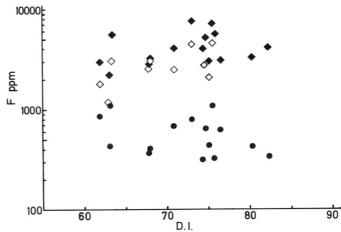

Figure 5. Behavior of fluorine in the Inagawa granites, Ryoke Belt. Closed circle: Rocks. Closed diamond: Biotites. Open diamond: Hornblendes.

diversity of fluorine content is thought to be due to some interaction between the granitic magmas and the surrounding rocks.

ACKNOWLEDGMENTS

The author wishes to thank T. Nozawa for critically reading the manuscript and for useful suggestions.

REFERENCES CITED

Aramaki, S., Hirayama, K., and Nozawa, T., 1972, Chemical composition of Japanese granites, Part 2, Variation trends and average composition of 1200 analyses: Journal of the Geological Society of Japan, v. 78, p. 39–49.

Deer, W. A., Howie, R. A., and Zussman, J., 1963, Rock forming minerals, v. 2, Chain silicates: Longmans, London, 379 p.

Goto, H., Kanisawa, S., and Katada, M., 1977, Zinc, lead, copper, nickel, cobalt, chromium, vanadium and lithium contents of some Ryoke granites and their constituting biotites and hornblendes, central Japan: Geological Survey of Japan Bulletin, v. 28, p. 47–57 (in Japanese with English abstract).

Haslam, H. W., 1968, The crystallization of intermediate and acid magmas at Ben Nevis, Scotland: Journal of Petrology, v. 9, p. 84–104.

Honma, H., 1974, Chemical features of biotites from metamorphic and granitic rocks of the Yanai district in the Ryoke Belt, Japan: Journal of the Japanese Association of Mining, Petroleum and Economic Geologists, v. 69, p. 390–402.

Isomi, H., and Kawada, K., 1968, Correlation of basement rocks between both sides of Fossa Magna: Fossa Magna, p. 4–12 (in Japanese).

Kanaya, H., and Ishihara, S., 1973, Regional variation of magnetic susceptibility of the granitic rocks in Japan: Journal of the Japanese Association of Mining, Petroleum and Economic Geologists, v. 68, p. 211–224 (in Japanese with English abstract).

Kanisawa, S., 1969, On the Hitokabe granodioritic mass, Kitakami Mountainland: Journal of the Japanese Association of Mining, Petroleum and Economic Geologists, v. 62, p. 275–288 (in Japanese with English abstract).

———,1972, Coexisting biotites and hornblendes from some granitic rocks in southern Kitakami Mountains: Journal of the Japanese Association of Mining, Petroleum and Economic Geologists, v. 67, p. 332–344.

———,1974, Granitic rocks closely associated with the Lower Cretaceous volcanic rocks in the Kitakami Mountains, northeast Japan: Journal of the Geological Society of Japan, v. 80, p. 355–367.

———,1975, Chemical composition of hornblendes of some Ryoke granites, central Japan: Journal of the Japanese Association of Mining, Petroleum and Economic Geologists, v. 70, p. 200–211.

———,1976, Chemistry of biotites and hornblendes of some granitic rocks in the San-in zone, southwest Japan: Journal of the Geological Society of Japan, v. 82, p. 543–548 (in Japanese with English abstract).

———,1979, Content and behaviour of fluorine in granitic rocks, Kitakami Mountains, northeast Japan: Chemical Geology, v. 24, p. 57–67.

Kanisawa, S., Nedachi, M., and Ueno, H., 1975, Opaque minerals in the granitic masses, Kitakami Mountains (Part 1) (Abstract): Journal of the Japanese Association of Mining, Petroleum and Economic Geologists, v. 70, p. 127 (in Japanese).

Kato, Y., 1972, Petrology of the Orikabe granitic body, Kitakami Mountainland: Journal of the Japanese Association of Mining, Petroleum and Economic Geologists, v. 67, p. 50–59 (in Japanese with English abstract).

———,1974, Petrology of the Tabashine granitic body, Kitakami Mountains, northeastern Japan: Journal of the Japanese Association of Mining, Petroleum and Economic Geologists, v. 69, p. 417–425.

Kato, Y., and Tanaka, H., 1973, Petrology of the Kinkasan granitic body, Kitakami Mountains: Journal of the Japanese Association of Mining, Petroleum and Economic Geologists, v. 68, p. 395–403 (in Japanese with English abstract).

Kawano, Y., and Ueda, Y., 1967, K-A dating on the igneous rocks in Japan (VI) — Granitic rocks, summary — : Tohoku University Scientific Report, Serial III, v. 10, p. 65–76.

Leake, B. E., 1965, The relationship between composition of calciferous amphiboles and grade of metamorphism, in Pitcher, W. S., and Flinn, G. W., eds., Controls of metamorphism: New York, John Wiley and Sons, Inc., 368 p.

———,1968, A catalog of analyzed calciferous and subcalciferous amphiboles together with their nomenclature and associated minerals: Geological Society of America Special Paper, no. 98.

Murakami, N., 1969, Two contrastive trends of evolution of biotite in granitic rocks: Journal of the Japanese Association of Mining, Petroleum and Economic Geologists, v. 62, p. 223–248.

Oba, T., Yamamoto, M., and Oba, N., 1977, Chemical compositions of coexisting biotites and hornblendes from Okueyama granodiorite, Kyushu, Japan: Journal of the Japanese Association of Mining, Petroleum and Economic Geologists, v. 72, p. 433–442.

Onuki, H., and Tiba, T., 1964, Petrochemistry of the Ichinohe alkali plutonic complex, Kitakami Mountainland, northern Japan: Tohoku University Scientific Report, Serial III, v. 9, p. 123–154.

Shibata, H., Oba, N., and Shimoda, N., 1966, Bearing on aluminum in mafic minerals in plutonic and metamorphic rocks: Tokyo Kyoiku Daigaku Scientific Report, Ser. C, v. 9, p. 89–123.

Shimazaki, H., 1976, Granitic magmas and ore deposits (2); Oxidation state of magmas and ore deposits: Mining Geology, Special Issue, v. 7, p. 25–35 (in Japanese with English abstract).

Tanaka, H., 1975, Magnesium-iron distribution in coexisting biotite and hornblende from granitic rocks: Journal of the Japanese Association of Mining, Petroleum and Economic Geologists, v. 70, p. 118–124.

——, 1977, Petrochemistry of some Mesozoic granitic rocks in the northern Abukuma Mountains: Journal of the Japanese Association of Mining, Petroleum and Economic Geologists, v. 72, p. 373–382.

Tsusue, A., and Ishihara, S., 1974, The iron-titanium oxides in the granitic rocks of southwest Japan: Mining Geology, v. 24, p. 13–30 (in Japanese with English abstract).

Wones, D. R., and Eugster, H. P., 1965, Stability of biotite: Experiment theory, and application: American Journal of Science, v. 50, p. 1228–1272.

MANUSCRIPT ACCEPTED BY THE SOCIETY JULY 12, 1982

Geological Society of America
Memoir 159
1983

Strontium isotopic compositions of late Mesozoic and Early Tertiary acid rocks in Japan

H. Kagami
Institute for Thermal Spring Research, Okayama University
Misasa, Tottori-ken, 682-02, Japan

K. Shuto
Nishi-ikocho, Adachi-ku, Tokyo, 121, Japan

ABSTRACT

The source materials for late Mesozoic and Early Tertiary acid rocks in Japan are examined on the basis of Sr isotopic data. The following are considered as possible source materials: 1) ultrabasic rocks constituting the upper mantle, or basic materials derived from the upper mantle; 2) rocks of andesitic basalt composition probably constituting the uppermost mantle or the lower crust of continental or adjacent regions; 3) rocks of Precambrian or earliest Paleozoic age probably constituting the lower part of the granitic layer of the crust under the Japanese islands; 4) rocks, mainly sedimentary, constituting the Honshu geosyncline.

Comparison of the Sr isotopic data of these four materials with those of the acid rocks indicates that the first two sources are more probable.

INTRODUCTION

Late Mesozoic and Early Tertiary acid rocks are extensive in Japan. The absolute ages of these rocks have been determined using the Rb-Sr whole-rock method by many research workers since the mid-1960s. The range of initial Sr isotopic ratios (abbreviated to SrI hereafter) for acid rocks in Japan is from 0.7038 to 0.711 (Table 1) and forms a convenient framework for the discussion of petrogenesis as well as of age. The names of the granitic provinces are those of Nozawa (1977) and their locations are shown in Figure 1.

INITIAL ^{87}SR/^{86}SR ISOTOPIC RATIOS

In northeast Japan, the SrI for acid rocks is about 0.7040 in Kitakami Granitic Province, 0.7055 in Abukuma Province, and from 0.7060 to 0.7080 in Uetsu Province (Shibata and Ishihara, 1977). On the other hand, those of the Ryoke Province and southern border of the Chugoku Province of Southwest Japan range from 0.7070 to 0.711. Those values in the Chugoku Granitic Province decrease gradually from south to north (i.e., toward the Japan Sea) from about 0.7070 to 0.7050 (Ishizaka and Seki, 1975).

The distribution of SrI in Japan forms a somewhat systematic pattern as shown by Shibata and Ishihara (1977) (Fig. 2). Kagami (1977) pointed out that the SrI of igneous rocks in the Ryoke metamorphic belt and its surrounding area is the highest in Japan.

In addition to its distribution pattern, two important facts about SrI should be noted. One is that the standard deviations (σ-value) for the SrI of acid rocks in Japan are generally very low compared with those from other parts of the world. The former are less than 0.0020 (σ), most about 0.0005 (Table 1), and the latter range from 0.070 to 0.0001 (Kagami and others, 1976). The σ-value is thought to have geological significance; those igneous rocks with high σ-values are considered to be contaminated with upper crustal materials or to have passed through some other complex geological processes. On the other hand, the igneous rocks with low σ-values are considered to have been less affected by such geological events (Kagami and others, 1975, 1976; Kagami and Shuto, 1977).

The second attribute of SrI is that, except for some igneous rocks of mid- to Late Tertiary age, basic rocks that are regionally associated with acid rocks have the same SrI values as the acid rocks (Shibata and Ishihara, 1977).

PETROGENESIS OF ACID ROCKS IN JAPAN

Four source materials for the acid rocks are considered

TABLE 1. INITIAL SR ISOTOPE RATIOS AND AGES OF THE LATE TERTIARY AND EARLY MESOZOIC ACID

ROCKS OF JAPAN

Location	Age in m. y.	Initial Sr Ratio	Reference
Kitakami Granitic Province			
Ichinohe qtz-syenite	104	0.7041(0.0001)	Kubo(1977)
Qtz-diorite(s-mass)	104	0.7046(0.0001)	Kubo(1977)
Tono-Kesengawa Gr.	105	0.7039(0.0002)	Shibata1974)
Miyako Gr.	121	0.7039(0.0002)	Shibata and others(1976)
Taro Gr.	128	0.7038(0.0002)	Shibata and others(1976)
Abukuma Granitic Province			
Iritono Gr.	145	0.7035( )	Maruyama(1972)
Tabito Gr.	150	0.7055(0.0003)	Maruyama(1972)
Ryoke Granitic Province			
Chubu District			
Habakawa Gr.	59	0.7088(0.0004)	Kagami(1973)
Kadoshima Gr.	76	0.7102(0.0003)	Kagami(1973)
Inagawa Gr.	78	0.7089(0.0004)	Kagami(1973)
Seinaiji Gr.	82	0.7098( )	Yamada.Shibata(1970)
Ikuta Gr.	84	0.7082(0.0001)	Kagami(1973)
Tenryukyo Gr.	117	0.7075(0.0009)	Kagami(1973)
Kamihara Gr.	167	0.7073(0.0013)	Kagami(1973)
Yanai District			
Towa Gr.	76	0.7083(0.0002)	Shigeno.Yamaguchi(1976)
Gamano Gr.	82	0.7078(0.0005)	Shigeno.Yamaguchi(1976)
Murotsu Gr.	96	0.7075(0.0003)	Shigeno.Yamaguchi(1976)
Kitaoshima Gr.	178	0.7070(0.0007)	Shigeno.Yamaguchi(1976)
Hida Granitic Province			
Kamioka Qp.	86	0.7061(0.0005)	Seki(1972)
Kamioka Gr.	143	0.7074(0.0016)	Seki(1972)
Funatsu Gr.	187	0.7048(0.0003)	Shibata et al.(1970)
Chugoku Granitic Province			
Chubu District			
Nohi Rhy.	78	0.7095(0.0020)	Okamoto et al.(1975)
Kinki District			
Koto Rhy.	72	0.7108(0.0005)	Seki.Hayase(1971)
Tenkadaiyama Rhy.	73	0.7100(0.0004)	Seki.Hayase(1974)
Harima Gr.	83	0.7056(0.0003)	Seki.Hayase(1971)
Nose Gr.	85	0.7064(0.0009)	Ishizaka(1971)
Aioi Vc.	119	0.7052(0.0004)	Seki.Hayase(1974)
Sanin District			
Tottori Gr.	67	0.7058(0.0004)	Hattori.Shibata(1974)
North Kyushu District			
Sawara Gr.	135	0.7046(0.0001)	Yanagi(1975)
Haki Gr.	161	0.7058(0.0001)	Yanagi(1975)
Itoshima Gr.	166	0.7041(0.0001)	Yanagi(1975)

Values in brackets following each SrI are standard deviations.

Gr - granite; Qp - quartz porphyry; Rhy - rhyolite; Vc - volcanics.

possible: 1) ultrabasic or basic rocks constituting the upper mantle; 2) rocks or andesitic basalt composition probably constituting the uppermost mantle or the lower crust of continental and adjacent regions; 3) rocks of Precambrian or earliest Paleozoic age (about 400 m.y. and older), probably constituting the lower part of the "granitic" layer of crust under the Japanese islands; 4) rocks, mainly sedimentary, constituting the Honshu geosyncline.

Knowledge of both Sr isotopic variation with time for the four possible source materials and SrI of the acid rocks is necessary for discussion of the source materials of the acid rocks. The values of the latter are shown in Table 1. The Sr isotopic variation with time is here termed the Sr evolution. In heterogeneous rocks of any region, there will be a Sr-evolution field rather than a line. As the description in detail of the Sr-evolution fields of the four materials is beyond the scope of this paper, it will only be summarized here. (For more detail, see Gorai and others, 1972, 1978; Kagami and others, 1975, 1976; Kagami and Shuto, 1977.)

The Sr-evolution field of the upper mantle (Fig. 3) is estimated from the SrI of Recent oceanic basalt and older ultrabasic rocks shown in Figures 4 and 5.

It is difficult to show the Sr-evolution field of the lower crust (Fig 3), because of lack of data. The writers have

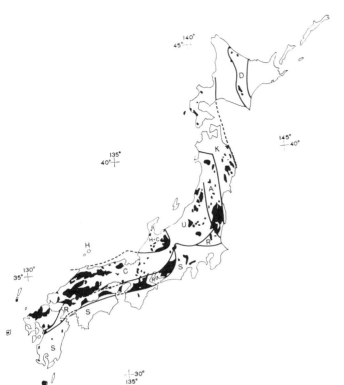

Figure 1. Distribution of granitic provinces in Japan. D: Hidaka, K: Kitakami, A: Abukuma, U: Uetsu, H: Hida, C: Chugoku, R: Ryoke, S: Shimanto.

estimated the field on the basis of SrI of continental basic rocks (Fig. 5).

The ages of Precambrian rocks are not well known in Japan. However, an age of about 2 b.y. was obtained from pebbles in a conglomerate at Kamiaso (Mino terrain) using the Rb-Sr whole-rock method (Shibata and Adachi, 1974), and an age of 1.8 b.y. based on the Pb^{207}/Pb^{206} method was reported from zircons in the Ryoke metamorphic rocks (Ishizaka, 1969). If these represent the most abundant rocks of the lower part of the "granitic" layer of the crust under the Japanese islands, the Sr-evolution field of these source rocks would be as shown in Figure 6. The field is estimated from the general trend of the Sr-evolution lines (Fig. 7) for each rock.

Rocks about 400 m.y. old have been reported from various places, especially from tectonic zones such as the Kurosegawa and Nagato belts. Their Sr-evolution field is determined in the same manner as for the Precambrian rocks (Fig. 8), and the Sr, evolution lines of these rocks are shown in Figure 9.

The rocks constituting the Honshu geosyncline are thought by many workers to overlie the older rocks mentioned above. The Sr-evolution field and lines are given in Figures 10 and 11, respectively.

To discuss the petrogenesis of acid rocks, it is necessary to compare the Sr evolution of the four materials mentioned

Figure 2. Regional variations in the SrI for igneous rocks of late Mesozoic and Early Tertiary age in Japan (after Shibata and Ishihara, 1978). Solid circles: gabbroic rock; open circles: granitic rock. The dotted lines indicate the middle Tertiary igneous rocks in the Outer Zone of southwest Japan.

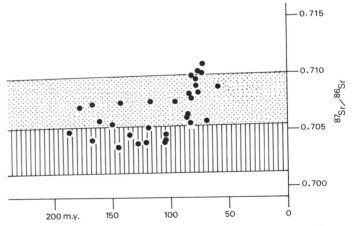

Figure 3. Sr-evolution fields of upper mantle and lower crust since 250 m.y. ago. The vertical striped pattern represents the Sr-evolution field of the upper mantle. The dotted pattern represents the field of lower crust. Solid circles indicate acidic rock.

above with the SrI of acid rocks. The SrI of acid rocks falls in the fields of upper mantle (material 1), uppermost mantle and lower crust (material 2), and 400-m.y.-old rocks (material 3), but not in the fields of the Precambrian rocks (material 3) or the Honshu geosyncline (material 4) (Figs. 3,

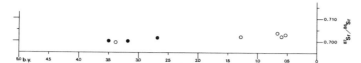

Figure 4. Relation between SrI and age of ultrabasic rocks (after Gorai and others, 1978). Solid circles: komatiite, open circles: ultrabasic inclusions.

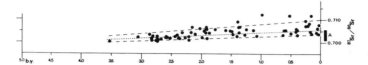

Figure 5. Relation between SrI and age of continental basic rocks. A: oceanic basalt of Recent age. Solid circles: continental basic rock. Dotted line: upper limit of distribution of ultrabasic rocks shown in Figure 4.

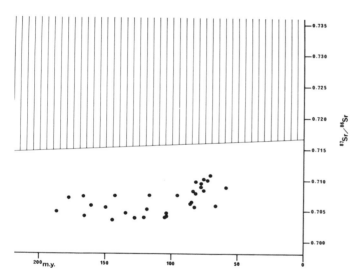

Figure 6. Sr-evolution field of Precambrian rocks. Vertical striped pattern indicates the lower part of Sr-evolution field of Precambrian rocks. Solid circles: acid rock.

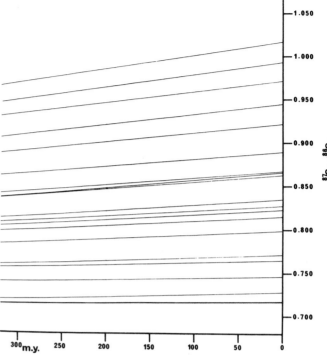

Figure 7. Sr-evolution lines of Precambrian rocks.

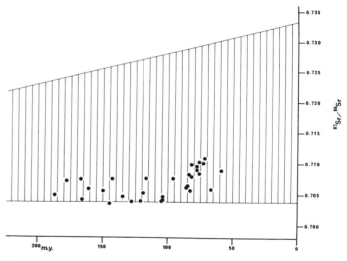

Figure 8. Sr-evolution field for 400-m.y.-old rocks. Solid circles: acid rock.

6, 8, 10). The last two should, therefore, be excluded as possible source materials of acid rocks because their Sr-evolution fields are situated in the uppermost part of that of the acid rocks.

If the 400-m.y.-old rocks overlie the Precambrian rocks, they must also be excluded as possible source materials for the acid rocks. Judging from the Rb/Sr ratios of Precambrian rocks and the 400-m.y.-old rocks, the former are more siliceous than the latter. Accordingly, the Precambrian rocks would melt at temperatures well below those of the 400-m.y.-old rocks, and the acid magma originating from either rock group must have a high SrI. On the other hand, the following case could also be possible. The Precambrian rocks are distributed only in a limited area and the 400-m.y.-old rocks immediately overlie the "basaltic" layer of the

crust in most parts of the Japanese islands. Consequently, a slight possibility exists that the 400-m.y.-old rocks could be the source materials of some acid rocks, but these acid rocks should then have a slightly higher SrI (Kagami, 1977).

In view of these considerations, the acid rocks in Japan are presumed to have been derived directly from the upper mantle or lower crust (materials 1 or 2).

However, another interpretation is possible. All the magma may have been derived initially from the upper

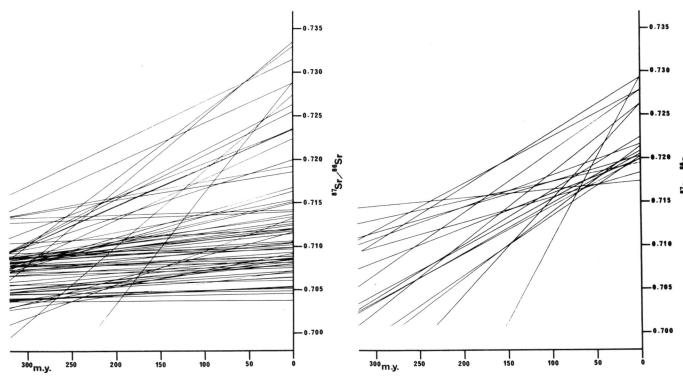

Figure 9. Sr-evolution lines for 400-m.y.-old rocks. The source of data is as follows: Shibata and others, 1970; Yamaguchi and Yanagi, 1970; Shibata, 1974; Ishihara and Yanagi, 1975; Yanagi, 1975.

Figure 11. Sr-evolution lines for rocks of the Honshu geosyncline. The source of data is as follows: Ueno and others, 1968; Kagami, 1973; Shigeno and Yamaguchi, 1976; Ueno and Ono, 1976.

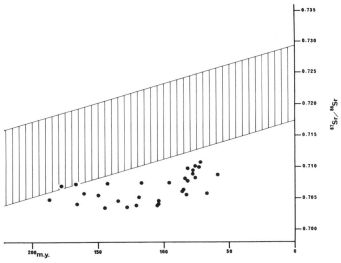

Figure 10. Sr-evolution field for rocks of the Honshu geosyncline. Solid circles: acid rocks.

mantle and much of it subsequently contaminated with upper crustal materials. Having passed through this process, the magma should have a SrI of 0.706 or more. But certain difficulties are encountered if all acid rocks with high SrI values are accounted for only on the basis of contamination.

Firstly, the distribution pattern of SrI in Japan is not irregular, and the regular pattern is difficult to explain by accidental geological events such as contamination. This fact is generally recognized for coeval acid rocks occurring in the circum-Pacific region (Le Couteur and Tempelman-Kluit, 1969; Kistler and Petterman, 1973; Armstrong and others, 1977).

Secondly, the acid rocks have very low σ-values. Thirdly, the basic rocks related to acid rocks have the same SrI. To explain the last two facts on the basis of contamination, one could assume that during the first stage magma derived from the upper mantle was throughly mixed with upper crustal materials, resulting in a homogeneous magma. Thereafter, basic materials and acid magma differentiated from the homogeneous magma. However, such homogenization is unlikely in the shallow and heterogeneous upper crust.

The writers do not deny totally the possibility of contamination (Kagami, 1977) but consider that it is impossible to explain all the acid rocks with values of 0.706 and over by contamination alone. The continental basic rocks of middle and late Mesozoic age are abundant, and most have high SrI (Fig. 4). Such rocks cannot be explained by contamination as pointed out by Faure and Powell (1972). We emphasize that the difference of SrI may reflect fundamentally different source materials.

Lastly, the distribution pattern of SrI may be related to variation in source material. It is possible to suppose that the variation of Rb/Sr ratios, or $^{87}Sr/^{86}Sr$, in the source materials is not random but has a certain pattern or horizontal or vertical variation. Zartman and others (1969) proposed that the ratio gradually decreases with increasing depth. According to this idea, acid rocks having high SrI were formed by melting at the higher levels of any source material (Kagami, 1977).

It is concluded from the Sr isotopic data that the late Mesozoic and Early Tertiary acid rocks of Japan are derived from either 1) the basic or ultrabasic rocks of the upper mantle, or 2) rocks of andesitic or basaltic composition constituting the uppermost mantle or lower crust.

REFERENCES CITED

Armstrong, R. L., Taubeneck, W. H., and Hales, P. O., 1977, Rb-Sr and K-Ar geochronometry of Mesozoic granitic rocks and their Sr isotopic composition, Oregon, Washington, and Idaho: Geological Society of America Bulletin, v. 88, p. 397–411.

Faure, G., and Powell, J. L., 1972, Strontium isotope geology: Springer-Verlag, Berlin, Heidelberg, New York, 188 p.

Gorai, M., Kagami, H., and Iizumi, S., 1972, Reexamination of the source material of granitic magmas: Journal of the Geological Society of Japan, v. 78, p. 549–559.

Gorai, M., Kagami, H., and Shuto, K., 1978, Geohistorical transition of the source materials of acid rocks: Magma, no. 52, p. 11–15 (in Japanese).

Hattori, H., and Shibata, K., 1974, Concordant K-Ar and Rb-Sr ages of the Tottori granite, Western Japan: Bulletin Geological Survey of Japan, v. 25, p. 157–173.

Ishizaka, K., 1969, U-Th-Pb ages of zircon of the Ryoke metamorphic terrain, Kinki district, Japan: Journal Japanese Association of Mining, Petroleum and Economic Geologists, v. 62, p. 191–197 (in Japanese with English abstract).

——, 1971, A Rb-Sr isotopic study of the Ibaragi granitic complex, Osaka, Japan: Journal of the Geological Society of Japan, v. 77, p. 731–740.

Ishizaka, K., and Seki, T., 1975, Zonal Sr isotopic variations in the Southwest Japan: Magma, no. 43, p. 10–12 (in Japanese).

Ishizaka, K., and Yanagi, T., 1975, Occurrence of oceanic plagiogranites in the older tectonic zone, Southwest Japan: Earth and Planetary Science Letters, v. 27, p. 371–377.

Kagami, H., 1973, A Rb-Sr geochronological study of the Ryoke granites in Chubu district, Central Japan: Journal of the Geological Society of Japan, v. 79, p. 1–10.

——, 1977, The origin of Ryoke granites — Considered from their initial strontium isotopic ratios: Monograph of the Association of Geological Collaborators of Japan, no. 20, p. 45–52 (in Japanese with English abstract).

Kagami, H., and Iizumi, S., 1972, A view on the source material of acid magmas — Considered from their strontium isotopic compositions: Journal Japanese Association of Mining, Petroleum and Economic Geologists, v. 67, p. 11–19 (in Japanese with English abstract).

Kagami, H., Shuto, K., and Gorai, M., 1975, Reexamination of the source material of acid igneous rocks, based on the selected Sr isotopic data: Journal of the Geological Society of Japan, v. 81, p. 319–330 (in Japanese with English abstract).

——, 1976, Consideration of the relation between the rock types of acid igneous rocks and their initial Sr isotopic ratios: Journal of the Geological Society of Japan, v. 82, p. 655–660.

Kagami, H., and Shuto, K., 1977, Sr isotope petrology: Association of Geological Collaborators of Japan, no. 21, 274 p. (in Japanese).

Kistler, R. W., and Peterman, Z. E., 1973, Variation in Sr, Rb, K, Na, and initial Sr^{87}/Sr^{86} in Mesozoic granitic rocks and intruded wall rocks in Central California: Geological Society of America Bulletin, v. 84, p. 3489–3512.

Kubo, K., 1977, A Rb-Sr isotopic study on the Ojika and Ichinohe gabbroic complexes in the Kitakami mountains, Northeast Japan: Journal

Japanese Association of Mining, Petroleum and Economic Geologists, v. 72, p. 412–418.

Le Couteur, P. C., and Tempelman-Kluit, D. J., 1976, Rb/Sr ages and a profile of initial Sr^{87}/Sr^{86} ratios for plutonic rocks across the Yukon crystalline terrane: Canadian Journal of Earth Sciences, v. 13, p. 319–330.

Maruyama, T., 1972, Geochronological study of the granites distributed over Gosaisho-Takanuki district, southern part of the Abukuma plateau: Magma, no. 31, p. 8–13 (in Japanese).

Nozawa, T., 1977, Plutonic rocks, *in* Tanaka, K., and Nozawa, T., eds., Geology and mineral resources of Japan: Geological Survey of Japan, p. 284.

Okamoto, K., Nohda, S., Masuda, Y., and Matsumoto, T., 1975, Significance of Cs:Rb ratios in volcanic rocks as exemplified by the Nohi rhyolite complex, Central Japan: Geochemical Journal, v. 9, p. 201–210.

Seki, T., 1972, A Rb-Sr geochronological study of porphyries in the Kamioka mining district, Central Japan: Journal Japanese Association of Mining, Petroleum and Economic Geologists, v. 67, p. 410–417.

Seki, T., and Hayase, H., 1971, Study of Rb-Sr isochron on late Mesozoic igneous activity of the inner belt of Southwest Japan: Abstracts, Annual Meeting of Geochemical Society of Japan, 15B10 (in Japanese).

——, 1974, Rb-Sr isochron of the Cretaceous acid volcanic rocks of Himeji district, Hyogo Prefecture, Japan: Mass Spectroscopy, v. 22, p. 55–59 (in Japanese with English abstract).

Shibata, K., 1974, Rb-Sr geochronology of the Hikami granite, Kitakami mountains, Japan: Geochemical Journal, v. 8, p. 193–207.

Shibata, K., Nozawa, T., and Wanless, R. K., 1970, Rb-Sr geochronology of the Hida metamorphic belt, Japan: Canadian Journal of Earth Sciences, v. 7, p. 1383–1401.

Shibata, K., and Adachi, M., 1974, Rb-Sr whole-rock ages of Precambrian metamorphic rocks in the Kamiaso conglomerate from Central Japan: Earth and Planetary Science Letters, v. 21, p. 277–287.

Shibata, K., Yanagi, T., and Hamamoto, R., 1976, Isotopic age of Miyako granite, Kitakami mountains: Abstracts, Annual Meeting of Geochemical Society of Japan, p. 58 (in Japanese).

Shibata, K., and Ishihara, S., 1977, @Sr^{87}/Sr^{86} initial ratios of gabbroic and granitic ratios in Japan: Magma, no. 49–50, p. 60–62 (in Japanese).

Shigeno, H., and Yamaguchi, M., 1976, A Rb-Sr isotopic study of metamorphism and plutonism in the Ryoke belt, Yanai district, Japan: Journal of the Geological Society of Japan, v. 82, p. 687–698 (in Japanese with English abstract).

Ueno, N., Ozima, M., and Ono, A., 1969, Geochronology of the Ryoke metamorphism — Rb-Sr, K-Ar isotopic investigation of the metamorphic rocks in the Ryoke metamorphic belt: Geochemical Journal, v. 3, p. 35–44.

Ueno, N., and Ono, A., 1976, A Rb-Sr study of Ryoke metasediments, Central Japan: Earth Science, v. 30, p. 268–271 (in Japanese with English abstract).

Yamada, N., and Shibata, K., 1970, Rb-Sr age of the granites in Ryoke belt, Central Japan [abstracts]: 77th Annual Meeting of the Geological Society of Japan, p. 309 (in Japanese).

Yamaguchi, M., and Yanagi, T., 1970, Geochronology of some metamorphic rocks in Japan: Eclogae Geologicae Helvetiae, v. 63, p. 371–388.

Yanagi, T., 1975, Rubidium-strontium model of formation of the continental crust and the granites at the Island Arc: Science Report, Kyushu University, Series D, Geology, no. 22, p. 37–98.

Zartman, R. E., and Wasserburg, G. J., 1969, The isotopic composition of lead in potassium feldspars from some 1.0-b.y.-old North American igneous rocks: Geochimica et Cosmochimica Acta, v. 33, p. 901–942.

MANUSCRIPT ACCEPTED BY THE SOCIETY JULY 12, 1982

PRINTED IN U.S.A.

Geological Society of America
Memoir 159
1983

Oxygen isotope and some other geochemical evidence for the origin of two contrasting types of granitic rocks of Japan

Hiroji Honma
Institute for Thermal Spring Research
Okayama University
Tottori-ken, Japan 682-02

ABSTRACT

There are two different groups of Late Cretaceous to Early Tertiary granitic rocks in Japan: (1) intrusions into non-metamorphic terranes, with $\sigma^{18}O$ of quartz ranging from 8.5 to 10.0 o/oo, and (2) granitic rocks of low P/T type metamorphic belts, with $\sigma^{18}O$ of quartz ranging from 11.5 to 14.5 o/oo.

Granitic rocks of the Chugoku non-metamorphic belt are only slightly (0.8 to 1.0 o/oo) enriched in ^{18}O compared with associated basic plutonic rocks having low $^{18}O/^{16}O$ ratios (whole-rock $\sigma^{18}O$: 6.2-1.5 o/oo). The paucity of ^{18}O enrichment is thought to be associated with magma crystallization and selective settling of certain minerals. Contribution of crustal rocks to the chemistry of granites is detected only locally and on a small scale.

In contrast, granite magmas in and around the Ryoke metamorphic belt are enriched in ^{18}O, whereas the surrounding metamorphic and migmatitic rocks are considerably depleted in ^{18}O compared with their original state. Other geochemical evidence, such as the NH_4 content and major element chemistry of biotites, also indicates a significant contribution by sedimentary rocks to the chemistry of granitic rocks of the Ryoke belt.

Characteristic geological and geochemical features of granitic rocks of the Ryoke belt may be related to the special geological situation of this belt. For example, the tectonic movement in the Ryoke belt may have enhanced the mobility of fluid and, thus, increased the interaction between granite magma and country rocks.

INTRODUCTION

The writer, with his co-workers, studied the oxygen isotope ratios of the Cretaceous granitic rocks of Japan. A preliminary investigation (Matsuhisa and others, 1972) revealed that the $^{18}O/^{16}O$ ratios of granites in the Ryoke and Abukuma low P/T type metamorphic belts are 2.0 to 3.0 o/oo higher than those of granites in non-metamorphic terranes. The detailed study of the rocks of the Chugoku province dealt with: (1) oxygen isotopic variation in basic to acidic plutonic rocks of the non-metamorphic terranes, (2) chemical and oxygen isotopic evidence for definite interaction between granite magmas and country rocks in the Ryoke metamorphic belt, and (3) systematic oxygen isotopic zonation in granitic rocks at the periphery of the Ryoke belt. The genetic relationship between the Ryoke and Chugoku granitic rocks will be discussed with special reference to the geological situation of the Ryoke belt.

$^{18}O/^{16}O$ RATIOS OF PLUTONIC ROCKS OF NON-METAMORPHIC TERRANES

According to Matsuhisa and others (1972), most granitic rocks emplaced in non-metamorphic terranes in Japan have uniformly low $^{18}O/^{16}O$ ratios (Fig. 1) with whole-rock $\sigma^{18}O$ values ranging from 7.9 to 8.5 o/oo. They could be assigned to the H1 group of granites of Taylor (1968).

$^{18}O/^{16}O$ ratios have been obtained for the major minerals from plutonic rocks of the Misasa-Okutsu area of the Chugoku province, which is located far from the Ryoke belt.

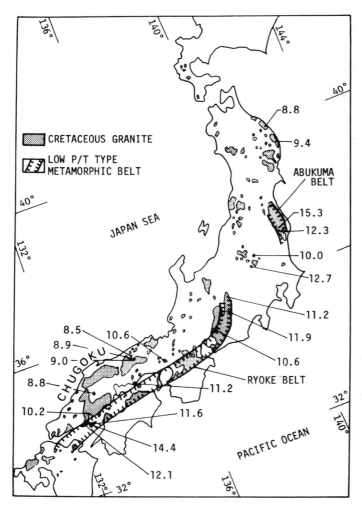

Figure 1. Variation of $\sigma^{18}O$ values of quartz in the Cretaceous granitic rocks of Japan. Data are from the present study and Matsuhisa and others (1972).

The plutonic rocks of the first cycle of volcano-plutonism in the area form, in order of emplacement, a gabbroic complex, a pluton (the Okutsu granodiorite), and a large granitic batholith (the Tottori batholith) — see Figure 2, after Yamada (1961).

Deuteric isotope exchange with meteoric water has greatly lowered the $^{18}O/^{16}O$ ratios of certain minerals from some parts of small and shallow intrusions. However, the primary $^{18}O/^{16}O$ ratios of rocks (or magmas) have been estimated from those of minerals which, by inspection of isotopic equilibrium, were considered to be free of ^{18}O depletion. Analyzed or estimated $^{18}O/^{16}O$ ratios of the representative rock types of the area are summarized in Figure 3.

Assuming that the hornblende gabbro and the enclosed small xenolithic blocks of anorthositic olivine norite (TA4) originated during an early stage of crystallization of the parental magma, it is suggested that the parental magma of

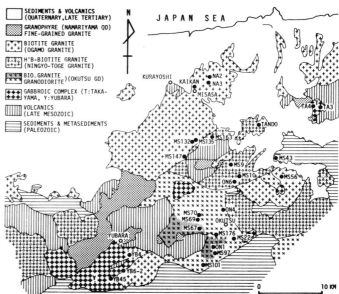

Figure 2. Schematic geology of the Misasa-Okutsu area (after Yamada, 1961) and localities of samples.

the gabbroic complex of this area was, with respect to oxygen, isotopically similar to "mantle materials." The magma of the gabbroic complex seems to have become slightly enriched in ^{18}O during its early crystallization because of the settling of low ^{18}O olivine, pyroxene, and magnetite coupled by flotation or suspension of the higher ^{18}O plagioclase. The high An content of the plagioclase cores and the high Fe/Mg ratio of pyroxene and olivine in this type of complex (Sasada, 1978) support the view that some differentiation did take place.

Granodioritic and granitic rocks are more enriched in ^{18}O than the associated basic rocks, but only slightly. This also suggests the mechanism of selective separation of certain minerals. Matsuhisa and others (1973) showed a regular ^{18}O enrichment inward in a zoned pluton of the Chugoku belt.

It seems unlikely from the above data that sedimentary rocks with high $^{18}O/^{16}O$ ratios played any significant role in the general chemical variation of the magma of the area.

In contrast to the gabbroic complex and Okutsu grano-

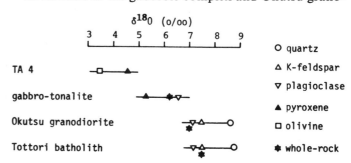

Figure 3. Generalized $\sigma^{18}O$ values of minerals and estimated whole-rock $\sigma^{18}O$ values, showing variation of $^{18}O/^{16}O$ ratios among representative rock-types of the Misasa-Okutsu area.

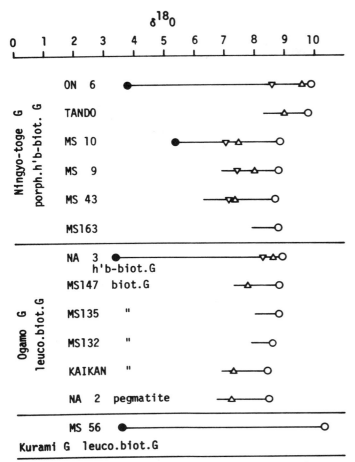

Figure 4. $\sigma^{18}O$ values of minerals from rocks of the Tottori batholith. Data for NA3 are from Matsuhisa and others (1972).

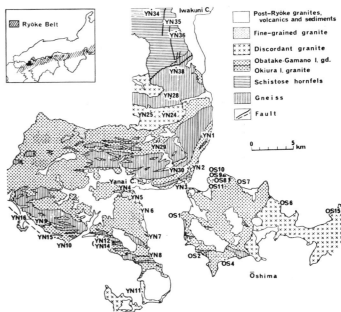

Figure 5. Schematic geology of the Yanai district (after Okamura, 1957) and sample localities.

diorite, local ^{18}O enrichment was detected in parts of the Tottori batholith (Fig. 4). The Ningyo-toge granite, a porphyritic hornblende-biotite granite, includes many minute mafic clots and basic xenoliths and often grades into the Ogamo granite. The Ningyo-toge granite is thought to have been formed from the Ogamo granite magma which had been contaminated by some basic materials rich in ^{18}O. The Kurami granite (MS56) is petrographically much the same as the Ogamo granite but differs in some respects. The $^{18}O/^{16}O$ ratio of its quartz is compatible with the presence of garnet in this granite, and the absence of magnetite suggests contamination by sedimentary rocks.

INTENSE INTERACTION BETWEEN GRANITE MAGMA AND COUNTRY ROCKS IN THE YANAI DISTRICT OF THE RYOKE BELT

The central part of the area (Fig. 5) is occupied by a gneiss-granite complex consisting mainly of "older" granites, "younger" granites, and siliceous banded gneiss with subordinate pelitic gneiss and amphibolite (Okamura, 1957).

The older granites are gneissose and form lenticular masses of varying size which grade into the country rocks. Among them, the Obatake and Gamano (O-G) coarse-grained granodiorites are most widely distributed. Chemical and Sr isotope studies suggest that the main facies of this granodiorite was formed from magma originating at depth (Honma, 1974a, b; Shigeno and Yamaguchi, 1976). On the other hand, judging from the wide distribution of ghost structures of sediments and abundant thin relic bands of siliceous gneiss, the main portion of the Okiura (OKI) fine-grained biotite granite may be essentially metasomatic and partly anatectic in origin. The younger granites form discordant intrusions.

OXYGEN ISOTOPIC EVIDENCE

The $^{18}O/^{16}O$ ratios of pelitic rocks in the marginal metamorphic zone decrease progressively with increasing metamorphic grade (Fig. 6). Advancing homogenization of the oxygen isotopic composition is the most characteristic feature of the gneiss-granite complex. Various types of granitic rocks are rather uniform isotopically, being about 3.0 to 4.0 o/oo enriched in ^{18}O compared to Cretaceous granitic rocks of non-metamorphic terranes in Japan. Most of migmatitic rocks are slightly higher in $^{18}O/^{16}O$ ratio than O-G granodiorite. It is clear that the isotopic homogenization is connected with chemical homogenization (i.e., granitization of metamorphic rocks) on both local and regional scales. However, no definite correlation exists between the isotopic and chemical compositions of rocks. This is most clearly shown by the large isotopic variation in

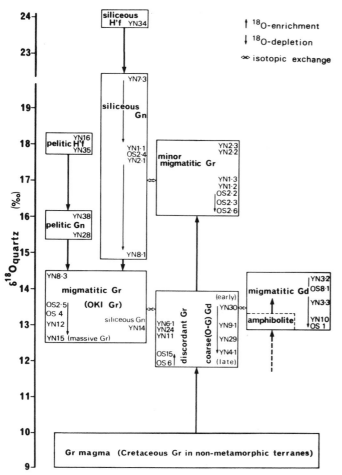

Figure 6. $\sigma^{18}O$ values of quartz showing oxygen isotopic variations within and among each rock-type of the Yanai district (Honma and Sakai, 1975).

the siliceous gneiss, indicating that the isotopic interaction between the granite and country rocks took place, in large part, through fluid media.

SUPPORTING EVIDENCE: CHEMICAL FEATURES OF BIOTITES

The relationship between composition and paragenesis defined by Nockolds (1947) for biotites from calc-alkaline igneous rocks holds for those from various types of rocks from high-grade metamorphic zone, including pelitic and siliceous gneisses, amphibolite, two types of the older granites, and the younger granites (Fig. 7). It was found that both the Al content and paragenesis of minerals are controlled by the chemistry of their host rocks. These facts suggest that all the biotites were formed under roughly equal P/T conditions.

The biotites from the main facies of O-G granodiorite follow the well-defined trend of biotites from the amphibolite without any significant variation in the relative content

of Al. On the other hand, unusually high Al content of biotites from some facies of OKI granite suggests a contribution from sedimentary gneiss to this granite.

Mg/Fe ratios of biotites and their host rocks are similar to and vary in parallel with each other, decreasing with increasing alkalis in the host rocks. This may be due to (1) interaction or exchange of materials between granite magma and rocks which existed prior to its intrusion and (2) crystallization of biotites at low oxygen fugacities, which is suggested by their low Fe^{3+}/Fe^{2+} ratios. Reduction of oxygen fugacity is likely to occur, owing to the reaction of oxygen with graphite in pelitic rocks (Miyashiro, 1964), but why such reduction took place intensely and uniformly among various kind of rocks of the Ryoke belt remains unanswered. All the characteristic features of the biotites from the Ryoke belt can be interpreted as being the result of

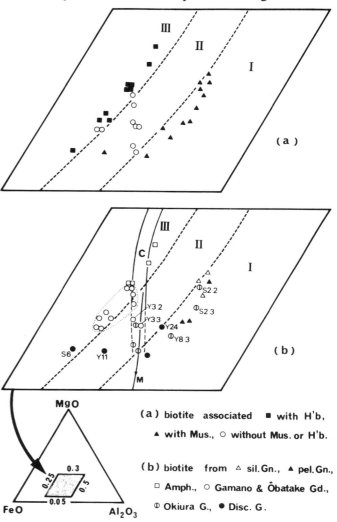

Figure 7. MgO + total iron as FeO - Al_2O_3 (weight per cent) for biotites classified by (a) paragenesis and (b) type of the host rocks. I, II, and III are the fields defined by Nockolds (1947) for biotites from calc-alkaline igneous rocks associated with muscovite, without muscovite or hornblende, and with hornblende, respectively.

intense interaction between the granite magma and country rocks (Honma, 1974b).

Correlation between $^{18}O/^{16}O$ *Ratio and Nh₄ Content of Biotites*

Recently, Itihara and Honma (in preparation) analyzed NH₄ in biotite from granitic and metamorphic rocks of Japan.

Figure 8 clearly shows a positive correlation between $^{18}O/^{16}O$ ratio and NH₄ content of biotites: for example, (1) biotites from granites of non-metamorphic terranes are the lowest in both $^{18}O/^{16}O$ ratio and NH₄ content, the latter averaging 22 p.p.m., (2) two of four biotites analyzed from granites occurring at the periphery of the Ryoke belt follow the general trend, (3) NH₄ contents of biotites from granitic rocks of the Ryoke belt are high, averaging at 67 p.p.m., and show generally an increase with increasing $^{18}O/^{16}O$ ratios, and (4) all biotites from the gneisses of the Ryoke belt are high in both NH₄ and $^{18}O/^{16}O$ ratio and plot on the extension of the variation trend for the granitic rocks.

It is well known that during diagenesis NH₄ forms as a result of decomposition of organic matter and is fixed (bound) or adsorbed in sediments. It is also known that biotite will fix NH₄ present, and fixed NH₄ in biotite is stable up to high temperatures.

Thus, one may be sure that high NH₄ content, as well as high $^{18}O/^{16}O$ ratio, indicates that the rock is genetically or chemically closely related to the sedimentary rocks.

OXYGEN ISOTOPIC ZONATION IN THE HIROSHIMA GRANITE COMPLEX

The Hiroshima granite complex is a member of the volcano-plutonic associations. It lies largely within the non-metamorphic Chugoku belt, but the southern boundary lies inside the margin of the Ryoke metamorphic belt (Fig. 9). Geological and structural evidence shows that the main facies of the complex reached their present position when the Ryoke metamorphism was almost completed (Nureki, 1960).

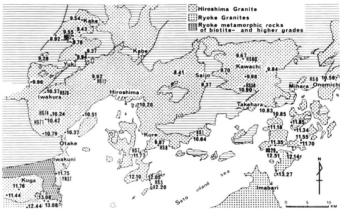

Figure 9. Generalized geological map of the Hiroshima region showing the variation of $\sigma^{18}O$ values of quartz in the main portion of the Hiroshima granite complex and adjacent Ryoke granites (Honma and Sakai, 1976).

Figures 9 and 10 indicate that $\sigma^{18}O$ values of quartz from rocks of the Hiroshima granite complex decrease progressively northward away from the Ryoke belt. Except for three samples (Fig. 11), the fractionation factors among coexisting minerals remain essentially constant, indicating that the isotopic zonation of quartz represents that of the magma which produced this complex. Oxygen isotopic data for rhyolitic crystal tuffs, the forerunners of the granite, support the view that the ^{18}O zonation originated at depth (Honma and Sakai, 1976).

The progressive southward enrichment in ^{18}O toward the Ryoke belt occurred without systematic change in the chemical composition of the Hiroshima magma, and the oxygen isotopic zonation is most likely a result of isotopic interaction between the granite magma and country rocks in and around the Ryoke belt through the fluid media. This

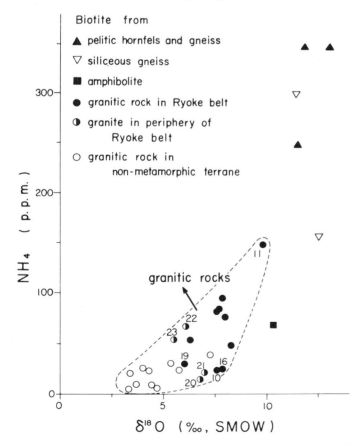

Figure 8. Relationship between NH₄ contents and $\sigma^{18}O$ values of biotite from the Cretaceous granitic rocks of Japan and metamorphic rocks of the Ryoke belt (Itihara and Honma, in preparation).

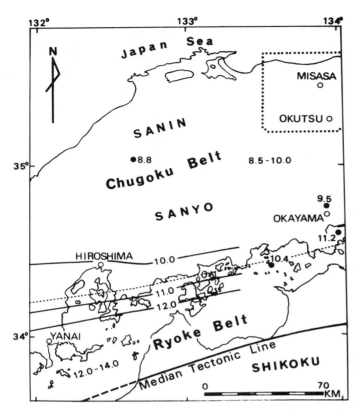

Figure 10. Map showing variation of $\sigma^{18}O$ values of quartz in granitic rocks of the Central Chugoku province. The dashed line indicates the approximate northern rim of the marginal metamorphic zone of the Ryoke belt.

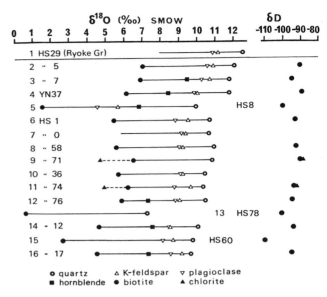

Figure 11. Oxygen and hydrogen isotopic compositions of minerals from rocks of the Hiroshima granite complex and adjacent Ryoke granite. For sample locations, see Figure 9. Abnormal oxygen isotopic fractionation among coexisting minerals and low D/H ratio of biotite (HS8, 60 and 78) are due to deuteric isotope exchange with meteoric water at late- or post-magmatic stages. Data are from Honma and Sakai (1976) and Honma (in preparation).

strongly points to the cognate origin of granite magmas in the Ryoke metamorphic and Chugoku non-metamorphic belts.

ORIGIN OF SPECIAL FEATURES OF THE RYOKE GRANITES

The Ryoke belt may be characterized geologically as the eastern front of the Late Mesozoic to Early Tertiary igneous activity on the eastern margin of the Asiatic continent (Yamada, 1966). Okamura (1957) and Nureki (1960) showed that structural deformation accompanied the plutono-metamorphism of this belt. Such tectonic movement should enhance the mobility of fluid and thus increase the rate of interaction between the granite magma and country rocks. Intense regional thermal metamorphism and chemical features of granitic rocks of this belt are thought to be the result of this intense interaction.

To account for the great width of the zone of ^{18}O enrichment around the Ryoke belt (Fig. 10, contour lines of 10.0 and 12.0 o/oo for $\sigma^{18}O$ of quartz lie at a distance of more than 30 km) and of the marginal metamorphism, the geosynclinal sediments must have subsided deeply preceding the Ryoke plutono-metamorphism, which was, in turn,

followed by orogenic movement. The following characteristic features of many of the "Older Ryoke" granites may be interpreted as having formed synkinematically at depth during the early stages of magmatic activity: (a) predominance of rocks of granodioritic and more basic compositions, (b) concordant intrusion and strong gneissosity, (c) regional thermal metamorphism, and presumed formation and emplacement of some metasomatic and anatectic granites, (d) little chloritization of biotite, and (e) lack of oxygen isotopic interaction with meteoric ground water.

Some of the "Younger Ryoke" granites may have been emplaced near the surface during the late-kinematic stages. As an example, the Towa granite of the Yanai district, which is, in the writer's opinion, a member of the Inagawa-type granite, one of the most widely distributed granites in the Ryoke belt in Chubu (Central) province, forms a large discordant pluton, varying from granodiorite to granite. The rock is, in general, strongly gneissose and parallel with the trend of the country rocks, while in some places it lacks gneissosity. Surrounding regional metamorphic rocks were poly-metamorphosed at the contact (Okamura, 1957). Biotites are generally chloritized moderately, and in acid facies significantly.

Certain features indicate that the Hiroshima granite was emplaced near the surface during the post-kinematic stage, such as: (a) intimate association with effusive forerunners, (b) discordant emplacement, (c) lack of gneissosity, (d) contact effect (poly-metamorphism) on the metamorphic rocks of the Ryoke belt, (e) widespread and significant chloritization of biotite, and (f) local infiltration of oxygen-isotopically light meteoric ground water.

The late- or post-kinematic emplacement of granites does not necessitate a late- or post-kinematic origin of their magma. The data presented in this paper suggest that these magmas were subjected to ^{18}O enrichment at depth in and around the "active" Ryoke belt and emplaced later at high levels.

ACKNOWLEDGMENTS

I wish to express my sincere thanks to H. Sakai and O. Matsubaya of this institute for their kind help in many aspects of this study. Discussions with N. Yamada and M. Sasada of the Geological Survey of Japan have been very useful.

REFERENCES CITED

Honma, H., 1974a, Major element chemistry of metamorphic and granitic rocks of the Yanai district in the Ryoke belt: Journal of the Japanese Association of Mining, Petroleum and Economic Geologists, v. 69, p. 193–204.

——,1974b, Chemical features of biotites from metamorphic and granitic rocks of the Yanai district in the Ryoke belt: Journal of the Japanese Association of Mining, Petroleum and Economic Geologists, v. 69, p. 390–402.

Honma, H., and Sakai, H., 1975, Oxygen isotope study of metamorphic and granitic rocks of the Yanai district in the Ryoke belt, Japan: Contributions to Mineralogy and Petrology, v. 52, p. 107–120.

——,1976, Zonal distribution of oxygen isotope ratios in the Hiroshima granite complex, Southwest Japan: Lithos, v. 9, p. 173–178.

Matsuhisa, Y., Honma, H. Matsubaya, O., and Sakai, H., 1972, Oxygen isotopic study of the Cretaceous granitic rocks in Japan: Contributions to Mineralogy and Petrology, v. 37, p. 65–74.

Matsuhisa, Y., Tainosho, Y., and Matsubaya, O., 1973, Oxygen isotope study of the Ibaragi granitic complex, Osaka, Southwest Japan: Geochemical Journal, v. 7, p. 201–213.

Miyashiro, A., 1964, Oxidation and reduction in the Earth's crust with special reference to the role of graphite: Geochimica et Cosmochimica Acta, v. 28, p. 717–729.

Nockolds, S. R., 1947, The relation between chemical composition and paragenesis in the biotite micas of igneous rocks: American Journal of Science, v. 245, p. 401–420.

Nureki, T., 1960, Structural investigation of the Ryoke metamorphic rocks of the area between Iwakuni and Yanai, Southwestern Japan: Journal of Science, Hiroshima University, (C), v. 3, p. 69–141.

Okamura, Y., 1957, Structure of the Ryoke metamorphic and granodioritic rocks of the Yanai district, Yamaguchi Prefecture: Journal of the Geological Society of Japan, v. 63, p. 684–697.

Sasada, M., 1978, Late Cretaceous-Paleogene intrusive rocks in the Yubara area, Okayama Prefecture, Southwest Japan: Geological Survey of Japan Bulletin, v. 84, p. 23–34.

Shigeno, H., and Yamaguchi, M., 1976, A Rb-Sr isotopic study of metamorphism and plutonism in the Ryoke belt, Yanai district, Japan: Journal of the Geological Society of Japan, v. 82, p. 687–698.

Taylor, H. P., 1968, The oxygen isotope geochemistry of igneous rocks: Contributions to Mineralogy and Petrology, v. 19, p. 1–71.

Yamada, N., 1961, Explanatory text of the geological map "Okutsu": Geological Survey of Japan.

——,1966, Nature of the late Mesozoic igneous activities in and around Southwest Japan: Earth Sciences (Chikyu Kagaku), v. 20, p. 53–58.

MANUSCRIPT ACCEPTED BY THE SOCIETY JULY 12, 1982

PRINTED IN U.S.A.

Geological Society of America
Memoir 159
1983

Mesozoic granitoids of northeast Asia

N. A. Shilo
Far Eastern Science Centre
Vladivostok, U. S. S. R.

A. P. Milov and A. P. Sobolev
Northeast Interdisciplinary Scientific
Research Institute
Magadan, U. S. S. R.

ABSTRACT

Most of the plutonic masses in northeastern Asia were emplaced in the late Mesozoic. They exhibit a wide range of compositions and, on the basis of ranges, the batholiths form two groups, one acidic with a narrow range from granodiorite to granite, and the other with a wide range extending from gabbro through diorite and tonalite to granodiorite and granite. The Verkhoyansk-Chukchi region, with its Preriphean crust, is characterized by the acidic series, as are miogeosynclinal regions in general. The gabbro-granite series is typical of eugeosynclines such as the northeast-trending Okhotsk-Chukchi volcanogenic belt.

The acidic series shows no close relationship with volcanic processes and is suggestive of low-temperature melts. The gabbro-granite series is closely related in space and time with volcanism (mainly andesitic) and is thought to be co-magmatic with the volcanic rocks. They imply derivation from relatively dry, high temperature magmas. Although the gabbro-granite series has $^{87}Sr/^{86}Sr$ ratios below 0.707, mantle sources are considered unlikely; they are accounted for by proposing a lower crust, granulite-eclogite source.

The Okhotsk-Chukchi volcanogenic belt is divided into an inner (continent side) zone and an outer (ocean side) zone. Plutonic rocks in the inner zone are potassic, whereas those in the outer zone are sodic. The boundary between the zones is thought also to mark the junction between sialic continental crust and crust transitional to the ocean basin.

INTRODUCTION

In northeastern Asia, the most vigorous granitoidal magmatism occurred during the late Mesozoic. In various parts of the region under consideration, granitoids differ regularly due to their distribution, composition and environment, association with volcanic processes, and spatial and time relationships with other types of plutonic rocks.

Batholiths composed largely of granodiorite, adamellite and granite, and plutons of the wider gabbro to granite range are identified fairly well. Among the latter, gabbro-plagiogranite, gabbro-diorite-tonalite, gabbro-diorite-granodiorite, gabbro-granodiorite-granite, and gabbro-syenite-alkali granite series are identified.

In the Verkhoyansk-Chukchi area (Fig. 1) with its Preriphean continental crust, the largest northeastern batholiths formed during the Late Jurassic and Early Cretaceous. These batholiths cover up to 7,000 km² and are composed predominantly of a granodiorite-granite complex.

YANA-KOLYMA MIOGEOSYNCLINAL SYSTEM

Within the Yana-Kolyma miogeosynclinal system, which is composed of Late Paleozoic to Jurassic terrigenous deposits (10 km thick), plutons form chains, linear belts, and orderly series concordant with the patterns of pre-existing

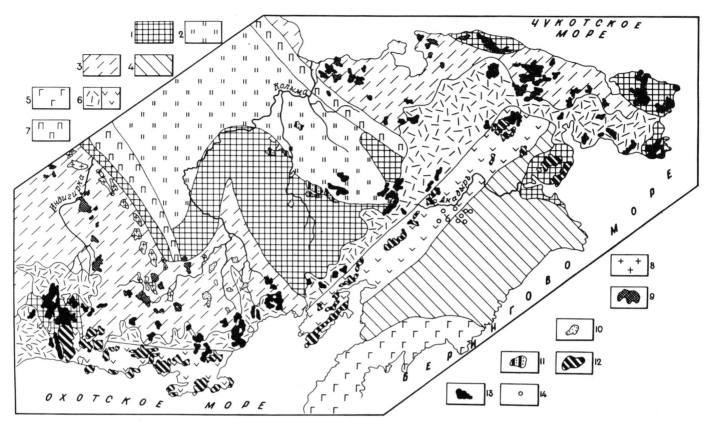

Figure 1. Locations of the Mesozoic intrusive complexes and series on the tectonic map of northeastern U. S. S. R. 1. Blocks with Precambrian basement and thin cover of Paleozoic and Mesozoic deposits. 2. Alazeya-Oloy eugeosynclinal system. 3. Yana-Kolyma and Chukchi systems within Inyali-Debin synclinorium. 4. Anadyr-Koryak Cenozoic eugeosynclinal system. 5. Olyutor-Kamchatka Cenozoic eugeosynclinal system. 6. Okhotsk-Chukchi volcanogenic belt: (a) outer zone and (b) inner zone. 7. Regenerated Late Mesozoic depressions. 8. Complex of peraluminous granites. 9. Granodiorite-granite complex. 10. Complex of leucocratic granites. 11. Gabbro-granodiorite-granite series (Oloy zone). 12. Gabbro-tonalite series of the Okhotsk-Chukchi volcanogenic belt's outer zone. 13. Gabbro-granodiorite-granite series of the Okhotsk-Chukchi volcanogenic belt's inner zone. 14. Gabbro-plagiogranite series.

folds and faults. These massifs are related to major zones of deep-seated faults, and the largest are bounded by districts with much different geologic histories. For example, the principal batholithic belt extends for some 2,500 km near the western border of the Alazeya-Oloy eugeosynclinal system and is confined to the Debin synclinorium with its thinned continental crust. Some rows of plutons accentuate the outlines of the Preriphean median massifs.

INYALI-DEBIN SYNCLINORIUM

In the Inyali-Debin synclinorium, massifs are lens-shaped and less than 10 km thick (Belyayev and others, 1968; Vaschilov, 1963). These massifs are elongated northwesterly.

The intrusives (Unit 8, Fig. 1) are composed of medium- to coarse-grained adamellite and granite characterized by the following: zoned plagioclase (An$^{25-50}$), microcline (albite, 15-20%), biotite (iron, 66-70%; alumina, 25-28%), muscovite,

and quartz. Also present are garnet, cordierite, andalusite, and sillimanite. The rocks contain significant ilmenite. Adamellite and granite are alumina enriched, and calcium and sodium depleted; sodium is subordinate to potassium (Fig. 2, Table 1).

The rocks under consideration were probably formed during the Late Jurassic (Zagruzina, 1975; Sobolev and Kolesnichenko, 1979).

In anticlinal uplifts and domes belonging to the Yana-Kolyma system, plutons (Unit 9, Fig. 1) form transversal, submeridianal and northwesterly trending rows composed of Early Cretaceous biotite-hornblende granodiorite, a-damellite, and granite ("calcic granitoids"). The minerals in these rocks are: plagioclase (An$^{26-40}$), alkali feldspar, biotite (iron, 56-70%; alumina, 14-21%), hornblende, and quartz. Sphene, zircon, and ilmenite as accessory minerals are typical of these rocks.

Granitoids are alumina enriched, and calcium and alkali

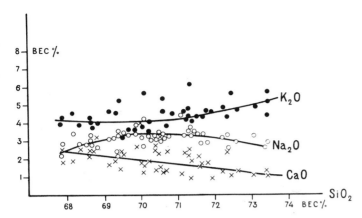

Figure 2. Alkali and calcium variation diagram for Mesozoic alumina granites of the Yana-Kolyma system.

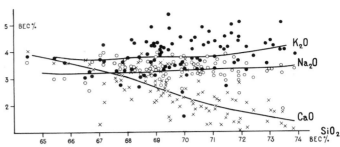

Figure 3. Alkali and calcium variation diagram for the Mesozoic granodiorite-granite complex of the Yana-Kolyma system.

depleted; potassium is slightly more abundant than sodium (Fig. 3).

An isolated lower Upper Cretaceous complex of leucocratic granite (Unit 10, Fig. 1) crops out in the southeastern part of the system (Sobolev, 1971, 1976). This complex comprises plutons which are elongated, faulted bodies trending along faults. The mineralogical and chemical compositions of the granite are normal; plagioclase is subordinate to alkali feldspar (ratios about 1:2.5-3). The minerals in these rocks are: albite-oligoclase (An $^{04-14}$), orthoclase, biotite (iron content, 80%; alumina content, 17-22%), and quartz. Accessory minerals are represented by

allanite, magnetite, monazite, and apatite; xenotime, cassiterite, and tourmaline are also found (Sobolev, 1975). The granite is typical of potassic varieties.

CHUKCHI MIOGEOSYNCLINE FOLDED SYSTEM

The granitoid batholiths (Unit 9, Fig. 1) of the Chukchi miogeosynclinal folded system are confined to anticlinal uplifts and junctions between northwesterly and northeasterly striking faults. Geological and geophysical evidence testifies to their being stratiform in section and up to 4-5 km thick. Potassium-argon ages indicate the granitoids are Early Cretaceous. The granitoids are localized in Triassic terrigenous or upper Paleozoic carbonate-terrigenous deposits, the latter being characteristic of uplifts.

TABLE 1. AVERAGE CHEMICAL COMPOSITIONS OF INTRUSIVE ROCKS FROM
THE VERKHOYANSK-CHUKCHI MIOGEOSYNCLINAL FOLDED SYSTEM

| | Yana-Kolyma system | | | | | Chukchi folded system | | | | | | | | | | | |
| | | | | | | western part | | | | | | eastern part | | | | | |
	1	2	3	4	5	6	7	8	9	10	11	12	13	14	15	16	17
SiO₂	68.40	71.59	67.53	71.87	75.30	56.02	60.34	65.66	68.00	70.16	74.27	54.49	60.84	65.11	67.81	70.51	74.18
TiO₂	0.56	0.33	0.51	0.27	0.15	0.85	0.70	0.56	0.48	0.35	0.21	1.04	0.69	0.57	0.48	0.35	0.23
Al₂O₃	15.45	14.66	15.86	14.56	12.92	15.47	16.06	15.62	15.01	14.75	13.55	17.72	16.85	15.98	15.47	14.73	13.56
Fe₂O₃	1.14	0.68	1.13	0.76	0.68	2.41	1.20	0.92	0.75	0.50	0.63	1.21	1.39	0.97	1.04	0.74	0.52
FeO	3.19	2.08	2.77	1.89	1.43	5.60	4.07	3.34	2.81	2.46	1.29	4.94	3.95	3.04	2.37	1.98	1.38
MnO	0.06	0.05	0.06	0.04	0.03	0.16	0.10	0.08	0.07	0.06	0.04	0.11	0.10	0.09	0.11	0.05	0.09
MgO	1.04	0.64	1.13	0.60	0.23	6.20	3.76	2.14	1.70	1.10	0.44	3.45	2.64	1.95	1.35	0.87	0.35
CaO	2.40	1.68	2.92	1.85	0.68	6.46	4.67	3.25	2.62	2.09	1.19	4.99	3.81	3.00	2.29	1.77	0.99
Na₂O	2.91	3.22	2.92	3.35	3.25	2.98	3.10	3.24	3.58	3.55	3.40	3.18	3.23	3.27	3.39	3.42	3.07
K₂O	3.94	4.36	3.36	4.09	4.69	2.44	3.63	3.49	4.15	4.35	4.62	5.14	4.43	4.62	4.11	4.56	4.64
No. An.	8	37	21	61	114	7	24	19	22	41	19	3	27	50	46	54	124

1,2 - high-alumina adamellites (1) and granites (2) from Inyali-Debin synclinorium; 3,4 - "calcareous" adamellites (3) and granites (4); 5 - leucocratic granites; 6,12 - diorites; 7,13 - quartz diorites; 8,14 - granodiorites; 9,15 - adamellites; 10,16 - granites; 11,17 - leucocratic granites.

The plutons comprise quartz diorite, hornblende-biotite granodiorite and granite, and biotite and leucocratic two-mica granite.

Quartz diorite is characterized by plagioclase (An^{35-60}) with hornblende (iron content 30-40%), biotite (iron content, 40-45%), alkali feldspar, and quartz. They constitute minor bodies measuring less than 0.5 km².

The mineralogy of the most common hornblende-biotite granite and granodiorite is represented by plagioclase (An^{20-45}), alkali feldspar (albite, 30-40%), hornblende (iron, 39-46%; alumina, 18-20%), biotite, and quartz. More sodic plagioclase (An^{18-36}) and higher contents of iron (82-92%) and alumina (19-30%) in biotite are typical of biotite and two-mica granites. Garnet (almandine) is common.

The accessory minerals include sphene, allanite, apatite, and ilmenite.

Potassium is significantly more abundant than sodium in all these rocks (Fig. 4, Table 1).

OLOY EUGEOSYNCLINAL ZONE

In the Oloy eugeosynclinal zone, granitoidal plutons (Unit 11, Fig. 1) are localized in Triassic-Jurassic volcano-sedimentary deposits and andesitic to basaltic rocks of presumably Late Jurassic to Early Cretaceous age.

The earliest in the Oloy zone are the Late Jurassic to Early Cretaceous plutons composed of the gabbro-syenite-granosyenite series. They commonly constitute small bodies (50-150 km²) controlled by deep-seated meridianally striking faults. The plutons were emplaced during three intervals, from oldest to youngest: gabbro and gabbro-diorite; syenite-diorite; and quartz syenite and granosyenite. Granodiorite and granite also occur with them.

The syenite-diorite-granosyenite complex is characterized by plagioclase (An^{15-40}) with alkali feldspar, amphibole, biotite, and quartz. Magnetite, apatite, and sphene are accessory minerals.

Sodium prevails noticeably over potassium throughout the whole rock series (Fig. 5).

The late Lower Cretaceous gabbro-granodiorite-granite series constitutes rather large batholith-like massifs (700-900 km²) which form a northwesterly striking row.

The plutons were emplaced during several phases, from oldest to youngest: gabbro and gabbro-diorite; quartz

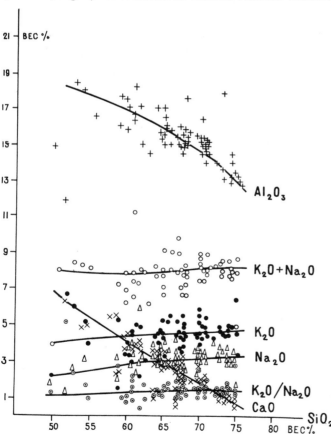

Figure 4. Variation diagram for average alkali, calcium, and aluminum values in the granodiorite-granite complex of the Chukchi folded system.

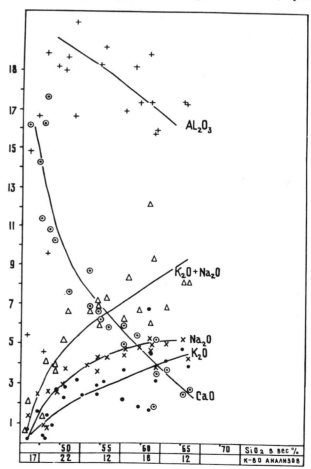

Figure 5. Variation diagram for average alkali, calcium, and aluminum values in the syenite-diorite-granosyenite complex of the Oloy folded system.

diorite and granodiorite; granite; and normal and alkaline granite.

Gabbro and gabbro-diorite are characterized by plagioclase (An^{70-80}) with hornblende, and monoclinic and rhombic pyroxene. Individual olivine grains are sometimes found in the gabbro.

Granodiorite and biotite-amphibole granite are characterized by the following minerals: plagioclase (An^{25-45}), alkali feldspar (albite component, 20-25%), hornblende (iron, 40-50%), low-iron biotite (iron, 40-48%), and quartz.

Normal and alkaline granites are commonly leucocratic with miarolitic cavities and micropegmatite textures; alkali feldspar predominates significantly over plagioclase. The mineral components are alkali feldspar (about 50% albite), albite-oligoclase (An^{05-15}), biotite (iron, 60%; alumina, 16-22%), individual hornblende grains, and quartz. Aegirine, riebeckite, and hastingsite are also found in alkaline granite.

The rocks are characterized by magnetite, zircon, apatite, and sphene as accessory minerals. Xenotime, tantalite, columbite, and fergusonite are also found. The rocks in this series range from potassic to sodic (Fig. 6, Table 2).

OKHOTSK-CHUKCHI VOLCANOGENIC BELT

The most important event during the early Late Cretaceous phase of granitoidal magmatism was the creation of the Okhotsk-Chukchi volcanogenic belt extending for some 3,000 km.

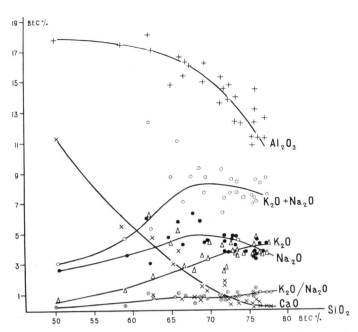

Figure 6. Variation diagram for average alkali, calcium, and aluminum values in the gabbro-granodiorite-granite series of the Oloy folded system: (a) Talalakh complex and (b) Khetakchan complex.

According to V. F. Bely (1977, 1978), two types of gabbro-granite series were being formed early in the development of the belt (Albian-Cenomanian). A continuous series, gabbro-

TABLE 2. AVERAGE CHEMICAL COMPOSITIONS OF INTRUSIVE ROCKS FROM THE EARLY CRETACEOUS COMPLEX OF THE OLOY ZONE

	1	2	3	4	5	6	7	8	9	10	11	12	13	14	15	16	17
SiO_2	50.85	49.69	53.40	53.36	61.77	62.44	64.06	65.63	66.62	66.29	67.89	70.80	71.53	70.67	75.44	75.08	75.89
TiO_2	1.11	2.28	0.77	1.84	0.59	0.53	0.49	0.53	0.42	0.41	0.60	0.41	0.29	0.28	0.16	0.21	0.13
Al_2O_3	17.93	15.91	19.47	15.88	17.37	17.60	17.06	16.58	16.53	16.82	15.82	15.00	15.17	15.35	12.92	13.33	11.30
Fe_2O_3	4.22	2.83	3.18	2.44	2.53	2.44	2.97	2.39	1.37	1.66	1.49	1.37	1.11	1.17	1.08	0.92	2.75
FeO	5.83	8.42	4.67	8.76	3.40	2.72	1.64	2.05	2.67	2.17	1.99	1.39	2.07	1.77	1.05	1.11	1.22
MnO	0.19	0.20	0.14	0.22	0.11	0.10	0.17	0.14	0.06	0.09	0.08	0.10	0.06	0.07	0.05	0.05	0.10
MgO	6.54	6.38	4.00	4.55	2.90	2.02	0.47	0.45	2.13	1.55	0.66	0.48	1.06	0.68	0.23	0.19	0.03
CaO	9.40	8.95	7.82	6.97	5.20	3.54	0.92	0.80	3.56	2.32	1.74	1.04	2.37	1.59	0.52	0.46	0.28
Na_2O	2.93	3.70	3.83	4.26	4.03	4.86	6.38	5.87	4.08	4.61	5.25	4.60	4.41	4.74	4.12	4.08	4.27
K_2O	0.94	0.78	2.69	1.15	2.06	3.65	5.76	5.25	2.55	4.06	4.37	4.46	1.91	3.65	4.40	4.41	4.00
No. An.	19	13	11	3	16	16	2	10	17	25	4	8	3	18	21	11	2

1,2 - gabbro; 3 - monzonite; 4 - syenite; 5 - quartz syenite-diorite, syenite-diorite; 6- quartz syenite-diorite; 7 - alkaline syenite; 8 - nordmarkite; 9 - granodiorite and granosyenite; 10,11 - granosyenite; 12,13,14 - normal granite; 15 - normal and alkali granite; 16 - alaskite granite; 17 - alkali granite.

Analyses nos. 1,3,5,6,9,10,13,14,15 by I.N. Kotlyar, V.F.Bely, A.P. Milov (1981); nos. 16 and 17 by Ye.F. Dylevsky (1980).

granodiorite-granite (Unit 13, Fig. 1), is typical of the belt's outer zone which is superimposed on the structures of mio- and eugeosynclines and ancient rigid massifs (Bely, 1977, 1978). The plutons composed of this rock series are commonly associated with basaltic andesite, andesite, dacite, and liparitic volcanic cover rocks. Some massifs transect Mesozoic structures as well as ancient massifs and form well-defined rows along the belt's strike. Plutons are located within major uplifts, and, in places, they surround certain volcano-tectonic depressions. The massifs underlie a few dozens to 1,500 km². They are presumably dome-like or stratiform in section.

Many intrusives are heterogeneous and show successive intrusions of gabbro and gabbro-diorite, diorite and quartz diorite, and granodiorite and granite (leucocratic normal and alkaline granite). Biotite-amphibole granodiorite (about 50%) and granite (about 40%) dominate the intrusive series.

Insignificant amounts of alkali feldspar and quartz are found in the gabbro and gabbro-diorite. These are predominantly amphibolic rocks. The main minerals are: plagioclase (An$^{56-75}$), hornblende, and monoclinic or rhombic pyroxene.

The granodiorite and biotite-amphibole granite are characterized by plagioclase (An$^{26-45}$), hornblende (iron, 40-50%; alumina, 7-12%), biotite (iron, 48-56%; alumina, 14-19%), alkali feldspar, and quartz.

The leucocratic granite consists of plagioclase (albite-oligoclase), alkali feldspar, biotite (iron, 60-70%; alumina, 16-21%), hornblende (iron, 40-45%), and quartz. In some places, these rocks grade into alkaline granite with riebeckite, aegirine, and arfvedsonite. In the granitoids under consider-

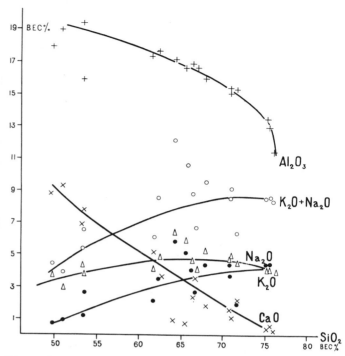

Figure 7. Variation diagram for average alkali, calcium, and aluminum values in the gabbro-granodiorite-granite series of the Okhotsk-Chukchi volcanogenic belt's outer zone.

ation, magnetite, apatite, sphene, allanite, and zircon are accessory minerals. Potassium and sodium values are roughly equal in the intrusive series (Fig. 7, Table 3).

The volcanogenic belt's inner zone is superimposed on the structures of the Koni-Taigonos tectonic-magmatic zone

TABLE 3. AVERAGE CHEMICAL COMPOSITIONS OF INTRUSIVE ROCKS FROM THE OUTER AND INNER ZONES OF THE OKHOTSK-CHUKCHI VOLCANOGENIC BELT

	outer zone						inner zone					
	1	2	3	4	5	6	7	8	9	10	11	12
SiO₂	52.08	56.90	60.92	65.95	70.90	75.40	48.71	54.86	61.60	65.71	69.76	75.27
TiO₂	1.31	0.95	0.80	0.95	0.35	0.17	0.84	0.86	0.67	0.52	0.34	0.17
Al₂O₃	18.80	18.30	17.50	16.50	15.22	13.30	17.48	18.41	16.89	16.17	15.20	13.27
Fe₂O₃	3.80	3.04	2.26	1.63	1.25	1.20	4.13	2.75	2.18	1.80	1.29	0.74
FeO	5.50	4.82	4.29	3.30	1.93	1.18	5.53	5.16	3.98	2.67	1.91	0.74
MnO	0.15	0.15	0.12	0.09	0.07	0.04	0.12	0.13	0.11	0.08	0.06	0.05
MgO	5.10	3.84	2.72	1.69	0.82	0.30	8.30	4.07	2.66	2.01	1.14	0.53
CaO	8.35	6.53	4.78	3.35	1.81	0.73	10.64	7.80	5.60	4.10	3.08	1.11
Na₂O	3.27	3.50	3.84	3.74	3.82	3.66	2.37	3.48	3.64	3.88	4.01	3.84
K₂O	1.50	1.94	2.70	3.27	3.85	4.20	0.48	0.81	1.53	2.16	2.41	3.89
No. An.	74	29	93	186	207	224	27	21	29	38	29	7

1,7 - gabbro; 2,8 - diorite; 3,9 - quartz diorite; 4,10 - granodiorite; 5,11 - granite; 6,12 - leucocratic granite.

TABLE 4. AVERAGE CHEMICAL COMPOSITIONS OF GRANITOID ROCKS FROM THE OKHOTSK-CHUKCHI VOLCANOGENIC BELT OUTER ZONE AND THE ANADYR-KORYAK FOLDED SYSTEM

	1	2	3	4	5	6	7	8	9	10
SiO_2	50.35	55.41	60.54	65.41	68.30	75.47	46.72	58.26	66.69	73.87
TiO_2	1.35	1.08	0.81	0.57	0.39	0.16	0.33	0.44	0.36	0.27
Al_2O_3	18.30	18.07	16.80	16.24	15.37	12.93	16.55	16.02	13.64	12.51
Fe_2O_3	3.48	4.00	1.81	1.81	1.44	1.13	6.42	6.34	3.16	1.75
FeO	5.87	3.56	3.73	2.81	1.99	2.09	2.63	2.78	1.91	1.31
MnO	0.15	0.10	0.12	0.09	0.07	0.03	0.07	0.06	0.08	0.05
MgO	3.29	3.55	2.51	1.46	1.04	0.27	8.94	4.14	2.91	1.21
CaO	9.11	6.91	4.19	2.83	2.35	0.42	11.95	6.72	4.23	2.21
Na_2O	3.19	3.72	4.32	4.42	4.14	3.96	2.12	4.86	4.14	4.71
K_2O	1.18	1.85	2.71	3.24	3.29	4.35	0.33	0.46	0.38	0.38
No. An.	5	12	37	41	28	25	47	6	15	47

1 - gabbro, 2 - diorite, 3 - quartz syenite-diorite, 4 - quartz monzonite, 5 - granosyenite, 6 - alkali and subalkali granite, 7 - gabbro, 8 - quartz diorite, 9 - granodiorite, 10 - plagiogranite.

("andesite" geosyncline of V. F. Bely), and it was the site for the development of a gabbro-tonalite-plagiogranite series (Unit 12, Fig. 1). Plutons, as large as 1,000 km², are associated with subaerial accumulations of volcanic rocks having basaltic to andesitic compositions. Massifs are arranged in markedly linear rows which are subordinate to the general northeasterly strike of the structures incorporated by the Okhotsk-Chukchi volcanogenic belt. They were formed under epizonal conditions and are laccolithic in shape. Biotite-hornblende tonalite and hornblende diorite dominate this series, forming up to 80%. Geochronological data suggest an Early Cretaceous (Aptian-Albian) age for these rocks.

In this zone, plutons were emplaced during several phases: gabbro and gabbro-diorite; diorite, granodiorite and tonalite; granodiorite and granite; and plagiogranite and leucocratic granite.

The gabbro is bipyroxene, commonly olivine-bearing, with plagioclase corresponding to labradorite and bytownite (An^{64-78}).

The tonalite and quartz diorite have the following mineral composition: plagioclase (An^{26-54}), hornblende (iron, 40-44%), biotite (iron, 52-54%; alumina, 12-16%), alkali feldspar, and quartz. In the plagiogranite and granite, plagioclase is represented by oligoclase (An^{18-24}), alkali feldspar

(10-22%), biotite (iron, 55%), and quartz. Rare hornblende grains are present. Magnetite, apatite, zircon, and rare sphene are accessories. Sodium prevails over potassium in all of the rock series (Fig. 8, Table 4). The intrusive series from the belt's outer and inner zones have a complicated relationship with the volcanic rock, the plutons having formed subsequent to the volcanic rocks.

During the later stage of the belt's evolution, a sodic series of syenite, diorite, and alkali granite was created. Intrusives of this series are irregularly distributed and occur in both zones of the belt. They are small (90-100 km²) and seem to be stocklike in shape. Hornblende syenite-diorite, syenite, and granosyenite, as well as alkali granite, are the commonest rocks. K-Ar age determinations made on these rocks point to their Late Cretaceous origin.

Syenite and granosyenite are characterized by oligoclase (An^{18-26}) with alkali feldspar, hornblende, biotite, and quartz. Magnetite, zircon, and apatite are accessory minerals. Sodium is markedly predominant over potassium in these rocks (Fig. 9, Table 4).

ANADYR-KORYAK EUGEOSYNCLINAL SYSTEM

The Anadyr-Koryak eugeosynclinal system features widely developed Late Jurassic-Cretaceous volcano-sedimen-

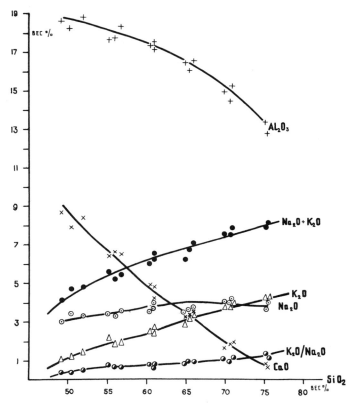

Figure 8. Variation diagram for average alkali, calcium, and aluminum values in the gabbro-tonalite-plagiogranite series of the Okhotsk-Chukchi volcanogenic belt's inner zone.

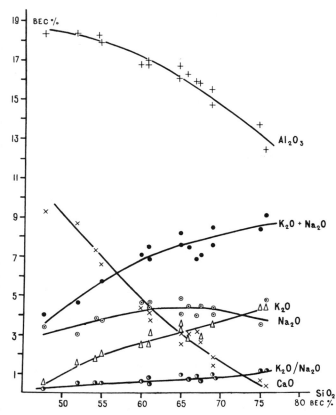

Figure 9. Variation diagram for average alkali, calcium, and aluminum values in the syenite-diorite-alkali granite series of the Okhotsk-Chukchi volcanogenic belt's outer zone.

tary deposits. The granitoids (Unit 14, Fig. 1) discussed earlier are absent from this system. Rare and small bodies of the gabbro-plagiogranite series are associated with Alpine-type ultramafic complexes.

The intrusives are mostly platelike bodies commonly having tectonic contacts with the host rocks. Geological evidence indicates a post-Valanginian to pre-Cenomanian age for these rocks.

The layered gabbros are characterized by plagioclase (An^{55-88}), hornblende, monoclinic and rhombic pyroxene, and olivine.

The plagiogranite of the second intrusive phase in the massifs consists of the following minerals: plagioclase (An^{18-26}), hornblende (iron, 47-60%; alumina, 10-15%), biotite (iron, 55%), and quartz. Micropegmatite, quartz, and plagioclase intergrowths are common. All these rocks are typical sodium entities (Table 4).

DISCUSSION

The above considerations imply that the distribution pattern of plutonic rocks in Northeast U.S.S.R. reflects the general circum-Pacific zonal character of the magmatism but is also related to the regional tectonic zonality.

The gabbro-granite series and granodiorite-granite complexes were created in different geotectonic environments; the former are typical of eugeosynclines and the Okhotsk-Chukchi volcanogenic belt, whereas the latter occur in miogeosynclines.

The granodiorite-granite complexes of the miogeosynclinal zones constitute epizonal plutons which do not show any close relationship with volcanic processes. The early plutonic phases are not preceded by gabbro or diorite. The mineral composition of the granitoids suggests moderately aqueous and comparatively low-temperature melts (Sobolev and Kolesnichenko, 1979; Milov and Sobolev, 1979).

The gabbro-granite series is closely related in space and time with products of volcanic eruptions (mainly andesitic rocks), and probably are comagmatic with them. Volcanism commonly preceded the emplacement of the intrusives. These granitoids were derived from low-water, high-temperature magmas.

The identified petrologic features of the granitoids are reflected in the initial strontium ratios. In rocks from the granodiorite-granite complexes, the $^{87}Sr/^{86}Sr$ ratios are 0.709 or more but are below 0.707 in rocks belonging to the gabbro-granite series.

The differences in the isotopic strontium values and the

petrologic features seem best accounted for by the melting of crustal substrata of different compositions, being "granite-metamorphic" in the case of higher initial strontium ratios and "granulite-eclogitic" for low ratios. Mantle sources are considered to be unlikely, the fact being supported by the absence of granitoids in oceanic crust areas.

The information about granitoids in Northeast U.S.S.R. makes possible the identification of I- and S-types of granitic rocks (White and Chappel, 1977) or the equivalent, magnetite and ilmenite types of Ishihara (1977). However, plutonic rocks which are transitional between these two types are certainly present, i.e., granodiorite-granite complexes of the Yana-Kolyma and Chukchi folded systems, which reveal numerous similarities with the Caledonian granitoids of Donegal (Pitcher and Berger, 1972).

CONCLUSIONS

In summary, the authors should mention that the sodic to potassic gabbro-granite series and the potassic granodiorite-granite complexes did not form southeast of the boundary separating the outer and inner zones of the Okhotsk-Chukchi volcanogenic belt. The sodic gabbro-granite series persisted through every stage of development of the structures in the inner zone of the Okhotsk-Chukchi belt. The zone's boundary is considered to be a line demarcating areas having transitional and continental crust types.

Widespread manifestations of plutonic rocks in the belt's outer zone result from partial degradation (melting) of the continental crust type.

REFERENCES CITED

Bely, V. F., 1977, The stratigraphy and structures of the Okhotsk-Chukchi volcanogenic belt: Nauka Publ., Moscow.

Bely, V. F., 1978, The formations and tectonics of the Okhotsk-Chukchi volcanogenic belt: Nauka Publ., Moscow.

Belyayev, I. V., Belyayeva, D. N., and Migovich, I. M., 1968, Igneous rocks and the structure of the abnormal physical fields in Northeastern U. S. S. R.: Abstracts, 1st Regional Petrographic Meeting, Magadan.

Ishihara, S., 1977, The magnetite-series and the ilmenite-series granitic rocks: Mining Geology, v. 27, p. 293–305.

Milov, A. P., and Sobolev, A. P., 1979, Biotite as an indicator of the conditions under which granitoids were formed (as exemplified by the major Mesozoic complexes): *in* Magmatic and metamorphic complexes of Northeastern U. S. S. R., Magadan.

Pitcher, W. S., and Berger, A. R., 1972, The geology of Donegal: a study of granite emplacement and unroofing: Wiley Interscience, London, 435 p.

Sobolev, A. P., 1971, Some features peculiar to tin-bearing biotite granites: Geology of Ore Deposits, n. 3.

Sobolev, A. P., 1975, Accessory minerals in Late Mesozoic granitoids of the Northeastern U. S. S. R.: Transactions, Northeastern Interdisciplinary Scientific Research Institute, Magadan, issue 6.

Sobolev, A. P., 1976, On the problem of relationships between tin mineralization and magmatism in Northeastern U. S. S. R.: Transactions, Northeastern Interdisciplinary Scientific Research Institute, Magadan, issue 69.

Sobolev, A. P., and Kolesnichenko, P. P., 1979, Mesozoic granitoidal complexes in the southern Yana-Kolyma folded system: Nauka Publ., Moscow.

Vashchilov, Yu. Ya., 1963, Deep-seated faults in the southern Yana-Kolyma folded zone and the Okhotsk-Chukchi volcanic belt and their role in the emplacement of granitic intrusions during the development of the structures (based on the geophysical data): Sovetskaya Geologiya, n. 4.

White, A. J. R., and Chappel, B. W., 1977, Ultrametamorphism and granitoid genesis: Tectonophysics, v. 43, p. 7–22.

Zagruzina, I. A., 1975, Geochronology of the Mesozoic granitoids from the Northeastern U. S. S. R.: Nauka Publ., Moscow.

MANUSCRIPT ACCEPTED BY THE SOCIETY JULY 12, 1982

Geological Society of America
Memoir 159
1983

Calc-alkaline plutonism along the Pacific rim of southern Alaska

Travis Hudson
U.S. Geological Survey
1209 Orca Street
Anchorage, Alaska 99501

ABSTRACT

Field, petrology, and age data on southern Alaskan plutonic rocks now enable the delineation of eight calc-alkaline plutonic belts. These belts of plutons or batholithic complexes are curvilinear to linear and trend parallel or subparallel to the continental margin. The belts represent the principal loci of emplacement for plutons of specific ages, and although there is spatial or temporal overlap in some cases, they are, more commonly, spatially and temporally distinct. Intermediate lighologies, such as quartz diorite, tonalite, and granodiorite, dominate in most of the belts, but granodiorite and granite characterize one. The belts are of Mesozoic or Cenozoic age, and plutonism began in six of them at about 195, 175, 120, 75, 60, and 40 m.y. ago; age relations in two are poorly known. Recognition of the belts is important for future studies of regional geology, tectonism, and magmatism along the Pacific rim of southern Alaska.

INTRODUCTION

Southern Alaska is a part of the Pacific rim that is now the site of major plate convergence and may have been so during earlier intervals of the Mesozoic and Cenozoic (Plafker, 1969, 1972, p. 921). Calc-alkaline plutonism has played a major role in the evolution of this continental margin, but data necessary for a relatively complete overview have only recently become available. These data enable the delineation of eight major calc-alkaline plutonic belts in southern Alaska. This paper outlines the distribution of these belts and summarizes available data concerning their geologic settings, petrologic characteristics, and age relations.

Southern Alaska (Fig. 1) is here considered to include the Aleutian and Alaska Ranges, all mainland areas to the south, all Gulf of Alaska islands, the St. Elias Mountains and adjacent coastal belt, and the Haines-Glacier Bay-Chichagof Island area of northern southeastern Alaska. Contiguous parts of Canada between the eastern Alaska Range and the Haines area are also included. The northern boundary of this 500,000 km² region approximately coincides with the Denali fault system from about longitude

153° W. eastward to Chatham Strait and with the north flank of the Aleutian and Alaska Ranges west of longitude 153° W.

This region of complex Phanerozoic geology is divisible into several geologic terranes (generalized in Fig. 1) composed primarily of marine sedimentary and volcanic rocks. The terranes were developed and aggregated through processes of sedimentation, magmatism, and tectonism that accompanied episodes of plate subduction and transform faulting. They include tectonostratigraphic sequences that may be only partly related to the evolution of presently nearby parts of the North American craton (Jones and others, 1972, 1976, 1977). Some plutonic belts are restricted to particular geologic terranes but others are not.

Petrologic nomenclature throughout the paper corresponds to that suggested by the International Union of Geological Sciences' Subcommission on the Systematics of Igneous Rocks (Geotimes, October 1973). Correlation of stratigraphic and absolute ages follows the time scale of Van Eysinga (1975). All radiometric ages used in this report have

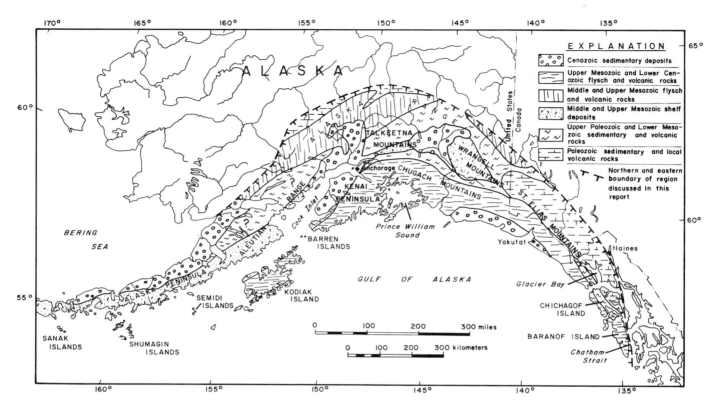

Figure 1. Index map and generalized geologic terranes of southern Alaska (modified from Beikman, 1978).

been determined by the potassium-argon (K-Ar) method, and single-mineral K-Ar ages are considered to be minimum ages.

PLUTONIC BELTS

A plutonic belt is a region in which plutonic rocks of specific ages are known or are reasonably expected to occur. Belts of different ages may be spatially distinct or overlap. They represent the loci of principal pluton emplacement and do not necessarily include all plutons of specific ages (Kistler, 1974, p. 409). On the basis of available information, eight Mesozoic and Cenozoic calc-alkaline plutonic belts can be delineated in southern Alaska. The distribution, geologic setting, petrologic characteristics, and age relations of each belt are summarized below. The belts are informally named after geographic areas or localities nearby.

Kodiak-Kenai belt

The Kodiak-Kenai belt (Fig. 2) trends northeasterly for 300 km along the western coast of the Kodiak Archipelago (Connelly and Moore, 1977), through the Barren Islands (Cowan and Boss, 1978), to the southern Kenai Peninsula (Carden and others, 1977; Donald L. Turner, personal commun., 1978). Extensions of the belt along trend are

likely, especially to the northeast to areas along the northern flank of the Chugach Mountains (Hudson, Plafker, and Lanphere, 1977, p. 169).

The plutons intrude a diverse assemblage of sedimentary

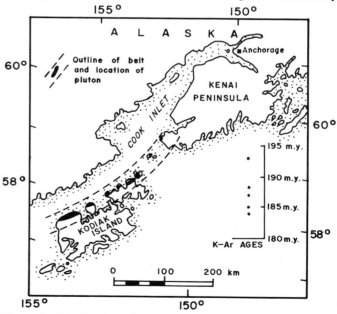

Figure 2. Distribution of plutons and summary of K-Ar ages (hornblende) for the Kodiak-Kenai plutonic belt.

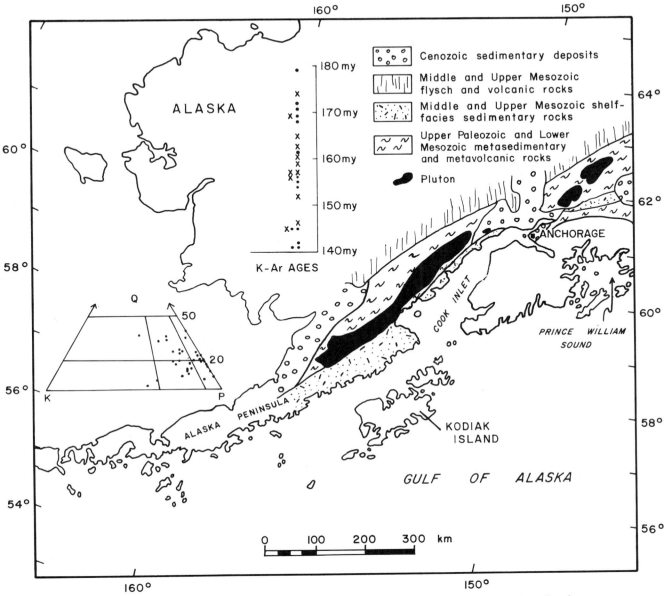

Figure 3. Distribution of plutons and nearby geologic terranes, summary of modal data (Reed and Lanphere, 1973), and summary of K-Ar ages (•, single mineral; ×, concordant mineral pair) for the Aleutian Range-Talkeetna Mountains plutonic belt.

ar.d metamorphic rocks that are part of a complicated upper Paleozoic and lower Mesozoic terrane that parallels the continental margin throughout southern Alaska and is structurally bound to the south by the Border Ranges fault (Plafker, 1969; MacKevett and Plafker, 1974; Plafker and others, 1976). Plutons are elongate parallel to regional trends and locally fault-bounded. Intrusive contacts are sharp to migmatitic, and diking and thermal metamorphism of nearby country rocks are locally well developed. Petrologic data are scarce, but hornblende (± biotite) diorite and quartz diorite appear to be characteristic lithologies; some chemical data have been reported by Hill and Morris (1977).

Plutons were emplaced in the belt during the earliest Jurassic and possibly Late Triassic; some intrude Upper Triassic (Norian) sedimentary rocks (including volcanogenic sandstone and tuff), and single-mineral (hornblende) K-Ar ages range from 184 to 193 m.y. (Fig. 2).

Aleutian Range-Talkeetna Mountains belt

The Aleutian Range-Talkeetna Mountains belt contains plutons that are almost continuously exposed in a narrow and linear, northeast-trending belt that appears to comprise two separate segments (Fig. 3). The southern segment is exposed from near Becharof Lake on the Alaska Peninsula northeast 500 km to near Tyonek on the western side of Cook Inlet (Reed and Lanphere, 1972). Reed and Lanphere (1973, p. 2593) suggested that this segment may extend unexposed for an additional 480 km or more to the southwest of Becharof Lake. The shorter (about 150 km) but similar northern segment is exposed in the Talkeetna Mountains, where Jurassic plutonic rocks make up the eastern one-third of the Talkeetna Mountains batholithic complex (Csejtey and others, 1978).

The northern and southern segments are both locally emplaced against Lower Jurassic sedimentary and volcanic rocks to the east and are, in turn, locally intruded by Upper Cretaceous-Lower Tertiary plutonic rocks to the west. Triassic or older metamorphic rocks form roof pendants and parts of complex border zones, and the southern segment is fault bounded for long distances along its eastern margin. Individual plutons are generally concordant to regional trends, and contacts are sharp to gradational with migmatitic and inclusion-rich zones locally developed. Foliated biotite hornblende quartz diorite, tonalite, and granodiorite dominate in the belt, but gabbro, diorite, and some monzonitic rocks are present (Fig. 3).

Plutons of the belt locally intrude Lower Jurassic (Sinemurian to Toarcian) marine sedimentary and volcanic rocks of the Talkeetna Formation, and boulders derived from the plutonic belt are present in the Upper Jurassic (lower Oxfordian) Naknek Formation (Grantz and others, 1963; Detterman and others, 1965). K-Ar ages (Fig. 3) from the southern segment (Reed and Lanphere, 1972, 1973) range from 142 to 179 m.y., with concordant mineral pairs being between 155 and 175 million years. In the northern segment, the range of K-Ar ages is similar (142-173 m.y., Grantz and others, 1963; Turner and Smith, 1974; Csejtey and others, 1978), and although dated mineral pairs are mostly discordant, they do include the youngest concordant age in the belt (145 m.y.). The field and K-Ar data show that the age of plutonism in the Alaska Peninsula-Talkeetna Mountains belt is Middle and Late Jurassic.

Tonsina-Chichagof belt

The Tonsina-Chichagof belt (Fig. 4) includes all plutons of known or probable Jurassic age located in or along the northern side of the eastern Chugach and St. Elias Mountains, as well as Jurassic plutons of Chichagof Island. The belt extends southeastward for about 800 km from near Tonsina at the northern side of the Chugach Mountains (Beikman, Holloway, and MacKevett, 1977), through parts of the Wrangell and St. Elias Mountains (Campbell and Dodds, 1978; Hudson and others, 1977), probably through parts of Glacier Bay National Monument, and to western Chichagof Island (Loney and others, 1975). Spatial continuity and temporal correlation of at least parts of this belt

with Jurassic belts to the west are possible but cannot be demonstrated at present.

Plutons of the Tonsina-Chichagof belt are mostly emplaced in greenschist to amphibolite facies, upper Paleozoic and lower Mesozoic metasedimentary and metavolcanic rocks that form the upper plate of the Border Ranges fault (MacKevett and Plafker, 1974; Plafker and others, 1976) but also intrude older Paleozoic rocks in parts of the St. Elias Mountains (Campbell and Dodds, 1978) and probably in Glacier Bay National Monument (Fig. 4). The plutons are commonly elongate, parallel to regional trends, and have sharp to migmatitic and gradational boundaries. Foliated biotite hornblende quartz diorite, tonalite, and granodiorite are characteristic of the belt.

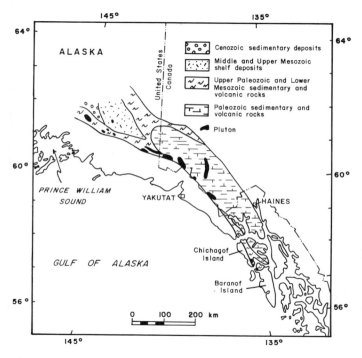

Figure 4. Distribution of plutons and host geologic terranes for the Tonsina-Chichagof plutonic belt.

Age relations in the belt are incompletely known. Plutons locally intrude Upper Triassic sedimentary rocks and are unconformably overlain by Lower Cretaceous sedimentary rocks. Biotite from a granodiorite clast in the Kotsina Conglomerate (Middle or Late Jurassic) is dated at 157 ± 6 m.y., and a microdiorite dike intruding this conglomerate is dated at 142 ± 5 m.y. (Grantz and others, 1966). Jurassic K-Ar ages from the belt range upward to 164 m.y., and two mineral pairs have yielded concordant ages at about 140 and 151 m.y. (R. B. Campbell, written commun., 1977; Loney and others, 1975). The age data show that a major Jurassic plutonic event (mostly Late? Jurassic) occurred in eastern southern Alaska.

Nutzotin-Chichagof belt

The Nutzotin-Chichagof belt (Fig. 5) consists of northern and southern segments that differ in some general field and petrologic characteristics. The northern segment extends from near the intersection of the Denali and Totshunda faults in the Nutzotin Mountains (Richter and others, 1975) 475 km southeast to near lower Dezadeash Lake in Yukon Territory, Canada (Campbell and Dodds, 1978). The southern segment extends 375 km south from near Mush Lake in Yukon Territory (Campbell and Dodds, 1978), through British Columbia (Watson, 1948), through parts of Glacier Bay National Monument and the Haines area (MacKevett and others, 1971; MacKevett, Robertson, and Winkler, 1974), to Chichagof Island (Loney and others, 1975).

Plutons of the northern segment commonly intrude Upper Jurassic-Lower Cretaceous flyschoid rocks of the Nutzotin Mountains sequence (Richter, 1976) and correlative rocks of the Dezadeash Formation in Canada (Eisbacher, 1976). They locally intrude upper Paleozoic and Triassic sedimentary and volcanic rocks that flank the Upper Jurassic-Lower Cretaceous flysch terrane to the west (Richter, 1976; Campbell and Dodds, 1978). These plutons are generally widely spaced, isolated intrusive complexes that are irregularly shaped and have sharp discordant contacts. Thermal aureoles are commonly well developed. In Alaska, the plutons are commonly composite, are petrologically variable, and display zonal relations; gabbro, diorite, quartz monzodiorite, quartz diorite, tonalite, granodiorite, and granite are present (Fig. 5). Some plutons are associated with porphyry-type copper and molybdenum mineralization.

Plutons of the southern segment intrude diverse Paleozoic metasedimentary and metavolcanic rocks. Contact relations vary from sharply discordant to gradational, migmatitic, and concordant. Many plutons are elongate and grossly conformable to regional structure. They are apparently more uniform petrologically than those to the north and characteristically contain foliated biotite and hornblende-bearing dioritic to tonalitic rocks. Granodiorite is also present.

In the northern segment, the Alaskan plutons locally intrude sedimentary and volcanic rocks of the Lower Cretaceous (Valanginian? to Barremian) Chisana Formation and were apparently the source of granitic cobbles incorporated in Upper(?) Cretaceous continental sedimentary rocks (Richter, 1976). K-Ar ages (Fig. 5) in the northern segment range from 105 to 117 m.y., with four biotite-hornblende mineral pairs yielding concordant dates between 111 and 114 m.y. (Richter and others, 1975; R. B. Campbell, written commun., 1977). K-Ar ages in the southern segment include several from the Haines area that range from 105 to 119 m.y. and include two concordant biotite-hornblende pairs at about 107 m.y. and 110 m.y. (MacKevett and others, 1974). Single mineral K-Ar ages reported from Chichagof Island range from 103 to 114 m.y. (Loney and others, 1975). The age data throughout the belt indicate the time of plutonism was late Early Cretaceous, between about 105 and 120 m.y. ago.

Alaska Range-Talkeetna Mountains belt

The Alaska Range-Talkeetna Mountains belt (Fig. 6) is about 700 km long and extends from near Nonvianuk Lake in the Aleutian Range, northeast through the southern Alaska Range (Reed and Lanphere, 1969, 1972, 1973) and the Talkeetna Mountains (Csejtey and others, 1978), to the central Alaska Range (Smith and Lanphere, 1971; Turner and Smith, 1974). The belt, as now exposed, is broader than any other; it is about 100 km wide in the southern Alaska Range and about 175 km wide between the Talkeetna Mountains and the central Alaska Range.

The southeastern parts of the belt contain elongate, somewhat continuous, and regionally concordant batholithic complexes that intrude Mesozoic and upper Paleozoic metamorphic rocks as well as plutons of the Aleutian Range-Talkeetna Mountains belt. These batholithic complexes are dominantly foliated biotite hornblende quartz

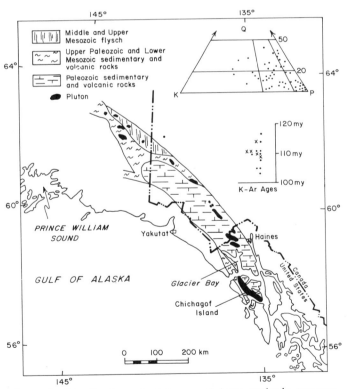

Figure 5. Distribution of plutons and host geologic terranes, summary of modal data (Alaska plutons in northern segment only, Richter and others, 1975), and summary of K-Ar ages (•, single mineral; ×, concordant mineral pair) for the Nutzotin-Chichagof plutonic belt.

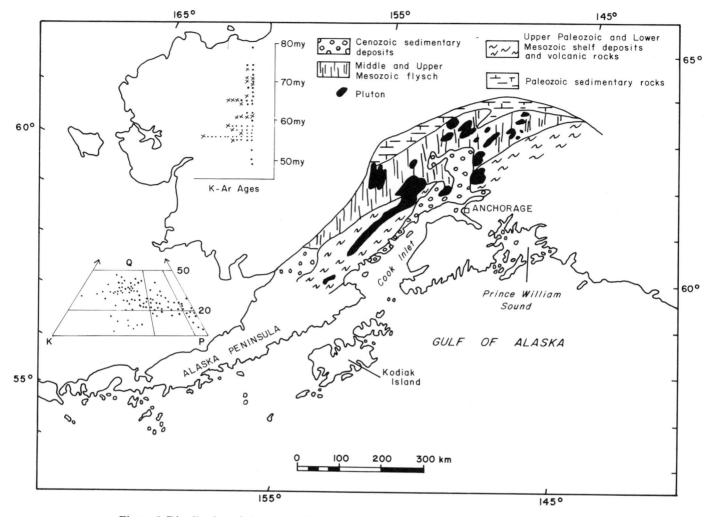

Figure 6. Distribution of plutons and host geologic terranes, summary of modal data (Reed and Lanphere, 1973; Reed and Nelson, 1977), and summary of K-Ar ages (•, single mineral; ×, concordant mineral pair) for the Alaska Range-Talkeetna Mountains plutonic belt.

diorite, tonalite, and granodiorite. The northwestern parts of the belt contain many irregular plutons, scattered through the central Alaska Range, that sharply and discordantly intrude a wide variety of Mesozoic and Paleozoic sedimentary, volcanic, and metamorphic rocks (Fig. 6). They mostly contain granodiorite and granite.

The age of plutonism in the belt is well established by K-Ar dating. K-Ar ages range from 50 to 74 m.y. (Fig. 6) and include concordant mineral pairs between 50 to 73 m.y. (Reed and Lanphere, 1972, 1973; Csejtey and others, 1978; Smith and Lanphere, 1971; Turner and Smith, 1974). The K-Ar data alone do not discriminate separate plutonic episodes, but the more mafic rock suites (quartz diorite to tonalite) are ≥ 62 m.y. old (Reed and Lanphere, 1973, p. 2599) and are found mainly in the southeastern parts of the belt. Most plutons were emplaced during the latest Cretaceous and Paleocene.

Yakutat-Chichagof belt

Plutons of the Yakutat-Chichagof belt (Fig. 7) have only recently been recognized on a regional scale (Hudson and others, 1977, p. 169), and only preliminary data are available concerning them; their age and field relations are poorly known. They have been identified over a distance of about 300 km in the eastern Gulf of Alaska near Yakutat, in the Lituya Bay area, and on Chichagof Island.

The plutons form small stock and larger elongate bodies enclosed in Cretaceous flyschoid rocks of the eastern Gulf of Alaska continental margin (Plafker and others, 1977). They contain foliated biotite hornblende quartz diorite, and tonalite that is characteristically sheared and altered. Alteration consists of extensive saussuritization, sericitization, and prehnite veining and replacement of biotite. In some places, the enclosing rocks are thermally metamorphosed

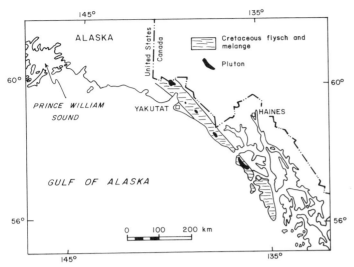

Figure 7. Distribution of plutons and host geologic terrane of the Yakutat-Chichagof plutonic belt.

(Hudson, Plafker, and Lanphere, 1977, p. 165), and on the basis of these relations, the belt is provisionally considered to be Cretaceous or Tertiary in age.

One small Jurassic pluton in the Yakutat area is apparently incorporated as an exotic block in melange of the coastal flysch belt (Hudson and others, 1977). This relation suggests the possibility that some of the plutons in the Yakutat-Chichagof belt may also be allochthonous and older than their enclosing rocks.

Sanak-Baranof belt

The Sanak-Baranof belt (Fig. 8) is over 2,000 km long and parallel to the continental margin throughout the Gulf of

Alaska region (Hudson, Plafker, and Lanphere, 1977, p. 169–170). It is defined by a series of discrete plutons that are present on Sanak Island (Moore, 1974), the Shumagin Islands (Grantz, 1963; Moore, 1974), the Semidi Islands (Burk, 1965), on central Kodiak Island (Moore, 1967), the Kenai Peninsula (Tysdal and Case, 1979), in the eastern Chugach Mountains (T. Hudson and G. Plafker, unpubl. data; Hudson and others, 1979), the eastern Gulf of Alaska (Hudson and others, 1977), and on western Chichagof and Baranof Islands (Loney and others, 1975).

Plutons throughout the belt intrude an upper Mesozoic-lower Tertiary accretionary prism of flyschoid rocks (Plafker and others, 1977) that lies between the Border Ranges fault (MacKevett and Plafker, 1974; Plafker and others, 1976) and the present continental margin. They include stocks and larger bodies, and although many are elongate and grossly conformable to regional structural trends, they commonly have sharp and locally discordant contacts. Distinct thermal aureoles are well developed adjacent to many plutons, but migmatitic and gradational contact zones are associated with some more deeply eroded plutons. The plutons characteristically contain biotite granodiorite and granite but include some biotite tonalite (Fig. 8). Hornblende is absent except in certain marginal zones that have complex contact relations with mafic volcanic rocks. Some white mica and garnet are present.

The plutons intrude strongly deformed Cretaceous and Paleocene rocks; K-Ar ages are almost all between 47 and 60 m.y. (Fig. 8). Prince William Sound apparently divides the belt between a slightly older 58-60 m.y. western limb (Moore, 1974; Kienle and Turner, 1976; Karlstrom and Ball, 1969, p. 29; Tysdal and Case, 1979) and a younger 47-52 m.y. eastern limb (Plafker and Lanphere, 1974; Hudson and

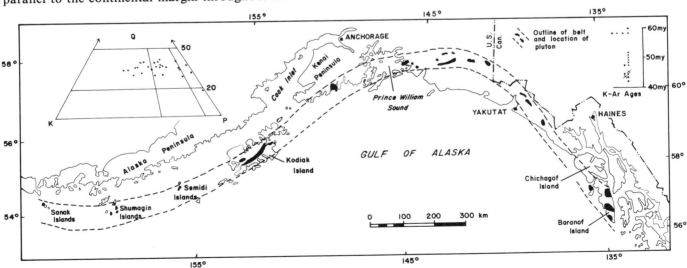

Figure 8. Distribution of plutons, summary of modal data (Hudson, Plafker, and Lanphere, 1977, and Hudson, unpubl. data), and summary of K-Ar ages (•, single mineral; ×, concordant mineral pair) for the Sanak-Baranof plutonic belt. All plutons are located within an upper Mesozoic (Cretaceous) and Lower Cenozoic flysch terrane (Fig. 1).

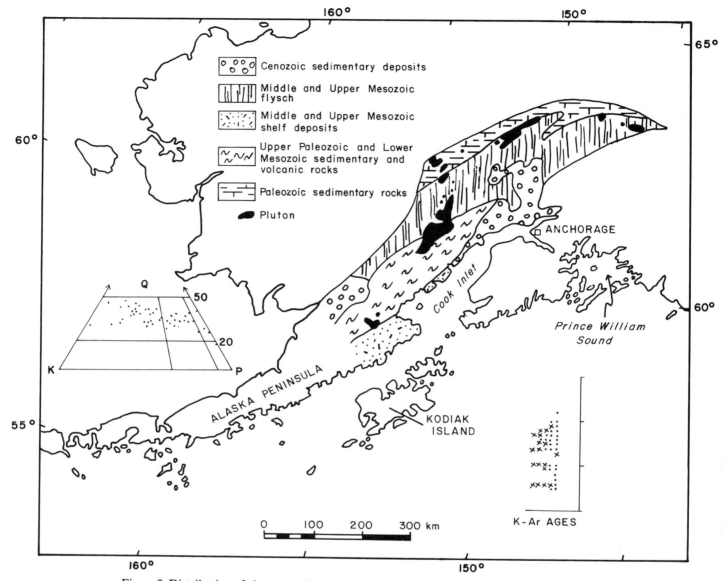

Figure 9. Distribution of plutons and host geologic terranes, summary of modal data (Reed and Lanphere, 1973), and summary of K-Ar ages (•, single mineral; ×, concordant mineral pair) for the Alaska Range plutonic belt.

others, 1977; Loney and others, 1975). One pluton on Baranof Island that is probably part of the belt is apparently about 43 m.y. old. The available data indicate that plutonism was approximately synchronous during the early Eocene (47-51 m.y.) in the eastern segment and during the late Paleocene (58-60 m.y.) in the western segment.

Alaska Range belt

Plutons of the Alaska Range belt (Fig. 9) are scattered through the southern and central Alaska Range in a curvilinear trend at least 600 km long from Lake Clark northeastward to the vicinity of the Richardson Highway (Reed and Lanphere, 1969, 1972, 1973; Turner and Smith, 1974). The belt extends southeasterly from Lake Clark to the Aleutian Peninsula, but plutons in this part have been identified only near Nonvianuk Lake (Reed and Lanphere, 1973, p. 2603-2604). The Alaska Range belt spatially overlaps the central and northern parts of the Late Cretaceous to Early Tertiary Alaska Range-Talkeetna Mountains belt.

Country rocks of the Alaska Range belt include extensive areas of Mesozoic flysch and volcanic rocks, as well as Paleozoic carbonate and clastic rocks. Plutons form a large elongate intrusive complex (the Merrill Pass sequence, Reed and Lanphere, 1973, p. 2601-2603) and many smaller isolated and irregular bodies. Many of the plutons are epizonal or hypabyssal, and some are spatially associated with probable cogenetic subaerial volcanic rocks. Lithologic variation is well developed in the belt, and rocks range from biotite-hornblende quartz diorite to muscovite-bearing granite (Fig. 9).

Age relations within the Alaska Range belt have been determined primarily by K-Ar dating; the K-Ar ages range from 25 to 41 m.y., with concordant mineral pairs between 26 and 38 m.y. (Fig. 9). These ages are probably a good measure of the emplacement ages in the belt and establish it as Oligocene — the youngest major plutonic belt now identifiable in southern Alaska.

Other Plutonic Rocks

The eight plutonic belts include the great bulk of plutonic rocks in southern Alaska, but others that are of local and, possibly, regional significance include: (1) upper Paleozoic rocks ranging from diorite to granite that form several plutons in the Wrangell-St. Elias Mountains (Hudson and others, 1977; Campbell and Dodds, 1978); (2) undated layered gabbro complexes in the St. Elias Mountains (Rossman, 1963; Plafker and MacKevett, 1970); (3) scattered granitic rocks in the Aleutian and Alaska Ranges that yield mid-Cretaceous K-Ar ages (Reed and Lanphere, 1972; Turner and Smith, 1974), but for which field and age relations are incompletely known (many granitic plutons in the central and eastern Alaska Range north of the area discussed in this report are mid-Cretaceous, Richter and others, 1975); and (4) Tertiary intrusive rocks that are spottily distributed but widespread in southern Alaska and mostly undated. They include many small plutons and dikes of early Miocene age in the Yakutat-St. Elias area (Hudson and others, 1977) and Oligocene plutons in the Prince William Sound area (Lanphere, 1966).

SUMMARY

The available data on southern Alaska plutonic rocks show that plutonism was widespread but characterized by specific loci of emplacement for plutons of specific ages. These loci define eight plutonic belts that together represent most of plutonic rocks in southern Alaska. The plutonic belts are defined by curvilinear to linear trends of exposed plutons or batholithic complexes that are now broadly parallel or sub-parallel to the continental margin. The lengths of the belts, based on known exposures and, in all cases, minimum lengths, range from 300 to 2,000 km, but most are between 650 and 850 km. Approximate widths of the belts range from 15 to 175 km, but most are between 30 and 100 km. Many of the belts are spatially distinct, but overlap of a younger belt on an older one is known in the Alaska Range, the Yakutat-St. Elias area, and probably the

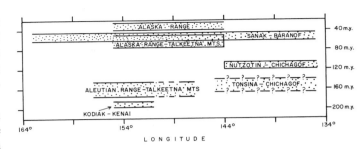

Figure 10. Diagrammatic summary of space-time relations of southern Alaska plutonic belts. The Yakutat-Chichagof belt is not included because of its poorly known age relations. Short dashes indicate where the continuity of a belt is uncertain. The indicated range in age for a belt is not necessarily constant along its length.

Glacier Bay-Haines-Chichagof area of southeastern Alaska.

Petrologically, the plutonic belts are dominantly of intermediate character — quartz diorite, tonalite, and granodiorite are by far the most abundant rock types. Gabbro and monzonitic rocks are locally a part of some intermediate-composition belts, and granite is an important part of the belts in the Alaska Range. Granodiorite and granite dominate in the Lower Tertiary Sanak-Baranof belt — a belt that is petrologically distinct in this respect.

Plutonism in southern Alaska is almost entirely of Mesozoic and Cenozoic age. Most of the belts are temporally distinct, but there is known overlap in ages between the belts containing Upper Jurassic and Lower Tertiary plutons (Fig. 10). The timing of pluton emplacement in six of the belts is moderately well known. Intrusive episodes apparently began in them at about 195, 175, 120, 75, 60, and 40 m.y. ago, and they were as short as about 10 m.y. and as long as about 30 m.y.

CONCLUSIONS

The available data on plutonic rocks in southern Alaska are incomplete, and many problems in the timing, distribution, and origin of plutonism in this region remain to be solved. This report provides an overview that should help to define important regional problems dealing with the spatial continuity and temporal correlation of the many southern Alaskan plutonic belts, as well as their relation to other regional rock units and structures. It also provides an initial framework for tectonic reconstructions of the Mesozoic and Cenozoic history of the region. Calc-alkaline plutonism has played a major role in the evolution of the tectonically complicated southern Alaskan continental margin, and most (Hudson, 1979), but not all (Hudson and others, 1979) of this plutonism may be directly related to episodes of plate convergence and subduction.

ACKNOWLEDGMENTS

The report is based on compilations of geologic relations, petrology, and K-Ar data available in the literature. R. B. Campbell shared unpublished K-Ar data from the St. Elias Mountains. Constructive technical reviews of an earlier draft of the report were provided by George Plafker and Marvin A. Lanphere.

REFERENCES CITED

Beikman, H. M., 1978, Preliminary geologic map of Alaska: U.S. Geological Survey Map, scale 1:2,500,000.

Beikman, H. M., Holloway, C. D., and MacKevett, E. M., Jr., 1977, Generalized geologic map of the eastern part of southern Alaska: U.S. Geological Survey Open-file Report 77-169-B, 1 sheet, scale 1:1,000,000.

Burk, C. A., 1965, Geology of the Alaska Peninsula-Island arc and continental margin (Part 1): Geological Society of America Memoir 99, 250 p.

Campbell, R. B., and Dodds, C. J., 1978, Operation Saint Elias, Yukon Territory: Current Research, Part A, Geological Survey Canada, Paper 78-1A, p. 35-41.

Carden, J. R., Connelly, W., Forbes, R. B., and Turner, D. L., 1977, Blueschists of the Kodiak Islands, Alaska: An extension of the Seldovia schist terrane: Geology, v. 5, n. 9, p. 529-533.

Connelly, W., and Moore, J. C., 1977, Geologic map of the northwest side of the Kodiak Islands, Alaska: U.S. Geological Survey Open-file Report 77-382.

Cowan, D. S., and Boss, R. F., 1978, Tectonic framework of the southwestern Kenai Peninsula, Alaska: Geological Society of America Bulletin, v. 89, p. 155-158.

Csejtey, B., Jr., Nelson, W. H., and others, 1978, Reconnaissance geological map and geochronology, Talkeetna Mountains quadrangle, northern part of Anchorage quadrangle, and southwest corner of Healy quadrangle, Alaska: U.S. Geological Survey Open-file Report 78-558-A.

Detterman, R. L., Reed, B. L., and Lanphere, M. A., 1965, Jurassic plutonism in the Cook Inlet region, Alaska: U.S. Geological Survey Professional Paper 525-D, p. D16-D21.

Eisbacher, G. H., 1976, Sedimentology of the Dezadeash flysch and it's implications for strike slip faulting along the Denali fault, Yukon Territory and Alaska: Canadian Journal of Earth Sciences, v. 13, n. 11, p. 1495-1513.

Geotimes, 1973, Plutonic rocks—classification and nomenclature recommended by the IUGS Subcommission on the Systematics of Igneous Rocks: Geotimes, v. 18, n. 10, p. 26-30.

Grantz, A., 1963, Aerial reconnaissance of the outer Shumagin Islands, Alaska: U.S. Geological Survey Professional Paper 475-B, p. B106-B109.

Grantz, A., Thomas, H., Stern, T. W., and Sheffey, N. B., 1963, Potassium-argon and lead-alpha ages for stratigraphically bracketed plutonic rocks in the Talkeetna Mountains, Alaska: U.S. Geological Survey Professional Paper 475-B, p. B56-B59.

Grantz, A., Jones, D. L., and Lanphere, M. A., 1966, Stratigraphy, paleontology, and isotopic ages of upper Mesozoic rocks in the southwestern Wrangell Mountains, Alaska: U.S. Geological Survey Professional Paper 550-C, p. C39-C47.

Hill, M. D., and Morris, J. D., 1977, Near-trench plutonism in southwestern Alaska, abs.: Geological Society of America, Abstracts with Programs, Cordilleran Sect., Sacramento, p. 436-437.

Hudson, T., 1979, Mesozoic plutonic belts of southern Alaska: Geology,

v. 7, p. 230-234.

Hudson, T., Plafker, G., and Lanphere, M. A., 1977, Intrusive rocks of the Yakutat-St. Elias area south-central Alaska: U.S. Geological Survey Journal of Research, v. 5, n. 2, p. 155-172.

Hudson, T., Plafker, G., and Peterman, Z. E., 1979, Paleogene anatexis along the Gulf of Alaska margin: Geology, v. 7, p. 573-577.

Jones, D. L., Irwin, W. P., and Ovenshine, A. T., 1972, Southeastern Alaska — a displaced continental fragment?: U.S. Geological Survey Professional Paper 800-B, p. B211-B217.

Jones, D. L., Pessangno, E. A., Jr., and Csejtey, B., Jr., 1976, Significance of the Upper Chulitna Ophiolite for the late Mesozoic evolution of southern Alaska: Geological Society of America, Abstracts with Programs, v. 8, p. 385-386.

Jones, D. L., Silberling, N. J., and Hillhouse, J., 1977, Wrangellia — a displaced terrane in northwestern North America: Canadian Journal of Earth Sciences, v. 14, n. 11, p. 2565-2577.

Karlstrom, T. N., and Ball, G. E., eds., 1969, The Kodiak Island refugium: its geology, flora, fauna, and history: Ryerson Press, Toronto, 262 p.

Kienle, J., and Turner, D. L., 1976, The Shumagin-Kodiak batholith — a Paleocene magmatic arc:, in Short notes on Alaska Geology, 1976: Alaska Division of Geological and Geophysical Surveys, Geology Report 51, p. 9-11.

Kistler, R. W., 1974, Phanerozoic batholiths in western North America: Annual Review of Earth and Planetary Sciences, v. 2, p. 403-418.

Lanphere, M. A., 1966, Potassium-argon ages of Tertiary plutons in the Prince William Sound region, Alaska: U.S. Geological Survey Professional Paper 550-D, p. D195-D198.

Loney, R. A., Brew, D. A., Muffler, L. J. P., and Pomeroy, J. S., 1975, Reconnaissance geology of Chichagof, Baranof, and Kruzof Islands, southeastern Alaska: U.S. Geological Survey Professional Paper 792, 105 p.

MacKevett, E. M., Jr., Brew, D. A., Hawley, C. C., Huff, L. C., and Smith, J. G., 1971, Mineral resources of Glacier Bay National Monument, Alaska: U.S. Geological Survey Professional Paper 632, 90 p.

MacKevett, E. M., Jr., and Plafker, G., 1974, The Border Ranges fault in south-central Alaska: U.S. Geological Survey, Journal of Research, v. 2, n. 3, p. 323-329.

MacKevett, E. M., Jr., Robertson, E. C., and Winkler, G. R., 1974, Geology of the Skagway B-3 and B-4 quadrangles, southeastern Alaska: U.S. Geological Survey Professional Paper 832, 33 p.

Moore, G. W., 1967, Preliminary geologic map of Kodiak Island and vicinity, Alaska: U.S. Geological Survey Open-file Report 271.

Moore, J. C., 1974, Geologic and structural map of the Sanak Islands, southwestern Alaska: U.S. Geological Survey Miscellaneous Geological Investigations, Map I-817.

Plafker, G., 1969, Tectonics of the March 27, 1964 Alaska earthquake: U.S. Geological Survey Professional Paper 543-I, 74 p.

——, 1972, Alaskan earthquake of 1964 and Chilean earthquake of 1960: Implications for arc tectonics: Journal of Geophysical Research, v. 71, p. 901-925.

Plafker, G., Jones, D. L., Hudson, T., and Berg, H. C., 1976, The Border Ranges fault system in the Saint Elias Mountains and Alexander Archipelago: U.S. Geological Survey Circular 722, p. 14-16.

Plafker, G., Jones, D. L., and Pessangno, E. A., Jr., 1977, A Cretaceous accretionary flysch and melange terrane along the Gulf of Alaska margin: U.S. Geological Survey Circular 751-B, p. B41-B43.

Plafker, G., and Lanphere, M. A., 1974, Radiometrically dated plutons cutting the Orca Group: U.S. Geological Survey Circular 700, p. 53.

Plafker, G., and MacKevett, E. M., Jr., 1970, Mafic and ultra-mafic rocks from a layered pluton at Mount Fairweather, Alaska: U.S. Geological Survey Professional Paper 700-B, p. B21-B26.

Reed, B. L., and Lanphere, M. A., 1969, Age and chemistry of Mesozoic ant Tertiary plutonic rocks in south-central Alaska: Geological Society of America Bulletin, v. 80, p. 23-44.

Reed, B. L., and Lanphere, M. A., 1972, Generalized geologic map of the Alaska-Aleutian Range batholith showing potassium-argon ages of the plutonic rocks: U.S. Geological Survey Map MF-372.

——,1972, Generalized geologic map of the Alaska-Aleutian Range batholith showing potassium-argon ages of the plutonic rocks: U.S. Geological Survey Map MF-372.

——,1973, Alaska-Aleutian Range batholith: Geochronology, chemistry and relation to circum-Pacific plutonism: Geological Society of America Bulletin, v. 84, p. 2583–2610.

Richter, D. H., 1976, Geologic map of the Nabesna quadrangle, Alaska: U.S. Geological Survey Map I-932.

Richter, D. H., Lanphere, M. A., and Matson, N. A., Jr., 1975, Granitic plutonism and metamorphism, eastern Alaska Range, Alaska: Geological Society of America Bulletin, v. 86, p. 819–829.

Smith, T. E., and Lanphere, M. A., 1971, Age of the sedimentation, plutonism, and regional metamorphism in the Clearwater Mountains region, central Alaska: Isochron/West, n. 2, p. 17–20.

Rossman, D. L., 1963, Geology and petrology of two stocks of layered gabbro in the Fairweather Range, Alaska: U.S. Geological Survey Bulletin 1121-F, 50 p.

Turner, D. L., and Smith, T. E., 1974, Geochronology and generalized geology of the central Alaska Range, Clearwater Mountains, and northern Talkeetna Mountains: Alaska Division of Geological and Geophysical Surveys, Open-file Report 72, 11 p.

Tysdal, R. G., and Case, J. E., 1979, Geological map of the Seward and Blying Sound quadrangles, southern Alaska: U.S. Geological Survey Miscellaneous Investigations Series, Map I-1150, scale 1:250,000.

Van Eysinga, F. W. B., 1975, Geological time table: Elsevier Scientific Publishing Company, Amsterdam.

Watson, K. DeP., 1948, The Squaw Creek-Rainy Hollow area, northern British Columbia: British Columbia Department of Mines and Petroleum Resources Bulletin 25, 74 p.

MANUSCRIPT ACCEPTED BY THE SOCIETY JULY 12, 1982

Geological Society of America
Memoir 159
1983

Intrusive rocks and plutonic belts
of southeastern Alaska, U.S.A.

David A. Brew and Robert P. Morrell*
U.S. Geological Survey
Menlo Park, California 94025

ABSTRACT

About 30 percent of the 175,000-km² area of southeastern Alaska is underlain by intrusive igneous rocks. Compilation of available information on the distribution, composition, and ages of these rocks indicates the presence of six major and six minor plutonic belts.

From west to east, the major belts are: the Fairweather-Baranof belt of early to mid-Tertiary granodiorite; the Muir-Chichagof belt of mid-Cretaceous tonalite and granodiorite; the Admiralty-Revillagigedo belt of porphyritic granodiorite, quartz diorite, and diorite of probable Cretaceous age; the Klukwan-Duke belt of concentrically zoned or Alaskan-type ultramafic-mafic plutons of mid-Cretaceous age within the Admiralty-Revillagigedo belt; the Coast Plutonic Complex sill belt of tonalite of unknown, but perhaps mid-Cretaceous, age; and the Coast Plutonic Complex belt I of early to mid-Tertiary granodiorite and quartz monzonite.

The minor belts are distributed as follows: the Glacier Bay belt of Cretaceous and(or) Tertiary granodiorite, tonalite, and quartz diorite lies within the Fairweather-Baranof belt; layered gabbro complexes of inferred mid-Tertiary age lie within and are probably related to the Fairweather-Baranof belt; the Chilkat-Chichagof belt of Jurassic granodiorite and tonalite lies within the Muir-Chichagof belt; the Sitkoh Bay alkaline, the Kendrick Bay pyroxenite to quartz monzonite, and the Annette and Cape Fox trondhjemite plutons, all interpreted to be of Ordovician(?) age, together form the crude southern southeastern Alaska belt within the Muir-Chichagof belt; the Kuiu-Etolin mid-Tertiary belt of volcanic and plutonic rocks extends from the Muir-Chichagof belt eastward into the Admiralty-Revillagigedo belt; and the Behm Canal belt of mid- to late Tertiary granite lies within and next to Coast Plutonic Complex belt II. In addition, scattered mafic-ultramafic bodies occur within the Fairweather-Baranof, Muir-Chichagof, and Coast Plutonic Complex belts I and II. Palinspastic reconstruction of 200 km of right-lateral movement on the Chatham Strait fault does not significantly change the pattern of the major belts but does bring parts of the minor mid-Tertiary and Ordovician(?) belts closer together.

The major belts are related to the stratigraphic-tectonic terranes of Berg, Jones, and Coney (1978) as follows: the Fairweather-Baranof belt is largely in the Chugach, Wrangell (Wrangellia), and Alexander terranes; the Muir-Chichagof belt is in the Alexander and Wrangell terranes; the Admiralty-Revillagigedo belt is in the Gravina and Taku terranes; the Klukwan-Duke belt is in the Gravina, Taku, and Alexander terranes; the Coast Plutonic Complex sill belt is probably between the Taku and Tracy Arm terranes; and the Coast Plutonic Complex belts I and II are in the Tracy Arm and Stikine terranes.

Present address: Amax Exploration, Inc., 4704 Harlan, Denver, Colorado 80212

Significant metallic-mineral deposits are spatially related to certain of these belts, and some deposits may be genetically related. Gold, copper, and molybdenum occurrences may be related to granodiorites of the Fairweather-Baranof belt. Magmatic copper-nickel deposits occur in the layered gabbro within that belt. The Juneau gold belt, which contains gold, silver, copper, lead, and zinc occurrences, parallels and lies close to the Coast Plutonic Complex sill belt; iron deposits occur in the Klukwan-Duke belt; and porphyry molybdenum deposits occur in the Behm Canal belt.

The Muir-Chichagof belt of mid-Cretaceous age and the Admiralty-Revillagigedo belt of probable Cretaceous age are currently interpreted as possible magmatic arcs associated with subduction events. In general, the other belts of intrusive rocks are spatially related to structural discontinuities, but genetic relations, if any, are not yet known. The Coast Plutonic Complex sill belt is probably related to a post-Triassic, pre-early Tertiary suture zone that nearly corresponds to the boundary between the Tracy Arm and Taku terranes. The boundary between the Admiralty-Revillagigedo and Muir-Chichagof belts coincides nearly with the Seymour Canal-Clarence Strait lineament and also is probably a major post-Triassic suture.

INTRODUCTION

About 30 percent of the 175,000-km² area of southeastern Alaska (Fig. 1) is underlain by intrusive igneous rocks. Within the last few years, several large parts of the region have been mapped, most in reconnaissance fashion, but some parts in considerable detail. These areas, together with earlier compilations by Brew, Loney, and Muffler (1966), Hutchison, Berg, and Okulitch (1973), Souther, Brew, and Okulitch (1974), Beikman (1975), and Brew (1975), provided the intrusive-rock distribution and composition information summarized and interpreted here. The present authors reexamined all original sources of the data to prepare a 1:1,000,000-scale compilation (Brew and Morrell, 1979a, 1980) that the reader should refer to for specific sources of data. Almost all available radiometric ages from the region have been compiled by Wilson, Dadisman, and Herzon (1979); the present authors have freely interpreted that information in this report. In general, the age assignments in this report are based on extrapolation from potassium-argon-dated bodies to undated but lithologically and structurally similar bodies. Age interpretations in the Coast Plutonic Complex and vicinity are currently being reevaluated as uranium-lead dates on zircons from selected bodies become available (J. G. Arth and J. G. Smith, oral communs., 1978, 1979). Hudson's (1979) interpretation of the Mesozoic plutonic belts of southern Alaska extends somewhat into southeastern Alaska; in general, his and the present authors' interpretations are compatible in the areas of overlap.

This report contains the following components: (1) a series of maps showing the distribution and composition of intrusive rocks of different ages; (2) a comparable series of maps showing how these rocks fall into six major and six minor belts; (3) a table (Table 1) summarizing the isotopic

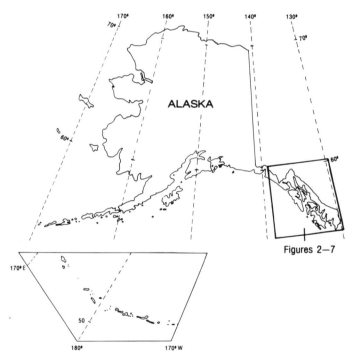

Figure 1. Index map of southeastern Alaska, showing area covered by Figures 2 through 7. See Figure 7B for individual place names.

ages, compositions, mineralogy, tectonic associations (Berg and others, 1978), metamorphic characteristics, and metallogenic associations of the belts; (4) brief comments on the different groupings of intrusive rocks; and (5) a discussion of several important general problems.

The compositional terms used here are those selected by Brew and Morrell (1979a) to provide a manageable general classification scheme that does not misrepresent any of the information taken from the original diverse sources. The authors first attempted to use the International Union of Geological Sciences (IUGS) classification (Streckeisen, 1973) in their compilation but found that many original sources did not provide enough information to permit its proper use. Reluctantly, a fivefold classification of granitoid rocks was adopted, as follows. Those containing less than 10 percent quartz are classified as alkalic (they are actually mostly syenite); and those containing greater than 10 and less than 50 percent quartz are subdivided according to potassium and plagioclase feldspar content: alkali granite contains less than 10, granite (and peralkaline granite) between 10 and 35, quartz monzonite between 35 and 65, and granodiorite greater than 65 percent plagioclase. The calc-alkalic part of the scheme is modified from Bateman (1961); alkalic, alkali granitic, peralkaline granitic, mafic, and ultramafic rock types also are included. The scheme unfortunately combines diorite, quartz diorite, tonalite, and granodiorite into a single category. Because of the different original classification schemes, granite and quartz monzonite may be incorrectly depicted in some map areas. The authors are aware that this general classification has serious shortcomings and fully expect that any future versions of these maps will incorporate new information from bodies that are now poorly known, so that IUGS classification can be applied.

The term "Coast Plutonic Complex" is used throughout this report both in a general sense and as part of the names for three different plutonic belts. The term was proposed by Douglas and others (1970) for the British Columbian and southeastern Alaskan parts of the unique batholithic complex that forms the core of the western Cordillera of North America. The term "Coast Plutonic Complex" is roughly synonymous with the terms "Coast Crystalline Belt" (Roddick, 1966), "Coast Range batholithic complex" (Brew and Ford, 1978), and "Coast Range Batholith" (Buddington and Chapin, 1929).

In this report, the intrusive rocks of the Coast Plutonic Complex are broken into four plutonic belts of different ages and compositions; the term "Coast Plutonic Complex" is used as a part of the names of three of those belts to emphasize the geographic unity of the complex. The three belts are: (1) The Coast Plutonic Complex sill belt, (2) the Coast Plutonic Complex belt I, and (3) the Coast Plutonic Complex belt II. The fourth belt, unnamed, is volumetrically much less significant than the other three. These, and all other, geologic names in this report are informal.

Although the information and interpretations given in this report will definitely be improved upon as more details become available, they probably define the major features of intrusive rocks in the region. The authors note that Buddington and Chapin (1929) anticipated several of the belts discussed here.

INTRUSIVE ROCKS OF TERTIARY AGE

The distribution and interpretation of intrusive rocks of known Tertiary age and those of Cretaceous and (or) Tertiary age are shown on the accompanying maps (Fig. 2); the rocks are described further in Table 1 (cols. 1-6). The Cretaceous and/or Tertiary age assignments for rocks of the Glacier Bay belt and Coast Plutonic Complex belt II are temporary in that both include either undated plutons or plutons with discordant ages that from field relations appear to be younger than nearby Cretaceous bodies and older than nearby Early Tertiary bodies; further studies will refine this assignment.

Some of these plutonic belts are related to tectonostratigraphic terranes, other belts, and structural features in ways that are not obvious from either the maps or the table: (1) much of the Fairweather-Baranof belt roughly parallels the Tarr Inlet suture zone (Brew and Morrell, 1978), which is currently interpreted to be a manifestation of the Wrangell (Wrangellia) terrane (Berg, Jones, and Coney, 1978; Brew and Morrell, 1979b); (2) the Kuiu-Etolin belt (Brew and others, 1979) is unusual in that it cuts across the Alexander, Gravina, and Taku terranes; (3) along much of their length, the Coast Plutonic Complex belts I and II are tightly constrained on the west by the mid-Cretaceous(?) Coast Plutonic Complex tonalite sill (Figs. 3A, 3B); (4) those belts are also continentward of the Cretaceous(?) Admiralty-Revillagigedo belt throughout all of its extent (Figs. 3A, 3B); and (5) the Behm Canal belt is of particular interest because it contains the important Quartz Hill molybdenite deposit east of Ketchikan, Alaska, as well as the Burroughs Bay molybdenite deposit.

Although the 200 km of right-lateral movement on the Chatham Strait fault (Ovenshine and Brew, 1972) may partly or completely predate the Kuiu-Etolin belt, palinspastic reconstruction, nevertheless, brings the belt into approximate alignment with the southeastern extension of the newly recognized Tkope volcanoplutonic belt in British Columbia, about due west of Skagway, Alaska (Campbell and Dodds, 1979).

INTERMEDIATE AND FELSIC INTRUSIVE ROCKS OF CRETACEOUS AND CRETACEOUS(?) AGE

Intrusive rocks of Cretaceous age are probably the most common in southeastern Alaska (Fig. 3; cols. 7-9, Table 1). The granodiorite shown as Cretaceous(?) age (lower half, Fig. 3A) is undated isotopically and could be Paleozoic or Mesozoic. Likewise, plutons interpreted as Cretaceous in the middle part of the Muir-Chichagof belt are also undated and could be older.

Two belts, the Muir-Chichagof and Admiralty-Revillagigedo, are subdivided into sections labeled I and II: section I contains abundant plutons and section II sparser plutons.

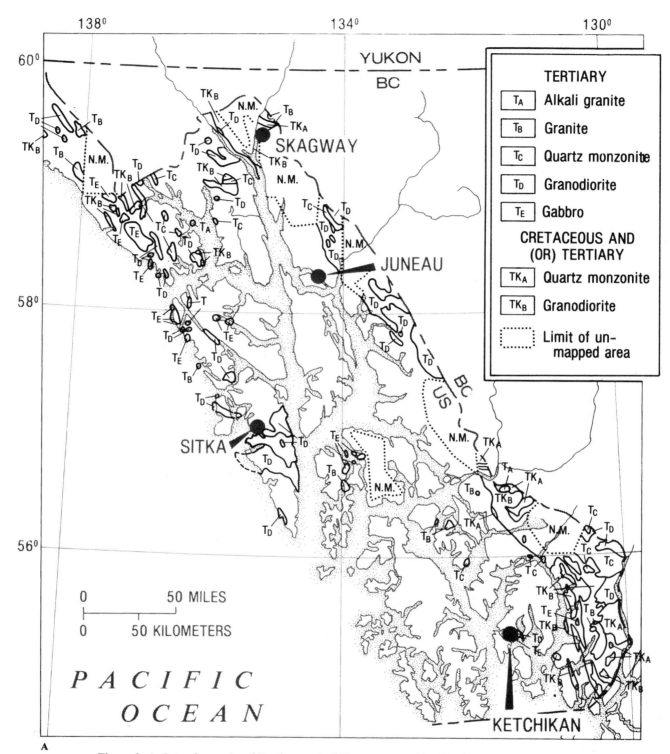

Figure 2. A, Intrusive rocks of Tertiary and of Cretaceous and/or Tertiary age, southeastern Alaska. B (facing page), Plutonic belts of Tertiary and of Cretaceous and/or Tertiary age, southeastern Alaska.

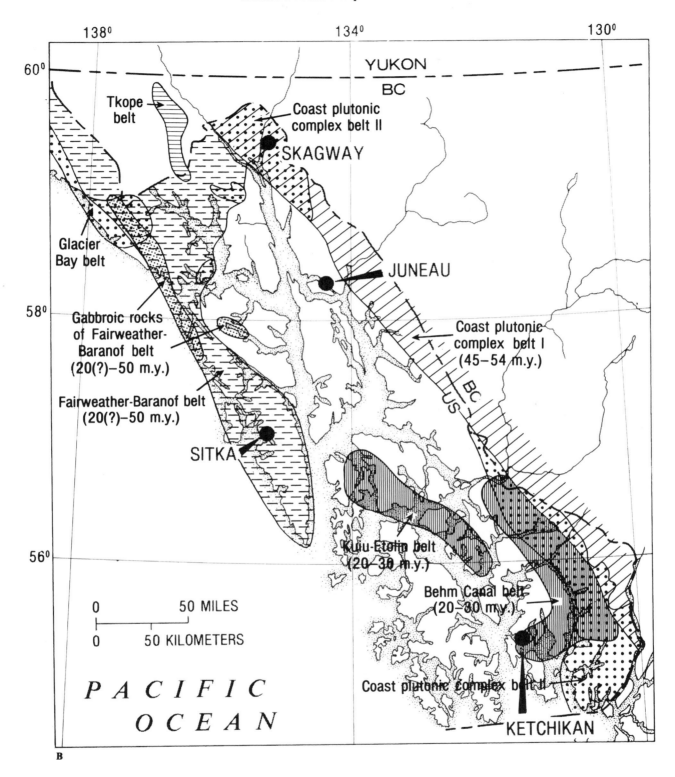

B

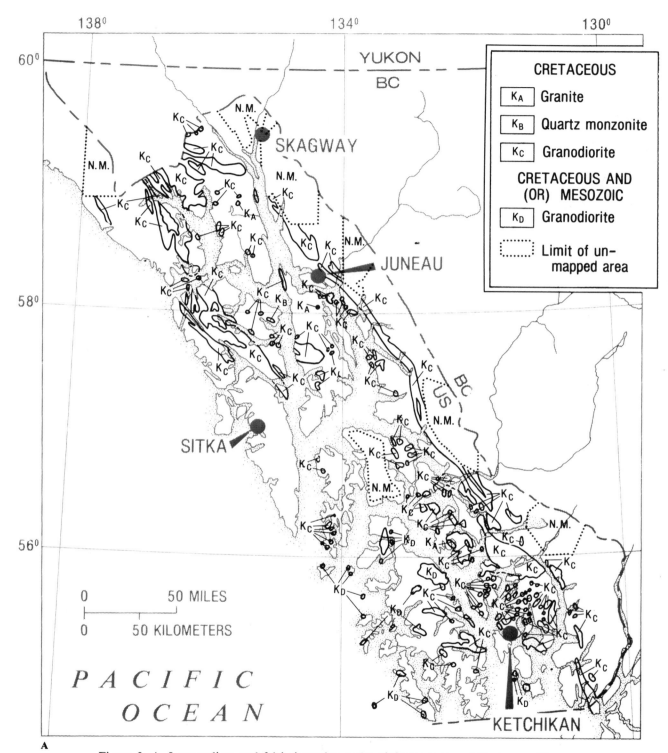

Figure 3. A, Intermediate and felsic intrusive rocks of Cretaceous and Cretaceous(?) age, southeastern Alaska. B (facing page), Plutonic belts of intermediate and felsic intrusive of Cretaceous age, southeastern Alaska.

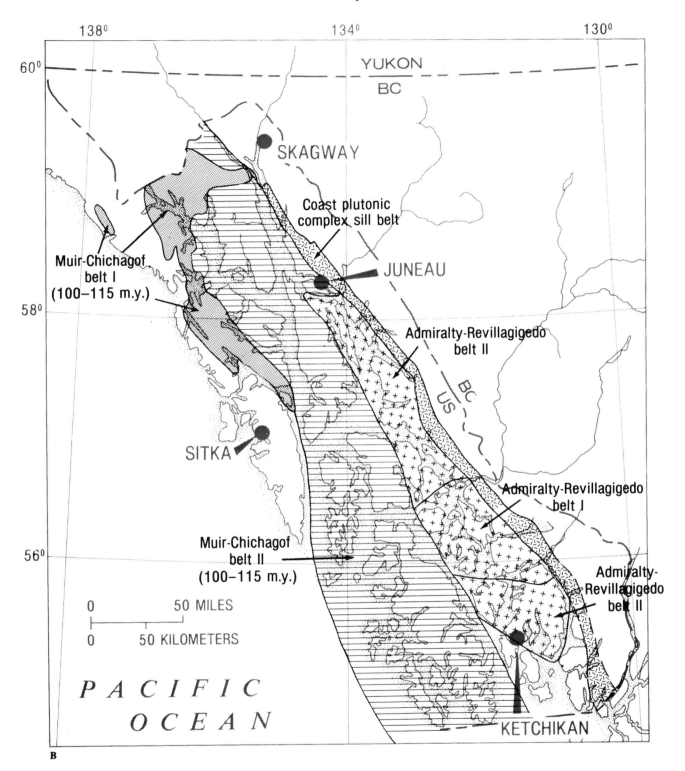

B

TABLE 1. CHARACTERISTICS OF INTRUSIVE ROCKS IN THE PLUTONIC BELTS OF SOUTHEASTERN ALASKA.

AGE	TERTIARY				TERTIARY	
	1	2	3	4	5	6
Name of belt or intrusive body (figure number reference)	Kuiu-Etolin belt* fig. 2	Behm Canal belt* fig. 2	Fairweather-Baranof belt** fig. 2	Coast plutonic complex belt I** fig. 2	Glacier Bay belt fig. 2	Coast Plutonic Complex belt II fig. 2
Isotopic age range (K-Ar method unless otherwise indicated).	Mid-Tertiary.	20-30 m.y. on biotite and approx. concordant biotite-hornblende.	Yakutat and Baranof areas: 1.20-30 m.y. on approx. concordant biotite and hornblende. 2.~40-50 m.y. on approx-concordant biotite/hornblende and biotite/muscovite. Glacier Bay area: 27-38 m.y. on biotite and muscovite.	45-54 m.y. on approx. concordant biotite/hornblende.	Discordant mid-Tertiary ages.	Discordant mid-Tertiary ages.
Dominant composition or compositional range.	1.Granite to quartz monzonite. 2.Gabbro sills near Keku Strait.	1.Alkali granite 2.Granite 3.Gabbro to granodiorite.	1.Tonalite to quartz monzonite. 2.Narrow "sub-belt" of gabbro-norite intrusives.	Granodiorite to quartz monzonite.	Granodiorite, tonalite, and quartz diorite.	Tonalite to quartz monzonite. Migmatite Orthogneiss
Primary characterizing and accessory minerals.	Biotite (olivine and clinopyroxene)	Biotite Minor pyrite and/or molybdenite. (Augite, hypersthene, hornblende and biotite in gabbro).	Intermediate rocks: Biotite with variable amounts of hornblende, garnet, muscovite, magnetite, sulfides, apatite. Rare clinopyroxene. Gabbro-norite: Olivine, pyroxene with variable amounts of hornblende, biotite, sulfides, magnetite, ilmenite.	Hornblende, biotite, sphene; locally garnet-bearing.	Hornblende, biotite, and local sphene and magnetite.	Biotite, hornblende with localized magnetite and sphene. K-feldspar phenocrysts are abundant in some units.
Host tectonostratigraphic terranes.+	Southern Craig Admiralty Gravina Taku	Taku Tracy Arm	Northern Craig Chugach Wrangell	Tracy Arm Stikine	Chugach Northern Craig	Tracy Arm
Foliation and metamorphic characteristics of plutons and country rocks	Unfoliated Plutonic contacts cut regional foliation trends. Extent of contact metamorphism is uncertain.	Unfoliated. Plutonic contacts cut regional foliation trends. Extent of contact metamorphism is uncertain.	Unfoliated except locally near contacts. Cross-cutting to conformable contact relations with respect to country rocks. Well-developed contact metamorphic aureoles. Typically surrounded by stockwork migmatite.	Unfoliated. Typically bordered by migmatite phase in Tracy Arm area.	Slightly foliated. Cross-cutting to conformable contact relations with respect to country rocks. Thermal aureoles developed (at least). Some plutons are extensively deformed and altered.	Weakly foliated to gneissic. Cross-cutting to conformable contact relations with respect to country rocks. Irregular compositional and textural variations. Intrude sillimanite to kyanite-bearing schist and gneiss.
Metallogenic associations++	Has not been assessed. Tungsten geochem anomalies associated with volcanics on Zarembo Island.	Porphyry molybdenum (Quartz Hill, Groundhog Basin?, Burroughs Bay) U-Th? prospect (Cone Mountain).	Cu-Ni-sulfide magmatic segregation deposits in gabbro-norite (Brady Glacier, Bohemia Basin, Mirror Harbor). Au-Ag-Cu-Pb, Zn quartz-sulfide veins (Chichagof-Sitka). Porphyry copper deposit (Margerie Glacier). Porphyry molybdenum deposit (the Nunatak).	Porphyry(?) Cu-Mo deposits and Mo-Ag deposits. Polymetallic(?) skarn deposits.		Magnetite skarn (Bradfield River prospect).
Remarks	Intrusive rocks generally miarolitic. Plutons associated with dike swarms and volcanics on Kuiu and with rhyolitic volcanics on Zarembo Island.	Intrusive rocks commonly miarolitic. Plutons associated with rhyolitic volcanics at Cone Mountain and with rhyolite porphyry dike swarm at Groundhog Basin.	Gabbros in the Fairweather Range are layered.	This belt may include many of the intrusive rocks that have been included in the Tertiary and/or Cretaceous Coast Plutonic Complex belt II which have not been adequately dated.	Distinguished from the Tertiary bodies of the Fairweather-Baranof belt primarily by the development of a weak foliation.	Heterogeneity of these units and discordance of isotopic dates make age determination ambiguous.
References	Berg and others, 1976; Brew and others, unpub. data, 1979; Muffler, 1967	Berg and others, 1978; Berg and others, 1977; Elliott and Koch, unpub. data, 1979; Hudson and others, 1979b.	Brew and others, 1978; Brew and Sonnevil, unpub. data, 1979; Hudson and others, 1979a; Loney and others, 1975; MacKevett and others, 1974.	Berg and others, 1978; Berg and others, 1977; Brew and others, 1977; Elliott and Koch, unpub. data, 1979; Ford and Brew, 1977a,b; Hudson and others, 1979b; MacKevett and others, 1974; Smith, 1977.	Brew and others, 1978; Brew and Sonnevil, unpub. data, 1979; Hudson and others, 1977	Berg and others, 1978; Berg and others, 1977; Elliott and Koch, unpub. data, 1979; Hudson and others, 1979b.

*Minor belt
**Major belt
+Tectonostratigraphic terranes are those defined by Berg, Jones, and Coney, 1978. Terminology modified slightly here.
++Berg, H. C., 1979, was an additional source of information regarding metallogenic associations.

	CRETACEOUS				MESOZOIC	
7	8	9	10	11	12	13
Muir-Chichagof belts I and II** fig. 3	Admiralty-Revillagigedo belts I and II** fig. 3	Coast plutonic complex sill belt fig. 3	Klukwan-Duke belt fig. 4	Ultramafic rocks, Baranof Island fig. 4	Ultramafic rocks, Admiralty Island fig. 4	Ultramafic rocks, Tracy Arm fig. 4
100-115 m.y. on approx. concordant biotite-hornblende. Pb-α ages on zircon of 110 m.y. and 150 m.y. (± 20 m.y.)	74-84 m.y. on nearly concordant biotite-hornblende reset to discordant ages to east by 50 m.y. event in Coast Plutonic complex.	110 m.y.? Discordant early Tertiary ages on biotite and hornblende.	100-110 m.y. determined by analysis of degree of concordance of biotite/hornblende with respect to proximity to younger granitic intrusive rocks	Mesozoic?	Mesozoic	Cretaceous or older.
Granodiorite, tonalite, quartz diorite, diorite, and gabbro. Minor monzonite, quartz monzonite (e.g. in Copper Mt. pluton on Prince of Wales Island).	Granodiorite, quartz diorite, diorite.	Tonalite	Dunite, pyroxenite, hornblendite, gabbro.	Wehrlite, serpentinite.	Serpentinite or serpentinized peridotite	Peridotite, dunite, pyroxenite, and minor gabbro and hornblendite
Hornblende, biotite, sphene, apatite, sulfides. Pyroxene generally rare but abundant with uralite in gabbro.	Biotite, garnet, hornblende. Most bodies in belt II are characterized by plagioclase phenocrysts. Belt II bodies commonly lack plagioclase phenocrusts but have more hornblende.	Hornblende (typically euhedral), biotite, sphene magnetite. Rare garnet or augite in plutons grouped in this belt.	Olivine, clinopyroxene, hornblende, magnetite, biotite, serpentine.	Olivine, clinopyroxene, chromite with secondary serpentine, magnetite, and talc-carbonate alteration.	Antigorite, talc, carbonate.	Pyroxene, hornblende, olivine, biotite. Secondary anthophyllite and tremolitic amphibole. Serpentine rare or absent.
Craig Wrangell Admiralty Chugach (west of Fairweather fault)	Taku Gravina	Western edge of Tracy Arm	Craig Admiralty Annette Gravina Taku	Wrangell(?)	Admiralty (Retreat Group)	Tracy Arm
Moderate to strong foliation. Contact zones are commonly stockwork migmatites. Contact metamorphism to amphibolite or hornblende-hornfels facies.	Belt II stocks are typically zoned with stronger foliation (and more felsic composition) near contacts and most are less metamorphosed than surrounding country rocks. Belt I plutons are strongly foliated and have both concordant and cross-cutting contact relations.	Moderate to strong foliation parallel to contacts. Sill truncates regional metamorphic isograds near Juneau.	Locally sheared or mylonitized. Sharp to gradational contact relations with country rocks. Uncertain contact metamorphic effects, but some definite aureoles.	Pervasive foliation. Serpentinization (to antigorite) was probably under greenschist facies conditions.	Sheared.	Strong foliation parallel to contacts and regional foliation. No contact metamorphic aureoles. In gneiss of almandine-amphibolite facies. Metamorphosed to (at least) greenschist facies.
W-Cu-Ag-Au quartz-sulfide veins in altered granitic rocks (Reid Inlet) W-Cu-Ag-Au-Zn skarn deposits (Highland Chief) Cu-Zn-Mo-Ag-Au skarn deposits where intruding Wales Group (Copper Mt. pluton).		Au quartz veins west of sill (Juneau gold belt). Cu-Zn, Au-Cu, and Zn-Pb sulfide mineralization in lenses and pods and disseminated to west of sill (Groundhog Basin Zn-Pb, Glacier Basin Zn-Pb, Berg Basin Pb-Zn, Tracy Arm Cu-Zn prospect, Sumdum Cu-Zn prospect).	Magnetite cumulate deposits.	Low-grade, impure chromite (Red Bluff Bay peridotite).		No anomalous metallic concentrations.
Belts I and II are of the same apparent age and compositional range. Belt I consists of large bodies with contacts concordant to foliation. Belt is widest in northern Glacier Bay National Monument, narrows north and south. Belt II consists mostly of small, scattered plugs.	Belt I bodies are large, strongly foliated, and either non-porphyritic or slightly porphyritic. Belt II bodies are generally small, isolated porphyritic stocks.	Sill is semi-continuous between Berners Bay and the Stikine River. South of the Stikine River, the typical sill lithology is found sporadically in heterogeneous gneiss and orthogneiss. Shown as granodiorite on MF-1048.	Rhythmically-layered, concentrically zoned, Alaskan-type intrusives. Zoning is generally ultramafic to gabbroic from core to periphery of intrusive.	Form two en echelon belts on Baranof Island. Age relation with country rock uncertain; bodies may be tectonically emplaced. Bodies resemble zoned ultramafics of Klukwan-Duke belt rather than harzburgitic alpine peridotites.	Ore body is crudely layered.	Many bodies in or near contact with marble. Ni-Cr content suggest primary igneous origin.
Brew and others, 1978; Brew and Sonnevil, unpub. data, 1979; Lathram and others, 1959; Lathram and others, 1965; MacKevett and others, 1974; Muffler, 1967; Turner and others, 1977.	Berg and others, 1976; Berg and others, 1978; Brew and others, unpub. data, 1979; Elliott and Koch, unpub. data, 1979.	Berg and others, 1978; Brew and others, 1977; Elliott and Koch, unpub. data, 1979; Ford and Brew, 1977a,b; Gault and others, 1953; Brew and others, 1976.	Berg and others, 1976; Berg and others, 1977; Irvine, 1974; Lanphere, 1968; Lanphere and Eberlein, 1966; Lathram and others, 1959; Lathram and others, 1965; MacKevett and others, 1964; Taylor, 1967.	Loney and others, 1975.	Lathram and others, 1965.	Brew and others, 1977; Grybeck and others, 1977.

A G E	JURASSIC	TRIASSIC AND/OR JURASSIC		LATE PALEOZOIC(?)	PENNSYLVANIAN
	14	15	16	17	18
Name of belt or intrusive (figure number reference)	Chilkat-Chichagof belt*	Bokan Mountain Granite	Texas Creek granodiorite	Art Lewis Glacier pluton	Intrusive pluton near Klawak
	fig. 5	fig. 5	fig. 5	fig. 6	fig. 6
Isotopic age range (K-Ar method unless otherwise indicated)	145-165 m.y. on hornblende and approx. concordant biotite hornblende. Pb-$_\alpha$ ages on zircon are 160 m.y. and 180 m.y. ($\pm$ 20 m.y.)	180-190 m.y. on reibeckite. Pb-$_\alpha$ age of 240 $\pm$ 30 m.y. on zircon.	200-206 m.y. on hornblende.	Ages of 225 m.y. and 136 m.y. on hornblende.	276 $\pm$ 8 m.y. on biotite.
Dominant composition or compositional range.	Tonalite to quartz monzonite (more basic variants are minor).	Peralkaline granite.	Granodiorite.	Diorite, quartz diorite.	Syenite.
Primary characterizing and accessory minerals.	Hornblende, biotite, sphene, epidote, and apatite.	Reibeckite, acmite, zircon, xenotime, fluorite, uranothorite, cordierite, euhedral quartz phenocrysts.	Euhedral phenocrysts of hornblende and K-feldspar, biotite, sphene.	Hornblende(?), biotite(?), apatite, magnetite, pyrite.	Biotite, hornblende.
Host tectonostratigraphic terranes*$^+$	Northern Craig Wrangell	Southern Craig	Stikine	Hubbard terrane (equivalent to Craig?)	Southern Craig
Foliation and metamorphic characteristics of plutons and country rocks.	Foliated. Narrow contact metamorphic aureole to hornblende-hornfels facies. Steep contacts with country rocks.	Locally developed cataclastic texture. Albitized aureole in surrounding Ordovician(?) intrusives.	Locally developed cataclastic texture and shear zones.	Foliated. Both gradational and sharp, crosscutting contacts. Intrudes amphibolite, subordinate marble, mica schist.	
Metallogenic associations.*$^+$		Uranium-thorium-REE deposits: 1.Primary segregation enhanced by hydrothermal concentration (Ross-Adams). 2.Syngenetic deposits in pegmatite and aplite dikes. 3.Open space filling and replacement epigenetic hydrothermal deposits. 4.Interstices of clastic sedimentary rocks.	Polymetallic quartz veins in sheared zones of Texas Creek Granodiorite. Volcanogenic Cu-Pb-Zn-Ag-Au deposits in volcaniclastics of Hazelton(?) Group.		
Remarks	Locally porphyritic. Concentric zoning of Kennel Creek pluton on Chichagof Island.	Roughly 3 square miles in area.		May correlate with Mt. Hubbard pluton with hornblende ages of 279 and 284 m.y. High zirconium and strontium content.	
References	Lathram and others, 1959; Loney and others, 1975.	Lanphere and others, 1964; MacKevett, 1963.	Berg and others, 1978; Berg and others, 1977; Byers and Sainsbury, 1956; Smith, 1977.	Hudson and others, 1977.	Churkin and Eberlein, 1975.

*Minor belt
**Major belt
$^+$Tectonostratigraphic terranes are those defined by Berg, Jones, and Coney, 1978.
$^+$-Berg, H. C., 1979, was an additional source of information regarding metallogenic associations.

ORDOVICIAN(?)				PRECAMBRIAN
SOUTHERN SOUTHEASTERN ALASKA BELT				
19	20	21	22	23
Sitkoh Bay complex	Kendrick Bay complex	Prince of Wales mafics/ultramafics	Annette pluton	Ruth Bay intrusive
fig. 6	fig. 6	fig. 6	fig. 6	fig. 6
Minimum age of 406 ± 16 m.y. on hornblende.	Minimum age of 446 ± m.y. on hornblende. Pb-α age of 510 ± 60 m.y. on zircon.	Minimum age of 430-440 m.y. on hornblende.	Minimum age of 416 ± 12 m.y. on hornblende.	Minimum age of 730 m.y., U-Pb date on zircon.
Syenite, syenodiorite, trondhjemite. (Lesser amounts of a wide variety of compositions).	1.Quartz diorite to diorite. 2.Quartz monzonite to granodiorite. 3.Alaskite. 4.Minor syenite.	1.Pyroxenite. 2.Gabbro.	Trondhjemite with minor leuco-granite, quartz monzonite, and quartz diorite.	Trondhjemite.
Hornblende, biotite with localized nepheline, sodalite, cancrinite, and sodic pyroxene.	Hornblende, biotite, augite. Rare garnet and aegerine in syenite.	Augite, uralitic hornblende, biotite magnetite, apatite, pyrite and ilmenite in gabbro.	Muscovite, minor biotite and hornblende. Secondary chlorite, epidote.	Not published.
Northern Craig	Southern Craig	Southern Craig	Annette	Southern Craig
	Poor foliation except near contacts and in gneissic quartz monzonite unit. Alkalic rocks appear to have intruded calcic rocks but many contacts are gradational. Albitized in part. Cu-Au quartz-carbonate veins Au-Ag in massive pyrite in marble septum intruded by quartz diorite. Gold-bearing calcite veins.	Locally sheared. Pyroxenite intruded by Ordovician(?) quartz monzonite or syenite. Intrusive breccia contact zones. Gabbro in gradational contact with Ordovician(?) diorite. Possible concentrations of magnetite.	Mild cataclastic texture in core; mylonitic schist, gneiss, and breccia near periphery. Baked contact zone few cm wide.	Pre-intrusive greenschist facies metamorphism. Thermal event reset K-Ar ages of metamorphic rocks in Early Ordovician. Undeformed and unmetamorphosed.
Each body conspicuously heterogeneous. Clasts of syenite probably from this complex occur in: 1.Point Augusta Fm. (Silurian). 2.Kennel Creek Limestone (Silurian and/or Devonian). 3.Cedar Cove Fm. (Middle and Upper Devonian).	Inclusions of gabbro, amphibolite, gneiss, and schist in diorite.	Contact relations suggest ultramafic rocks to be oldest intrusive rocks in Bokan Mt.-Kendrick Bay area. So their minimum age is considered 446 m.y.		In Wales Group.
Lanphere and others, 1965; Loney and others, 1975.	Lanphere and others, 1964; MacKevett, 1963; Turner and others, 1977.	MacKevett, 1963; Turner and others, 1977.	Berg, 1972; Berg and others, 1978.	Churkin and Eberlein, 1977; Turner and others, 1977.

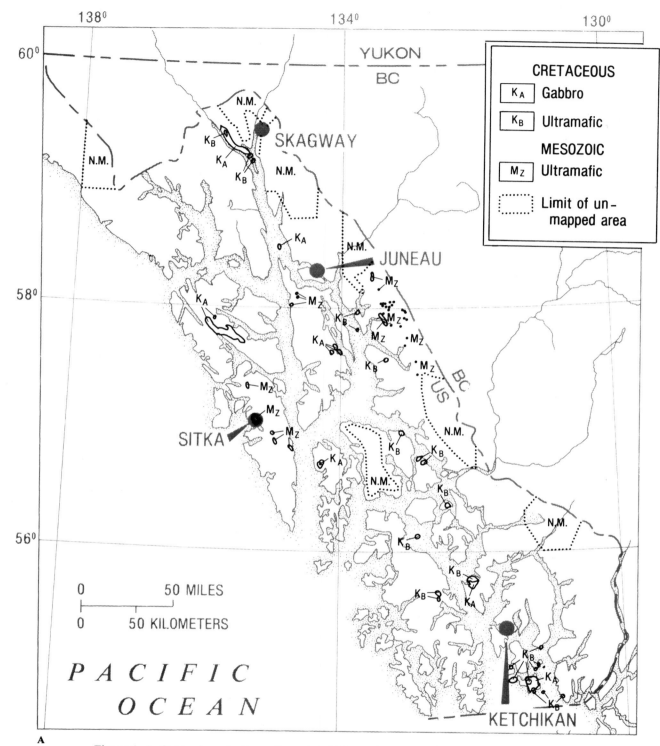

Figure 4. A, Gabbroic and ultramafic rocks of Mesozoic age, southeastern Alaska. B (facing page), Gabbroic and ultramafic belts of Mesozoic age, southeastern Alaska.

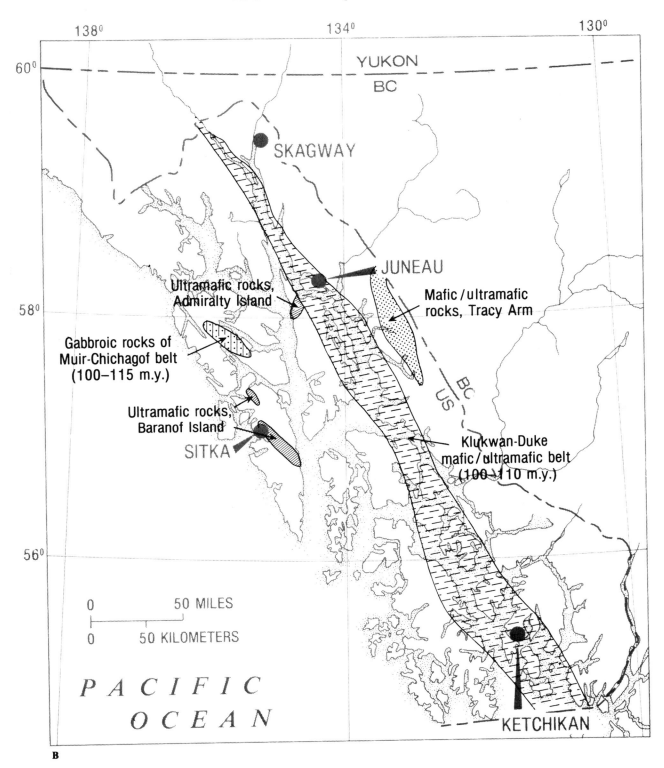

B

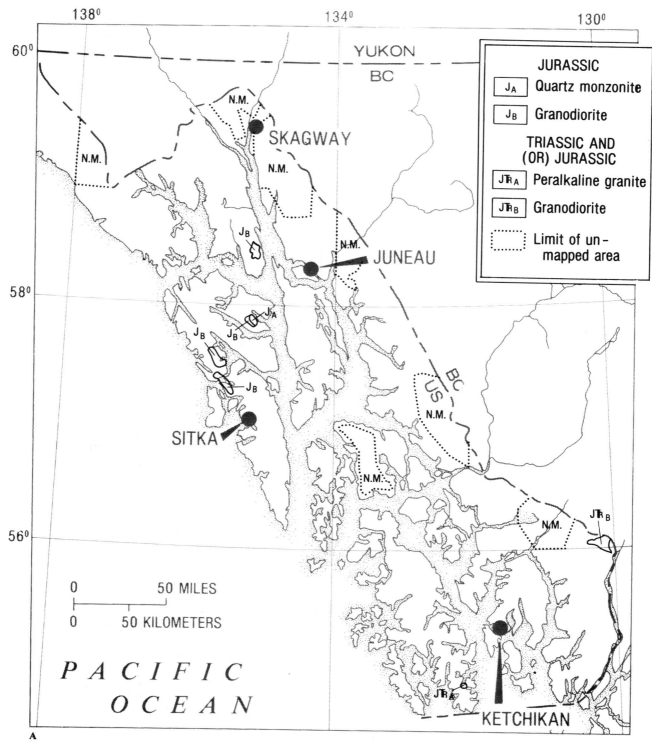

Figure 5. A, Intrusive rocks of Jurassic age, southeastern Alaska. B (facing page), Plutonic belts and isolated intrusions of Jurassic age, southeastern Alaska.

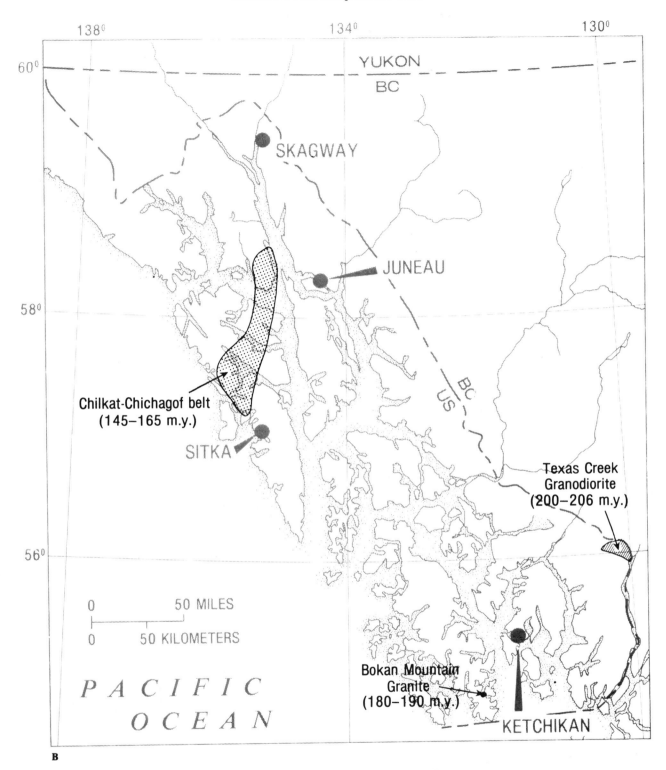

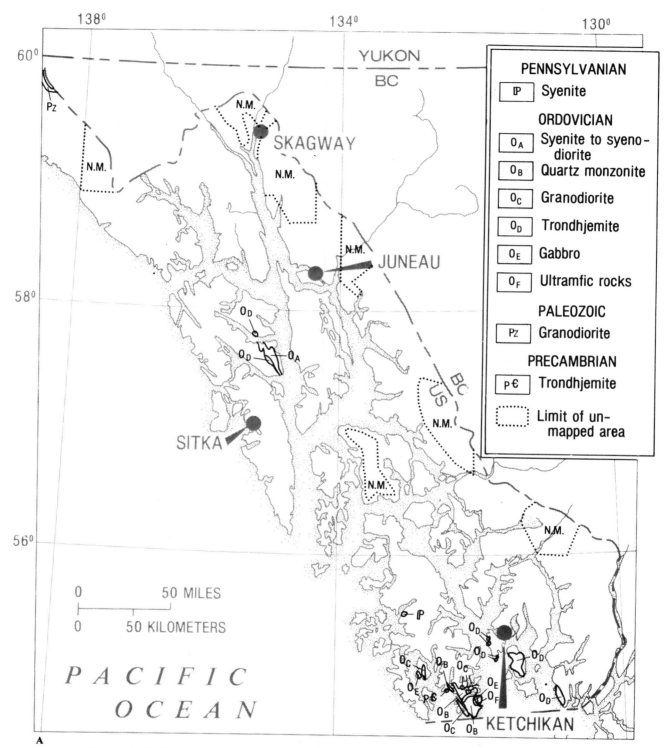

Figure 6. A, Intrusive rocks of Precambrian and of Paleozoic age, southeastern Alaska. B (facing page), Plutonic belts of Precambrian and of Paleozoic age in southeastern Alaska.

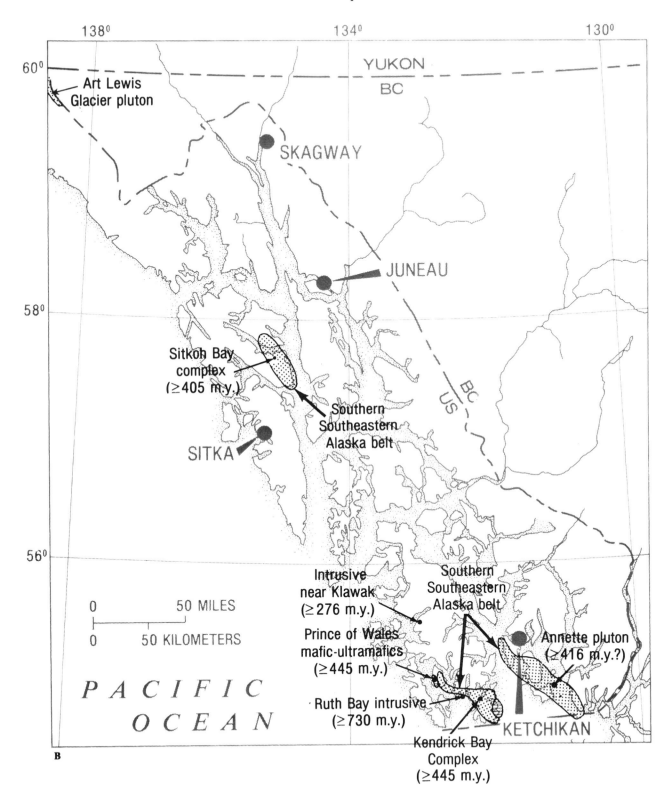

138° 134° 130°

60°

Art Lewis
Glacier pluton

YUKON
BC

SKAGWAY

JUNEAU

58°

Sitkoh Bay
complex
(≥405 m.y.)

BC
US

Southern
Southeastern
Alaska belt

SITKA

56°

Intrusive
near Klawak
(≥276 m.y.)

Southern
Southeastern
Alaska belt

Prince of Wales
mafic-ultramafics
(≥445 m.y.)

Annette pluton
(≥416 m.y.?)

Ruth Bay intrusive
(≥730 m.y.)

Kendrick Bay
Complex
(≥445 m.y.)

KETCHIKAN

0   50 MILES

0   50 KILOMETERS

P A C I F I C

O C E A N

B

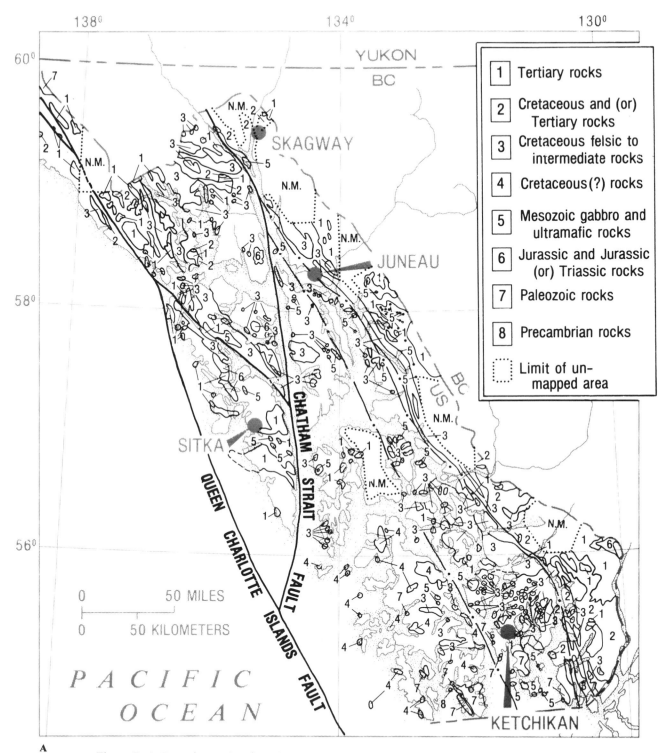

Figure 7. A, Intrusive rocks of southeastern Alaska. B (facing page), Major plutonic belts of southeastern Alaska.

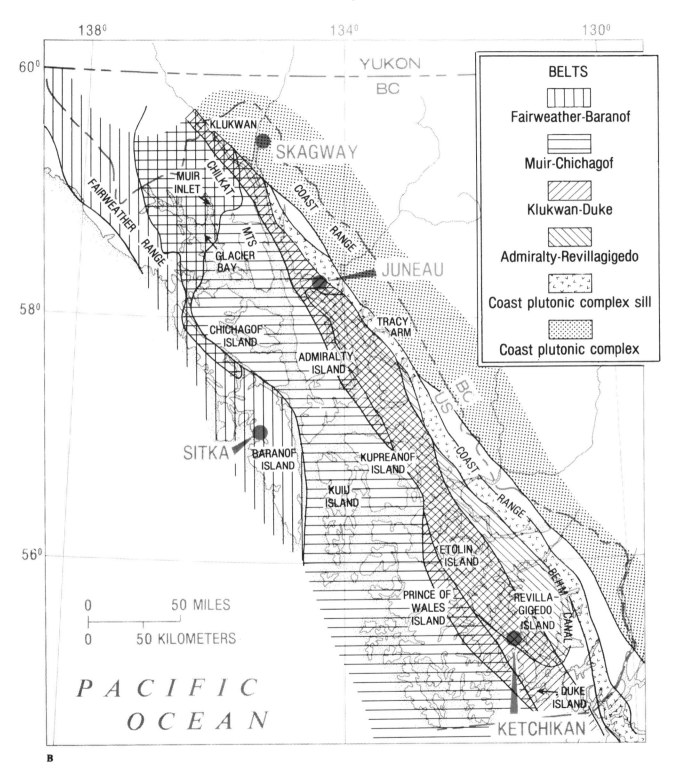

B

These two belts join at the Seymour Canal-Clarence Strait lineament, whose location is revised from that shown by Brew and Ford (1978). These two belts, together with the three Coast Plutonic Complex sill belts, cover almost all of southeastern Alaska except the Chugach and part of the Wrangell terranes in the west (Brew and Morrell, 1979b, c). Brew and Ford (1978) discussed the origin of the Coast Plutonic Complex sill and suggested that it was emplaced during mid- or Late Cretaceous time along an important structural discontinuity. The sill becomes discontinuous to the south where it adjoins section I of the Admiralty-Revillagigedo belt but appears to re-establish itself near the Alaska-Canadian border southeast of Ketchikan, Alaska; it continues southeast in Canada as the Quottoon pluton (Hutchison, 1982). The Muir-Chichagof belt more or less corresponds to the Nutzotin-Chichagof belt of Hudson (1979).

Reconstruction of the 200 km of right-lateral movement on the Chatham Strait fault increases the overall width of the Muir-Chichagof belt and makes its strike northwest.

GABBROIC AND ULTRAMAFIC ROCKS OF MESOZOIC AGE

A variety of gabbroic and ultramafic rocks of Mesozoic age occurs in southeastern Alaska (Fig. 4; cols. 10-13, Table 1). A Cretaceous age has been established for some of the bodies, including the concentrically zoned or Alaskan-type mafic-ultramafic complexes which form the only major belt. Hudson (1979) did not include these rocks in his analysis of Mesozoic plutonic belts.

INTRUSIVE ROCKS OF JURASSIC AGE

Relatively few intrusive rocks of Jurassic age and of Triassic and/or Jurassic age are known in southeastern Alaska (Fig. 5; cols. 14-16, Table 1). Many more Jurassic bodies may exist but are so similar petrographically to the Cretaceous bodies that they have not been recognized. Only one belt, the Chilkat-Chichagof, has been defined; its southern end coincides with the Tonsina-Chichagof belt of Hudson (1979), but to the north, for some reason, the Tonsina-Chichagof belt excludes radiometrically dated Jurassic bodies in the Chilkat Range and northeastern Chichagof Island, and also includes a large area in which no Jurassic plutons have been identified.

INTRUSIVE ROCKS OF PALEOZOIC AND PRECAMBRIAN AGES

Intrusive rocks of Paleozoic and Precambrian ages constitute a diverse group of plutons (Fig. 6; cols. 17-23, Table 1). Although the age of the Annette pluton is uncertain and the plutonic complexes differ in composition, a widespread

intrusive event of probable Ordovician age clearly occurred (M. A. Lanphere, oral commun., 1978). Palinspastic reconstruction of 200 km of right-lateral movement on the Chatham Strait fault (Ovenshine and Brew, 1972) brings these Ordovician(?) complexes significantly closer together and aligns them in a 160-km-long east-west-trending area near the U.S.-Canadian border.

The limited available isotopic evidence that reveals unusual bodies like the Pennsylvanian syenite and Precambrian trondhjemite on Prince of Wales Island suggests that more such plutons may be present, though unrecognized.

DISCUSSION

The distribution, composition, and age information presented by Brew and Morrell (1980) and included here has been synthesized to give an interpretation of the plutonic belts in southeastern Alaska. Taken as a whole (Fig. 7A), the situation is complex; the major belts alone present a complicated picture (Fig. 7B).

No simple time-space distribution of the plutonic belts of southeastern Alaska exists. However, some observations are appropriate to the major and several of the minor belts:

1. The Kuiu-Etolin (20-30 m.y.), Behm Canal (20-30 m.y.), and, perhaps, Tkope belts are not obviously related to any single tectonic element. The present authors suggest that they are related to either (a) mid- to Late Tertiary vertical movements (Brew and others, 1966; Brew, 1968) that led to local formation of continental sedimentary basins, or (b) tension associated with large-scale strike-slip movements. In either case, deep fractures provided conduits for magmas from below.

2. The Fairweather-Baranof belt (20(?)-50 m.y.) and Coast Plutonic Complex belt I (45-54 m.y.) flank the dominant Cretaceous plutonic belts of southeastern Alaska on the west and east, respectively. The Tertiary belts are grossly similar but have some important differences. Hudson, Plafker, and Peterman (1977) argued that the plutons in the Fairweather-Baranof belt are anatectic and derived from the thick accretionary prism of Cretaceous turbidites in the same area. Their hypothesis does not account for the gabbroic rocks that apparently form part of the same belt (although the gabbros are largely associated with metavolcanic rocks and not the metaturbidites), nor does it apply to the Coast Plutonic Complex belt because of its diverse host rocks. Brew, Johnson, and others (1978) suggested that oblique subduction directed to the north in Early and mid-Tertiary time was an unlikely cause for the Fairweather-Baranof belt, but they had no alternative hypothesis to offer.

3. The Muir-Chichagof, Admiralty-Revillagigedo, Klukwan-Duke, and Coast Plutonic Complex sill belts form the plutonic spine of southeastern Alaska; an understanding of their significance depends on deciphering their mutual

age relations, but these age relations are still uncertain. All these belts are presently thought to be nearly contemporaneous (100-115 m.y.) except for the slightly younger Admiralty-Revillagigedo belt (74-84 m.y.), whose plutons deform the foliation that penetrates the Coast Plutonic Complex sill. The Klukwan-Duke belt of concentrically zoned, mafic-ultramafic plutons generally adjoins, but slightly overlaps, the Muir-Chichagof belt. These belts are interpreted as being closely related, the mafic-ultramafic rocks as possibly representing the specific "roots" of the volcanic piles that make up much of the Gravina terrane (Berg and others, 1972; Irvine, 1973), and the foliated granitic rocks as possibly representing the general base of the magmatic arc. The Coast Plutonic Complex sill was interpreted (Brew and Ford, 1978) as having been emplaced at about the same time along a pre-existing structural discontinuity that may have separated the Taku and Tracy Arm terranes of Berg, Jones, and Coney (1978). The slightly younger Admiralty-Revillagigedo belt may represent another magmatic arc or a late phase of the 100-115-m.y. plutonic episode just described. A preliminary lead-uranium age of 140 m.y. on a pluton from the Admiralty-Revillagigedo belt raises the possibility that the belt is Jurassic rather than Cretaceous. Scattered, as yet unrecognized, Jurassic plutons may exist within both the Muir-Chichagof and Admiralty-Revillagigedo belts.

4. Jurassic plutonic rocks of the Alaska-Aleutian Range batholith have been extensively studied by Reed and Lanphere (1973), but their connection with the Chilkat-Chichagof belt, by way of the Tonsina-Chichagof belt of Hudson (1979), has not been established. Reed and Lanphere (1973) concluded that a magmatic arc with northward-dipping polarity was present, but Hudson (1979) suggested that the evidence for this interpretation is insufficient.

ACKNOWLEDGMENTS

We thank the technical reviewers, H. C. Berg and R. A. Sonnevil, and the technical editor, George Harach.

REFERENCES CITED

Bateman, P. C., 1961, Granitic formations in east-central Sierra Nevada, near Bishop, California: Geological Society of America Bulletin, v. 72, no. 10, p. 1521-1537.

Beikman, H. M., compiler, 1975, Preliminary geologic map of southeastern Alaska: U.S. Geological Survey Miscellaneous Field Studies Map MF-673, scale 1:1,000,000, 2 sheets.

Berg, H. C., 1972, Geologic map of Annette Island, Alaska: U.S. Geological Survey Miscellaneous Geologic Investigations Map I-684, scale 1:63,360, 8 p.

——, 1979, Significance of geotectonics in the metallogenesis and resource appraisal of southeastern Alaska: A progress report: Abstract prepared for Alaska Geological Society Meeting, April 1979.

Berg, H. C., Elliott, R. L., Koch, R. D., Carten, R. B., and Wahl, F. A., 1976, Preliminary geologic map of the Craig D-1 and parts of the Craig C-1 and D-2 quadrangles, Alaska: U.S. Geological Survey Open-File Report 76-430, scale 1:63,360.

Berg, H. C., Elliott, R. L., Smith, J. G., and Koch, R. D., 1978, Geologic map of the Ketchikan and Prince Rupert quadrangles, Alaska: U.S. Geological Survey Open-File Report 78-73-A, scale 1:250,000.

Berg, H. C., Elliott, R. L., Smith, J. G., Pittman, T. L., and Kimball, A. L., 1977, Mineral resources of the Granite Fiords wilderness study area, Alaska, *with a section on* Aeromagnetic data, by Andrew Griscom: U.S. Geological Survey Bulletin 1403, 151 p.

Berg, H. C., Jones, D. L., and Coney, P. J., 1978, Map showing pre-Cenozoic tectonostratigraphic terranes of southeastern Alaska and adjacent areas: U.S. Geological Survey Open-File Report 78-1085, scale 1:1,000,000, 2 sheets.

Berg, H. C., Jones, D. L., and Richter, D. H., 1972, Gravina-Nutzotin belt — tectonic significance of an upper Mesozoic sedimentary and volcanic sequence in southern and southeastern Alaska, *in* Geological Survey research 1972: U.S. Geological Survey Professional Paper 800-D, p. D1-D24.

Brew, D. A., 1968, The role of volcanism in post-Carboniferous tectonics of southeastern Alaska and nearby regions, North America: International Geological Congress, 23d, Prague, 1968, Section 2, Proceedings, p. 107-121.

——, 1975, Progress report on plutonic rocks of southeastern Alaska (abs.): Circum-Pacific Plutonism Project Meeting, 4th, Vancouver, Canada, 1975, p. 4.

Brew, D. A., Berg, H. C., Morrell, R. P., Sonnevil, R. A., Hunt, S. J., and Huie, C., 1979, The Tertiary Kuiu-Etolin volcanic-plutonic belt, southeastern Alaska, *in* Johnson, K. M., and Williams, J. R., eds., The United States Geological Survey in Alaska — accomplishments during 1978: U.S. Geological Survey Circular 804-B, p. B129-B130.

Brew, D. A., and Ford, A. B., 1978, Megalineament in southeastern Alaska marks southwest edge of Coast Range batholithic complex: Canadian Journal of Earth Sciences, v. 15, no. 11, p. 1763-1772.

Brew, D. A., Ford, A. B., Grybeck, D., Johnson, B. R., and Nutt, C. J., 1976, Key foliated quartz diorite sill along southwest side of Coast Range complex, northern southeastern Alaska, *in* Cobb, E. H., ed., The United States Geological Survey in Alaska: Accomplishments during 1975: U.S. Geological Survey Circular 733, p. 60.

Brew, D. A., Grybeck, D., Johnson, B. R., Jachens, R. C., Nutt, C. J., Barnes, D. F., Kimball, A. L., Still, J. C., and Rataj, J. L., 1977, Mineral resources of the Tracy Arm-Fords Terror wilderness study area and vicinity, Alaska: U.S. Geological Survey Open-File Report 77-649, 282 p.

Brew, D. A., Johnson, B. R., Grybeck, D., Griscom, A., Barnes, D. F., Kimball, A. L., Still, J. C., and Rataj, J. L., 1978, Mineral resources of the Glacier Bay National Monument wilderness study area, Alaska: U.S. Geological Survey Open-File Report 78-494.

Brew, D. A., Loney, R. A., and Muffler, L. J. P., 1966, Tectonic history of southeastern Alaska: Canadian Institute of Mining and Metallurgy Special Volume no. 8, p. 149-170.

Brew, D. A., and Morrell, R. P., 1978, Tarr Inlet suture zone, Glacier Bay National Monument, Alaska, *in* Johnson, K. M., ed., The United States Geological Survey in Alaska: Accomplishments during 1977: U.S. Geological Survey Circular 772-B, p. B90-B92.

——, 1980, Preliminary map of intrusive rocks in southeastern Alaska: U.S. Geological Survey Miscellaneous Field Studies Map MF-1048.

——, 1979a, Intrusive rock belts of southeastern Alaska, *in* Johnson, K. M., and Williams, J. R., eds., The United States Geological Survey in Alaska: Accomplishments during 1978: U.S. Geological Survey Circular 804-B, p. B116-B121.

——, 1979b, The Wrangell terrane ("Wrangellia") in southeastern Alaska; the Tarr Inlet suture zone with its northern and southern extensions, *in* Johnson, K. M., and Williams, J. R., eds., The United States Geological Survey in Alaska: Accomplishments during 1978: U.S. Geological Survey Circular 804-B, p. B121-B123.

——,1979c, Correlation of the Sitka Graywacke, unnamed rocks in the Fairweather Range, and Valdez Group, *in* Johnson, K. M., and Williams, J. R., eds., The United States Geological Survey in Alaska: Accomplishments during 1978: U.S. Geological Survey Circular 804-B, p. B123–B125.

Buddington, A. F., and Chapin, T., 1929, Geology and mineral deposits of southeastern Alaska: U.S. Geological Survey Bulletin 800, 398 p.

Byers, F. M., Jr., and Sainsbury, C. L., 1956, Tungsten deposits of the Hyder district, Alaska: U.S. Geological Survey Bulletin 1024-F, p. 123–140.

Campbell, R. B., and Dodds, C. J., 1979, Operation Saint Elias, British Columbia, *in* Current research, pt. A: Geological Survey of Canada Paper 79-1A, p. 17–20.

Churkin, M., Jr., and Eberlein, G. D., 1975, Geologic map of the Craig C-4 quadrangle, Alaska: U.S. Geological Survey Geologic Quadrangle Map GQ-1169, scale 1:63,360.

——,1977, Ancient borderland terranes of the North American Cordillera: Correlation and microplate tectonics: Geological Society of America Bulletin, v. 88, no. 6, p. 769–786.

Douglas, R. J. W., Gabrielse, H., Wheeler, J. O., Stott, D. F., and Belyea, H. R., 1970, Geology of western Canada, *in* Douglas, R. J. W., ed., Geology and economic minerals of Canada: Geological Survey of Canada Economic Geology Report 1, p. 365–488.

Ford, A. B., and Brew, D. A., 1977a, Preliminary geologic and metamorphic-isograd isograd map of northern parts of the Juneau A-1 and A-2 quadrangles, Alaska: U.S. Geological Survey Miscellaneous Field Studies Map MF-847, scales 1:125,000 and 1:31,680.

——,1977b, Truncation of regional metamorphic zonation pattern of the Juneau, Alaska, area by the Coast Range batholith, *in* Blean, K. M., ed., The United States Geological Survey in Alaska — accomplishments during 1976: U.S. Geological Survey Circular 751-B, p. B85–B87.

Gault, H. R., Rossman, D. L., Flint, G. M., Jr., and Ray, R. G., 1953, Some zinc-lead deposits of the Wrangell district, Alaska: U.S. Geological Survey Bulletin 998-B, p. 15–58.

Grybeck, D., Brew, D. A., Johnson, B. R., and Nutt, C. J., 1977, Ultramafic rocks in part of the Coast Range batholithic complex, southeastern Alaska, *in* Blean, K. M., ed., The United States Geological Survey in Alaska — accomplishments during 1976: U.S. Geological Survey Circular 751-B, p. B82–B85.

Hudson, T. L., 1979, Mesozoic plutonic belts of southern Alaska: Geology, v. 7, p. 230–234.

Hudson, T. L., Plafker, G., and Lanphere, M. A., 1977, Intrusive rocks of the Yakutat-St. Elias area, south-central Alaska: Journal of Research of the U.S. Geological Survey, v. 5, no. 2, p. 155–172.

Hudson, T. L., Plafker, G., and Peterman, Z. E., 1979a, Paleogene anatexis along the Gulf of Alaska margin: Geology, v. 7, p. 573–577.

Hudson, T. L., Smith, J. G., and Elliott, R. L., 1979b, Petrology, composition, and age of intrusive rocks associated with the Quartz Hill molybdenite deposit, southeastern Alaska: Canadian Journal of Earth Sciences, v. 16, p. 1805–1822.

Hutchison, W. W., 1982, Geology of the Prince Rupert-Skeena map area: Geological Survey of Canada Memoir 394, 116 p.

Hutchison, W. W., Berg, H. C., and Okulitch, A. V., 1973, Skeena River, British Columbia, geological map: Geological Survey of Canada Open-File Report 166, scale 1:1,000,000.

Irvine, T. N., 1973, Bridget Cove volcanics, Juneau area, Alaska; possible parental magma of Alaskan-type ultramafic complexes: Carnegie Institute of Washington Yearbook 72, p. 478–491.

——,1974, Petrology of the Duke Island ultramafic complex, southeastern Alaska: Geological Society of America Memoir 138, 240 p.

King, P. B., compiler, 1969, Tectonic map of North America: Washington, D.C., U.S. Geological Survey, scale 1:5,000,000.

Lanphere, M. A., 1968, Sr-Rb-K and Sr isotopic relationships in ultramafic rocks, southeastern Alaska: Earth and Planetary Science Letters, v. 4, no. 3, p. 185–190.

Lanphere, M. A., and Eberlein, G. D., 1966, Potassium-argon ages of magnetite-bearing ultramafic complexes in southeastern Alaska [abs.]: Geological Society of America Special Paper 87, p. 94.

Lanphere, M. A., Loney, R. A., and Brew, D. A., 1965, Potassium-argon ages of some plutonic rocks, Tenakee area, Chichagof Island, southeastern Alaska, *in* Geological Survey research 1965: U.S. Geological Survey Professional Paper 525-B, p. B109–B111.

Lanphere, M. A., MacKevett, E. M., Jr., and Stern, T. W., 1964, Potassium-argon and lead-alpha ages of plutonic rocks, Bokan Mountain area, Alaska: Science, v. 145, no. 3633, p. 705–707.

Lathram, E. H., Loney, R. A., Condon, W. H., and Berg, H. C., 1959, Progress map of the geology of the Juneau quadrangle, Alaska: U.S. Geological Survey Miscellaneous Geologic Investigations Map I-303, scale 1:250,000.

Lathram, E. H., Pomeroy, J. S., Berg, H. C., and Loney, R. A., 1965, Reconnaissance geology of Admiralty Island, Alaska: U.S. Geological Survey Bulletin 1181-R, p. R1–R48.

Loney, R. A., Brew, D. A., Muffler, L. J. P., and Pomeroy, J. S., 1975, Reconnaissance geology of Chichagof, Baranof, and Kruzof Islands, southeastern Alaska: U.S. Geological Survey Professional Paper 792, 105 p.

MacKevett, E. M., Jr., 1963, Geology and ore deposits of the Bokan Mountain uranium-thorium area, southeastern Alaska: U.S. Geological Survey Bulletin 1154, 125 p.

MacKevett, E. M., Jr., and Blake, M. C., Jr., 1963, Geology of the North Bradfield River iron prospect, southeastern Alaska: U.S. Geological Survey Bulletin 1108-D, p. D1–D21.

MacKevett, E. M., Jr., Berg, H. C., Plafker, G., and Jones, D. L., 1964, Preliminary geologic map of the McCarthy C-4 quadrangle, Alaska: U.S. Geological Survey Miscellaneous Geological Investigations Map I-423, scale 1:63,360 (1 inch to 1 mile).

MacKevett, E. M., Jr., Robertson, E. C., and Winkler, G. R., 1974, Geology of the Skagway B-3 and B-4 quadrangles, southeastern Alaska: U.S. Geological Survey Professional Paper 832, 33 p.

Muffler, L. J. P., 1967, Stratigraphy of the Keku Islets and neighboring parts of Kuiu and Kupreanof Islands, southeastern Alaska: U.S. Geological Survey Bulletin 1241-C, p. C1–C52.

Ovenshine, A. T., and Brew, D. A., 1972, Separation and history of the Chatham Strait fault, southeast Alaska, North America: International Geological Congress, 24th, Montreal, 1972, Section 3, Proceedings, p. 245–254.

Reed, B. L., and Lanphere, M. A., 1973, Alaska-Aleutian Range batholith: Geochronology, chemistry, and relation to circum-Pacific plutonism: Geological Society of America Bulletin, v. 84, no. 8, p. 2583–2610.

Roddick, J. A., 1966, Coast crystalline belt of British Columbia: Canadian Institute of Mining and Metallurgy Special Volume 8, p. 73–82.

Smith, J. G., 1977, Geology of the Ketchikan D-1 and Bradfield Canal A-1 quadrangles, southeastern Alaska: U.S. Geological Survey Bulletin 1425, 49 p.

Souther, J. G., Brew, D. A., and Okulitch, A. V., 1974, Iskut River, British Columbia, geologic map: Geological Survey of Canada Open-File Report 214, scale 1:1,000,000.

——,1979, Iskut River 1:1,000,000 geological atlas NTS 104, 114: Geological Survey of Canada Map 1418A.

Streckeisen, A. L., chairman, 1973, Plutonic rocks — classification and nomenclature recommended by the International Union of Geological Sciences (IUGS) Subcommission on the Systems of Igneous Rocks: Geotimes, v. 18, no. 10, p. 26–30.

Taylor, H. P., 1967, The zoned ultramafic complexes of southeastern Alaska, *in* Wyllie, P. J., ed., Ultramafic and related rocks: New York, John Wiley and Sons, p. 97–121.

Turner, D. L., Herreid, G., and Bundtzen, T. K., 1977, Geochronology of southern Prince of Wales Island, Alaska, *in* Short notes on Alaskan geology — 1977: Alaska Division of Geological and Geophysical Surveys Geologic Report 55, p. 11–16.

Wilson, F. H., Dadisman, S. V., and Herzon, P. L., 1979, Map showing radiometric ages of rocks in Alaska; part I, southeastern Alaska: U.S. Geological Survey Open-File Report 79-594, scale 1:1,000,000.

MANUSCRIPT ACCEPTED BY THE SOCIETY JULY 12, 1982

Geological Society of America
Memoir 159
1983

Geophysical review and composition of the Coast Plutonic Complex, south of latitude 55° N

J. A. Roddick

Geological Survey of Canada
100 West Pender St., Vancouver, B.C., V6B1R8

ABSTRACT

The continental margin is unmistakably indicated by the steep decline in Bouguer values from 150 mgal to 30 mgal from west to east across the continental slope. A decline of almost equal magnitude marks the western side of the Coast Plutonic Complex and is possibly an expression of a former continental margin. The "seismic" crust appears to differ from the "gravity" crust in most places, but this is best accounted for by the lack of constraints on the interpretation of both types of data. The depth to the Moho discontinuity beneath the western margin of the Coast Plutonic Complex is thought to be between 23 and 27 km, and beneath the eastern margin, about 33 km. The Moho appears to deepen easterly beneath the Intermontane Belt. Heat production in the plutonic rocks is fairly constant across the Coast Plutonic Complex. A strong magnetic anomaly over the western margin of the Coast Plutonic Complex is the most striking feature revealed by geophysical data. Its source is large and deep, a mass at least 250 km long, about 50 km wide, and possibly as much as 40 km deep. It is thought to be a near-surface segment of lower crust rather than an intrusive body or a zone mineralized with magnetite.

Modal study of about 8000 specimens of plutonic rock from the southern Coast Mountains (south of latitude 52° N) confirms that the average plutonic rock of the Coast Plutonic Complex is a quartz diorite. When the specimens are divided, each according to its source in one of three belts parallel with the regional trend, a marked quartz deficiency is revealed, not in the western zone where expected, but in the axial belt; quartz diorite is seen to be relatively more abundant in the eastern, rather than in the western, belt. Also contrary to expectations, granodiorite becomes progressively less abundant from west to east. Most common plutonic rock types were found to become denser and more mafic-rich from east to west.

About 600 chemical analyses from the Coast Plutonic Complex between latitudes 51° and 54° N showed that the average composition of the plutonic rock of the Coast Plutonic Complex is nearly identical to that of typical continental crust in SiO_2, FeO, TiO_2, and P_2O_5, and similar in MgO and CaO. The Coast Plutonic Complex, however, is markedly richer in Al_2O_3, somewhat richer in Na_2O, and distinctly poorer in K_2O. Chemically, the volcanic equivalent to the average plutonic rock in the Coast Plutonic Complex is a tholeiitic andesite.

The specimens for chemical analyses were selected so as to form three widely spaced cross-sections of the Coast Plutonic Complex. The most conspicuous feature shown by the chemical cross-sections is a marked depression in SiO_2 values near, but not at, the western edge of the Complex. The chemical analyses show no systematic increase in SiO_2 or K_2O from west to east across the Coast Plutonic Complex.

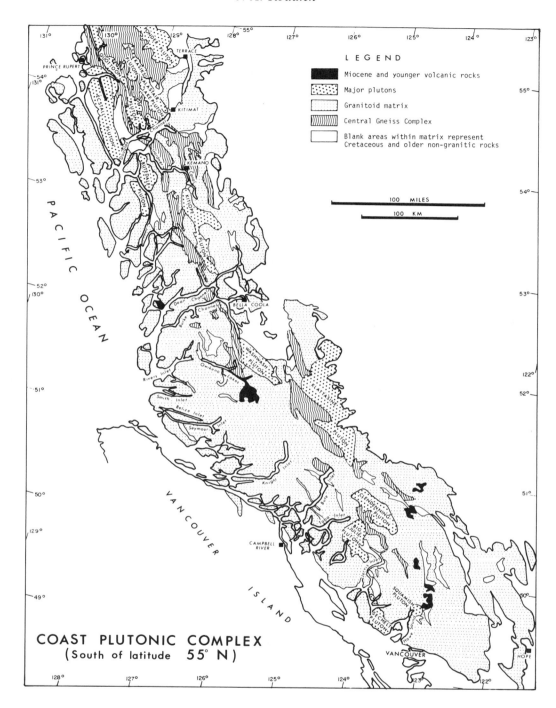

Figure 1. Sketch map of the Coast Plutonic Complex south of latitude 55°N. Three cross-section lines on which the chemical analyses were projected are labeled: DC = Douglas Channel, BC = Bella Coola, OW = Owikeno. Each cross-section line has a thickened segment near its western end marking the silica dip on which the variation plots shown in Figure 7 are keyed. Two northwest-trending lines divide the southern Coast Plutonic Complex into three belts, western (W), axial (A) and eastern (E) which are used for the analyses of model data shown in Tables 1, 2, 3 and 4.

INTRODUCTION

As the initial reconnaissance of the Coast Plutonic Complex south of latitude 55° N nears completion, a review of the main findings seems appropriate. The data base is already large, and any new data will probably form only a small addition. The major conclusions should be valid, but because most individual conclusions rest upon only a small part of the data, new insights and the ripple effects of unexpected discoveries may impair previous judgments.

This paper comprises a brief general summary of the Complex, a review of the sparse geophysical data, and previously unpublished information on its lithologic and chemical composition.

The Coast Plutonic Complex, a long (1700 km), narrow (100 km) belt dominated by intermediate and basic plutonic rock, forms one of the largest and probably the most basic large plutonic terrane in the world (Fig. 1). The Complex is regarded as a unit that extends to the Moho; that is, it forms the crust rather than an entity within the crust. Were one to view the Coast Plutonic Complex from space, stripped of its cover, it would resemble a belt of irregularly layered and veined megagneiss, consisting of narrow, often disrupted, bands of metavolcanic and metasedimentary rock, gneisses and heterogeneous foliated plutonic rock, and swellings and patches of more homogeneous, less foliated granitoid rocks representing plutons. As the general features of the belt were summarized in Roddick and Hutchison (1972, 1974) and in other publications listed therein, only a sketchy outline of the more important aspects is presented here.

The strata flanking the Coast Plutonic Complex consist mainly of Mesozoic volcanic and sedimentary rocks, although lower Paleozoic rocks occur in southeastern Alaska, in the San Juan Islands, and in the adjoining northern Cascade Mountains. Upper Paleozoic rocks are also known in those places, as well as on Vancouver Island and along the eastern flank of the Coast Mountains.

Although rare in rocks older than Late Jurassic, all major systems associated with the Coast Plutonic Complex contain some granitic debris, indicating that the belt was the site of at least scattered plutonic activity in pre-Mesozoic time, but the nature of the belt then is not known. It may have been at least partly emergent since late Carboniferous time, if the Pennsylvanian plants found at the southern end of the Coast Mountains (Monger, 1966) had an extensive distribution.

Lack of Lower Triassic strata in the Coast Mountains, marked by a disconformity between the Permo-Carboniferous Chilliwack Group and the Upper Triassic-Jurassic Cultus Formation (Monger, 1970, p. 9), and the angular unconformity between Middle Triassic sediments and Upper Paleozoic beds east of the Coast Mountains near the United States border constitute indirect evidence of Early Triassic orogenic activity (Little and Thorpe, 1965). Late Triassic time was one of major basaltic volcanic activity from Vancouver Island in the west to the Interior Plateau in the east.

A significant change occurred in the Early Jurassic when basaltic volcanism gave way to andesitic volcanism on Vancouver Island. Although adjacent areas indicate plutonism and uplift in the northern part of the Coast Plutonic Complex in Late Triassic time, the first major burst of intrusive activity came in the Jurassic, and the belt became a major positive feature flanked on the east by two troughs, which in mid-Jurassic time broke up into three successor basins. They display a gradual transition from marine deposition in the Jurassic and Early Cretaceous to non-marine deposition in the mid-Cretaceous, accompanied by an increasing volume of granitoid debris derived from the Coast Plutonic Complex. Plutonic activity reached its height in the Cretaceous and continued into the early Tertiary but was followed by scattered plutonic intrusions, up to at least the end of Miocene time.

The previous paragraph, slightly modified from the guidebook to the southern end of the Coast Plutonic Complex (Roddick, Mathews, and Woodsworth, 1977, p. 3), employs the nebulous but conventional phrase "plutonic activity." Its connotation of intrusion and solidification of magma is far beyond the facts, and actually in opposition to those few that are not ambiguous. The phrase, more modestly interpreted, reflects that a number of isotopic ages indicate a rather profound cooling event that is thought to have followed soon after one or more intrusive events. The creation and evolution of plutonic rock were probably much longer processes, which were only terminated by intrusion.

Bodies of volcanic, sedimentary, and metamorphic rocks, ranging in age from at least as old as Devonian to mid-Cretaceous (Albian), are found as pendants within the Coast Plutonic Complex throughout its length. They form a somewhat larger proportion of the crystalline belt between latitudes 53° and 55° N than elsewhere. Most of these bodies are Mesozoic and are metamorphosed to greenschist or amphibolite grade, but a few are only slightly altered and, locally, contain fossils. In the southern Coast Mountains, volcanic rocks, mainly andesitic to basaltic, are more abundant than sedimentary strata in most pendants. Most of the pendants are long and narrow and in fault or intrusive contact with plutonic rocks. The pendants, especially those consisting mainly of greenstone, are commonly, but not everywhere, associated with dioritic complexes (see Geological Survey of Canada Open Files (maps) Nos. 480 and 611).

As the pendants within the Coast Mountains have yielded few fossils, most of the evidence bearing on the timing of plutonism comes from radiometric age determinations (mainly K-Ar). For the region between latitudes 52° and 55° N, the earlier pattern of three distinct belts changed as additional determinations were made. The eastern belt remains well defined in the 43 to 50 m.y. range and constitutes all of the Complex lying east of its central axis, but the two western belts have merged into a single belt with Early Cretaceous (and rare Jurassic) ages on the west through to Late Cretaceous ages on the east.

The isotopic age pattern in the southern Coast Mountains between latitudes 49° and 51° N is similar, but not identical, to the northern region. The eastern belt spans a greater range, 35 to 80 m.y., and the western transitional belt shows somewhat less scatter, ranging from 75 to 158 m.y.

About half of the 143 K-Ar age determinations in the Coast Mountains fall within the Late Cretaceous, and about one quarter fall within the Early Tertiary. These ages may be interpreted as probable minimum final cooling ages for large parts of the Coast Plutonic Complex, and they indicate at least that the western part cooled before the eastern part. Most of the 31 hornblende-biotite pairs are more or less concordant, with the hornblende age being grater than that of biotite in most places. The range of discordancy does not appear to be significantly greater to the west. In this respect, it differs from the extension of the Coast Plutonic Complex into southeastern Alaska as reported by Smith (1975).

When all of the K-Ar dates in the Coast Mountains are considered, only one substantial gap is revealed; it lies between 115 and 140 m.y. and covers pre-Aptian Early Cretaceous and the last stage of the Jurassic. The 90 to 110 m.y. dates, which are common in the Coast Mountains, are absent on Vancouver Island, whereas Late Jurassic dates, which are common on the Island, are represented only locally in the Coast Mountains.

INFERENCES FROM GEOPHYSICAL DATA

Although the topographic relief in the Coast Mountains is considerable (1500 to 3000 metres in most places), the great area of the region makes geological information essentially horizontal. For information about the third dimension, reliance must be placed on geophysical data. Unfortunately, such data are not abundant, and some appear conflicting. Also, the lack of restraints makes it exceedingly difficult to convert most geophysical data to useful information. The geophysical data indicate that the complexity at the surface, which consists mainly of steeply dipping to vertical elements, does not extend to depth (Betty and Forsyth, 1975, p. 203). In broad terms, this inference is probably true, as the tendency to homogenize horizontally (parallel with the isotherms) undoubtedly increases with depth, but even if surface complexity were preserved at depth, the ability of geophysical methods to resolve it seems questionable.

Depth of the Coast Plutonic Complex and Adjacent Belts

Seismic data (Forsyth and others, 1974) indicate a 26-km thick crust beneath the northern Insular Belt (Queen Charlotte Islands), which thickens slightly towards the western edge of the Coast Plutonic Complex. The Moho transition in the vicinity of the Skeena Arch (near DC section on Fig. 1) was found to be deepest east of the height of land in the Coast Mountains. However, Berry and Forsyth (1975, p. 197) stated, "In the Coast Plutonic Complex, the crust thickens from 23 km beneath Ripley Bay (127° 53′W 52° 25′N) to 33 km along its eastern edge. In fact, the 33-km contour follows the boundary between the Coast Plutonic Complex and the Intermontane Belt closely except in the south, where the contour continues in a southeastern direction where the boundary turns more to the south." They note, however, that the 33-km contour continues to follow the eastern physiographic boundary of the Coast Mountains very closely. An easterly-dipping (about 5°) Moho boundary beneath the Coast Mountains seems compatible with both greater uplift on the west and higher terrane on the east. The apparent westward tilt of the belt is caused by the inability of the thinner western crust to fully compensate for erosion by further isostatic uplift.

According to Berry and Forsyth (1975, p. 197), the crust of "the western Intermontane Belt is thicker than the Coast Plutonic Complex and, in fact, except for Vancouver Island (a dubious exception), has the thickest section of crust that is well defined by the data." When Berry and Forsyth combined their seismic data with Stacey's (1973) gravity data, they found an apparent mass deficiency in the Intermontane Belt as compared with the Coast Plutonic Complex. In other words, the crust of the Coast Plutonic Complex is denser or thinner than that of the Intermontane Belt, assuming no lateral variations in the density of the upper mantle. Berry and Forsyth, however, favoured a denser upper mantle under the Coast Plutonic Complex and, especially, under Vancouver Island.

The Vancouver Island problem was considered by Riddihough (1979), who noted that if the crust there were 50 km thick, as proposed on negative evidence by White and Savage (1965) and Tseng (1968), and the density of the crust were normal, a negative anomaly of about −300 mgal should exist, but actually a small positive anomaly is present and is part of an extensive pattern which extends southward into Washington and Oregon. The pattern is, moreover, typical of that on active margins elsewhere. The crustal thickness to the south has been seismically determined as being about 20 km (Berg and others, 1966). Riddihough concluded that beneath Vancouver Island the "seismic" crust is not coincident with the "gravity" crust and that normal density crust probably extends down 20 to 30 km, with denser material (3.2 to 3.3 g/cm³) beneath that. The seismic velocities (about 7.1 km/s) in the zone between 30 and 50 km deep are considered too low for mantle material, and they are compatible with an uncommon metamorphic rock consisting of plagioclase-clinopyroxene-garnet granulite (derived experimentally as the material which most closely satisfies the unusual shear/compressional velocity ratio).

Riddihough thought that water coming off the down-going plate could alter the overlying wedge of mantle and lower crust in such a way as to produce the proposed rock. With or without subduction, some such rock type must form in the lower crust simply in response to the normal temperature gradient, and although the rate of formation would be controlled by incorporated hydrous fluid, additional water from the subducted plate does not seem required.

For the region east of the Coast Mountains, Cumming and others (1979) showed, from a partly reversed seismic refraction profile, that the Moho discontinuity dips to the east from a depth of about 30 km east of the Coast Plutonic Complex near latitude 50° 30′N to about 40 km west of the southern part of the Rocky Mountain Trench.

Intra-crustal Discontinuity

Cumming and others (1979) found above the Moho a 12-km-thick layer of lower crust having a velocity of about 6.9 km/s. The upper surface of this layer (east of the Coast Mountains), like the Moho beneath the Coast Mountains, dips to the east. Above, it is a broad, low-velocity zone. Hales and Nation (1973), working in the northern (U. S.) Rocky Mountains, also noted the intra-crustal discontinuity. There, it appeared at a depth of about 21 km, below which crustal velocities of about 6.4 km/s were recorded. This intra-crustal, low-velocity zone break is known elsewhere as the Conrad discontinuity and the crust between it and the Moho as the Conrad layer. It is not everywhere present, and nowhere has its significance been satisfactorily explained.

The Conrad layer may coincide with Caner's (1971) conductive, hydrated lower crust, which he postulated to be centered at about 30 km depth. If the hydration is dependent upon the breakdown of fairly evenly distributed hydrous minerals, then the discontinuity should rise and fall with the temperature gradient, disappearing, perhaps, in regions of very low temperature gradients, such as the Rocky Mountains. Such crustal inferences cannot be made, however, if Dragert and Clarke (1977), who extended Caner's work, are correct. They concluded that his conductive zone, instead of being centered at about 30 km depth, is probably several tens of kilometres deeper. But more relevant is later magnetotelluric work which Dragert and others (1980, p. 166) conducted within the Coast Mountains near Pemberton. It showed a good conductive zone at a depth of about 20 km.

Crustal Temperature

South of the Canadian border, Blackwell (1969) showed that the Northern Rocky Mountains physiographic province is marked by higher than normal heat flow, whereas the Pacific Border province is normal. In Canada, this would translate into high heat flow between the Rocky Mountain Trench and the eastern side of the Coast Plutonic Complex, and normal heat flow for the Coast Plutonic Complex. This pattern is confirmed south of Kamloops, but in the central Intermontane Belt, heat flow drops to normal.

Heat flow measurements made in the inlets of the southern Coast Mountains by Hyndman (1976) showed low values (37 mW/m²) from the continental margin inland about 200 km (to the heads of the inlets). Stations at the heads of Bute and Jervis Inlets produced values of 63 mW/m², which are transitional to values of about 84 mW/m² characteristic of points farther inland.

Lewis (1976) measured the heat production from specimens of plutonic rock from the southern Coast Plutonic Complex. He found an average heat production of 0.79 W/m³ and a range extending from 0.50 W/m³ for quartz diorite to 1.3 W/m³ for beta granite. The values are distinctly lower than the 2.5 W/m³ average value which he obtained from the Cassiar and Nelson Batholiths, but across the Coast Mountains, heat generation (in contrast to heat flow) does not appear to vary.

Western Magnetic Anomaly

A large-scale magnetic anomaly (Fig. 2) marks the western edge of the Coast Plutonic Complex between latitudes 49° and 53°N (Coles and Currie, 1977). Intermediate and acid plutonic rocks show no major variations in magnetic susceptibility across the Coast Plutonic Complex, but diorite shows a distinct increase in susceptibility towards the west (op. cit., p. 1758). This concurs with modal data which show an increase in mafic-mineral content and density in plutonic rocks from east to west (Table 4). The Curie temperatures (565°C to 580°C) indicate that the magnetic phase is nearly pure magnetite. Depth to the Curie

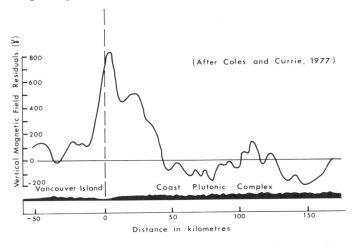

Figure 2. Western magnetic anomaly: the prominent positive anomaly is well shown by this northeast-trending flightline which passed over the northeast shoulder of Vancouver Island, Flight altitude ranged between 5.1 and 5.8 km above sea level.

isotherm has been estimated by Hyndman (1976) to be about 40 km in the western Coast Mountains and about 20 km in the eastern Coast Mountains, where the heat flow is higher. Lewis (1976) thought the depth to be considerably greater, as his calculations showed a temperature of only 300°C at 20 km beneath the Coast Mountains.

To reproduce the sharp western anomaly, Coles and Currie (1977) found it necessary in their model to assume that a narrow, highly magnetic body exists along the western edge of the Coast Plutonic Complex. This body would appear to be best represented by the gabbro-diorite complex on McCauley Island, which lies beneath the northern end of the anomaly. The gabbro is mostly coarse grained, black weathering, and rich in magnetite. The McCauley Island body is confined between two northwest-trending faults, allowing the speculation that the magnetic material was brought up as a horst. For a number of reasons, however, these faults are thought to be much younger than the emplacement of the gabbro-diorite complex and probably have mainly right-lateral, rather than vertical, displacement. A greater problem exists in most places farther southeast, where no such basic rocks are present (at the surface, at least) to account for the magnetic high. If the magnetic material is subsurface there, it must be correspondingly more magnetic than the McCauley Island surface material. It may be inferred, perhaps safely, that even where the surface rock is magnetic, the anomaly is caused mainly by subsurface material. Whether this material has approached the surface by faulting, plastic deformation, or magmatic intrusion is not clear. The material is thought to be basic rock from the lower crust rather than from the mantle, as the basic rock at the surface on the western side of the Coast Plutonic Complex is characterized by screens of amphibolite and carbonate from which it appears to have been derived and is cut by basaltic dykes, which clearly represent the mantle contribution.

Coles and Currie (1977) point out that the source body for the strong magnetic anomaly is necessarily large, at least 250 km long, 50 km wide, and perhaps as much as 40 km deep. A mass of that size is more likely to be a crustal segment than an intrusive body. Coles and Currie suggested that the anomaly might be accounted for by large magnetite deposits formed at the contact between the Coast Plutonic Complex and the Insular Belt, analogous to the iron deposits on Texada Island, but favoured extensive precipitation of magnetite by oxidizing hydrous fluids rising through the crust from a subducted slab. The enormous size of the anomaly, however, detracts from both possibilities. They observed also that, although Vancouver Island is in the region of low crustal temperature with respect to subduction geometry, permitting a deeper magnetization there than for the Coast Plutonic Complex, the magnetization on Vancouver Island seems to have a depth of only a few kilometres. They concluded that the deep crust beneath Vancouver Island is very different from that beneath the western Coast Plutonic Complex, and farther east the magnetic crust is thin. An intense magnetic anomaly west of the Sierra Nevada Batholith, the Great Valley anomaly (Zeitz, 1969), occupies an analagous position to that of the western anomaly in British Columbia.

Inferences from Paleomagnetism

Paleomagnetic work on the Upper Triassic (Karnian) Karmutsen Formation on Vancouver Island (Symons, 1971; Irving and Yole, 1972; Yole and Irving, 1980) showed that the island was far south of its present position relative to North America in the Late Triassic. The paleolatitude indicated is either 18°N or 18°S. Although the matter is unresolved, the latter is thought to be more consistent with other Cordilleran magnetizations. The movement of the western elements of the Cordillera from a region about 4900 km to the southwest is favoured by Yole and Irving (1980). Whether this migrating crustal segment included the Coast Plutonic Complex is uncertain; the idea received no definite support from geomagnetic studies conducted on the Coast Plutonic complex itself (Symons, 1974, 1977).

Paleomagnetic work on Eocene and Upper Cretaceous plutons in the Coast Plutonic Complex near Prince Rupert showed that the Eocene plutons have pole positions concordant with Eocene pole positions for cratonic North America, but that the more westerly Upper Cretaceous pluton yielded a pole position that is discordant with the established Upper Cretaceous pole positions. Symons (1974) accounted for the discordancy by proposing that the Upper Cretaceous pluton (Ecstall Pluton on Fig. 4) had been tilted about 20° to the west. Other plutons in the same region, whose magnetism was studied later (Symons, 1977), also required a western tilting (about 30°) to correct their discordance. Symons showed that such tilting brought the pole for the 136-Ma diorite on Gil Island close to the Late Jurassic position and the pole for the 102-Ma Stephens Island Pluton (west of Prince Rupert) close to the mid-Cretaceous position. It should be mentioned that Symons assumed, perhaps incorrectly, that a K-Ar date marks the cooling of a pluton through the Curie temperature.

The pole for the 144-Ma Banks Island (on Fig. 4) plutonic rocks, which are separated from the rest of the Coast Plutonic Complex to the east by two major northwest-trending faults, is concordant with the Jurassic reference pole and was, therefore, not affected by the tilting.

Symon's work indicates no northward translation of the Coast Plutonic Complex since Jurassic time, but Beck and Noson (1972) concluded that the Cascade Mountains have been shifted northward about 30° of latitude relative to the North American craton since the mid-Late Cretaceous and, according to Beske and others (1973), before the end of the Eocene. If all concerned are correct, problems are displacing problems. Although the Insular Belt and the Cascade Mountains may have shifted while the Coast Plutonic Complex remained stationary, the mechanics are not clear.

UPLIFT OF THE COAST PLUTONIC COMPLEX

East of the Coast Mountains in the Tyaughton Trough, sediments changed from marine to continental in mid-Cretaceous (Albian) time. The exclusion of the sea from that region is thought to have been the result of uplift of the Coast Plutonic Complex (Tiper, 1969, p. 60). Whether it became a high standing barrier at that time is not known, but conglomeratic beds containing abundant plutonic cobbles were deposited in the Jackass Mountain Group on the eastern flank of the Coast Plutonic Complex about then. By the end of the Cretaceous, a low-relief erosion surface had developed extending from Vancouver Island, across the Coast Mountains and Intermontane Belt. Plateau basalts were extruded on this surface in the Miocene. Since then, the eastern part of the Coast Mountains has been uplifted by more than 2 km, an amount inferred from high-standing remnants of the plateau basalts in the southern Coast Mountains. Coupled with this uplift was the depression of the Fraser Lowland and parts of the Strait of Georgia, where terrestrial sediments, thought to be Miocene, are preserved 200 to 1400 metres below sea level. On Vancouver Island between Nanaimo and Comox, the old erosion surface can be seen in several places to be dipping northeasterly under the Strait of Georgia, whereas on the mainland side, the opposite dip off the Coast Mountains is evident in many places. That the western margin of the Coast Plutonic Complex has been uplifted less than the eastern margin since Miocene time is also evident from the nearly horizontal continental basalt flows of that age near Bella Bella (Baer, 1973, p. 34). They lie on the western margin of the Coast Plutonic Complex and remain at sea level.

On the eastern flank of the southern Coast Mountains, deformation of Miocene lavas, dissection of a probable Miocene erosion surface, and a dramatic climatic change indicate a late Neogene uplift of the Coast Mountains. Fission track dates on apatite at sea level are 8 to 20 m.y. north of latitude 52° N and 35 to 45 m.y. south of that latitude, indicating a difference in uplift history between the two regions. From this and other data, Parrish (1981) concluded that the southern region underwent rapid orogenic uplift in the Eocene and slow uplift in the mid-Cenozoic, increasing slightly in the late Neogene.

Aside from its physiographic effects, the uplift of the Coast Plutonic Complex, or at least some components of it, can be roughly measured by certain metamorphic mineral assemblages and fluid inclusions. Such work by Hollister, Crawford, and their students indicates that the Coast Plutonic Complex has been differentially uplifted across its width. On the west, near Prince Rupert, uplift on the order of 27 to 30 km is indicated (Crawford and others, 1979). Farther east, but still west of Quottoon Pluton, uplift appears to be about 18 to 24 km (Lappin and Hollister, 1980). In the axial belt, between Quottoon and Alastair Lake Plutons, only 12 to 15 km of uplift are indicated (Kenah, 1979). Curiously, the temperatures determined show a reversed trend, from about 625° C on the western side of the complex to about 750° C in the axial zone.

Hollister (1975) found orthopyroxene at several places in the axial zone and concluded that the host rocks had reached the granulite-facies of metamorphism. Orthopyroxene commonly forms relict cores in cummingtonite crystals. This feature is found also in dioritic and gabbroic rocks in the western part of the Coast Plutonic Complex and, locally, in more acid rocks farther east. In southeastern British Columbia, the Adamant Pluton, described by Fox (1969), contains a core of hypersthene monzonite. The hypersthene in these rocks can be accounted for metamorphically by appealing to high temperatures (750° C to 850° C) generated deep in the crust. In the axial belt of the Coast Plutonic Complex, hypersthene is found in a limited area but clearly not contact related; it may mark only a permeable zone connected to deeper, hotter levels.

Much different are the largest areas of hypersthene occurrence, which are found near the western margin of the Coast Plutonic Complex; they are the basic complexes, such as that on McCauley Island and that forming the Denman diorite west of Big Julie Pluton in the Bute Inlet map-area (between latitudes 50° and 51° N). Basic plutonic rock, whether or not it contains hypersthene, may represent lower crust that has reached the surface more or less intact. More acid rocks, such as parts of the Central Gneiss Complex, Adamant Pluton, etc., which contain hypersthene, may or may not have originated in the lower crust, but it seems that during their trip to the surface they dwelt for long periods at moderate depths, where they were extensively altered — in fact, converted to the more acid composition typical of lesser depths.

LITHOLOGIC COMPOSITION OF THE COAST PLUTONIC COMPLEX

During the Coast Mountains Project of the Geological Survey of Canada, more than eleven thousand specimens were collected, representing the principal lithology at an equivalent number of points in the Coast Plutonic Complex south of latitude 52° N. In so far as they are truly representative of the Complex, their proportional abundances, shown in Table 1, approximate its overall lithologic composition.

Because of the access afforded by the intricate coastline, the sampling density is much higher for the western part of the region (accounting for nearly 65% of the specimens collected). To compensate for the sampling imbalance and to provide further insight into the complex, the region was

TABLE 1. CONSTITUENT ROCK CLASSES OF THE COAST PLUTONIC COMPLEX

	Western Belt	Axial Belt	Eastern Belt	Total S. CPC
Number of Spec	6,952	2,503	1,629	11,084
Plutonic	88.7%	78.3%	61.1%	77.7%[1]
non-migmatitic	(78.9)	(63.8)	(53.1)	(66.8)
migmatitic	(9.8)	(14.5)	(8.0)	(10.9)
Metavolcanic	4.8	7.2	13.7	8.0
Metasedimentary	4.5	11.1	15.7	9.8
Volcanic	1.6	3.2	7.2	3.7
Sedimentary	0.4	0.2	2.3	0.8
Area	24,800 Km²	21,400 Km²	17,400 Km²	63,600 Km²

[1]Percentages in this column are weighted according to the areas of the respective belts.

divided into western, axial, and eastern belts. The data were compiled separately for each belt, then combined and weighted according to the relative area of each belt. Easily eroded rocks, especially sedimentary and some metasedimentary rocks, are probably under-represented, as they form debris-covered saddles on the ridges and beaches on the shore. The relative proportions of the resistant rocks, however, are thought to be reasonably accurate.

Plutonic rocks are seen in Table 1 to underlie about three quarters of the Coast Plutonic Complex, but the proportion varies in the three belts, from more than 89% on the west to about 61% on the east. The axial belt, which is more than 78% plutonic, approaches the value for the combined belts. The western belt contains distinctly less metasedimentary, volcanic, and metavolcanic terrane than the other two. The axial belt contains the most migmatite (14%), reflecting the more extensive gneissic terrane there.

Slightly less than 8,000 specimens of non-migmatitic plutonic rock, each representing the principal rock-type at a single station, form the basis for Table 2. Surprisingly, quartz diorite, which is by far the most abundant plutonic rock and underlies about 40% of the region, becomes relatively more abundant towards the east, increasing by 8%. This trend appears to oppose the common circum-Pacific trend (or at least the perception of it), which would have quartz diorite decreasing eastward, giving way to more siliceous and K-feldspar-rich types. Tonalite (16.6%) and

diorite (15.1%) follow quartz diorite in overall abundance but, even if taken together, are subordinate to quartz diorite. Wholly unexpected is the distribution of tonalite; in the western and eastern belts, it forms about 20% of the plutonic rock, but in the axial belt only about 11%. Although the I.U.G.S. classification (Streckeisen, 1976), which has been used throughout, emphasizes this anomaly, a real and doubtlessly significant quartz deficiency exists in the axial belt (and is confirmed in Table 3).

As might be expected, diorite decreases eastward; but it is practically as abundant in the axial belt (17.8%) as in the western belt (18.0%), and it exhibits a marked decline in the eastern belt, where it forms only about 7.8% of the plutonic terrane. Map patterns farther north, between latitudes 52° and 55° N, indicate considerably less diorite in the axial belt and more in the western zone, but no equivalent statistical study has been done for that region.

Granodiorite distribution, like that of quartz diorite, is contrary to expectations. Instead of becoming more abundant to the east, it actually decreases one percentage point for each belt, 12.6%, 11.6%, and 10.6%, respectively. Quartz monzodiorite, however, shows an opposite trend, at a lower abundance level, from 7.2% in the western belt to 9.2% in the eastern belt.

Gabbro is predictably most abundant on the west (3.0%) and least abundant on the east (2.0%), but the difference is less than expected. Gabbro is more abundant than granite only in the western belt; it is subordinate to granite in the axial and eastern belts, as well as overall. Only the beta variety of granite is significant in the Coast Plutonic Complex, as alpha granite (more K-feldspar than beta granite) is virtually absent.

At 0.5% or less, quartz monzonite is not a significant plutonic rock-type south of latitude 52° N. Miscellaneous specimens, not statistically significant, include alpha granite, monzonite, quartz-rich granitoids, quartz syenite, and ultramafic rock. The last is very rare in the Coast Plutonic Complex, and the nearest large body of ultramafic rock lies east of the Coast Plutonic Complex in the Shulaps Range in northeastern Pemberton map-area.

An attempt has been made in Table 3 to show what rock type results when the modes for the plutonic rocks are averaged, first for each belt, then combined, with regard for the proportion of the Coast Plutonic Complex represented by each belt. Although differing slightly in modal compo-

TABLE 2. RELATIVE ABUNDANCES OF PLUTONIC ROCK TYPES IN THE COAST PLUTONIC COMPLEX SOUTH OF LATITUDE 52° N

	Western Belt	Axial Belt	Eastern Belt	Total S. CPC
Number of Spec	5,484	1,599	866	7,949
Qtz Diorite	36.1%	43.0%	44.1%	40.6%[1]
Tonalite	19.1	11.0	20.1	16.6
Diorite	18.0	17.8	7.8	15.1
Granodiorite	12.6	11.6	10.6	11.7
Qtz Monzodiorite	7.2	8.7	9.2	8.3
Gabbro	3.0	2.8	2.0	2.7
Beta Granite	2.4	3.9	2.7	3.0
Monzodiorite	0.6	0.7	0.1	0.5
Qtz Monzonite	0.5	0.2	0.4	0.4
Miscellaneous	0.5	0.3	3.0	1.1

[1]Percentages in last column are weighted according to the areas of the respective belts.

TABLE 3. AVERAGE MODAL COMPOSITION OF THE COAST PLUTONIC COMPLEX SOUTH OF LATITUDE 52° N

	Western Belt	Axial Belt	Eastern Belt	Total S. CPC
Number of Spec	5,484	1,599	866	7,949
Mafic Minerals	18.9	16.2	14.1	16.7
K-feldspar	4.7	5.5	5.5	5.2
Quartz	13.6	12.1	14.9	13.5
Plagioclase	62.8	66.2	65.5	64.6
Spec. Grav.	2.76	2.75	2.73	2.75

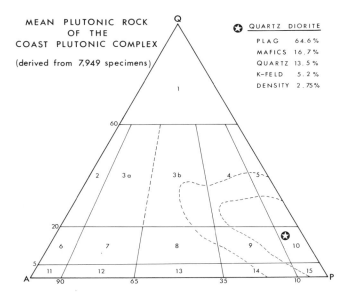

Figure 3. Mean plutonic rock of the Coast Plutonic Complex south of latitude 52°N. Calculated from the estimated modes of 7949 stained specimens. Inner dashed line encloses about 75% of specimens, and outer dashed line about 95%.

sition and density, the plutonic rocks resulting from the homogenization of each belt are quartz diorites, and the entire southern Coast Plutonic Complex could be fairly represented as quartz diorite, containing 16.7% mafic minerals (with hornblende dominant), 5.2% K-feldspar, 13.5% quartz, and 64.6% plagioclase. The product of this distillation (see Figure 3) has a specific gravity of 2.75. Table 3 also clearly shows the progressive decrease in mafic mineral content and density from west to east and the curious quartz deficiency in the axial belt.

Table 4 was constructed to determine whether the mean

TABLE 4. AVERAGE MODE AND SPECIFIC GRAVITY FOR EACH PLUTONIC ROCK TYPE IN THE COAST PLUTIONC COMPLEX SOUTH OF LATITUDE 52°N

	Belt	Specimens	Mafics	K-feld	Quartz	Sp. gr.
Qtz Diorite	W[1]	1,961	19.1(8.1)	2.1(2.2)	10.0(3.9)	2.77(.06)
	A[2]	644	15.6(7.5)	2.2(2.3)	9.4(3.9)	2.75(.05)
	E[3]	369	15.0(6.5)	2.2(2.5)	10.1(4.1)	2.74(.05)
Tonalite	W	1,043	17.1(7.9)	2.2(2.1)	23.9(6.2)	2.74(.05)
	A	164	14.7(7.3)	2.6(2.2)	22.6(4.7)	2.72(.05)
	E	173	13.2(5.5)	2.0(2.0)	22.9(5.6)	2.73(.09)
Diorite	W	982	25.6(9.2)	0.8(1.5)	1.3(1.3)	2.82(.05)
	A	267	23.0(8.3)	0.5(1.3)	1.6(1.4)	2.83(.05)
	E	65	23.9(7.7)	0.2(1.0)	1.7(1.4)	2.82(.06)
Granodiorite	W	687	12.2(6.5)	12.1(4.5)	25.6(6.4)	2.70(.18)
	A	174	10.7(7.1)	14.1(5.0)	24.4(6.0)	2.69(.10)
	E	89	8.5(4.4)	13.3(5.2)	27.0(7.2)	2.67(.04)
Qtz Monzodiorite	W	391	14.6(8.2)	12.0(4.3)	12.5(3.7)	2.71(.05)
	A	130	12.9(6.7)	12.6(3.6)	12.7(3.8)	2.70(.05)
	E	77	11.2(5.7)	12.8(4.7)	12.7(3.7)	2.69(.04)
Gabbro	W	164	37.7(15.8)	0.0(0.2)	0.3(0.7)	2.98(.07)
	A	42	39.3(12.6)	0.1(0.4)	0.7(1.0)	3.10(.07)
	E	17	33.9(9.6)	0.1(0.3)	0.7(0.9)	3.01(.10)
Beta Granite	W	130	8.0(6.2)	27.9(5.6)	29.5(6.4)	2.64(.03)
	A	58	6.8(4.1)	29.6(4.8)	29.9(6.8)	2.64(.03)
	E	23	6.0(4.0)	28.0(5.5)	32.2(8.7)	2.65(.04)
Monzodiorite	W	34	23.2(9.9)	12.8(5.0)	2.0(1.4)	2.75(.05)
	A	10	24.4(6.6)	12.2(3.3)	1.3(1.2)	2.78(.03)
	E	1	13.0	12.0	2.0	2.74
Qtz Monzonite	W	27	12.6(10.1)	34.3(7.5)	12.1(4.1)	2.68(.05)
	A	4	7.5(5.3)	33.0(5.7)	11.8(3.3)	2.64(.01)
	E	3	5.5(6.4)	34.5(10.6)	14.0(1.5)	2.66(.08)

[1]W - Western Belt; [2]A - Axial Belt; [3]E - Eastern Belt.
Standard deviations in parentheses.

mineralogical composition of individual plutonic rock-types, quartz diorite, for example, change across the Coast Moutains. Significant variations are evident. The mean mafic mineral content is lower in the eastern than in the western belt for each of the nine plutonic rock-types considered. It is intermediate in the axial belt except for the minor lithologies, gabbro and monzodiorite, for which the mafic minerals reach their maximum abundance in the axial belt. Mean specific gravity values show a variation parallel with that of the mafic minerals.

For most rock types, K-feldspar does not show a significant variation across the Coast Mountains, with the curious exception of diorite, for which the K-feldspar, although not abundant, shows a persistent decline to the east. In contrast, quartz in diorite increases from west to east. In the three most abundant rocks other than diorite, namely, quartz diorite, tonalite, and granodiorite, minimum values for quartz are exhibited in the axial belt.

Table 5 was derived from Table 4 by weighting the average modes of the three belts according to the area of each, so as to obtain values for the main plutonic rock types across the entire width of the Coast Plutonic Complex.

TABLE 5. AVERAGE MODE AND SPECIFIC GRAVITY FOR THE MAIN PLUTONIC ROCK TYPES IN THE COAST PLUTONIC COMPLEX SOUTH OF LATITUDE 52°N

	Tot.sp.	Mafics	K-feld	Quartz	Plag	Sp.gr.
Qtz Diorite	2,974	16.8	2.2	9.8	71.2	2.76
Tonalite	1,380	15.2	2.4	23.2	59.2	2.74
Diorite	1,314	24.3	0.5	1.5	73.7	2.83
Granodiorite	950	10.7	13.1	25.6	50.6	2.69
Qtz Monzodiorite	598	14.0	12.4	12.6	61.0	2.70
Gabbro	223	37.2	0.1	0.5	62.2	3.03
Beta Granite	211	7.0	28.5	30.4	34.1	2.65
Monzodiorite	45	23.8	12.5	1.7	65.0	2.76
Qtz Monzonite	34	8.9	33.9	12.5	44.7	2.66

Derived from average values for each belt (as shown in Table 4) and weighted according to the relative area of each belt.

CHEMICAL COMPOSITION OF THE COAST PLUTONIC COMPLEX

About 600 chemical analyses have been completed for the Coast Plutonic Complex between latitudes 51° and 54°N. The specimens analyzed were selected so as to form sections across the Coast Plutonic Complex, which from north to south are herein referred to as Douglas Channel, Bella Coola, and Owikeno cross-sections (Figs. 4, 5, and 6). Their respective locations appear on Figure 1.

With the minor exception involved in the determination of volcanic equivalents, the analyses will be examined without the use of norms and other derivative values, which have as their theoretical basis crystallization under equilibrium conditions according to certain rules. Neither the equilibrium conditions nor the rules, which pertain to magmas, seem to be applicable to the extremely heterogeneous Coast Plutonic Complex. Owing to this hetero-

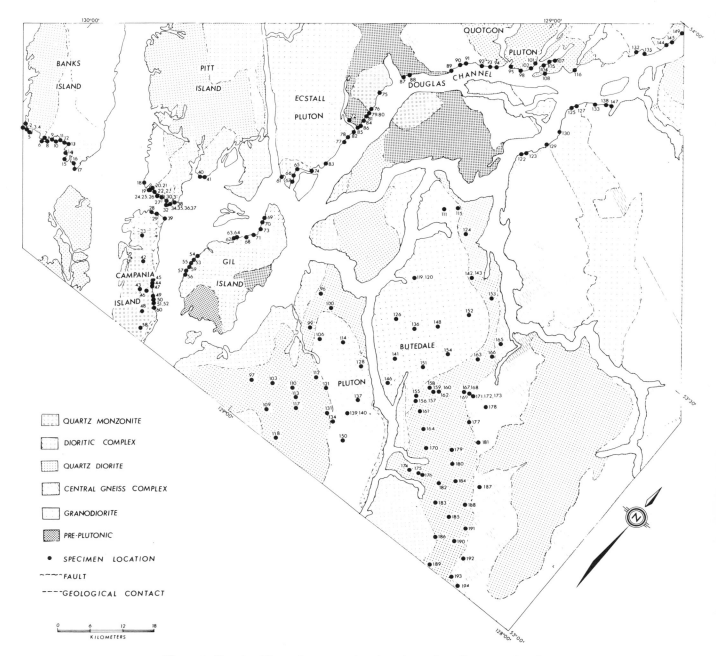

Figure 4. Douglas Channel cross-section; locations of specimens analyzed.

geneity and to the limited significance of individual analyses, the data are grouped according to pluton or rock-unit, and mean values are used to display major trends across the Coast Mountains.

Chemical Comparison with Typical Continental Crust

To obtain an average chemical composition for the Coast Plutonic Complex, the averages for each cross-section were obtained by weighting the average values for each plutonic unit according to its average width measured across the trend of the belt (projected to the line of the section). The three sections were then combined to produce the average values for the Coast Plutonic Complex, shown in Table 6, where they are compared with average values for continental crust as calculated by Ronov and Yaroshevsky (1969). Unexpectedly, the two sets of values are practically identical for SiO_2, FeO, TiO_2, and P_2O_5. Similar values are evident for MgO and CaO, the Coast Plutonic Complex being slightly poorer in the former and slightly richer in the latter. The Coast Plutonic Complex is appreciably richer in Al_2O_3 (17% versus 15%, approximately) than average continental

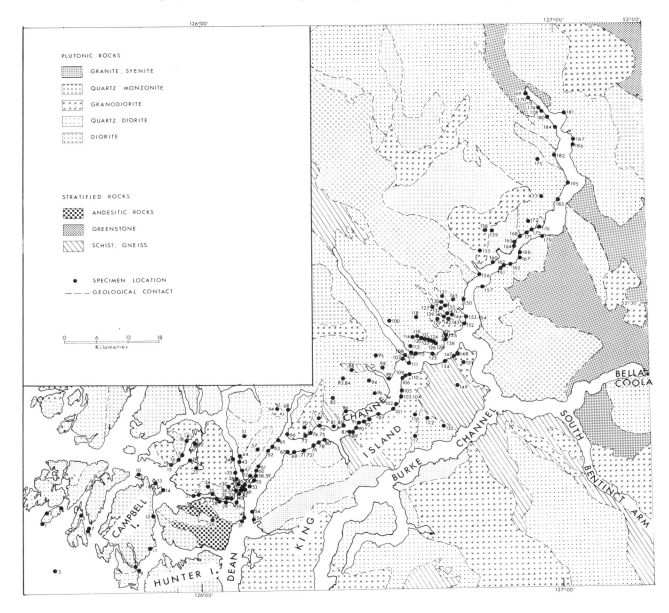

Figure 5. Bella Coola cross-section; locations of specimens analyzed.

crust; in fact, it is equivalent to oceanic basalt in that oxide.

The Coast Plutonic Complex is truly remarkable in its high soda content (4.37%), which is unmatched by any major rock component of the earth's crust. In contrast, its potash content (1.82%) is distinctly intermediate, being much higher than oceanic igneous material (0.18%) and much lower (about one third lower) than typical continental crust (2.86%). The ferric oxide content of the Coast Plutonic Complex (1.75%) is also about one-third less than that of average continental crust (2.48%).

The average chemical composition of the Coast Plutonic Complex, as well as the average value for each of the three sections, is equivalent to tholeiitic andesite (average series) according to the classification of Irvine and Baragar (1971).

The classification program, which produces norms as a byproduct, calculated normative quartz as slightly less than 10% and, as expected, no normative olivine.

Comparison of the Chemical Sections

Comparison of the mean values for the three sections, listed in Table 7 from north to south, shows them to be closely similar, but within their limited range, most of the major oxides, except the alkalis, show a progressive change. From north to south, SiO_2 and Al_2O_3 increase, whereas Fe_2O_3, FeO, MgO, and CaO decrease. The alkali pattern, however, is different; the central section is higher in Na_2O and lower in K_2O. Whether any of these variations is

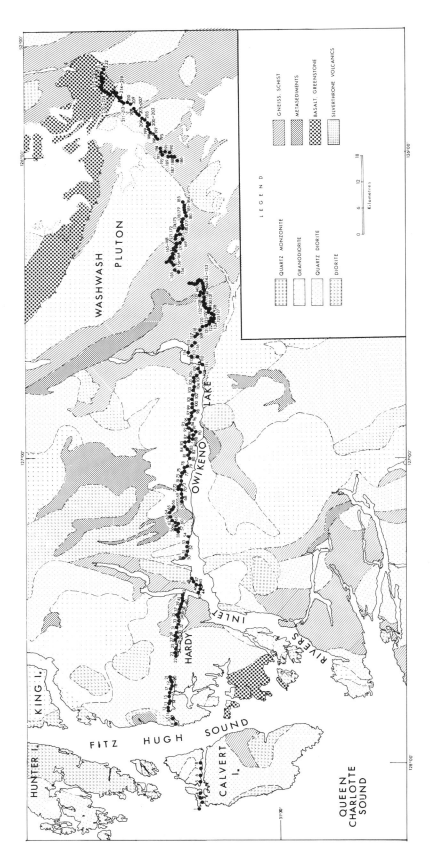

Figure 6. Owikeno cross-section; locations of specimens analyzed.

TABLE 6. AVERAGE CHEMICAL COMPOSITION OF THE COAST PLUTONIC COMPLEX COMPARED
WITH THAT OF AVERAGE CONTINENTAL CRUST

	Coast Plutonic Complex[1]	Continental Crust[2]	% Diff.
SiO_2	60.15	60.22	- 0.1
Al_2O_3	16.98	15.18	12
Fe_2O_3	1.75	2.48	-29
FeO	3.72	3.77	- 1
MgO	2.97	3.06	- 3
CaO	5.85	5.51	6
Na_2O	4.37	2.97	47
K_2O	1.82	2.86	-36
TiO_2	0.72	0.73	- 1
P_2O_5	0.24	0.24	0
MnO	0.12	0.14	-14
S	0.03	0.04	-25
H_2O	1.00	1.38	-28

[1]From this study (470 analyses used).
[2]From Ronov and Yaroshevsky (1969).

sufficiently great to be significant is doubtful.

Oxide Variation across the Coast Plutonic Complex

To determine whether any significant variation is present across the Coast Plutonic Complex, average values of principal oxides for individual plutons and other map-units are plotted in Figure 7. The most conspicuous feature is a marked dip in SiO_2 values near, but not at, the western margin of the Coast Plutonic Complex for each section. The silica dip is centered on Gil Island in the Douglas Channel section, on Campbell Island in the Bella Coola section, and on eastern Hardy Inlet in the Owikeno section. This dip in the most abundant oxide provided the control point for the lateral positioning of the sections in Figure 7.

As would be expected, the silica dip coincides with highs in CaO, FeO, MgO, and H_2O. It coincides also with a dip in Na_2O values, although in the Owikeno section, the low is

TABLE 7. AVERAGE COMPOSITIONS OF THREE CROSS-SECTIONS OF THE COAST
PLUTONIC COMPLEX

	Douglas Channel (125 analyses)	Bella Coola (119 analyses)	Owikeno (226 analyses)
SiO_2	59.07	60.13	61.25
Al_2O_3	16.44	17.14	17.36
Fe_2O_3	2.01	1.80	1.44
FeO	4.01	3.60	3.54
MgO	3.33	2.84	2.75
CaO	6.26	5.84	5.45
Na_2O	4.23	4.64	4.23
K_2O	1.88	1.75	1.83
TiO_2	0.77	0.70	0.70
P_2O_5	0.26	0.23	0.23
MnO	0.12	0.11	0.12
S	0.04	0.00	0.06
H_2O	1.10	1.09	0.80

Mean values for the cross-sections were derived from mean values
for plutonic units by weighting the latter according to the widths
of the plutonic units.

much broader, extending across three map-units. K_2O behaves somewhat differently, in that its low average value coincides with the silica dip only in the Douglas Channel section, whereas it lies in the westernmost map-unit of the Bella Coola section and forms a broad low in the Owikeno section similar to Na_2O. More inconsistent is Al_2O_3, which matches the silica dip in the Douglas Channel section but reaches its highest average value at the silica dip in the Owikeno section. In the Bella Coola section, Al_2O_3 does not distinguish itself, being neither notably high or low.

The variation plots do not show a clear increase in SiO_2 and K_2O from west to east, and the present data do not confirm the "systematic increase" in K_2O reported by Smith and others (1979, p. 346). Nevertheless, an increase in SiO_2 and K_2O towards the east is possibly a real trend and might be evident if sufficiently large masses, such as whole belts, were represented. The oxide variation patterns shown by individual sections are clearly too dependent on chance location to be of much significance to the belt as a whole, but if confirmed by other sections, the major variations must be accounted for in models of the Coast Plutonic Complex.

Chemical Correlations

One hundred and twenty-five selected analyses from the Douglas Channel section were checked for correlation against each other and against the distance measured perpendicular to the trend of the Coast Plutonic Complex from an arbitrary point to the west (Table 8).

The correlation with distance is generally poor. Even the best, K_2O (0.15), is not statistically significant at the 0.05 probability level. Correlation between SiO_2 and distance is an insignificant 0.03. In Table 9, these correlations are compared with those of Smith and others (1979), whose specimens came partly from the Douglas Channel area and partly from the Prince Rupert area to the north. Their correlations are also weak, but, for K_2O and SiO_2, much higher than the present author found. Also of interest are their small negative correlations for Na_2O and TiO_2, for which the present author found small positive correlations.

The correlations between oxides are, more or less, as would be expected. The greatest correlation is the inverse one between SiO_2 and CaO (–0.94), whereas that between SiO_2 and Na_2O is surprisingly small (0.10). The strongest correlation involving Al_2O_3 is that with K_2O (–0.56), which is the only major component with which Al_2O_3 has an inverse correlation (see Table 10). Almost half of the major oxides have their greatest correlation (invariably negative) with SiO_2, reflecting its comparative abundance in all plutonic rocks.

CONCLUSIONS

Although much is known about the Coast Plutonic

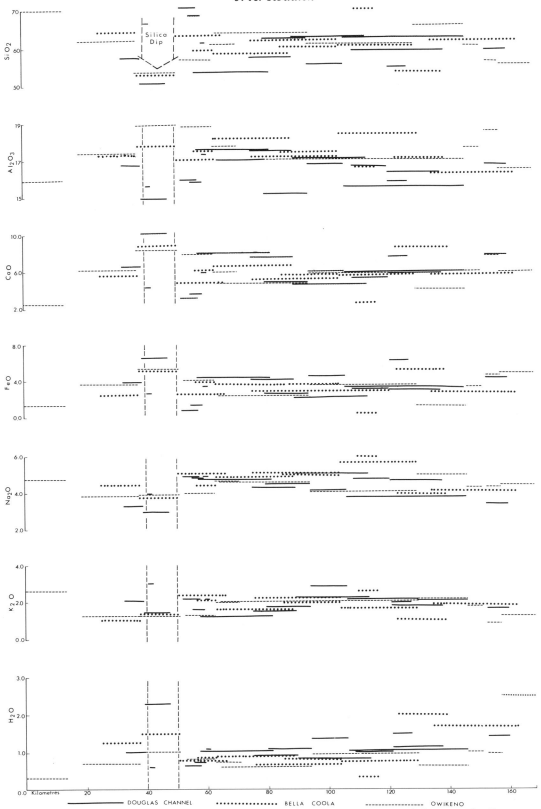

Figure 7. Variation of selected oxides (averaged for plutons or map-units) across the Coast Plutonic Complex. X-axis measures distance across the CPC but because the plots are keyed to the silica dip, it measures the distance from the western margin only for the longest section (Owikeno).

TABLE 8. CORRELATION MATRIX FOR 125 SELECTED[1] ANALYSES FROM DOUGLAS CHANNEL CROSS-SECTION

	SiO_2	Al_2O_3	Fe_2O_3	FeO	CaO	MgO	Na_2O	K_2O	TiO_2	P_2O_5	MnO	CO_2	H_2O
SiO_2													
Al_2O_3	-0.48												
Fe_2O_3	-0.78	0.48											
FeO	-0.86	0.16	0.59										
CaO	-0.94	0.50	0.76	0.80									
MgO	-0.86	0.10	0.61	0.82	0.82								
Na_2O	0.10	0.19	-0.14	-0.28	-0.21	-0.26							
K_2O	0.59	-0.56	-0.55	-0.42	-0.67	-0.41	-0.16						
TiO_2	-0.88	0.26	0.68	0.88	0.77	0.78	-0.16	-0.46					
P_2O_5	-0.64	0.37	0.52	0.54	0.56	0.47	0.10	-0.46	0.57				
MnO	-0.80	0.35	0.76	0.80	0.83	0.69	-0.29	-0.51	0.71	0.55			
CO_2	-0.08	-0.05	0.05	0.03	0.01	0.03	0.01	0.14	0.03	-0.05	-0.04		
H_2O	-0.65	0.01	0.48	0.64	0.55	0.73	-0.24	-0.25	0.64	0.32	0.50	0.18	
Dist	0.03	-0.12	-0.09	0.05	-0.11	-0.01	0.08	0.15	0.07	0.10	-0.14	0.07	0.11
Mean	61.04	16.71	1.80	3.42	5.59	2.68	4.47	2.03	0.65	0.23	0.11	0.05	0.97
St Dev	7.95	1.39	1.17	1.94	2.44	2.18	1.15	0.90	0.38	0.14	0.53	0.14	0.44

[1]From a total of 192 analyses. Most rejects had totals beyond acceptable limits (100 ± 2), a few represented rare rock types, such as hornblendite, etc.

TABLE 9. COMPARISON OF DISTANCE - OXIDE CORRELATIONS FOR TWO SETS OF CHEMICAL ANALYSES IN THE DOUGLAS CHANNEL REGION[1]

	SiO_2	TiO_2	Al_2O_3	MgO	CaO	Na_2O	K_2O	P_2O_5
Smith and others (1979) (122 analyses)	0.13	-0.07	-0.10	-0.08	-0.22	-0.16	0.38	0.11
This paper (125 analyses)	0.03	0.07	-0.12	-0.01	-0.11	0.08	0.15	0.10

[1]Distance measured at right angles to the northwesterly trend of the Coast Plutonic Complex from a point west of it. Only the value of 0.38 for K_2O is significant at the 0.05 probability level.

TABLE 10. STRONGEST CORRELATIONS BETWEEN OXIDES

CaO-SiO_2 (-0.94)	Fe_2O_3-SiO_2 (-0.78)
SiO_2-CaO (-0.94)	H_2O-MgO (0.73)
TiO_2-SiO_2 (-0.88), FeO (0.88)	K_2O-CaO (-0.67)
FeO-TiO_2 (0.88)	P_2O_5-SiO (-0.64)
MgO-SiO_2 (-0.86)	Al_2O_3-K_2O (-0.56)
MnO-FeO (0.83)	Na_2O-MnO (-0.29)

Complex, the problems increase as the data accumulate, and any simplistic history of this enigmatic belt is easily disproven. The dominant characteristics of the Coast Plutonic Complex are its heterogeneity and its basicity. The heterogeneity of the plutonic rocks is thought to indicate a long history prior to their emplacement, and the basicity is interpreted as reflecting origin in the lower crust. The latter implies that erosion has exposed levels in the Coast Plutonic complex deeper than those common for more acidic plutonic terranes.

If basic plutonic rocks, indeed, represent lower crustal material and, consequently, more uplift, then the Moho discontinuity would be expected at shallower depths beneath the Coast Plutonic Complex than beneath a more acid plutonic terrane, such as the Sierra Nevada Batholith, which occupies a similar position with respect to the North American craton. The seismic data confirm the fact, if not the reason.

The magnetotelluric indication of a conductive zone centered about 20 km beneath the surface in the Coast Plutonic Complex is thought to be of particular significance. It is interpreted here as the hydrated zone which must be present in the crust at levels where hydrous minerals are breaking down (currently, perhaps) and producing an intergranular fluid. This, in essence, is the active zone of the crust, or zone of plutonism, and is susceptible to both chemical change and plastic mass mobility. Such zones probably have long lives but eventually should become stable when the hydrous minerals are largely destroyed, and the resultant hydrous fluid escapes to higher levels.

The western magnetic anomaly is difficult to account for

because of the huge dimensions of the mass that are required to produce it. It can be modeled in different ways, none of which seem very convincing. Further work, however, may produce a model that is in good accord with known geological, seismic, and gravity data. Models involving segments of lower crust brought to high levels by faulting seem the most promising.

If homogenized, the plutonic rocks in the Coast Mountains would produce a quartz diorite. The analysis of the several thousand available modes, however, showed other features, which were both unpredicted and surprising. The abundance of tonalite in the western and eastern belts of the Coast Plutonic Complex and its scarcity in the axial belt result in a relative quartz deficiency in the axial belt. The relative abundance of quartz diorite increases from west to east, whereas that of granodiorite decreases. Such phenomena are contrary to other circum-Pacific plutonic terranes, although most must be judged from chemical rather than modal data.

Chemical analyses of plutonic rocks have a long history of enlarging the data base without significantly increasing the understanding of these rocks. The difficulty is that such analyses represent products obtainable in a variety of ways, and too often, the only valid conclusion is that the analyses are roughly compatible with some partly understood concept.

The dominant feature shown by each of the three chemical sections is the major dip in SiO_2 values near the western margin. That the rocks are more basic at the silica dip was evident, without chemical analyses, from the geological maps. The phenomenon is real and coincides with the northern part of the western magnetic high, previously described, which has the characteristics of an anomaly produced by a deep-seated, rather than a shallow, mass. The fact that the silica dip lies near, but not at the western edge of the Coast Plutonic Complex, makes it more difficult to interpret in terms of contamination of granitic magma by oceanic basalt, which would require the most basic rock to be nearest the ocean basin. If fusion played only a minor role in the creation and emplacement of components of the Coast Plutonic Complex, the interpretation of the silica dip would be different but, perhaps, no simpler. The silica dip may reflect the presence of rocks that once formed a lower part of the crust than did the adjoining rocks. If so, the silica dip marks a belt of relatively greater uplift.

REFERENCES CITED

Baer, A. J., 1973, Bella Coola-Laredo Sound map-areas, British Columbia: Geological Survey of Canada Memoir 372, 122 p.

Beck, M. E., and Noson, L., 1972, Anomalously low paleolatitudes in Cretaceous granitic rocks of the Stevens Pass area, central Washington, U. S. A.: Nature, v. 235, p. 11–13.

Berg, J. W., Tembly, L. T., McKnight, W. R., Sarmah, S. K., Souders, R., Thiruvathukal, J. V., and Vossler, D. A., 1966, Crustal refraction profile, Oregon Coast Range: Bulletin of the Seismological Society of America, v. 56, p. 1357–1362.

Berry, M. J., and Forsyth, D. A., 1975, Structure of the Canadian Cordillera from seismic refraction and other data: Canadian Journal of Earth Sciences, v. 12, p. 182–208.

Beske, S. J., Beck, M. E., and Noson, L., 1973, Paleomagnetism of the Miocene Grotto and Snoqualmie Batholiths, central Cascades, Washington: Journal of Geophysical Research, v. 78, p. 2601–2608.

Blackwell, D. D., 1969, Heat flow determinations in northwestern United States: Journal of Geophysical Research, v. 74, p. 992–1007.

Caner, B., 1971, Quantitative interpretation of geomagnetic depth-sounding data in western Canada: Journal of Geophysical Research, v. 76, p. 7202–7216.

Coles, R. L., and Currie, R. G., 1977, Magnetic anomalies and rock magnetizations in the southern Coast Mountains, British Columbia; possible relation to subduction: Canadian Journal of Earth Sciences, v. 14, p. 1753–1770.

Crawford, M. L., Kraus, D. K., and Hollister, L. S., 1979, Petrologic and fluid inclusion study of calc-silicate rocks, Prince Rupert, British Columbia: American Journal of Science, v. 279, p. 1135–1159.

Cumming, W. B., Clowes, R. M., and Ellis, R. M., 1979, Crustal structure from a seismic refraction profile across southern British Columbia: Canadian Journal of Earth Sciences, v. 16, no. 5, p. 1024–1040.

Dragert, H., and Clarke, G. K. C., 1977, A detailed investigation of the Canadian Cordillera geomagnetic transition anomaly: Journal of Geophysics, v. 42, p. 373–390.

Dragert, H., Law, L. K., and Sule, P. O., 1980, Magnetotelluric soundings across the Pemberton volcanic belt, British Columbia: Canadian Journal of Earth Sciences, v. 17, p. 161–167.

Forsyth, D. A., Berry, M. T., and Ellis, R. W., 1974, A refraction survey across the Canadian Cordillera at 54° N: Canadian Journal of Earth Sciences, v. 11, p. 533–548.

Fox, P. E., 1969, Petrology of Adamant Pluton, British Columbia: Geological Survey of Canada, Paper 67-61, 101 p.

Hales, A. L., and Nation, J. B., 1973, A seismic refraction survey in the northern Rocky Mountains; more evidence for an intermediate crustal layer: Geophysical Journal of the Royal Astronomical Society, v. 35, p. 381–399.

Hollister, L. S., 1975, Granulite facies metamorphism in the Coast Range Crystalline Belt: Canadian Journal of Earth Sciences, v. 12, p. 1953–1955.

Hyndman, R. D., 1976, Heat flow measurements in the inlets of southwestern British Columbia: Journal of Geophysical Research, v. 81, p. 337–349.

Irvine, T. N., and Baragar, W. R. A., 1971, A guide to the chemical classification of the common volcanic rocks: Canadian Journal of Earth Sciences, v. 8, p. 523–548.

Irving, E. and Yole, R. W., 1972, Paleomagnetism and the kinematic history of mafic and ultramafic rocks in fold mountain belts: in Irving, E., editor, The ancient oceanic lithosphere: Earth Physics Branch, Ottawa, v. 42, no. 3, p. 87–95.

Kenah, C., 1979, Mechanisms and physical conditions of emplacement of the Quottoon Pluton, British Columbia: PhD thesis, Princeton University, Princeton, N. J.

Lappin, A. R., and Hollister, L. S., 1980, Partial melting in the Central Gneiss Complex, near Prince Rupert, British Columbia: American Journal of Science, v. 280, p. 518–545.

Lewis, T. J., 1976, Heat generation in the Coast Range Complex and other areas of British Columbia: Canadian Journal of Earth Sciences, v. 13, p. 1634–1642.

Little, H. W. and Thorpe, R. I., 1965, Greenwood (82E/2) map-area: in Report of Activities, Geological Survey of Canada, Paper 65-1, p. 56–60.

Monger, J. W. H., 1966, Stratigraphy of the type area of the Chilliwack Group, southwestern British Columbia: University of British

Columbia, PhD thesis.

——, 1970, Hope map-area, west half, British Columbia: Geological Survey of Canada, Paper 69-47, 75 p.

Parrish, R., 1981, Cenozoic uplift history of the Coast Mountains of British Columbia: Geological Association of Canada, Cordilleran Section Meeting, Programme and Abstracts, p. 30.

Riddihough, R. P., 1979, Gravity and structure of an active margin, British Columbia and Washington: Canadian Journal of Earth Sciences, v. 16, p. 350–363.

Roddick, J. A., and Hutchison, W. W., 1972, Plutonic and associated rocks of the Coast Mountains of British Columbia: 24th International Geological Congress, Field Excursion Guidebook A04.

——, 1974, Setting of the Coast Plutonic Complex, British Columbia: Pacific Geology, v. 8, p. 91–108.

Roddick, J. A., Mathews, W. H., and Woodsworth, G. J., 1977, Southern end of the Coast Plutonic Complex: Geological Association of Canada, Fieldtrip Guidebook 9, 29 p.

Ronov, A. B., and Yaroshevsky, A. A., 1969, Chemical composition of the earth's crust: *in* Hart, P. J. (editor), The earth's crust and upper mantle: American Geophysical Union, Geophysical Monograph 13, p. 37–57.

Smith, J. G., 1975, K-Ar evidence for timing of metamorphism and plutonism in the Coast Mountains near Ketchikan, Alaska: Abstr., in Intrusive rocks and mineralization of the Canadian Cordillera: Geological Association of Canada, Cordilleran Section Meeting, p. 21.

Smith, T. E., Riddle, C., and Jackson, T. A., 1979, Chemical variation within the Coast Plutonic Complex of British Columbia between lat. 53° and 55° N: Geological Society of America Bulletin, Part 1, v. 90, p. 346–356.

Stacey, R. A., 1973, Gravity anomalies, crustal structure and plate tectonics

in the Canadian Cordillera: Canadian Journal of Earth Sciences, v. 10, p. 615–628.

Streckeisen, A. L., 1976, To each plutonic rock its proper name: Earth Science Review, v. 12, p. 1–33.

Symons, D. T. A., 1971, Paleomagnetic notes on the Karmutsen basalts, Vancouver Island, British Columbia: Geological Survey of Canada, Paper 71-24, p. 9–24.

——, 1974, Age and tectonic implications of paleomagnetic results from plutons near Prince Rupert, British Columbia: Journal of Geophysical Research, v. 79, p. 2690–2697.

——, 1977, Paleomagnetism of Mesozoic plutons in the westernmost Coast Complex of British Columbia: Canadian Journal of Earth Sciences, v. 14, p. 2127–2139.

Tipper, H. W., 1969, Mesozoic and Cenozoic geology of the northeast part of Mount Waddington Map-area (92N), British Columbia: Geological Survey of Canada, Paper 68-33, 103 p.

Tseng, K., 1968, A new model for the crust in the vicinity of Vancouver Island: MSc thesis, University of British Columbia, Vancouver, B. C.

White, W. R. H., and Savage, J. C., 1965, A seismic refraction and gravity study of the earth's crust in British Columbia: Bulletin of the Seismological Society of America, v. 55, p. 463–485.

Yole, R. W., and Irving, E., 1980, Displacement of Vancouver Island: paleomagnetic evidence from the Karmutsen Formation: Canadian Journal of Earth Sciences, v. 17, p. 1210–1228.

Zietz, I., 1969, Aeromagnetic investigation of the earth's crust in the United States: *in* Hart, P. J. (editor), The earth's crust and upper mantle: American Geophysical Union, Geophysical Monograph 13, p. 404–414.

Manuscript Accepted by the Society July 12, 1982

PRINTED IN U.S.A.

Geological Society of America
Memoir 159
1983

The Idaho batholith and associated plutons, Idaho and Western Montana

Donald W. Hyndman
Department of Geology,
University of Montana, Missoula, Montana 59812

ABSTRACT

The 39,000-km^2 Idaho batholith lies 600 km east of the present Pacific coastline and east of the Columbia River basalt plateau. The batholith is Late Cretaceous in age and is emplaced immediately east of the Triassic Seven Devils volcanic arc, an apparently allochthonous terrane which may be part of the recently recognized "Wrangellia terrane" of western Canada and southern Alaska. Locus of a Late Cretaceous subduction zone related to the Idaho batholith is not yet defined but must lie west of the Seven Devils arc.

Country rocks of the Idaho batholith are Proterozoic Belt metasediments and pre-Belt basement orthogneisses. Pre-batholithic, sillimanite-zone, regional dyna-mothermal metamorphism, apparently Jurassic or Cretaceous in age, is broadly concentric to the northern half of the batholith, extending for a few to several kilometers beyond the contact.

The Idaho batholith is dominantly medium-grained, massive to moderately foliated, muscovite-biotite granite and granodiorite. Gneissic tonalite, rich in biotite and hornblende, forms a 12-16 kilometer-wide western border zone of the batholith, and tonalite or trondhjemite form satellitic plutons for 50-70 kilometers to the west. Granodiorite makes up a 10-20 kilometer-wide border zone against exposed country rocks of most of the batholith and surrounds the voluminous granite of the batholith interior. Such granodiorite has not been documented adjacent to the pre-Belt basement rocks of the northwest-trending Salmon River arch, which divides the Idaho batholith into a northern Bitterroot lobe and a southern Atlanta lobe. Granodiorite may have formed a broad shell over most of the batholith which may have originally extended across the deeply eroded Salmon River arch.

The borders of a few large separate intrusions have been partly documented, especially in the interior of the Bitterroot lobe, but for most of the batholith, separate major bodies are as yet unknown. Foliation in the western tonalitic border zone of both lobes dips 50 to 70 degrees eastward under the batholith. To the east in the main body of the batholith, the foliation weakens and gradually arches to nearly horizontal in the interior. Southwestern and northeastern border zones of the deeper northern part of the batholith are marked by large alternating sheets of granitic and high-grade country rocks. Large, tabular or contorted inclusions and nebulous schlieren are abundant towards the interior.

The north-trending Bitterroot dome of the northeastern Bitterroot lobe appears to have formed a mushroom-shaped diapir into the country rocks, then rose isostatically in response to eastward unloading of the 6000 km^2 Sapphire tectonic block. The base of the flanks of this part of the batholith has been mapped on the southwest and northeast.

Chemically, the Idaho batholith appears related to the volumetrically minor "sodic series" recognized by Tilling in the Boulder batholith to the east. Radiometric ages in both batholiths appear to be similar. The hornblende-bearing tonalites of the western border and more-mafic satellitic plutons to the west, north, and northeast of the Idaho batholith appear to have the "I-type", "magnetite-series" mineralogy of Chappell and White and of Ishihara, respectively. The muscovite-bearing main units of the batholith appear to have "S-type" and probably "ilmenite-series" mineralogy.

Sources for magmas forming the tonalitic western border zones of the Idaho batholith were probably mafic-rich rocks of the upper mantle or of subducted oceanic crust or young continental margin volcanic rocks. Magmas forming the granodiorite-granite main body of the batholith wre probably derived by partial melting of Precambrian continental basement rocks. The main body of the Bitterroot Lobe of the batholith appears to have been emplaced at a depth of 15 to 20 km, and much of the magma was probably generated at not much greater depth.

Metallic mineral deposits are essentially absent from the deeper Bittrroot lobe. Correlation of magnetite-series granitic rocks and Cu-Mo mineralization and ilmenite-series granitic rocks with Sn-W mineralization can be partly documented for the Atlanta lobe and for plutons east of the batholith, but data are sketchy and, in part, appear contradictory.

INTRODUCTION

The Idaho batholith occupies much of central Idaho and part of western Montana in the northern Rocky Mountains of the United States. It lies between latitudes 43° N and 47° N and mainly between longitudes 114° W and 116° W. About 390 kilometers long in a north-south direction and averaging about 100 kilometers wide, it occupies an area of about 39,000 km² (Fig. 1). Thus, it is comparable in size to the Sierra Nevada batholith of California and somewhat smaller than the Coast Range plutonic complex of western Canada. In contrast to their 150 to 200 kilometer distance from the present coastline, the Idaho batholith lies more than 600 kilometers from the coast. This somewhat anomalous position is at the eastern end of the Columbia arc which is marked by Mesozoic ocean-floor ultramafic rocks, mafic volcanics, and metamorphosed sedimentary rocks in the Blue Mountains of central Oregon. The arc encloses a broad area of Miocene Columbia River basalts and no rocks older than Tertiary age.

The Blue Mountains province extends northeastward to the Seven Devils arc-volcanic pile (Mv in Figure 1) at the western flank of the Idaho batholith. The Triassic Seven Devils arc is cut by east-dipping thrust faults and consists of basalt and andesite submarine lavas and agglomerates (Vallier, 1977). Gabbroic rocks in the lowest part of the section may be solidified magma chambers or, in part, possibly ocean-floor material. The latter certainly occurs farther west in the Blue Mountains region. The structural and lithologic makeup of the Seven Devils volcanics suggests involvement in Mesozoic subduction at the continental margin (Talbot and Hyndman, 1975; Hamilton, 1976, 1978; Hyndman, 1979).

The juxtaposition of this Triassic volcanic arc against Precambrian crustal rocks to the east around the Idaho batholith, with no intervening Paleozoic rocks, suggests that the Seven Devils–Blue Mountains block may have been moved into place from elsewhere (Hamilton, 1976). The recent recognition of identical overall stratigraphic section, unconformities, fossils, and structure of the Seven Devils complex with that of the "Wrangellia terrane" of western-most Canada and southern Alaska suggests a similar history (Jones and others, 1977). Paleomagnetic evidence that basalts in Alaska may have originated near the paleo-equator suggests that the Seven Devils terrane also moved from far to the south (Hillhouse, 1977).

East of the Seven Devils arc-volcanic pile are the extensive Proterozoic Belt metasedimentary rocks and the underlying pre-Belt basement rocks, into which the Idaho batholith was emplaced (Fig. 1). The batholith is in two geographically distinct parts, separated by regionally metamorphosed Belt and pre-Belt basement gneisses forming the southeast-trending Salmon River arch (Armstrong, 1975a). The 14,000 km² northern or Bitterroot lobe is elongate northwest-southeast. The 25,000 km² southern or Atlanta lobe is elongate north-south.

Country rocks of the Bitterroot lobe are Belt schists and paragneisses and probable pre-Belt basement orthogneisses. The Belt rocks consist of muddy siltstones, quartz-rich sandstones, and calcareous arkosic silts. Generally concordant to the batholith, these units are regionally metamorphosed to the amphibolite facies within several or many kilometers of the main plutons. Country rocks of the Atlanta lobe on the north and northeast are Belt and pre-Belt basement schists and gneisses; those on the east are Devonian to Permian marine clastics which show eastward thrust faulting away from the batholith. Country rocks along the western contact are metamorphosed submarine

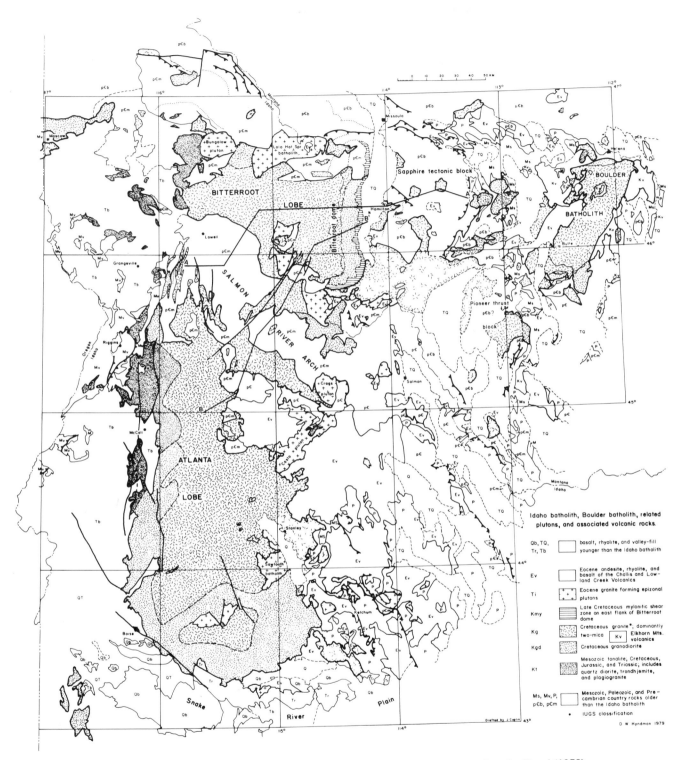

Figure 1. Geological map of the Idaho batholith. Modified from compilations by Bond (1978), Alt and Hyndman (1976), Lemoine and others (1978), and many other references listed herein.

volcanic rocks and basalts of probable Permo-Triassic age (cf. Bond, 1978). Large expanses of the country rock contacts of the Atlanta lobe are obscured on the west, south, and east by younger rocks. On the west are the Miocene Columbia River flood basalts and Pliocene stream and lake sediments; on the south are Pliocene welded tuffs and Pleistocene flood basalts of the Snake River Plain; and on the east are Eocene intermediate to silicic fragmental rocks and flows of the Challis volcanics (Bond, 1978).

The age of the Idaho batholith is poorly constrained by the nature of its contacts with other rocks. It intrudes Triassic volcanics on the west and Pennsylvanian-Permian marine sediments on the east. Upper Triassic or Lower Jurassic shales on the west are metamorphosed, presumably by the batholith or a pre-batholith event. The Idaho batholith is overlain or intruded, in part, by Eocene Challis volcanics and related plutons. Its geological age is, therefore, Mesozoic and probably Jurassic or Cretaceous. Thirty reasonable radiometric ages on the main phases of the Idaho batholith range from 85 to 66 million years in the Bitterroot lobe and from 117 to 62.6 million years in the Atlanta lobe. But ages greater than 95 million years are only in tonalites on the western fringe of the batholith. These ages include zircon U-Pb, biotite and hornblende Rb-Sr and K-Ar, and sphene fission-track determinations. The Late Cretaceous age of emplacement (?) and cooling, therefore, seems generally well established. A few Early Cretaceous, Jurassic, and Triassic dates have been obtained on smaller plutons west of the batholith. No clearcut age progression is apparent across the batholith itself.

A group of younger and generally shallower plutons is emplaced in a broad irregular belt through the eastern half of the Idaho batholith and a short distance to the east. Many of these plutons appear related to the Eocene Challis volcanics, which are centered on the eastern border of the Atlanta lobe. Reliable radiometric dates on the epizonal plutons in this group range from 49 to 39 million years (cf. Bennett, 1980).

The main Late Cretaceous phases of the Idaho batholith consist of granite (IUGS classification, Streckeisen, 1976) and granodiorite with some tonalites along the western-most border (Figs. 1, 2). In a general way, granodiorite is more common towards the borders of each lobe, granite towards the interior. But this generalization does not hold everywhere. The epizonal Eocene granites are character-istically pink quartz syenites and granites with miarolitic cavities, indicating emplacement at shallow levels.

GRANITIC ROCKS

The Upper Cretaceous granodiorite-granite main phase of the Idaho batholith dominates the granitic rocks. The Eocene alkali feldspar-rich granitic suite occupies only 3 or 4 percent of the area of granitic rocks.

Variation in mineralogical composition of the Idaho batholith is poorly known. A few areas, especially near the western and northern borders, have been mapped in some detail, but most of the batholith is either known only in broad reconnaissance, especially along main roads, or not known at all.

The generalized distribution of granitic lithologies in the Idaho batholith, Boulder Batholith, and satellite plutons is shown on the geological map (Fig. 1). It seems apparent that the main masses of the Atlanta and Bitterroot lobes of the Idaho batholith are biotite granite. A discontinuous zone 10 to 30 kilometers wide, which is dominantly granodiorite, borders much of the Idaho batholith. It is not known to occur along much of the northeastern margin of the Atlanta lobe and is absent from parts of the southwestern border of the Bitterroot lobe, across the Precambrian basement rocks of the Salmon River arch described by Armstrong (1975a). But much of this area is poorly accessible, and even such general aspects of the lithology of the batholith are not well known. The westernmost 12 to 16 kilometers of the Atlanta lobe are largely foliated tonalite or trondhjemite, as are many of the small satellite plutons for 50 to 70 kilometers west of both lobes of the batholith (cf.: Hamilton, 1962; Taubeneck, 1971). The gneissic tonalite border zone con-tinues around the southern border of the Idaho batholith southwest of the Snake River Plain near Silver City as demonstrated by Taubeneck (1971). Steep foliation in the gneissic border zone parallels the inferred country-rock contact.

The granodiorite-granite suite of the Atlanta lobe is also medium-grained and massive to weakly foliated and con-tains minor biotite and lesser muscovite, with scattered megacrysts of K-feldspar. Most individual plutons or size-able areas average 36 - 61 percent plagioclase, less than 1 - 25 percent alkali feldspar, and 20 - 32 percent quartz (see Table 1; Fig. 2). Tonalitic rocks along the western border contain less K-feldspar and more plagioclase. Biotite (3 - 16% of the rock) and muscovite (0.1 - 1.4%) are the most common minor minerals. Hornblende is absent except in tonalitic rocks or satellite plutons of the western border zone of the lobe.

A comparison of the average modes for the Atlanta lobe (Table 1) and the Bitterroot lobe (Table 2) suggests that tonalitic rocks are more abundant in the Atlanta lobe and granites less so (see also Figs. 2, 3). However, although tonalite is probably more abundant in the Atlanta lobe, the number of separate tabulations is largely a function of the availability of modal analyses from the western border of the lobe. And the single tabulation reflects the paucity of modal analyses rather than the area of granite exposure. In the Bitterroot lobe the lesser apparent tonalite may reflect a real difference between the two lobes as discussed further below.

The granodiorite-granite suite of the Bitterroot lobe is a

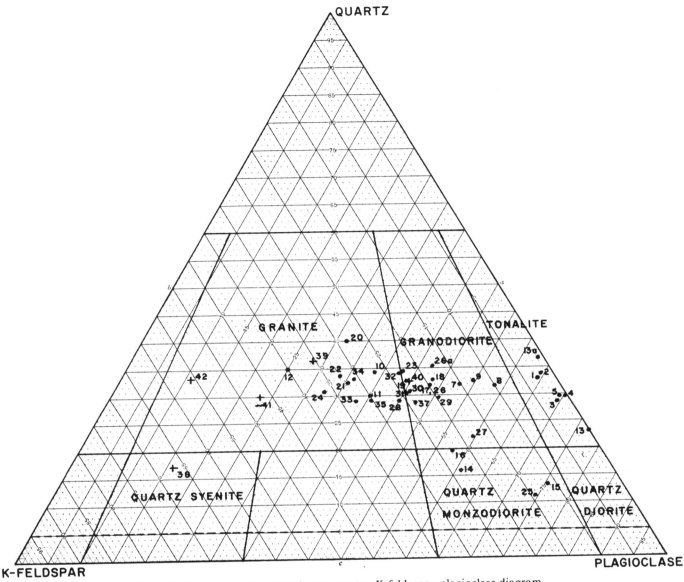

Figure 2. Average modes plotted on a quartz - K-feldspar - plagioclase diagram.

medium-grained, weakly foliated granitic rock with biotite and lesser muscovite and scattered megacrysts of K-feldspar (Hyndman and Williams, 1977). Most individual plutons or sizeable areas average 30 - 56 percent plagioclase, 16 - 31 percent alkali feldspar, and 19 - 38 percent quartz (Table 2, Figure 3). Biotite, averaging 3 - 10% and muscovite are the most common minor minerals. Hornblende occurs only in tonalitic or granodioritic border-zone rocks or satellite plutons which also contain more biotite. Streaky, diffuse-to-faint, slightly more-mafic schlieren are moderately common in some areas of the main body of the Bitterroot lobe, especially near sheets or septa of country-rock schists.

A broad belt of Eocene epizonal plutons 100 to 150 kilometers wide extends in a northerly direction through the eastern part of the Atlanta lobe and to its east, and through

the Bitterroot lobe of the batholith. These rocks are medium- to coarse-grained, massive, biotite granites and quartz syenites. The Eocene plutons are characterized by abundant K-feldspar which imparts a pink color to most rocks, miarolitic cavities containing crystals of smoky quartz and, in some cases, alkali feldspar and a well-defined vertical joint system. These intrusions contain two-to-three-times greater radioactivity from U, Th, and K^{40} compared with the Cretaceous main-phase granitic rocks as measured by gamma ray spectrometry (Swanberg and Blockwell, 1973; W. B. Strowd, *in* Bennett, 1980; Hyndman, unpubl. data). They appear to be related to eruption of voluminous Eocene Challis volcanic rocks, which overlap much of the eastern border of the Atlanta lobe.

Individual large plutons have not been delineated in the

TABLE 1. AVERAGE AND RANGE OF MODES AND CHEMICAL ANALYSES FOR THE

	(1)	(2)	(3)	(4)	(5)
Unit and Location	batholith SW of Snake R. Plain gneissic border zone	gneissic border zone of SW Atlanta lobe near Snake R. plain, ∿15 km wide	quartz diorite of Council Mtn. and Deserette, W of Atlanta lobe	quartz diorite W of Donnelly pluton; gneissic W border atl. lb.	Donnelly pluton, western Atlanta lobe
Rock name:	tonalite	tonalite	tonalite	tonalite	tonalite
Quartz	28.1 (18.3-32.9)	28.8 (22.1-32.8)	23.5 (8.2-33.4	25.8 (21.4-33.5)	20.1 (17.9-22.4)
Plagioclase (An)	54.9 (49.1-58.6) (n.d.)	54.9 (45.9-59.7 (n.d.)	57.2 (52.0-65.3) (n.d.)	61.4 (58.4-64.5) (n.d.)	47.3 (42.9-52.4) (n.d.)
K-feldspar	2.2 (0 - 3.6)	1.46 (0 - 10.6)	1.6 (0 - 10.1)	0.36 (0 - 0.7)	0.85 (0 - 2.3)
megacrysts:	in a few areas	no	no	no	no
Muscovite	0.6 (0 - 1.0)	0.8 (0 - 2.0)	{0 in Council Mtn. {0.7 in Deserette	0.1 (0 - 0.4)	--
Biotite (chlorite)	12.8 (9.4-14.9)	12.8 (9.0-19.7)	8.0 (4.4-13.6)	9.7 (6.2-11.3)	16.0 (14.4-17.6)
Hornblende	1.0 (0 - 4.8)	0.6 (0 - 5.8)	{13.9 in Council Mtn. {0 in Deserette	1.7 (0 - 5.7)	14.7 (12.4-17.9)
Apatite Zircon Sphene Monazite	} 0.35	} 0.5 (0.1-1.5)	} 0.61 (0 - 1.3)	} 0.5 (0 - 0.9)	} 0.52 (0.2-1.0)
Magnetite/ Ilmenite	--	0.2 (0 - 1.0)	0.17 (0 - 0.5)	0.13 (0 - 0.5)	--
Epidote Allanite			2.2 (0.7 - 5.3)	--	
Other					0.5(0-2. Augite)
No.averaged	5	15	13	6	6
Reference	Taubeneck, 1971	Taubeneck, 1971	Taubeneck, 1971	Taubeneck, 1971	Taubeneck, 1971

SiO_2					
TiO_2					
Al_2O_3					
Fe_2O_3					
FeO					
MnO					
MgO					
CaO					
Na_2O					
K_2O					
H_2O+					
CO_2					
P_2O_5					
No.averaged					
Reference					

ATLANTA LOBE OF THE IDAHO BATHOLITH

(6)	(7)	(8)	(9)	(10)	(11)	(12)
trondhjemite of N. Atlanta lobe; gneissic W border zone	batholith interior SW of Snake River Plain	Cascade-type granodiorite, SW Atlanta lobe	granodiorite of Atlanta lobe, including "Atlanta type"	granite of Atlanta lobe	granite of SW Atlanta lobe	granite of N Atlanta lobe
tonalite	granodiorite	granodiorite	granodiorite	granite	granite	granite
abundant	29.4 (27.5-32.2)	28.3 (26.9-30.)	30.4 (5.9-39)	32.(20 - 37)	30.(20.-43.)	35 (25-50)
abundant (∿20)((15-?))	48.7 (45.0-52.5) (n.d.)	51.9 (50.8-53.) (25)	50.8 (40.-69.2) (29)((25-35))	35.(27 - 43) (25)((25. 25))	40.(25.-57.) (16)	25 (5-50) (25)
a few (0 - ?)	13.9 (10.7-19.2)	8.8 (7 - 11.2)	12.1 (0 - 21.9)	25. (21 - 30)	30.(12-42)	40 (20-45)
no	yes	yes	some areas yes others no	yes	yes	no
in many	1.4	0.67 (0.4-1.0)	0.71 (0 - 4.7)	0.7 (0 - 2.8)	(<1.-6.)	2-3 (0-5)
yes (5-12)	6.4 (4.5-8.3)	9.1 (8 - 10.9)	7.8	4.7 (2 - 10)	4. (2. - 8.)	1-2 (0-5)
				±	±	
a little	--	--	1.6	± (minor)	--	
		0.6	(0-0.3)		t	--
		0.005 }0.05 (0.2-0.7)	t (0-0.01) }1.0	t (0.003-0.005)	t	t
	0.15	0.15	± (0 - 2)	±	--	--
			±	t (0.005-0.025)		
	0 (0 - 0.1)	0.2 (0 - 0.3)	0.2 0 - 1)	t	t	t
up to 2.	1.				--	
	0.015			t (0 - t)	t	
minor garnet				±garnet with musc.	t- garnet	t- garnet
	4	3	22	9	20-25	20
Hamilton, 1962	Taubeneck, 1971	Taubeneck, 1971; Larsen & Schmidt, 1958	Larsen & Schmidt, 1958; Taubeneck, 1971; Shenon and Ross, 1936	Larsen & Schmidt, 1958; Ross, 1934	Anderson &	Otto, 1978
46.46		68.80	67.8 (65.23-71.99)	70.42(68.42-74.53)		
1.50		0.61	0.63(0.23-0.91)	0.36(0.07-0.55)		
13.27		15.84	15.76(14.17-16.23)	14.71(14.06-15.29)		
2.60		1.03	0.80(0.02-1.60)	0.74(0.21-0.97)		
7.20		2.27	2.83(1.91-3.63)	1.93(1.09-2.62)		
0.20		0.06	0.08(0.05-0.12)	0.07(0.06-0.08)		
7.32		0.91	1.53(1.16-2.28)	0.63(0.22-1.21)		
8.20		4.00	3.56(1.97-4.24)	2.21(1.28-2.81)		
1.68		3.57	3.68(3.13-4.21)	3.61(3.23-3.97)		
1.71		1.99	2.72(2.02-2.94)	4.05(3.36-4.60)		
3.25		0.49	0.61(0.15-0.88)	0.85(0.35-1.47)		
5.50			0.07(0.05-0.25)	0.20		
0.25		0.18	0.17(0.08-0.28)	0.12(0.09-0.16)		
1		1	7	9		
Shenon & Ross, 1936		Larsen & Schmidt, 1958	Larsen & Schmidt, 1958; Shenon & Ross, 1936	Larsen & Schmidt, 1958; Ross, 1934; Lindgren, 1900		

D. W. Hyndman

TABLE 2.   AVERAGE AND RANGE OF MODES AND CHEMICAL ANALYSES FOR THE

Unit and Location	(13) NW Bitterroot lobe and satellites	(13a) satellites NW of Bitterroot lobe	(14) White Sand Creek stock, NE of Bitterroot lobe	(15) Skookum Butte stock NE of Bitterroot lobe	(16) N border of NE Bitterroot lobe (main body)	(17) E border of NE Bitterroot lobe
Rock name:	tonalite	tonalite	qtz monzodiorite	qtz monzodiorite	qtz monzodiorite	granodiorite
Quartz	16.1 (13.3-22.1)	30.1 (22.3-29.)	15 (5 - 20)	12 (10 - 20)	19 (13 - 25)	29 (16-43)
Plagioclase (An)	54.1 (51.4-55.4) (38) ((32-38))	51.5 (36.7-61.0) (33) ((27-37))	57 (39-72) ((12-37))	68 (50 - 83) ((20-45))	56 (40 - 79) ((18-33))	44 (27-69) (28)((20-38))
K-feldspar	0	0.7 (0 - 2.2)	22 (t - 39)	10 (t - 36)	22 (1 - 45)	18 (t-49)
megacrysts:	no	no	yes	yes	yes	±
Muscovite	--	2.5 (0 - 3.3)	t	--	t	1 (0 - 7)
Biotite (chlorite)	13.4 (0-14.7) (0.55)((o-2.2))	9.1 (3.0-11.7) (0.9)((0-2.1))	3 (1 - 5)	5 (4 - 7)	3 (t - 5)	6 (1 - 20)
Hornblende	13.4 (9-17.9)	2.4 (0 - 12)	3 (0 - 27)	5 (t - 8)	--	--
Apatite	0.5 (0.4-0.7)	0.2 (0.03-04)	t	t	t	t
Zircon		(0.05)	--	t	t	t
Sphene	0.2 (? - 0.6)	0.06(0 - 0.2)	t	t	t	t
Monazite						
Magnetite/ Ilmenite	0.9 (0.6 - 1.8)	0.2/ 0.08	t	t	t	t
Epidote	2.2 (0 - 3.5)	2.8 (0 - 4.2)	t	t	--	0-t
Allanite			t	t	t	
Other	0.1 calcite	0.1 calcite				± t sillimanite
No.averaged	4	5	8	5	12	45
Reference	Hietanen, 1962	Hietanen, 1962; Larsen & Schmidt, 1958	Nold, 1968	Nold, 1968	Nold, 1968	Chase, 1968

SiO_2	60.22 (57.3-64.05)	68.52 (65.42-75.52)				67.8 (63.3-73.5)
TiO_2	0.71 (0.53-0.85)	0.46 (0.27-0.72)				0.3 (0.2 - 0.6)
Al_2O_3	17.81 (17.24-18.22)	16.38 (12.58-18.90)				15.4 (14.4-16.7)
Fe_2O_3	1.69 (1.27-2.07)	0.58 (0.29-1.30)				2.7 (as Fe_2O_3)
FeO	3.64 (2.55-4.29)	2.21 (1.15-3.52)				(1.3-5.3)
MnO	0.09 (0.08-0.11)	0.04 (0.02-0.08)				
MgO	3.07 (2.06-4.26)	1.37 (0.80-2.58)				1.0 (0.6 - 1.5)
CaO	6.41 (5.95-7.11)	4.37 (2.68-5.19)				2.2 (1.5 - 3.5)
Na_2O	3.99 (3.71-4.44)	4.06 (2.77-5.00)				4.6 (3.9 - 5.4)
K_2O	1.06 (0.17-1.47)	1.15 (0.55-1.51)				3.6 (1.6 - 5.1)
H_2O+	0.90 (0.54-1.16)	0.50 (0.37-0.63)				0.3 (0.2 - 0.6)
CO_2	0.04 (0.02-0.1)	0.01 (0 - 0.04)				
P_2O_5	0.24 (0.17-0.30)	0.16 (0.03-0.28)				
No.averaged	4	5				9
Reference	Hietanen, 1962	Hietanen, 1962; Larsen & Schmidt, 1958				Chase, 1968

BITTERROOT LOBE OF THE IDAHO BATHOLITH

(18) mylonitic "detachment zone" in NE Bitterroot lobe	(19) Five Island complex, SW Bitterroot lobe	(20) Black Canyon complex, SW border of Bitterroot lobe	(21) Boulder Creek pluton, SW Bitterroot lobe	(22) Tom Beal Park granite, Bitterroot lobe	(23) Tom Beal Park granite near nebulite zone Bitterroot lobe	(24) main body of Bitterroot lobe
granodiorite	granod.-granite	granod.-granite	granod.-granite	granod.-granite	granod.-granite	granite
28 (19 - 37)	29.1(18.2- 35)	35.0(22.- 75)	30.8(25 - 36.5)	30.8(16.4-35.)	31.8(27 - 35)	29
41 (11 - 57)	43.6(20 - 60)	33.6(5 - 57)	38.0(25 - 50.)	36.1(30.-61.1)	40.3(25 - 56)	31
(27)((18-35))	(24)((20-26))	(22)((21-27))	(22)((20-27))	(23)((20-29))	(23)((20-29))	(26)
16 (t - 56)	22.3(t - 50)	25.3(13 - 45)	27.0(9 - 40)	28.1(8.5- 40)	21 (t - 40)	35
±	(0-3)	(0-3)	(2-3)	1.3 (0 - 3)	no	yes
3 (t - 9)	0.3 (0 - 0.5)	1.3(0 - 3.3)	0.09(0.03-0.14)	1.0 (0.2-3.2)	1 (0 - 4)	
10 (5 - 16)	4.8(1 - 10)	4.4(1 - 10)	4.9(1 - 9.6)	3.2 (t - 7)	3 (1 - 8)	4
	(0.14)((0-0.24))	(0.1)((0.04-0.2))	(t)((0.03-0.17))	(0.6)((0-2.5))	±t(0-3)	
	--	--	-- (0 - 2)	--	--	
t	t (0-t)	t	t (0 - 0.3)	±t (0-0.3)	t (0 - t)	
t	t (0-t)	t	t	±t (0-0.05)	t (0 - t)	
t	--	t	t (0 - 0.05)	±t (0-0.03)	± t (0 - t)	
	--	--	--	--		
t	-- (0 - t)	t (0 - 1)	t (0-0.3)	0.2(0-0.32)	t (0 - 0.1)	
--	-- (0 - t)	±t (0 - 0.05)	-- (0 - t)	--	-- (0 - t)	
	-- (0 - t)	±t (0 - t)	-- (0 - 0.16)	--	(0 - t)	
--	±t sillimanite	±t sillimanite	--	--	±t sillimanite	
13	8	15	11	55	23	1
Chase, 1968	Williams, 1977 Hyndman, unpubl.	Williams, 1977 Hyndman, umpbl.	Williams, 1977; Hietanen, 1963 Hyndman, unpubl.	Williams, 1977 Hyndman, umpbl.	Williams, 1977 Hyndman, unpubl.	Larsen and Schmidt, 1968
67.7	68.8(67.0-70.83)	72.96	69.95(67.0 -71.37)	70.93(66.5-73.3)	71.97	71.66
0.2	0.19(0.12-0.25)	0.21	0.23(0.17- 0.27)	0.21(0.17-0.24)	0.20	0.22
15.2	15.87(15.35-16.3)	15.48	15.94(15.56-16.5)	15.82(15.2-17.15)	15.52	15.48
2.0 (as Fe_2O_3)	0.38	0.91	0.66(0.58- 0.72)	0.61(0.17-0.24)	0.21	0.56
	1.04	0.67	0.83(0.70-0.92)	0.70(0.58-0.95)	0.73	1.45
	0.03(0.03-0.04)	0.02	0.03 (0.02- 0.04)	0.03 (0.02-0.05)	0.01	0.04
1.5	0.39(0.3 - 0.46)	0.43	0.42 (0.38- 0.46)	0.40 (0.12-0.48)	0.46	0.47
1.8	2.01(1.74-2.25)	1.74	2.04 (1.70- 2.72)	1.92 (1.46-2.31)	1.89	1.97
3.5	4.43(4.33-4.5)	4.05	4.56 (3.82- 4.57)	4.42 (3.82-4.16)	4.47	4.34
3.6	3.61(3.3 -4.04)	3.98	3.55 (2.55- 4.05)	3.60 (2.90-4.16)	3.61	3.24
0.5	0.36	0.48	0.39 (0.32- 0.44)	0.40 (0.21-0.65)	0.55	0.19
			0.02 (n.d.- 0.02)			
	0.06	0.06	0.04 (0.04- 0.05)	0.06 (0.04-0.11)	0.05	0.04
1	3	1	4	15	1	1
Chase, 1968	Williams, 1977 Hyndman, unpubl.	Hyndman, unpubl.	Williams, 1977; Hietanen, 1963 Hyndman, unpubl.	Williams, 1977 Hyndman, unpubl.	Hyndman, unpubl.	Larsen and Schmidt, 1958

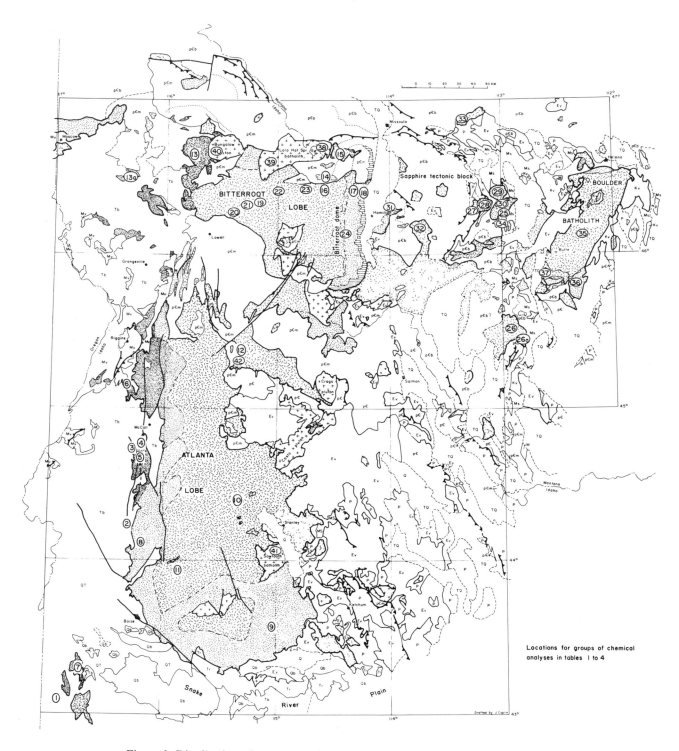

Figure 3. Distribution of average modes and chemical analyses in the Idaho batholith.

TABLE 2a. AVERAGE CHEMICAL ANALYSES FOR MAIN LITHOLOGIC UNITS
OF THE IDAHO BATHOLITH (FROM TABLES 1 AND 2)

	Tonalites of Western Border (5)	Granodiorites of both Atlanta and Bitterroot Lobes (19)	Granites of both Atlanta and Bitterroot Lobes (17)
SiO_2	57.48	67.86	70.24
TiO_2	0.89	0.42	0.29
Al_2O_3	16.90	15.60	15.29
Fe_2O_3	1.87	0.83	0.72
FeO	4.35	2.76	1.79
MnO	0.11	0.07	0.16
MgO	3.92	1.12	0.52
CaO	6.77	2.77	2.10
Na_2O	3.53	4.19	6.18
K_2O	1.21	3.17	3.80
H_2O+	1.37	0.44	0.74
CO_2	1.13	0.07	0.18
P_2O_5	0.24	0.17	0.11

main Late Cretaceous bodies of the Atlanta lobe and only partly in the Bitterroot lobe. Although broad areas of granite can be separately delineated from areas of granodiorite (Fig. 1), it is not clear that the areas mark separate plutons. Nor is it clear that separate intrusions do not make up individual areas of granite or of granodiorite. Exposure is poor over broad areas because of deep weathering and brush or forest cover. In other areas, accessibility is limited by lack of roads, as in wilderness areas. Even where exposure is fairly good, the mineralogy and texture show only subtle variation from place to place.

The Atlanta and Bitterroot lobes of the Idaho batholith seem to be obvious candidates for separately intruded bodies or composite bodies. They seem distinctly separated by the Precambrian basement of the northwest-trending Salmon River arch and by a broad area of probably late Precambrian Belt sedimentary rocks metamorphosed to the sillimanite zone of the upper amphibolite facies (Fig. 1). But such sillimanite-zone regional metamorphic rocks imply removal of many kilometers of rock overlying them at the time of metamorphism. It is not clear whether granitic rocks originally overlay this broad, high-grade area in the manner argued by Hamilton and Myers (1967) for part of New England, and, therefore, whether the two lobes of the batholith were originally connected over a broad area. Large masses of granitic rock of the Atlanta and Bitterroot lobes are now separated in one area by only a few kilometers near the Salmon River. More-mafic tonalites, representative of the western border zones of the batholith, and granodiorites, characteristic of most phases within several kilometers of intruded country rocks, appear to be absent over much of the contact areas near the Salmon River arch.

A belt of granodiorite, 10 to 20 kilometers wide, seems

fairly well documented almost completely around the Idaho batholith, except across the Salmon River arch and along the eastern side of the Atlanta lobe, where the border zone is buried under Eocene volcanic rocks. Uncertainties remain in sizeable areas in the eastern part of both lobes. The origin of such a shell of granodiorite around a dominantly granite core is unknown. One possibility is that a broad zone of assimilation of more calcic, less potassic rocks lies below the present level of exposure with considerable mixing inward or downward from the roof. Another is that broad border zones could have crystallized early from a molten region about the size of the whole batholith. A third is that early formed, more-granodioritic magmas occupying the present area of the batholith were followed by voluminous granites emplaced into the central part of the batholith. None of these possibilities is without difficulties. In any case, the granodioritic shell may have originally marked the roof of the batholith as well. If so, drier, granodioritic magmas would reasonably have risen to shallower levels before reaching the solidus curve, whereas the wetter granite magmas would reach the solidus curve at deeper levels. Granodioritic rocks concentrated around the outer margins and across the roof of the Idaho batholith would also account for the absence of such rocks bordering the deep levels of exposure of the Salmon River arch but would not preclude the former continuation of the batholith across the Salmon River arch (Fig. 7).

It seems possible that separate major intrusions occur in the southwestern part of the Bitterroot lobe where Williams (1977) has mapped the Black Canyon complex (loc. 20*), Boulder Creek pluton (loc. 21), and the Five Island complex (loc. 19) separate from the main-phase Tom Beal Park granite (loc. 23). The former three units, each 7 to 9 kilometers across, trend northwestward parallel to the overall trend of the lobe. The enormous Tom Beal Park granite is 18 kilometers across at the level of the Lochsa River canyon and more than 30 kilometers across at higher elevations (Williams, 1977; Hyndman, unpubl. data). This size is comparable to that of the Tuolumne Intrusive Series and other large plutons in the Sierra Nevada batholith (Bateman and others, 1963; Bateman and Chappell, 1978). The Boulder Creek pluton appears lithologically similar to the Tom Beal Park granite and may be closely related in time and source magma. As neither the northwestern nor southeastern limits of either mass have been mapped, the two masses may or may not be separate intrusions.

It also seems possible that the main body of the Atlanta lobe consists of only one or a few enormous plutons. Anderson (1952) argues in favor of multiple emplacement of discrete masses to form the Idaho batholith. Clearly, there are some sizeable separate masses in the foliated quartz

---

*Numbers in parentheses refer to locations on Figure 3 and to average modes and chemical analyses in Tables 1-4.

TABLE 3. AVERAGES AND RANGES OF MODES AND CHEMICAL ANALYSES FOR SATELLITIC

Unit and Location	(25) Racetrack pluton, SW of Deer Lodge, MT	(26) Pioneer batholith NW of Dillon, MT	(26a) Mount Torrey batholith NW of Dillon, MT	(27) Philipsburg batholith Bimetallic stock, E of Philipsburg, MT	(28) Dora Thorn pluton, E of Philipsburg, MT	(29) Royal stock SE of Drummond, MT	(30) Mt. Powell batholith, W of Deer Lodge, MT
Rock name:	qtz monzodiorite	granodiorite	granodiorite	granodiorite	granod.- granite	granodiorite	granod.- granite
Quartz	7 (t - 25)	27.9 (22.8-34.1)	30	16 (8.-32.2)	25.6 (15-45)	26.4 (17.5-35.1)	28.1 (22.5-32.9)
Plagioclase (An)	47 (15 - 65)	42.4 (31.5-51.0)	40 ((25-35))	43.4 (46.2-82.0) (50)((40-57)) cores	40. (29-65) (42)((36-44)) cores	45.4 (38.2-53.1) (32.8)((27-36))	42.0 (37.9-50.1) (28)((22-42))
K-feldspar	9 (0 - 60)	17.1 (4.3-29.0)	15	13. (4 - 31.2)	23. (9.5-43)	17.6 (5.2-27.1)	21.4 (14.5-31.3)
megacrysts:	[only in K-spar-rich, fol. rocks]			no	[in more K-spar-rich rocks]	no	yes
Muscovite	-- (0 - minor)	tr	t	--	± t	1.6 (0.6-3.2)	4.5 (2.1-7.6)
Biotite	7 (t - 20)	8.0 (4.3 -15.4)	10	12. (∿3∿12)	5.6 (3.-12.)	7.6 (5.1-10.9)	3.2 (1.0-7.1)
(chlorite)	1 (t - 6)	0.6 (0 - 1.2)	(t)	(t)	(1) (< 1-3)		
Hornblende	∿26 (t - 50)	1.7 (0 - 4.3)	3	12. (∿3∿30)	2.2 (0 - 10)	--	--
Apatite	1 (1 - 2)	tr	t	t	t (t)	t	t
Zircon		tr	--	t	t (0 - t)	± t	± t
Sphene	1 (1 - 3)	0.4 (0 - 0.8)	t	t	t (t - 1)	t	± t
Monazite					± t		
Magnetite/ Ilmenite	1 (t - 4)	tr	t	2	1.6 (t - 3)	t	t
Epidote	1 (0 - 3)	tr	--	t	0.5 (0 - 1)	± t	± t
Allanite		tr			± t		± t
Other							
No. averaged		4		50	88	159	242
Reference	Hawley, 1975	Zen and others, 1975	Willis, 1978	Hyndman and others, 1976	Hyndman and others, 1976	Allen, 1966; Hyndman and Silverman, unpub.; Benoit, 1972	Hyndman and Silverman, unpub.; Benoit, 1972

	(25)	(26)		(27)	(28)	(29)	(30)
SiO_2	56.93 (49.3-64.28)	68.2 (64.7-72.8)		59.04 (55.81-60.89)	69.42 (68.13-70.93)	67.6 (66.4-69.2)	72.3 (68.8-75.3)
TiO_2	0.98 (0.72-1.22)	0.37 (0.21-0.59)		0.79 (0.66 - 0.97)	0.40 (0.29 - 0.50)	0.34 (0.27-0.42)	0.17 (0.08-0.24)
Al_2O_3	17.88 (14.79-20.98)	16.0 (15.1-16.4)		18.44 (16.99-20.49)	15.77 (15.05-16.28)	17.5 (16.7-18.3)	16.0 (14.6-18.2)
Fe_2O_3	8.51 (6.13-10.68) (as FeO)	1.8 (0.86-2.8)		6.51 (5.12 - 7.73) (as FeO)	3.39 (2.66 - 4.06) (as FeO)	0.85 (0.6-1.0)	0.43 (0.1-1.2)
FeO		1.8 (0.76-2.9)				2.16 (1.8-2.5)	1.5 (1.2-1.9)
MnO		0.08 (0.08-0.10)					
MgO	3.97 (1.78-6.60)	1.1 (0.43-1.8)		2.84 (2.18 - 3.56)	1.30 (0.88 - 1.61)	1.01 (0.8-1.2	0.5 (0.2-0.7)
CaO	7.42 (3.67-10.44)	3.7 (2.2 -4.5)		5.89 (4.65 - 6.57)	3.10 (2.30 - 3.78)	3.95 (3.6-4.2)	2.2 (1.0-3.6)
Na_2O	3.72 (1.27-6.46)	3.3 (3.0 -3.8)		3.85 (3.47 - 4.86)	3.31 (3.10 - 3.51)	3.39 (3.2-3.6)	3.6 (3.2-4.1)
K_2O	2.23 (0.69-4.38)	2.9 (2.6-3.5)		2.41 (1.97 - 2.83)	3.82 (2.48 - 4.36)	3.38 (3.1-3.7)	3.6 (2.6-4.2)
H_2O+		0.82 (0.75-0.96)					
CO_2		0.03 (0.02-0.04)					
P_2O_5		0.17 (0.14-0.21)					
No. averaged	11	4		8	9	8	17
Reference	Hawley, 1975	Zen and others, 1975		Hyndman and others, 1982	Hyndman and others, 1982	Benoit, 1972	Benoit, 1972

dioritic border phases, and separate satellitic plutons both west and east of the Atlanta and Bitterroot lobes. But the separate intrusions noted by Anderson are small (at most, several square kilometers), in part, dike-like bodies related to the Eocene Challis volcanics (cf. Reid, 1963, p. 13). The pre-Challis "white quartz monzonite" of Ross (1934) and Reid (1963), a somewhat finer-grained phase slightly younger than the main phase of the batholith, forms bodies of several square kilometers. The presence of large plutons, separately emplaced in the main bodies of the Atlanta and Bitterroot lobes, seems likely, considering the existence of innumerable intrusions in the otherwise similar Sierra Nevada batholith (e.g., Bateman and others, 1963). But poor exposures and few detailed studies in the huge interior of the batholith have so far inhibited documentation of such separate intrusions.

The internal structure of the granitic rocks is somewhat variable. Anderson (1942, 1952), Schmidt (1964), and Taubeneck (1971) described the gneissic tonalite which forms the western border zone of the Atlanta lobe. It is 10 to 20 kilometers wide and strongly foliated parallel to the country-rock contact and to the country-rock foliation. The

PLUTONS EAST OF THE BITTERROOT LOBE AND FOR THE BOULDER BATHOLITH

(31) Willow Creek stock, E of Hamilton, MT	(32) Upper Rock Creek pluton, Sapphire Range, MT	(33) Garnet stock NE of Missoula, MT	(34) Lost Creek stock, N of Anaconda, MT	(35) Butte quartz monzonite, betw. Butte & Helena	(36) Boulder batholith — older satellites: Burton Park, Rader Creek, Unionville	(37) younger satellites; "sodic series": Climax Gul., Hell Cany., Donald
granod.- granite	granod.- granite	granite	granite	granite	granodiorite	granod.- granite
29.1 (11.3-51.1)	30.8 (19.9-43.4)	26.5 (29.8, 23.1)	31.3 (19.6-40.6)	26.1 (14.-35.)	(14-29)	26.(18-40)
43.7 (22.9-63.0) (26)((23-31))	39.1 (20.0-53.0) ((23-48))	35.1 (34.3, 35.9)	34.0 (21.9-54.7) (30)((20-32))	36.7 (22.-53.) (38)((40-58)) cores	(44-62) ((45-63)) cores	43.8 (31-50) ((20-40))
23.3 (8.3-44.1)	21.2 (8.1-40.0)	29.9 (28.9, 30.8)	29.4 (8.8 - 37.1)	27.5 (14.-42.)	(13-36)	21.9 (14-32)
yes		no		in many rocks	no?	yes
4. (0.4 - 9.5)	0.6	0.1 (0.1, 0.2)	± t	--	--	<1
	5.2 (0 - 10.0) (t)	4.3 (5.1, 3.6) (t)	5.3 (1.4-16.9)	5.5 (0.5-13.0)	7 - 22 subequal	6.
	0.5 (0 - 4.0)	3.4 (1.4, 5.3)	--	3.8 (0 - 11.)		1 - 2 (up to 3.5)
	t	t (t - 0.1)	±		t	t
	± t	t	±	1.2 (t - 4.0)	t	t
	t (0 - t) (± t)	0.1 (0.2, 0.1)	±		t (t)	t
t	0.6 (t - 1.4)	0.5 (1.1, 1.0)	± hematite		t	t
	± t	t	--			
				0.5 diopside(0-7)	± t tourmaline	± diopside
45 Presley, 1970	19 Pederson, 1976	2 Brenner, 1968	18 Winnegar, 1970	57 Becraft & others, 1963; Smedes, 1966; Weed, 1912; Knopf, 1957; Smedes & others, 1968	Tilling, 1964; Tilling and others, 1968	15 Tilling, 1964; Tilling, & others, 1968; Smedes, 1966; Smedes & others, 1968; Mathews & others, 1977
				64.6 (62. -71.5)	64.6 (54.5-71.)	70.0 (65.2-72.5)
				0.5 (0.23- 0.73)		0.36 (0.26-0.54)
				15.3 (14.2-17.2)		15.2 (14.8-15.6)
				1.94 (1.0-4.0)		1.4 (0.89-1.77)
				2.23 (0.68-3.3)		1.6 (0.92-2.66)
				0.08 (0.03-0.15)		0.08 (0.06-0.10)
				1.90 (0.65-2.7)		1.12 (0.51-2.20)
				4.01 (2.1 -4.7)	4.5	2.33 (1.8-4.14)
				2.99 (2.62-4.2)	3.5	3.59 (3.09-4.0)
				4.06 (3.3 -5.2)	2.9 (2.-5.2)	3.60 (2.6 -4.1)
				0.67 (0.26-1.4)		0.68 (0.57-0.80)
				0.18 (0.08-0.40)		0.13 (0.09-0.19)
				31 Tilling, 1974; Beacraft & others, 1963; Smedes, 1966; Weed, 1912; Smedes & others, 1968; Klepper & others, 1957	Tilling, 1973	15 for SiO$_2$, CaO, Na$_2$O, K$_2$O; 3 for others. Tilling, 1973

foliation in the tonalites over extensive areas dips 60 to 70 degrees eastward, that is, under the batholith. The gneissic border zone apparently grades to more massive granodiorites and granites in the interior of the batholith. In some areas, at least, such as along the northern border of the Snake River Plain, the foliation maintains its general northerly strike, but dips flatten to 25 and 30 degrees (Taubeneck, 1971). There are little data for the eastern Atlanta lobe, but Reid (1963) indicates that a well-developed foliation dips 45 to 70 degrees east in part of the Sawtooth area. Most rocks of the main "inner facies" of the lobe are massive or weakly foliated (cf. Anderson, 1942, 1952; Ross, 1934). In a few areas, especially near the borders in the Hailey and Casto areas, foliation is conspicuous.

In the Bitterroot lobe, mineral foliation in the tonalite at the northwestern end of the lobe is characteristically parallel to country-rock contacts and to bedding in the metasedimentary country rocks. It trends generally somewhat west of north and dips eastward about 50 degrees near the contact, but only 15 to 25 degrees farther away. The accompanying mineral lineation plunges 30 to 45 degrees eastward (Hietanen, 1962, 1963). This fabric extends for 10

TABLE 4. AVERAGES AND RANGES OF MODES AND CHEMICAL ANALYSES FOR
EPIZONAL EOCENE PLUTONS IN THE IDAHO BATHOLITH AREA

	(38)	(39)	(40)	(41)	(42)
Unit and Location	Lolo Hot Springs batholith	Indian Grave Pk. -Horshoe Lake. complex (W pt. of Lolo H.S. bathol.	Bungalow pluton, N of Bitterroot lobe	Sawtooth batholith	small plutons of N Atlanta lobe, Sheepeater Pk. area, S of Salmon R.
Rock name:	qtz.syenite granite	granite	granite-granodiorite	granite	granite
Quartz	17 (15-25)	35 (30-40)	30.6 (26.0, 35.2)	yes	30
Plagioclase (An)	16 (0-25) ((5-31))	27 (25-30) (21)((20-24))	42.5 (48.8, 36.3) (12) ((18, 6))	∿1/3 of felds. ((2-32))	10 (10 - 15) (22)((20-28))
K-feldspar	66 (59-80)	34 (30-37)	21.4 (18.2, 24.5)	∿2/3 of felds.	50 (45 0 55)
megacrysts:	no				no
Muscovite	--	--	1.6 (2.24, 0.90)	yes	t
Biotite (chlorite)	1 (t - 2)	4 (3 - 5)	4.3 (5.19, 3.40)	yes (yes)	5 (5 - 10)
Hornblende	--	--	--	yes	(1 - 4)
Apatite	t	t	0.08 (0.08, 0.08)	yes	
Zircon	t	t		yes	t
Sphene	--	t (0 - t)	--	yes	t
Monazite					
Magnetite/ Ilmenite	t	t	0.24 (0.36, 0.12)	yes/ yes	t
Epidote	--				
Allanite	t	t (0 - t)	--	yes	
No. averaged	6	3	2		3
Reference	Nold, 1968	Williams, 1977	Larsen & Schmidt, 1958; Hietanen, 1963	Reid, 1963	Otto, 1978
SiO_2	71.32		76.35 (77.43, 75.28)		
TiO_2	0.35		0.10 (0.06, 0.14)		
Al_2O_3	14.09		12.81 (12.47, 13.14)		
Fe_2O_3	0.16		0.56 (0.22, 0.89)		
FeO	0.91		1.16 (0.84, 1.47)		
MnO	0.05		0.06 (0.04, 0.07)		
MgO	0.52		0.20 (0.13, 0.27)		
CaO	1.08		0.7 (0.51, 0.89)		
Na_2O	4.15		6.6 (6.84, 6.36)		
K_2O	4.98		5.29 (5.22, 5.35)		
H_2O+	0.60		0.13 (0.19, 0.08)		
CO_2			0.1 (0.1, --)		
P_2O_5	0.08		0.03 (0.04, 0.03)		
No. averaged	1		2		
Reference	Hyndman, unpubl.		Larsen & Schmidt, 1958; Hietanen, 1963		

to 12 kilometers to the east, gradually giving way to more massive tonalite. The eastern contact dips eastward 50 to 60 degrees against a large inlier of metasedimentary country rocks. Farther east, the main-phase granite of the Bitterroot lobe is characteristically massive but locally weakly foliated (Hietanen, 1963, p. D 11). About 40 kilometers farther southeast, the foliation in the granite is weak to moderately strong. Foliation in the central Bitterroot lobe appears to describe a major dome or arch plunging northwest parallel to the trend of the lobe. Dips are about 80 degrees northeast near the contact, about 45 degrees southwest about 10 kilometers into the batholith, and essentially horizontal or 20 to 25 degrees northwest parallel to the trend of the lobe about 18 kilometers in from the contact. Nebulous schlieren and large, generally planar inclusions of country-rock schists and gneisses parallel with the foliation are widespread and fairly abundant in this southwestern part of the lobe. The remaining 23 kilometers across the Tom Beal Park granite (loc. 23) in the narrowest part of the lobe show gradually increasing dips to essentially vertical at the northeastern contact. Thus, foliation in the central Bitterroot lobe appears to describe a major dome or arch plunging northwest parallel to the trend of the lobe.

Farther east in the Bitterroot lobe, along the northerly part of the boundary between Idaho and Montana, is the northward-trending Bitterroot dome (Chase and Talbot, 1973). The Bitterroot dome is 100 kilometers long and about 50 kilometers across. Foliation on the western flank is weak, gradually increasing in intensity to very strong on the eastern flank, which dips about 25 degrees east (Hyndman, 1979). A penetrative mineral and slickenside lineation plunges east-southeastward in the surface of foliation. The lineation maintains its eastward trend across the dome, even at the south end of the dome, where the foliation dips southward. Thus, rise of the dome must post-date formation of the lineation.

Eocene granitic plutons, such as the Sawtooth batholith, Casto pluton, and Lolo Hot Springs batholith, are massive and otherwise lacking in internal structure.

Texture of the granitic rocks varies most prominently with rock composition. As previously noted, the more mafic tonalites of the western border zones are strongly foliated, the less mafic granodiorites and granites of the main mass of both the Atlanta and Bitterroot lobes are moderately to very weakly foliated, and the smaller, shallower Eocene plutons are massive.

The texture of most rocks is hypidiomorphic granular. Plagioclase, biotite, and where fairly abundant, hornblende, are generally subhedral. Quartz and K-feldspar are most commonly anhedral and where less abundant, interstitial to other minerals. Where more abundant, K-feldspar forms larger anhedral grains or megacrysts enclosing all of the other minerals and replacing them along the K-feldspar boundaries.

K-feldspar megacrysts away from the contacts of the Bitterroot lobe are commonly 1 to 2 centimeters across and, in some places, up to 5 centimeters. In most areas, they show fair crystal form and make up a few percent of the rock (Williams, 1977; Nold, 1968; Chase, 1968; Hietanen, 1963). No K-feldspar megacrysts are present in most of the Black Canyon complex although they may amount to about 5 percent locally. They form an average of 2 to 3 percent and up to 5 percent in the Boulder Creek pluton and the Five Island complex. The largest pluton in the center of the lobe, the Tom Beal Park granite, averages about 5 percent K-feldspar megacrysts; they range from about 1 to 10 percent and may be as long as 3 centimeters.

Similarly, in the Atlanta lobe, K-feldspar megacrysts are commonly 1 to 2 centimeters across and, in some areas, up to 5 centimeters or more. In most places, they amount to a few percent of the rock (Larsen and Schmidt, 1958; Anderson, 1942, 1952; Reid, 1963; Taubeneck, 1971; Olson, 1968). Locally, in the southwestern part of the lobe, between the South Fork of the Payette River and the Boise River, they make up to 10 to 30 percent of the rock over large areas (Anderson and Rasor, 1934, p. 293). In the gneissic western border zone, K-feldspar megacrysts are rare (Ross, 1934, p. 36). In the northeastern-most Atlanta lobe, west of the confluence of the Middle Fork of the Salmon River and the main Salmon River, nearly half of the granitic rock contains conspicuous megacrysts of K-feldspar averaging 5 centimeters and up to 15 centimeters long (Cater and others, 1973, p. 19).

Contacts of most of the granitic rocks are sharp on the scale of handspecimen or outcrop. Western contacts of the Atlanta lobe are mostly faulted against Permo-Triassic volcanics or buried under Miocene Columbia River basalts. Where preserved, the contacts between trondhjemite and country rocks to the west are semiconcordant and form contact migmatites up to one kilometer wide, or they are extremely irregular and crosscutting in detail (Hamilton, 1962). The contact between the gneissic border zone and the more felsic internal facies is sharp, the latter locally showing apophyses into the border zone (Anderson, 1952). Much of the eastern margin of the Atlanta lobe is buried under Eocene Challis volcanics. In the Casto 30'-quadrangle, the contact is apparently discordant and has mildly metamorphosed the country rocks (Ross, 1934).

The northwestern border of the Bitterroot lobe is largely concordant but to some extent discordant, especially where granitic rocks intertongue with the country rocks. In most places, it is gradational over a few centimeters to hundreds of meters; elsewhere, it is sharp (Hietanen, 1963, p. 9). For about 9 kilometers into the batholith (the Black Canyon complex of Williams, 1977), along the southwestern flank of the Bitterroot lobe, the contact zone consists of steep sheets of somewhat foliated granite and granodiorite from 0.5 to 2 kilometers thick, separated by somewhat thinner concordant sheets of metasedimentary country rocks (Williams, 1977; Hyndman, unpubl. data). Along the northern edge of the Bitterroot lobe in the northern Bitterroot dome, the contact is concordant on the scale of an outcrop. The granitic rock forms a broad series of large sills with intervening sheets of metamorphic rock. The zone is less well-exposed but resembles that of the southwestern border zone. A complicating factor between the granitic sheets and the main body of the lobe is a large mass of about 80 km² of probable pre-Belt basement orthogneiss exposed deep in the Lochsa River canyon, which may represent rocks underlying the batholith. The White Sand Creek stock (loc. 14), a nearby satellitic pluton, also exhibits contacts characterized by concordant sheets of granodiorite with intervening gneiss over a zone a few hundred meters wide (Nold, 1968, p. 43, 49). The Skookum Butte stock (loc. 15), a satellite about 20 kilometers farther northeast, is sharply bounded and concordant in some places, discordant in others. It has produced a contact-metamorphic aureole about 2.5 kilometers wide. A short distance farther east, in the northern part of the Bitterroot dome, the border zone of the main mass of the Bitterroot lobe forms a concordant series of sill-like granitic bodies, one centimeter to more than 30 meters wide. These sills locally truncate the foliation. Numerous apophyses penetrate the country rocks (Chase, 1973, p. 10).

The Eocene granitic plutons are sharply bounded and show a variety of evidence indicating emplacement at shallow depth, including miarolitic cavities and common, single feldspar mineralogy. The Sawtooth batholith contains pegmatite sheets mostly emplaced "at low angles parallel to tension joints that would form in the upper part of the batholith" (Reid, 1963, p. 24). The coarse-grained Lolo Hot Springs batholith is characterized by abundant miarolitic cavities, except in a chilled, medium-grained border zone 30 to 60 meters wide (Nold, 1974). The batholith apparently vented near its center to form the Crooked Fork rhyolite plug. The discordant Bungalow pluton is marked by a granite porphyry along most of its contacts with the country rocks (Hietanen, 1963, p. 9-10). The Eocene plutons and their associated mineral deposits were reviewed by Bennett (1980).

Dikes of pegmatite and aplite are widespread in the Idaho batholith. The composition of the pegmatite is characteristically much like that of the related granitic rock or somewhat more felsic and higher in alkalis. The characteristics are those expected of differentiation of the magma of the related pluton.

In the Atlanta lobe, granodioritic and granitic pegmatites are more common than aplites and are more abundant at the roof and in the marginal zones of the batholith. Hornblende and biotite quartz dioritic pegmatites are abundant in parts of the marginal facies. Granodiorite and granite pegmatites related to the inner facies of the batholith occur in the outer parts of the inner facies and in bordering parts of the marginal facies. Most pegmatite dikes are sharply bounded, but some grade into the surrounding granite or into associated aplite (Anderson,1942, p. 1115-1116; Larsen and Schmidt, 1958, p. 11; Reid, 1963, p. 18). In the Sawtooth area of the eastern Atlanta lobe, pegmatite dikes mostly trend northeast and dip steeply southeast (Reid, 1963). Pegmatite dikes are mostly near horizontal in the western part of the Atlanta lobe and variable farther north in the central part (Olson, 1968). Aplite dikes are abundant in a few areas, especially in and near some mining districts, near Idaho City, the Atlanta district, and the Casto area. They are tabular, sharply bounded, and generally steeply inclined.

In the Bitterroot lobe, granodiorite and granite pegmatites also are more abundant near the border zones of the batholith. Satellite plutons of tonalite west of the Bitterroot lobe in the Orofino area are cut by abundant dikes of biotite quartz dioritic pegmatite. Local patches of hornblende quartz dioritic pegmatite apparently formed by segregation in the tonalite (Hietanen, 1962, p. 57). Biotite quartz plagioclase pegmatite is abundant along the contact zones of tonalite where it grades to its gneissic tonalite border facies in the western margin of the Bitterroot lobe. Large masses of granite pegmatite occur along many contacts of coarser granite, against metamorphic country rocks, and locally against the tonalite border zone (Hietanen, 1963, p. 25). A section northeastward through the main mass of the Bitterroot lobe shows an average of about 15 percent (1 to 50%) pegmatite and lesser aplite dikes in the Black Canyon complex near the southwestern margin. The Boulder Creek pluton to the northeast averages 2 to 3 percent (0 to 10%) pegmatite, with the higher values concentrated near its margins. The Five Island complex farther northeast averages about 5 percent (1 to 10%) pegmatite. The Tom Beal Park granite, occupying the main mass of the central Bitterroot lobe, averages only about 1 percent (0 to 10%) pegmatite. For 18 kilometers in from the southwestern border, the pegmatite and aplite dikes characteristically dip 10 to 35 degrees northwesterly, cutting at high angles across the steeply dipping sheets of granitic and metasedimentary rocks. The pegmatite dikes are apparently controlled by the orientation of gently dipping shear surfaces on which slickenside striae plunge down-dip to the northwest. Pegmatites of the batholith locally penetrate the country rocks, cutting the metamorphic pegmatites therein (Chase, 1973, p. 10), and, in places, extensively penetrate roof rocks above the batholith.

Pegmatite and aplite dikes are also common in the epizonal Tertiary plutons. In the Sawtooth batholith, most pegmatite dikes dip at low angles, not parallel, to prominent joint sets, which probably formed later. A few near-vertical dikes trend northwest, north, and northeast parallel to major joint sets (Reid, 1963, p. 19). In the Bungalow pluton at the north end of the Idaho batholith, pegmatite dikes 10 to 20 centimeters wide are common (Hietanen, 1963). In the

Lolo Hot Springs batholith, sharply bounded aplite dikes 1 centimeter to 3 meters thick are moderately common. Less common are pegmatite pods up to 1 meter or so long and 30 centimeters across (Nold, 1968, p. 54).

RADIOMETRIC AGES

Radiometric ages on the Idaho batholith and surrounding areas (Fig. 4) have been reviewed by Armstrong (1975b) and Armstrong, Taubeneck, and Hales (1977). These and related events are plotted for comparison in Table 5. They indicate that the bulk of the Atlanta lobe cooled between 100 and 75 million years ago and the bulk of the Bitterroot lobe cooled between 80 and 70 million years ago. Reasonable ages in the granodiorite-granite interior of the Atlanta lobe include 13 dates between 62.6 and 95 m.y. Those from tonalitic rocks of the western marginal facies and nearby satellites include nine dates between 65 and 117 m.y. Reasonable ages in the granodiorite-granite main phase of the Bitterroot lobe include only a single 66 m.y. zircon date from the north-eastern margin of the lobe (loc. 17; Chase, Bickford, and Tripp, 1978) and a single 81 m.y. sphene fission-track date from the Skookum Butte stock (loc. 15) just north of the lobe (Ferguson, 1975). Tonalite and tonalite orthogneiss from the northern border zones of the Bitterroot lobe have yielded four reasonable ages ranging from 70 to 85 m.y. (Grauert and Hoffman, 1973; Reid and others, 1973; Ferguson, 1975; Chase and others, 1978).

Other radiometric ages from the southeastern third of the Atlanta lobe and the bulk of the Bitterroot lobe range from 43 to 46 and 37 to 56 m.y., respectively. These younger dates on granitic rocks of apparent Late Cretaceous age have apparently been reset by an Eocene event (cf. Armstrong, 1974). Explanations include contact metamorphic heating by large unrecognized Eocene plutons in the two lobes of the batholith (Swanberg and Blackwell, 1973), anomalously high heat flow or uplift or chemical alteration related to Eocene Challis volcanic and plutonic activity (Armstrong, 1974; Williams, 1977, p. 93), or uplift and cooling of deep-seated granitic rocks sufficient to retain argon in the mica mineral lattice beginning in Eocene time (Ferguson, 1975).

A more detailed analysis of the effect of the Eocene plutons on the Late Cretaceous rocks shows that the oxygen isotopes in the Cretaceous granitic rocks vary with distance from the Eocene plutons (Criss and Taylor, 1978).

In the western half of the Atlanta lobe, west of 115° 35′ $\delta^{18}O$ is greater than +8 with $\delta D \approx -120$, indicating a very low water/rock ratio (<0.1). To the east, $\delta^{18}O$ is variable, ranging from about –8 to +8 with D = –130 to –170, indicating strong exchange with meteoric/hydrothermal water. $\delta^{18}O$ is lowest in chloritized rocks near the Eocene plutons where the water/rock ratio is at least 1 to 10. In detail, $\delta^{18}O$ decreases outward from the relatively unaltered core of the Sawtooth batholith (about 20 × 20 kilometers) to

a low $\delta^{18}O$ "moat" forming an oval about 60 km north-south and 40 km east-west. The 5-to-15-kilometer wide moat may be the eroded equivalent of a large caldera ring-fracture system and the locus of greatest hydrothermal circulation and alteration (Criss and Taylor, 1978). Such alteration may explain the Eocene radiometric ages so widespread over large areas of the Late Cretaceous batholith.

CHEMISTRY

Average and range of chemical analyses from various parts of the Atlanta and Bitterroot lobes of the Idaho batholith are given in Tables 1 and 2, respectively. Their locations are shown in Figure 3. All available published and unpublished chemical analyses from the batholith cluster narrowly for several major oxides — SiO_2 ranges from 63.3 to 74.5 percent; Al_2O_3 from 14.06 to 17.15 percent; and MgO from 0.12 to 2.28 percent. The only exception to these values and those outlined below is one tonalite (no. 6) from the Atlanta lobe, which has high H_2O+ and CO_2 values, indicating that it is strongly altered. Available analyses are too few to permit firm distinction between the chemistry of the two lobes, but the average Na_2O is apparently a little lower in the Atlanta lobe. Units in the Atlanta lobe average 3.47 to 3.68 percent Na_2O, compared with 3.99 to 4.56 percent Na_2O in the Bitterroot lobe. Average analyses for plutons east of the Bitterroot lobe, for the Boulder batholith, and for Eocene plutons in the Idaho batholith are given in Tables 3 and 4.

A plot of average K_2O against SiO_2 for each unit, as was done by Tilling (1973) for the Boulder batholith, shows, in a general way, the characteristic correlation of increasing K_2O with SiO_2 (Fig. 5). All average values for units in the Idaho batholith (units 1 - 24) and those east of the Bitterroot lobe of the batholith (units 25 - 34) fall in Tilling's *sodic series* or along the plotted trend of the sodic-series field. The Butte quartz monzonite of his *main series* is appreciably richer in K_2O for the same percent SiO_2 (by about 1%) than any of the units of the Idaho batholith or its satellitic plutons. A plot of the average values of K_2O, Na_2O, and CaO for each unit (cf. Tilling, 1973) also shows this separation of more Na_2O-rich Idaho batholith compositions, satellitic plutons, and sodic series of the Boulder batholith from the more K_2O-rich main series of the Boulder batholith (Fig. 6).

Tilling (1973, 1974) demonstrated for the Boulder batholith that the "main series" and "sodic series" can be distinguished by K_2O or Na_2O versus SiO_2, K_2O or Na_2O versus CaO, Rb or Sr versus SiO_2, initial Pb^{206}/Pb^{207} ratios versus age, and mean K_2O or K_2O/SiO_2 versus age. The increase in K_2O from about 2.8 to 4.0 percent with decreasing age from about 77 to 69 million years in the sodic series of the Boulder batholith cannot accurately be compared with rocks of the Idaho batholith because of the paucity of reliable radiometric ages and insufficient data on

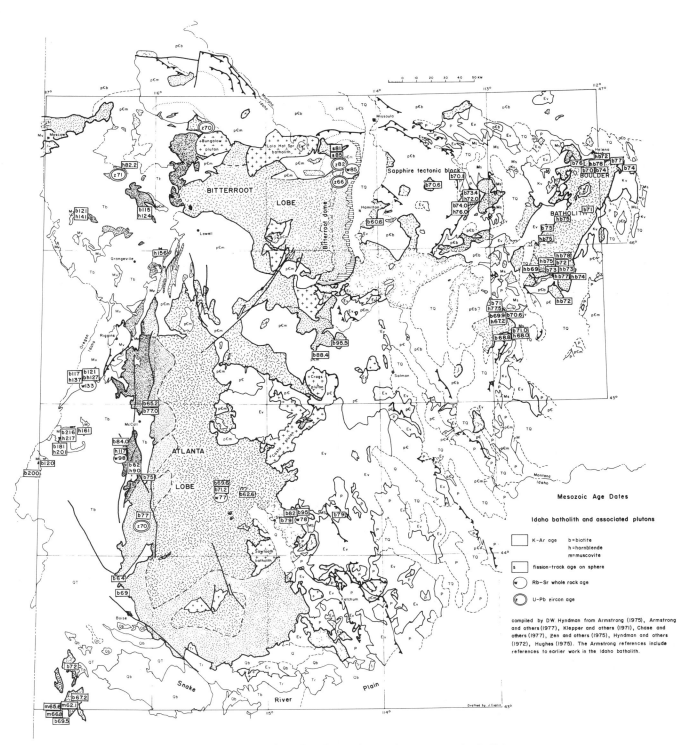

Figure 4. Distribution of radiometric ages in the Idaho batholith.

TABLE 5. TIMING OF LATE CRETACEOUS AND EARLY TERTIARY PLUTONIC, VOLCANIC, AND STRUCTURAL EVENTS
FOR THE IDAHO AND BOULDER BATHOLITHS AND RELATED AREAS

	PLUTONIC EVENTS		VOLCANIC EVENTS	STRUCTURAL & OTHER EVENTS
	Idaho batholith	East of Idaho batholith		

Chart (ages in millions of years, 40 to 100):

- 40
- Epizonal plutons (Idaho batholith, ~42–49)
- Challis Volcanics (~44–48)
- brecciation of volcanics on top of mylonitic zone (~48)
- 50 — Lowland Creek Volcanics (~49)
- 60
- 70 — Idaho batholith; Pioneer batholith; Miner's Gulch, Henderson stocks; E. / W. Philipsburg batholith; Boulder batholith
- Isostatic rebound; Thrusts bounding Sapphire block; Bitterroot mylonitic zone
- early Elkhorn Mtns. Volcs.
- 80 — Racetrack diorite (?)
- quartz diorite orthogneiss (~80–85)
- 90
- regional metamorphism
- 100 (?)

Note: For Idaho batholith, X = sample in or near Bitterroot lobe; ● = sample in or near Atlanta lobe.
Lowland Creek volcanics from Smedes and Thomas, 1965.

the extent and chemical characteristics of individual plutons in the latter. However, the plausibility of a correlation of K_2O with age in the Idaho batholith is suggested by the abundance of K_2O values in the range 2.8 to 4.0 percent and age determinations in the range 81 to 69 million years ranges similar to those in the sodic series of the Boulder batholith (Fig. 5, Table 5).

Two chemically distinct magma series, each with a different source or sources in the lower crust or upper mantle, and operating during a small span of time in a restricted area, have been argued by Tilling (1973). He suggests that the source areas may be separated either vertically or laterally. The comparisons made above support the suggestion that such source areas differ laterally, areas west of the Boulder batholith producing magmas which are distinctively more sodic than those of the main series of the Boulder batholith. If the source(s) are in the lower crust or upper mantle as suggested by Doe and others (1968) and

Tilling (1973), it seems possible that the lower crust or upper mantle of the area of the Idaho batholith are more sodic than those of the Boulder batholith to the east.

Chappell and White (1974) distinguish S-type granitic rocks thought to be derived from sedimentary source rocks from I-type granitic rocks thought to be derived from igneous source rocks that have not been through surface weathering processes. I-type granitic rocks have relatively high Na_2O/K_2O, higher Ca, Al, Fe_2O_3/FeO, and lower Fe, Mg, Sc, V, Cr, Co, Ni, Cu, Zn, Ba, Rb, Th, La, Ce, and Y than S-types with comparable SiO_2 (Chappell and White, 1974; Hine and others, 1978). Mineralogically, I-type granitic rocks contain biotite + hornblende ± sphene ± magnetite ± ilmenite; S-type contains biotite ± muscovite ± cordierite ± garnet ± ilmenite. The I- and S-types correlate to some extent with "magnetite-series" and "ilmenite-series" granitic rocks (Ishihara, 1977; Takahashi and others, 1980), respectively, the Boulder batholith being magnetite-series

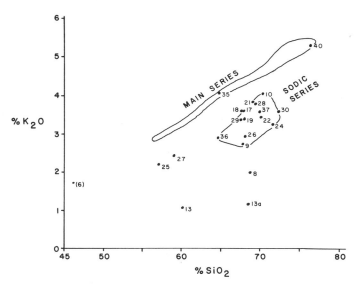

Figure 5. K_2O versus SiO_2 for averages of units of the Idaho and Boulder batholiths, with outlines of Tilling's (1973) "main series" and "sodic series" for the Boulder batholith.

and the Idaho batholith being ilmenite-series granitic rocks (S. Ishihara, verbal communication, 1978). The I-type also contains more plagioclase, less quartz, and commonly less than 66% SiO_2 in contrast to S-type granitic rocks. Using the mineralogical criteria in particular (Fig. 2), quartz diorite and tonalite of the western border of the Atlanta and Bitterroot lobes are I-type granitic rocks, as are the White Sand Creek stock (loc. 14), Skookum Butte stock (loc. 15), Racetrack pluton (loc. 25), Mount Torrey batholith (loc. 26a), Philipsburg batholith (locs. 27-28), Garnet stock (loc. 33), and the Boulder batholith (locs. 35-37). In contrast, the bulk of both the Atlanta and Bitterroot lobes consists of S-type granitic rocks, as do the Royal stock (loc. 29), Mount Powell batholith (loc. 30), and Willow Creek stock (loc. 31), which lie between the Bitterroot lobe and the Boulder batholith. A few plutons are ambiguous using these criteria, and the high Na_2O/K_2O of many of the S-type units of the Idaho batholith does not fit the distinction used by Hine and others (1978, Fig. 3).

An independent evaluation of igneous versus sedimentary source rocks for the Idaho batholith and related plutons is not generally available. It seems likely, however, that the I-type tonalitic rocks of the western border zone of the Idaho batholith are derived by partial melting either of oceanic lithosphere in the vicinity of a subduction zone or of metavolcanic rocks along the continental margin. The dominantly S-type granitic rocks to the east are emplaced into old rocks of the continental margin, including Belt metasedimentary rocks and pre-Belt basement rocks, which, in significant part at least, appear to be orthogneiss. The main I/S boundary, then, comes close to the main suture between oceanic crust on the west and continental crust on the east. If the orthogneisses are representative of much of

the source for the bulk of the Idaho batholith, the igneous versus sedimentary distinction may not hold in this region.

I-type granitic rocks forming satellitic plutons and the Boulder batholith clearly lie on the continental side of the oceanic/continental suture. They include satellites of the northeastern Bitterroot lobe, Boulder and Mount Torrey (loc. 26a) batholiths, and many plutons of the intervening Sapphire tectonic block. It is not independently clear that these plutons have igneous source rocks. It seems possible that the I/S-type distinction is not a universal criterion for igneous versus sedimentary source rocks. However, it may still reflect a difference in composition of the source rocks.

Ishihara (1977) suggests that magnetite-series granitic rocks were generated at deep levels of the upper mantle and lowest crust and have not interacted with C-bearing minerals, whereas lower oxygen-fugacity ilmenite-series granitic rocks were generated in the middle to lower continental crust and mixed with C-bearing metamorphic and sedimentary rocks at various stages. That the I-type and abundant magnetite-series granitic rocks are concentrated on the oceanic side of the inferred ocean-continent suture, in rocks having characteristics compatible with subduction, suggests that they were generated in the lowermost crust or upper mantle. The more-felsic compositions, S-type and presumably ilmenite-series characteristics, and continental-type country rocks of the bulk of the Atlanta and Bitterroot lobes of the Idaho batholith suggest derivation from the continental crust, in agreement with Ishihara's proposal.

Strontium-isotope ratios also provide some indication of source rock for the batholith. Low $^{87}Sr/^{86}Sr$ ratios ranging from 0.70362 to 0.70406 are confined to the tonalitic or quartz dioritic rocks of the western border zone of the batholith (Armstrong, Taubeneck, and Hales, 1977). They obtained a $^{87}Sr/^{86}Sr$ initial ratio of 0.70356 on the Council Mountain quartz diorite near McCall, Idaho, and an age of about 70-100 m.y. suggests that the initial ratio for the other samples is comparably low. Such low ratios indicate probable derivation from subducted oceanic crust or a mantle source without much contamination from old radiogenic continental crust. Or the source rock could be mafic-rich continental-margin rocks such as the Seven Devils volcanics, themselves recently derived from oceanic crust or mantle. However, not all of the tonalitic rocks have low ratios. The Donnelly quartz diorite, a few kilometers southeast of the Council Mountain quartz diorite, has ratios of 0.70848 to 0.70890 and an initial ratio of 0.70825. Farther east, the granodiorite and granite of the main body of the Idaho batholith have $^{87}Sr/^{86}Sr$ ratios ranging from 0.70702 to 0.71336 (Armstrong and others, 1977; Chase and others, 1978). The single available $^{87}Sr/^{86}Sr$ initial ratio of 0.7103 from the main body is from the northeastern corner of the Bitterroot Lobe of the batholith (Chase, Bickford, and Tripp, 1978). These higher ratios of greater than 0.706 from the main body of the batholith support derivation from old

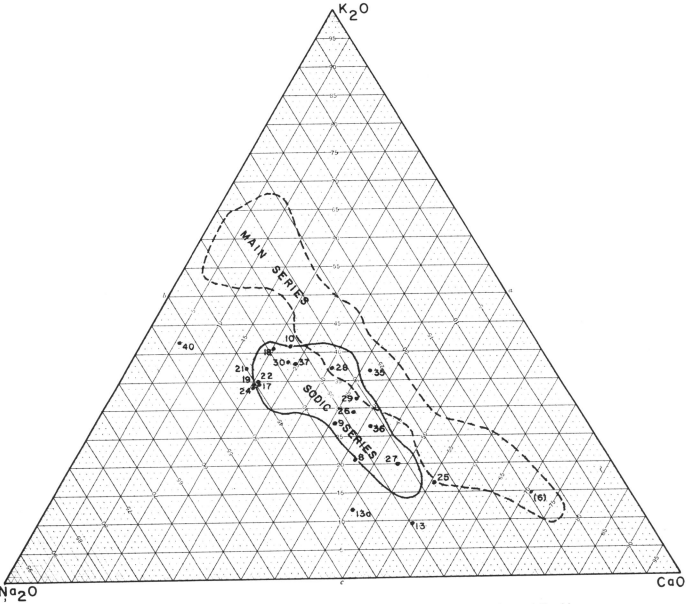

Figure 6. K_2O-Na_2O-CaO ternary diagram for averages of units of the Idaho and Boulder batholiths with outlines of Tilling's (1973) "main series" and "sodic series" for the Boulder batholith.

continental crust. The anomalously high initial ratio from the Council Mountain quartz diorite could have been influenced by contamination by old continental crust into which it is apparently emplaced, as suggested by Armstrong and others (1977).

The pink epizonal plutons of Tertiary age have I-type characteristics. Particularly distinctive is the higher oxidation state exhibited by the common pink coloration of the potassium fledspar. Hornblende in rocks of the Sheepeater Peak area is also indicative, but muscovite in the Bungalow pluton suggests that some S-type rocks may be included in the group.

METALLIC MINERAL DEPOSITS AND GRANITE TYPE

The characteristics of I-type (magnetite-series) granitic rocks compared with S-type (ilmenite-series) granitic rocks outlined above suggest a correlation with copper-molybdenum versus tin-tungsten mineral deposits, respectively, as suggested by Ishihara (1977). Little work has been done on the Idaho batholith with such a correlation in mind, and an attempt using published information shows only moderate correspondence. Most significant mineral occurrences in Idaho have been compiled in Idaho Bureau of Mines and

Geology Special Report No. 1 (1964), and those in western Montana have been compiled by Lange (1977). Excluding the western tonalitic border, for which metallic mineral deposits are not recorded, the Atlanta lobe shows dominantly S-type mineralogy and some W mineralization (units 9, 10). But it also contains Cu-Mo deposits (units 10, 11). I-type plutons east of the Bitterroot lobe (units 26, 27, 28, 33) show strong aeromagnetic highs (Johnson and others, 1965) and presumably contain magnetite; they also contain Cu and some Mo mineralization. S-type plutons (units 29, 30, 31) show very modest or no magnetic high, probably contain little or no magnetite, and unit 30 contains some Sn-W mineralization, but at least units 29 and 30 contain Cu mineralization. The I-type Boulder batholith is magnetically moderate to very strong and contains primarily Cu-Mo mineralization as expected, but the Butte quartz monzonite (unit 35) also contains a little tungsten mineralization. The Atlanta lobe, Boulder batholith, and plutons east of the Bitterroot lobe also contain deposits of Au, Ag, Pb, Zn, Sb, and U, with no apparent correlation to type of granitic rock. The poor degree of correlation may indicate that such correlation is not universal or that the lack of recognition of separate plutons, reconnaissance nature of mineral and chemical studies, and generalized, rather than specific, correlation between mineralization and nature of the directly associated granitic rocks blur the distinction between granitic rock types in the Atlanta lobe and associated plutons.

The absence of metallic mineral deposits recorded for either I- or S-type granitic rocks of the Bitterroot lobe is in striking contrast to other plutons of the region. There seems to be little bulk mineralogical or chemical reason for this absence, but the Bitterroot lobe and adjacent Atlanta lobe across the Salmon River arch appear to have been emplaced at deeper levels and in higher-grade country rocks than other plutons of the region. It may be that the relatively incompatible elements that form metallic mineral deposits have migrated to higher levels than those exposed in most of the Bitterroot lobe. The original upper parts of the eastern half of the lobe over the present Bitterroot dome slid eastward to form the Sapphire tectonic block with its shallower granitic plutons and low-grade country rocks; the upper parts of the western half of the lobe may have moved northwestward.

The Eocene plutons are characterized by much greater background radiation than the Cretaceous plutons, apparently contributed by K^{40}, U, and Th. Many of these plutons are associated with disseminated molybdenum, gold, tungsten, copper, uranium, and in one case, tin (Bennett, 1980).

STRUCTURES RELATED TO THE IDAHO BATHOLITH

The regional distribution of Precambrian country-rock units of at least the northern Idaho batholith appears to parallel the northwestward trend of the Salmon River arch. The northern or Bitterroot lobe of the batholith approximately follows this orientation, possibly lying in a complex synclinorium (Hyndman and Williams, 1977; Hyndman, 1979). The east-southeastward extension of the Bitterroot lobe (Figs. 1, 7) may have been in large part carried or injected eastward with the Sapphire tectonic block described by Hyndman and others (1975) and Hyndman (1980). The Bitterroot lobe, during its emplacement, may have been oriented northwestward parallel to the major country-rock trend. As noted above, the country rocks of the Atlanta lobe are buried to a major extent by younger volcanic rocks and,

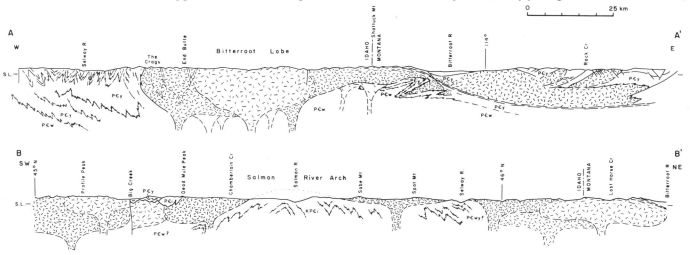

Figure 7. Cross-sections of the Idaho batholith and its country rocks with interpretation to the base of the bathlith. No vertical exaggeration. A-A': Across the Bitterroot Lobe of the Idaho batholith. B-B': Across the Salmon River Arch. Ki = Cretaceous granitic rocks; KP€i = orthogneiss, age uncertain; P€i = Precambrian igneous rocks; P€y = Late Precambrian metasedimentary rocks; P€w = Early Precambrian metamorphic rocks.

especially on the east, disrupted by thrust faults and folds. But the trends of Phanerozoic country-rock units, thrust faults, and post-batholith block-fault-controlled valleys are northwestward and suggest a dominant northwestward structural control for the Atlanta lobe as well. The lack of a northwestward orientation of features west of the Atlanta lobe is probably related to the major suture which juxtaposed the Seven Devils - Blue Mountains block, a presumed part of the "Wrangellia terrane."

Several stages of mesoscopic structures are found in the border zones of the Idaho batholith. The first stage, F_1, associated with moderate-to-high-grade regional dynamothermal metamorphism of Belt Supergroup rocks is locally preserved as similar-style isoclinal folds in bedding and metamorphic schistosity parallel to the axial planes of the folds. A second stage of deformation folded the schistosity, formed during F_1, in a moderately tight style, showing some characteristics of both similar and concentric folds. In some areas, the deformed micas of the first recognizable post-regional-metamorphism set of folds are themselves somewhat recrystallized with development of a new but weaker axial-plane schistosity. In most areas, at least two sets of post-metamorphic folds are recognized, but correlation between different areas is not yet documented. These fold events are reviewed by Greenwood and Morrison (1973), Chase (1973), Chase and Johnson (1977), Wiswall (1979), and Reid and others (1979).

Veins in amphibolite-facies regional metamorphic rocks are dominantly granitic and plot near the ternary minimum in the granite system (Chase and Johnson, 1977). Most are discordant, dilational, and probably injected. About five percent of the veins have mafic selvages and host-rock compositions appropriate to formation by anatexis or metamorphic deformation. Veins in greenschist facies regional metamorphic rocks farther from the batholith are nearly pure quartz or calcite, depending on the composition of the host rock. The veins are tabular, sharply bounded, and dilational and probably result from metamorphic differentiation.

Thrust faults west of the Idaho batholith dip eastward towards the batholith (Hamilton, 1963, 1969; Vallier, 1974). They may be related to pre-batholith subduction along the continental margin (Talbot and Hyndman, 1975; Hamilton, 1976; Hyndman, 1979) or to uplift and westward spreading of the Idaho batholith area (Scholten and Onasch, 1977).

North of the batholith, east- and east-southeast-trending faults and thrust-fault traces are complicated by complex stratigraphic variations (Harrison, 1972; Harrison and others, 1974a). But northward-asymmetric isoclinal folds (Reid, field trip, Belt Symposium, 1974) and flat thrust faults in country rocks (Harrison and others, 1974b) in the southwestern border zone of the Bitterroot lobe indicate mushrooming of the Idaho batholith infrastructure into its overlying suprastructure and gravitational spreading north-

ward and eastward (Hyndman and others, 1975; Talbot, 1977). Slickenside striae plunging northwestward on shear surfaces and thin mylonitic zones dipping gently northward and northwestward in the western part of the Bitterroot lobe suggest northwestward gravitational unroofing of this part of the lobe (Hyndman, unpubl. data). Emplacement of pegmatitic sheets along these shears and development of slickensides on the surface of these sheets suggest that much gravitational unroofing occurred during emplacement of granitic magmas of the Idaho batholith.

Eastward unroofing of the Bitterroot lobe of the Idaho batholith to form the suprastructural Sapphire tectonic block apparently occurred about 75 to 80 million years ago during late stages of consolidation of the Idaho batholith (Hyndman and others, 1975; Hyndman, 1980). This block, about 100 kilometers long, 70 kilometers wide, and 15 kilometers thick, apparently moved eastward about 60 kilometers across a prominent mylonitic "detachment" zone marking the eastern flank of the Bitterroot dome. The dome probably rose isostatically in response to unloading of the Sapphire tectonic block (Hyndman, 1980). Immediately south of the Sapphire tectonic block are flat thrust faults (Fraser and Waldrop, 1972) of the Pioneer thrust block which, although still poorly defined, also probably moved eastward from the Idaho batholith. A detachment surface for these thrusts has not been recognized.

Uplift and shedding of detritus from the Idaho batholith area during Cretaceous and Early Tertiary time occurred at about the time of eastward thrust faulting east of the Atlanta lobe of the batholith (Scholten, 1968; Ryder and Scholten, 1973; Scholten and Onasch, 1977). The temporal and regional association between these events suggests a cause and effect relationship between them. Scholten and co-workers suggest gravitational gliding away from the rising batholith. Lithologies of clasts in the Beaverhead Formation indicate that, by mid-Cretaceous time, about 12 kilometers of Paleozoic and late Precambrian Belt sedimentary rocks had been eroded from the Idaho batholith area (Scholten and Onasch, 1977).

METAMORPHISM AROUND THE IDAHO BATHOLITH

A broad passive burial metamorphism to the lower amphibolite facies affected Prichard Formation sedimentary rocks of the Belt Supergroup over much of northern Idaho and northwestern Montana, probably in Precambrian time (Norwick, 1972, 1977). Stratigraphic units overlying the Prichard are successively lower grade. Rocks marginal to the Bitterroot lobe are foliated and increase in grade toward the batholith to the sillimanite zone. This superimposed, high-grade, dynamothermal metamorphism is thought to be Jurassic or Cretaceous in age (Nold, 1974; Hyndman and Williams, 1977) because regional metamorphism, thought to be the same event, affects the Late Triassic Martin Bridge and Lucille formations north of Riggins, Idaho (Hamilton, 1963; Onasch, 1978) and because of a presumed genetic relationship between the regional metamorphism and generation of the Idaho batholith. Although the high-grade, regional dynamothermal metamorphism is crudely concentric to the Idaho batholith, the granite cuts across the metamorphic isograds over tens of kilometers (Nold, 1968; Hyndman and Williams, 1977). This discordance may be at low angle as viewed in three dimensions (Chase and Johnson, 1977). Regional metamorphic assemblages containing kyanite and, nearby, sillimanite surround the Bitterroot Lobe of the batholith to the southwest, northwest, and northeast (Reid and others, 1979; Hietanen, 1962; Wehrenberg, 1972). This suggests a considerable depth of metamorphism just prior to emplacement of the northern Idaho batholith. Diapiric rise of the batholith during emplacement probably dragged upward adjacent parts of the regional metamorphic environment, leaving the highest grade rocks spatially associated with borders of the batholith.

BASE OF THE IDAHO BATHOLITH

Regional metamorphism has raised the pre-batholithic rocks between the Atlanta and Bitterroot lobes to the sillimanite grade. According to Hamilton (1978), the general progression from granitic rocks in upland areas, through gneissic granites, to gneiss and migmatite in the deep canyons in this region suggests exposure of Shuswap-type metamorphic domes beneath the Idaho batholith (Hamilton, 1978). Hamilton was probably referring to the canyons in the Idaho Primitive area south of Salmon River, where he was mapping several years ago (Carter and others, 1973). Detailed mapping in the south-central part of the Bitterroot lobe north of the Salmon River demonstrates that the base of the Idaho batholith over a sizeable area is sharply bounded and nearly horizontal to gently undulating. Wiswall (1979) mapped the lower contact up a number of canyons tributary to the upper Selway River and demonstrated that a steeper contact under the Bitterroot lobe rises and flattens southwestward towards the Salmon River arch. Across the Bitterroot dome to the northeast, Chase (1977) has mapped a comparable near-horizontal base of the batholith over several kilometers. Thus, the southwestern and northeastern flanks of the Bitterroot dome apparently mushroomed during intrusion.

The general distribution of mapped areas of granite on the ridges overlying high-grade, regional metamorphic rocks over a 50-to-60-kilometer span on either side of the Salmon River arch and lack of prominent areas of granodiorite or tonalite near the contact suggest that the Idaho batholith may have been originally continuous completely across the Salmon River arch (see Fig. 7). Country rocks probably

overlying granite between the Bitterroot and Atlanta lobes are most obvious along the large canyons of the Selway River, Salmon River, and Big Creek (tributary to the Middle Fork of the Salmon River), all of which trend west or northwest generally parallel to the Salmon River arch and to the most prominent fold trend as outlined above. Orthogneiss probably belonging to pre-Belt basement is most prominently exposed along the Salmon River arch, along the upper Selway River valley, and northwest of the Bitterroot lobe (Hyndman and Williams, 1977), all areas probably under the base of the batholith at one time. The contact along the northwestern end of the Bitterroot lobe also dips 25 to 50 degrees eastward under the batholith (Hietanen, 1962). The only other large exposure of probable pre-Belt basement is along the canyon of the Lochsa River on the north-central border of the Bitterroot lobe. There also it is flanked by granite on the adjacent ridges. Such widespread exposures of probable pre-Belt basement adjacent to the northern half of the Idaho batholith suggest that it underlies all of the northern half of the batholith, at least, and may represent some of the source rocks from which the granitic magmas were generated.

DEPTH OF BATHOLITH EMPLACEMENT

The high metamorphic grade of the country rocks adjacent to the Bitterroot lobe of the Idaho batholith suggests that the batholith was emplaced at considerable depth, perhaps deeper than either the Coast Plutonic Complex of British Columbia or the Sierra Nevada batholith of California. Regional metamorphism of probable Jurassic or Cretaceous age has affected sedimentary country rocks to the sillimanite zone over broad areas — primarily the area between the Bitterroot and Atlanta lobes, and large areas at the northwestern end and northeastern corner of the Bitterroot Lobe. At least locally, in all three areas, the rocks contain kyanite. Migmatites, some of which appear anatectic, suggest high temperatures along with the required high pressures. The inferred pressure depends on the choice of conditions for the kyanite-sillimanite-andalusite invariant point, but 3.5 kbar or 13.3 kilometers from the data of Richardson and others (1969) and Holdaway (1971) appear most reasonable. Other experimental studies mostly infer somewhat higher pressures. At temperatures near the minimum-melting conditions for water-saturated granite, the kyanite-sillimanite boundary lies at about 6.5-7 kbar or 22-23 kilometers depth. At the minimum temperatures for water-undersaturated melts, assuming that all of the water is supplied by the breakdown of muscovite in the metamorphic rocks, the kyanite-sillimanite boundary lies at about 7.5 kbar or 25 kilometers depth (Hyndman, 1981).

Many or most of the sillimanite- and kyanite-bearing regional metamorphic rocks appear to be in the Prichard Formation of the Belt Supergroup. Summation of the stratigraphic section probably overlying the rocks at the time of metamorphism suggests between 13 and 21 kilometers of Belt section and perhaps an additional 5 kilometers of Paleozoic-Mesozoic section, unless the latter was removed prior to metamorphism or never deposited in the area of the batholith (Hyndman, 1980).

An estimate of depth of crystallization from within the granite itself is available from the widespread presence of minor muscovite with quartz and feldspars. Muscovite can crystallize directly from magma only at pressures greater than that for the intersection of the muscovite + quartz + Na-rich plagioclase-breakdown curve with the granite minimum-melting-temperature curve or at depths from about 5 to 17 kilometers (Hyndman, 1981). Muscovite in the Bitterroot Lobe is almost as coarse as the biotite, but much is secondary and replaces earlier-formed grains. It may be deuteric rather than magmatic. At near-magmatic conditions, its temperature of crystallization would still require pressures near the 5 kbar or 17 kilometers inferred above.

If the granitic magma of the Bitterroot Lobe rose diapirically through the metamorphic stratigraphic section, dragging deeper country rocks upward along its border zones, as seems to be the case, the depth estimates for granite emplacement based on those rocks may be too great. The depths inferred from the presence of muscovite in the granite itself should be approximately correct and are compatible with estimates from the other sources. Allowing for a modest degree of uncertainty in the data, it appears that the Bitterroot Lobe of the Idaho batholith crystallized at a depth of perhaps 15 to 20 kilometers in the continental crust. If partial melting to form the magma was initiated by release of water during the breakdown of muscovite with quartz and feldspar in the source area, the source of the magma was probably only a little deeper (Hyndman, 1981).

In contrast, the Eocene granite plutons, characterized by the presence of miarolitic cavities and spatially associated with penecontemporaneous volcanic ejecta, must have been emplaced at shallow depths — no more than 2-3 kilometers. Considerable removal of upper crustal rocks must have occurred in the period between latest Cretaceous emplacement of the Idaho batholith and about 50 m.y. ago when the Eocene plutons crystallized.

SUMMARY: TWO MAGMA SOURCES

Tonalitic rocks of the western side of the Idaho batholith may be slightly older and are compositionally and, to some extent, texturally distinct from granodiorites and granites of the main body of the batholith. Reasonable radiometric ages from the western tonalitic rocks of both lobes of the batholith range from 65 to 117 million years. Those from the granodiorite-granite main phase largely overlap, ranging from 62.6 to 95 million years.

Chemically (see Table 2a), the tonalitic rocks are rather similar to average andesite or diorite, except for slightly lower Al_2O_3 and total iron and except for lower K_2O of about two-thirds of the average value. In contrast to tonalitic rocks in some regions, they are much higher in SiO_2 and are much more evolved than tholeiitic basalts. Compared with the average granodiorite, that of the main body of the Idaho batholith contains more Na_2O, less CaO and MgO, and slightly less total iron. Compared with the average granite or adamellite, that of the Idaho batholith contains much more Na_2O and somewhat less MgO and total iron.

$^{87}Sr/^{86}Sr$ ratios are less than 0.704 in most of the tonalitic rocks of the western border zone of the batholith, suggesting derivation from subducted oceanic crust or upper mantle or from young continental margin rocks such as the Seven Devils volcanics, themselves recently derived from such sources. $^{87}Sr/^{86}Sr$ ratios from the granodiorite-granite main mass of the batholith are mostly in the range 0.707 to 0.711, suggesting derivation from older continental crust. The boundary between the two isotopic areas lies essentially, but not precisely, on the boundary between the western tonalitic rocks and the more-felsic main mass of the batholith.

The combination of slightly older tonalitic rocks having I-type characteristics similar in composition to average andesite, and having low $^{87}Sr/^{86}Sr$, suggests derivation by partial melting of subducted oceanic crust or of Phanerozoic basaltic to andesitic rocks of the continental margin. The combination of slightly younger sodic granodioritic to granitic rocks of the main mass of the batholith, having S-type characteristics, and having higher $^{87}Sr/^{86}Sr$, suggests derivation by partial melting of Precambrian continental basement rocks consisting largely of granitoid orthogneisses and possibly of some late Precambrian pelitic metasedimentary rocks of the Belt Supergroup. As noted by Armstrong and others (1977), the low $^{87}Sr/^{86}Sr$ ratios correlate well spatially with the occurrence of Paleozoic or Mesozoic eugeoclinal rocks, whereas the higher ratios correlate with known or inferred Precambrian rocks.

ACKNOWLEDGMENTS

I thank G. Wiswall, L. D. Williams, D. Alt, R. Chase, J. L. Talbot, and others for discussions of various topics related to the Idaho batholith and J. P. Wehrenberg for review of the manuscript. National Science Foundation grant EAR 76-84399 supported research on the Bitterroot lobe of the batholith.

REFERENCES CITED

Allen, J. C., Jr., 1966, Structure and petrology of the Royal stock, Flint Creek Range, central-western Montana: Geological Society of America Bull. vol 77, p. 291–302.

Alt, D., and Hyndman, D. W., 1976, Butte Sheet, 1:250,000: Montana Bureau of Mines and Geology, open-file geological map and index sheet.

Anderson, A. L., 1942, Endomorphism of the Idaho batholith: Geological Society of America Bulletin, vol. 53, p. 1099–1126.

——,1952, Multiple emplacement of the Idaho batholith: Journal of Geology, vol. 60, p. 255–265.

——,and Rasor, A. C., 1934, Composition of a part of the Idaho batholith in Boise County, Idaho: American Journal of Science, vol. 277, p. 287–294.

Armstrong, R. L., 1974, Geochronometry of the Eocene Volcanic-plutonic episode in Idaho: Northwest Geology, vol. 3, p. 1–15.

——,1975b, The Geochronometry of Idaho: Isochron/West, no. 14, p. 1–50.

——,1975a, PreCambrian (1500 m.y. old) rocks of Central Idaho - The Salmon River Arch and its role in Cordilleran sedimentation and tectonics: American Journal of Science, vol. 275-A, p. 437–467.

——,Taubeneck, W. H., and Hales, P. O., 1977, Rb-Sr and K-Ar geochronometry of Mesozoic granitic rocks and their Sr isotopic composition, Oregon, Washington, and Idaho: Geological Society of America Bulletin, vol. 88, p. 397–411.

Bateman, P. C., and Chappell, B. W., 1978, Crystallization, fractionation, and solidification of the Tuolumne Intrusive Series, Yosemite National Park, California: manuscript.

——,Clark, L. D., Huber, N. K., Moore, J. G., and Rinehart, C. D., 1963, The Sierra Nevada Batholith - A synthesis of Recent Work across the Central Part, U.S. Geological Survey Professional Paper 414-D, p. D1–D46.

Becraft, G. B., Pinckney, D. M., and Rosenblum, S., 1963, Geology and mineral deposits of the Jefferson City quadrangle, Jefferson and Lewis and Clark Counties, Montana: U. S. Geological Survey Professional Paper 428, 101 p.

Bennett, E. H., 1980, Granitic rocks of Tertiary age in the Idaho batholith and their relation to mineralization: Economic Geology, vol. 75, p. 277–288.

Benoit, W. R., 1972, Vertical zoning and differentiation in granitic rocks — central Flint Creek Range, Montana: M. S. thesis, University of Montana, Missoula, 53 p.

Bond, J. G., 1978, Geologic map of Idaho: Idaho Department of Lands, Bureau of Mines and Geology, Moscow, Idaho, 1:500,000.

Brenner, R. L., 1968, The geology of Lubrecht Experimental Forest: Lubrecht Series One, Montana Forest and Conservation Experiment Station, University of Montana, Missoula, 71 p.

Cater, F. W., Pinckney, D. M., Hamilton, W. B., Parker, R. L., Weldin, R. D., Close, T. J., and Zilka, N. T., 1973, Mineral Resources of the Idaho Primitive Area and Vicinity, Idaho: U. S. Geological Survey Bulletin 1304, 431 p.

Chappell, B. W., and White, A. J. R., 1974, Two contrasting granite types: Pacific Geology, vol. 8, p. 173–174.

Chase, R. L., 1968, Petrology of the Northeast Border of the Idaho batholith: Ph.D. dissertation, University of Montana, Missoula, 189 p.

——,1973, Petrology of the northeast Border Zone of the Idaho batholith: Montana Bureau of Mines and Geology, Memoir 43, 28 p.

——,1977, Structural evolution of the Bitterroot dome and zone of cataclasis, p. 1–24 *in* Mylonite detachment zone, eastern flank of Idaho batholith: Field Guide No. 1, Geol. Soc. America Rocky Mountain Sec., Ann. Mtg., Missoula, Montana.

——,Bickford, M. E., and Tripp, S. E., 1978, Rb-Sr and U-Pb isotopic studies of the northeastern Idaho batholith and border zone: Geological Society of America Bulletin, vol. 89, p. 1325–1334.

——,and Johnson, B. R., 1977, Border-zone relationships of the northern Idaho batholith: Northwest Geology, vol. 6-1, p. 38–50.

——,and Talbot, J.L., 1973, Structural evolution of the northeastern border zone of the Idaho batholith, western Montana: Geological Society of America Abstr. with Progr., vol. 5, no. 6, p. 471–472.

Criss, R. E., and Taylor, H. P., Jr., 1978, Regional $^{18}O/^{16}O$ and D/H

variations in granitic rocks of the southern half of the Idaho batholith and the dimensions of the giant hydrothermal systems associated with emplacement of the Eocene Sawtooth and Rocky Bar plutons: Geological Society of America, Abstr. with Progr., vol. 10, no. 7, p. 384.

Doe, B. R., Tilling, R. I., Hedge, C. E., and Klepper M. R., 1968, Lead and strontium isotope studies of the Boulder batholith, southwestern Montana: Economic Geology, vol. 63, p. 884–906.

Ferguson, J. A., 1975, Tectonic implications of some geochronometric data from the northeastern border zone of the Idaho batholith: Northwest Geology, vol. 4, p. 53–58.

Fraser, G. D., and Waldrop, H. A., 1972: Geologic map of the Wise River quadrangle, Silver Bow and Beaverhead Counties Montana: U. S. Geological Survey Map GQ 988.

Grauert, B., and Hofmann, A., 1973, Old radiogenic lead components in zircons from the Idaho batholith and its metasedimentary aureole: Carnegie Inst. Washington Yearbook, no. 72, p. 297–299.

Greenwood, W. R., and Morrison, D. A., 1973, Reconnaissance geology of the Selway-Bitterroot wilderness area: Idaho Bureau of Mines and Geology, Pamphlet 154, 30 p.

Hamilton, W., 1962, Trondhjemite in the Riggins quadrangle, western Idaho: U. S. Geological Survey, Professional Paper 450-E, p. 98–101.

——,1963, Metamorphism in the Riggins region, western Idaho: U. S. Geological Survey Professional Paper 436, 95 p.

——,1969, Reconnaissance geologic map of the Riggins quadrangle, west-central Idaho: U. S. Geological Survey, Misc. Geological Investigations Map I-579.

——,1976, Tectonic history of west-central Idaho: Geological Society of America, Abstr. with Progr., vol. 8, p. 387–379.

Hamilton, W., 1978, Mesozoic tectonics of the western United States: *in* Mesozoic paleogeography of the western United States, D. G. Howell and K. A. McDougall, editors, Pacific Section, Society of Economic Paleontologists and Mineralogists, Pacific Coast Paleogeography Symposium 2, p. 33–70.

——,and Myers, W. B., 1967, The nature of batholiths: U. S. Geological Survey Professional Paper 554-C, p. C1–C30.

Harrison, J. E., 1972, Precambrian Belt basin of northwestern United States: Its geometry, sedimentation, and copper occurrences: Geological Society of America, Bulletin, vol. 83, p. 1215–1240.

——,Griggs, A. B., and Wells, J. D., 1974a, Tectonic features of the Precambrian Belt basin and their influence on post-Belt structures: U.S. Geological Survey, Professional Paper 866, p. 1–15.

——,1974b, Preliminary Geologic Map of part of the Wallace 1:250,000 sheet, Idaho-Montana: U. S. Geological Survey open-file map OF-74-37.

Hawley, K. T., 1975, The Racetrack Pluton — A newly defined Flint Creek Pluton: Northwest Geology, vol. 4, p. 1–8.

Hietanen, A., 1962, Metasomatic metamorphism in western Clearwater County, Idaho: U. S. Geological Survey, Professional Paper 344-A, 116 p.

——,1963, Idaho batholith near Pierce and Bungalow, Clearwater County, Idaho: U. S. Geological Survey, Professional Paper 344-D, 42 p.

Hillhouse, J., 1977, Paleomagnetism of the Triassic Nikolai Greenstone, south-central Alaska: Canadian Journal of Earth Science, vol. 14, p. 2578–2592.

Hine, R., Williams, I. S., Chappell, B. W., and White, A. J. R., 1978, Contrasts between I- and S-type granitoids of the Kosciusko batholith: Journal of the Geological Society of Australia, vol. 25, p. 219–234.

Holdaway, M. J., 1971, Stability of andalusite and the aluminum silicate phase diagram: American Journal of Science, vol. 271, p. 97–131.

Hughes, G. H., Jr., 1975, Relationship of igneous rocks to structure in the Henderson-Willow Creek igneous belt, Montana: Northwest Geology, vol. 4, p. 15–25.

Hyndman, D. W., 1979, Major tectonic elements and tectonic problems along the line of section from northeastern Oregon to west-central Montana: Geological Society of America Map and Chart Series MC-28C, 11 p. + 1:250,000 strip map and cross section.

——,1980, Bitterroot dome - Sapphire tectonic block, an example of a plutonic-core gneiss-dome complex with its detached suprastructure, *in* Cordilleran Metamorphic Core Complexes: P. J. Coney, Max Crittenden, Jr., and G. H. Davis, eds., Geological Society of America Memoir, 153, p. 427-443.

——,1981, Controls on source and depth of emplacement of granitic magma: Geology, vol. 9, in press.

——,Obradovich, J. D., and Ehinger, R., 1972, Potassium-argon age determinations of the Philipsburg batholith: Geological Society of America Bulletin, vol. 83, p. 473-474.

——,Talbot, J. L., and Chase, R. B., 1975, Boulder batholith: A result of emplacement of a block detached from the Idaho batholith infrastructure?: Geology, vol. 3, p. 401-404.

——,Silverman, A. J., Ehinger, R., Benoit, W. R., and Wold, R., 1982, The Philipsburg batholith, western Montana: Mineralogy, petrology, internal variation, and evolution: Montana Bureau Mines and Geology, Memoir 49, 37 p.

Hyndman, D. W., and Williams, L. D., 1977, The Bitterroot lobe of the Idaho batholith: Northwest Geology, vol. 6-1, p. 1-16.

Ishihara, S., 1977, The magnetite-series and ilmenite-series granitic rocks: Mining Geology, vol. 27, p. 293-305.

Jones, D. L., Silberling, N. J., and Hillhouse, J., 1977, Wrangellia — A displaced terrane in northwestern North America: Canadian Journal of Earth Science, vol. 14, p. 2565-2577.

Johnson, R. W., Jr., Henderson, J. R., and Tyson, N. S., 1965, Aeromagnetic map of the Boulder batholith area, southwestern Montana: U. S. Geological Survey, Geophysical investigations map GP-538.

Klepper, M. R., Robinson, G. D., and Smedes, H. W., 1971, On the nature of the Boulder batholith of Montana: Geol. Soc. America Bull., vol. 82, p. 1563-1580.

——,Weeks, R. A., and Ruppel, E. T., 1957, Geology of the southern Elkhorn Mountains, Jefferson and Broadwater Counties, Montana: U. S. Geological Survey Professional Paper 292, 82 p.

Knopf, A., 1957, The Boulder bathylith of Montana: American Journal of Science, vol. 255, p. 81-103.

Lange, I. M., 1977, Metallic mineral deposits of western Montana: openfile map, Montana Bureau of Mines and Geology, Butte, Montana, 1:500,000.

Larsen, E. S. and Schmidt, R. G., 1958, A reconnaissance of the Idaho batholith and comparison with the southern California batholith: U. S. Geological Survey, Bull. 1070A, p. 1-33.

Lemoine, S., Enterline, T., Alt, D., and Hyndman, D. W., 1978, Dillon Sheet 1:250,000: Montana Bur. Mines and Geol., open-file geological map and index sheet.

Lindgren, W., 1900, The gold and silver veins of Silver City, DeLamar, and other mining districts in Idaho: U. S. Geol. Survey 20th Ann. Rept., pt. 3-B, p. 65-256.

Mathews, G. W., McClain, L. I., and Johanns, W. M., 1977, Petrogenetic aspects of the Hell Canyon pluton and its relation to the Boulder batholith, southwestern Montana: Northwest Geology, vol. 6-2, p. 77-84.

Nold, J. L., 1968, Geology of the northeastern border zone of the Idaho batholith, Montana and Idaho: Ph.D. dissertation, University of Montana, Missoula, 159 p.

——,1974, Geology of the northeastern border zone of the Idaho batholith: Northwest Geology, vol. 3, p. 47-52.

Norwick, S. A., 1972, The regional Precambrian metamorphic facies of the Prichard Formation of western Montana and northern Idaho: unpubl. Ph.D., dissertation, University of Montana, Missoula, 129 p.

——,1977, Precambrian amphibolite facies metamorphism in the Belt rocks of northern Idaho: Geological Society of America Progr. with Abstr., vol. 9, no. 6, p. 753.

Olson, H. J., 1968, The geology and tectonics of the Idaho porphyry belt from the Boise Basin to the Casto Quadrangle: Unpubl. Ph.D. dissert., Univ. of Arizona, 154 p.

Onasch, C. M., 1978, Multiple folding along the western margin of the Idaho batholith in the Riggins, Idaho area: Northwest Geology, vol. 7, p. 34-38.

Otto, B. R., 1978, Structure and petrology of the Sheepeater Peak area, Idaho Primitive Area, Idaho: unpubl. M. S. Thesis, University of Montana, Missoula, 68 p.

Pederson, R. J., 1976, Geology of the Upper Rock Creek drainage, Granite County, Montana: M. S. thesis, Montana College of Mineral Science and Technology, Butte, 237 p.

Presley, M. W., 1970, Igneous and metamorphic geology of the Willow Creek drainage basin, southern Sapphire Mountains, Montana: M. S. thesis, University of Montana, Missoula, 64 p.

Reid, R. R., 1963, Reconnaissance geology of the Sawtooth Range: Idaho Bureau of Mines and Geology Pamph. 129, 37 p.

Reid, R. R., Bittner, E., Greenwood, W. R., Ludington, S., Lund, K., Motzer, W., and Toth, M., 1979, Geologic section and road log across the Idaho batholith: Idaho Bureau Mines and Geology Inf. Circ. 34, 20 p.

——,Morrison, D. A., and Greenwood, W. R., 1973, The Clearwater orogenic zone: a relic of Proterozoic orogeny in central and northern Idaho: Belt Symposium, Dept. of Geology, University of Idaho, Moscow, p. 10-56.

Richardson, S. W., Gilbert, M. C., and Bell, P. M., 1969, Experimental determination of kyanite-andalusite and andalusite-sillimanite equilibria; the aluminum silicate triple point: American Journal of Science, vol. 267, p. 259-272.

Ross, C. P., 1934, Geology and ore deposits of the Casto quadrangle, Idaho: U. S. Geological Survey Bull. 854, 135 p.

Rupple, E. T., 1963, Geology of the Basin quadrangle, Jefferson, Lewis and Clark, and Powell Counties, Montana: U. S. Geological Survey Bull. 1151, 121 p.

Ryder, R. T., and Scholten, R., 1973, Syntectonic conglomerates in southwestern Montana: Their nature, origin, and tectonic significance: Geological Society of America Bull., vol. 84, p. 773-796.

Schmidt, D. L., 1964, Reconnaissance petrographic cross section of the Idaho batholith in Adams and Valley Counties, Idaho: U. S. Geological Survey Bull., 1181-G, 50 p.

Scholten, R., 1968, Model for evolution of Rocky Mountains east of the Idaho batholith: Tectonophysics, vol. 6, p. 109-126.

——,and Onasch, C. M., 1977, Genetic relations between the Idaho batholith and its deformed eastern and western margins: Northwest Geology, vol. 6-1, p. 25-37.

Shenon, P. J., and Ross, C. P., 1936, Geology and ore deposits near Edwardsburg and Thunder Mountain, Idaho: Idaho Bureau of Mines and Geology, Pamphlet 44, 45 p.

Smedes, H. W., 1966, Geology and igneous petrology of the northern Elkhorn Mountains, Jefferson and Broadwater Counties, Montana: U. S. Geological Survey Professional Paper 510, 116 p.

——,Klepper, M. R., and Tilling, R. I., 1968, Boulder batholith — A description of geology and road log: Geological Society of America, (Rocky Mtn. Sec., Bozeman, MT): Field Trip 3, 21 p.

——,and Thomas, H. H., 1965, Reassignment of the Lowland Creek Volcanics to Eocene age: Journal of Geology, vol. 73, p. 508-510.

Streckeisen, A., 1976, To each plutonic rock its proper name: Earth Science Reviews, vol. 12, p. 1-33.

Swanberg, C. A., and Blackwell, D. D., 1973, Areal distribution and geophysical significance of heat generation in the Idaho batholith and adjacent intrusions in eastern Oregon and Western Montana: Geological Society of America Bull., vol. 84, p. 1261-1282.

Takahashi, M., Aramaki, S., and Ishihara, S., 1980, Magnetite-series / ilmenite-series versus I-type / S-type granitoids: *in* Ishihara, S., and Takenouchi, S., eds., Granitic magmatism and related mineralization:

Mining Geology Special Issue No. 8, 247 p., p. 13–28.

Talbot, J. L, 1977, The role of the Idaho batholith in the structure of the northern Rocky Mountains, Idaho and Montana: Northwest Geology, vol. 6-1, p. 17–24.

——,and Hyndman, D. W., 1975, Consequence of subduction along the Mesozoic continental margin west of the Idaho batholith: Geological Society of America, Abstr. with Progr., vol. 7, p. 1290.

Taubeneck, W. H., 1971, Idaho batholith and its southern extension: Geological Society of America Bull. vol. 82, p. 1899–1928.

Tilling, R. I., 1973, The Boulder batholith, Montana: A product of two contemporaneous but chemically distinct magma series: Geological Society of America Bull., vol. 84, p. 3879–3900.

——,1974, Composition and time relations of plutonic and associated volcanic rocks, Boulder batholith region, Montana: Geological Society of America Bull. vol. 85, p. 1925–1930.

——,Klepper, M. R., and Obradovich, J. D., 1968, K-Ar ages and time span of the emplacement of the Boulder batholith, Montana: American Journal of Science, vol. 266, p. 671–689.

Vallier, T., 1974, A preliminary report on the geology of part of the Snake River Canyon, Oregon and Idaho: Oregon Department of Geology and Mineral Industries, Geol. Map Ser., GMS-6.

Vallier, T., 1977, The Permian and Triassic Seven Devils Group, western Idaho and northeastern Oregon: U. S. Geological Survey Bull. 1437, 58 p.

Weed, W. H., 1912, Geology and ore deposits of the Butte district, Montana: U. S. Geological Survey Professional Paper 74, p. 262.

Williams, L. D., 1977, Petrology and petrography of a section across the Bitterroot lobe of the Idaho batholith: Ph.D. dissertation, University of Montana, Missoula, 221 p.

Willis, G. F., 1978, Geology of the Birch Creek molybdenite prospect, Beaverhead County, Montana: M. S. Thesis University of Montana, Missoula, 74 p.

Winegar, R., 1970, The petrology of the Lost Creek stock and its relation to the Mount Powell batholith: M. S. Thesis, Univ. Montana, Missoula, 60 p.

Wiswall, G., 1979, Structure and petrology below the Bitterroot lobe of the Idaho batholith: Northwest Geology, vol. 8, p. 18–28.

Zen, E-an, Marcin, R. F., and Mehnert, H. H., 1975, Preliminary petrographic, chemical, and age data on some intrusive and associated contact metamorphic rocks, Pioneer Mountains, southwestern Montana: Geological Society of America Bull., vol. 86, p. 367–370.

MANUSCRIPT ACCEPTED BY THE SOCIETY JULY 12, 1982

Geological Society of America
Memoir 159
1983

A summary of critical relations in the central part of the Sierra Nevada batholith, California, U.S.A.

Paul C. Bateman
Branch of Field Geochemistry and Petrology
U.S. Geological Survey
345 Middlefield Road
Menlo Park, California 94025

ABSTRACT

The Sierra Nevada batholith was emplaced in Mesozoic time into strongly deformed, but weakly metamorphosed, strata of Precambrian, Paleozoic, and Mesozoic ages. The batholith is composed of many discrete plutons that are in sharp contact with one another or are separated by thin septa (screens) of older metamorphic or igneous rocks. The plutons can be grouped into formations and the formations into comagmatic suites. Within comagmatic suites, successively younger granitoids are commonly, but not invariably, more felsic, representing progressively lower temperature mineral assemblages. Compositional differences within suites have been attributed to crystal fractionation and differential upward movement of a parent magma produced during a single fusion event. However, recognition that initial $^{87}Sr/^{86}Sr$ increases in the younger and more felsic units of some suites suggests that the composition of the parent magma changed during emplacement, probably because of continuing incorporation of felsic, low-melting crustal rocks. The relative importance of crystal fractionation and of changing composition of the parent magma in the source region is not settled.

The granitoid suites are of Triassic, Jurassic, and Cretaceous ages. Triassic granitoids appear to be limited to a single extensive suite in the east-central part of the batholith. The Jurassic granitoids form a belt of scattered discrete plutons, which trends about N. 40° W. in the central Sierra Nevada. These granitoids are crossed by a N. 20° W. trending continuous belt of granitoids along the central Sierra Nevada. The optimum isotopic U-Pb age of the Triassic granitoid suite is about 210 m.y.; the Jurassic granitoids range in age from about 186 to 155 m.y. and the Cretaceous granitoids from about 125 to 88 m.y. Cretaceous granitoid suites are progressively younger eastward, but no pattern is apparent for the Jurassic granitoids. Few granitoids appear to have been emplaced in the central part of the Sierra Nevada batholith between 155 and 125 m.y. ago, an interval that includes the Late Jurassic Nevadan orogeny.

Quartz diorite, tonalite, plagiogranite, and gabbro predominate in the west, granodiorite and granite in the axial part of the batholith, and monzodiorite, monzonite, quartz monzonite, and granite in the east. The variations in the compositions of the granitoids across the central Sierra Nevada are independent of their ages. The most conspicuous chemical variation is an eastward increase in potassium. This change is accompanied by eastward increases in uranium, thorium, beryllium, rubidium, the oxidation ratio, total rare earths, and initial $^{87}Sr/^{86}Sr$, and by decreases in specific gravity and calcium.

Seismic P-wave velocities increase downward within the crust from 6.0 km/s near the surface to 6.4 km/s at intermediate depths, and to 6.9 km/s just above the seismic Moho. The Moho is depressed beneath the high Sierra Nevada to a depth of about 50 km but rises eastward to about 35 km and westward to about 20 km. The depression in the Moho follows the axis of the Cretaceous granitoids. The measured P-wave velocity in the upper mantle is 7.9 km/s.

Negative Bouguer gravity anomalies coincide with the axis of the Cretaceous granitoids, doubtless because of the increased thickness of the batholith. The residual magnetic intensity reflects the abundance of magnetite in the rock and is highest in the core of the batholith and lowest in the western part. Both heat flow and radioactive heat generation in the exposed rocks increase eastward to the crest of the range. Farther east, heat flow continues to increase at a slower rate, and heat generation decreases. Heat-flow measurements range from a little more than 0.4 H.F.U. to about 1.8 H.F.U. A plot of heat-flow measurements against heat-generation measurements made at the same sites is linear and shows that, at zero heat generation, the heat flow would be 0.4 H.F.U. This amount of heat probably represents the mantle contribution to the heat flow in the Sierra Nevada.

INTRODUCTION

The Sierra Nevada (Snowy Range) is a magnificient mountain range that extends northerly through eastern California for more than 400 km, from 35° N. lat. to 40° N. lat. (Fig. 1). The range is asymmetric, having a long, gentle, western slope and a short, steep, eastern escarpment that culminates in the highest peaks. It ranges in width from 80 to 100 km and in altitude from near sea level along the western edge to more than 4,000 m along the crest. Much of the western slope is covered with conifer forests, but the crestal region (the "high Sierra") rises above timber line, and the western foothills are occupied by grasslands and oak woodlands. The western foothills were the site of the famous California gold rush of 1849, which attracted adventurers from all parts of the world. Although many people still think of the Sierra Nevada as a rich metalliferous region, only tungsten was being mined in significant amounts in 1981. Now the Sierra Nevada is valued chiefly for its forests, which support a flourishing lumbering industry, for its rivers, which are the principal source of water and an important source of power for the cities and rich farmlands of California, and for the opportunities it offers for outdoor recreation.

Geologically, the Sierra Nevada is a huge block of the earth's crust that has broken free on the east along the Sierra Nevada fault system and been tilted westward, chiefly during Cenozoic time. It separates the heavily populated, lush Pacific coastal region from the sparsely populated, arid and undrained Great Basin. The range is a formidable barrier to both travel and to the passage of moisture eastward from the Pacific Ocean. Early settlers traveling in wagons feared being marooned in high passes by snow storms, and, even today, only a few roads cross the range.

During the winter, polar-front cyclones from the Pacific expand adiabatically as they move eastward up the western slope and cool below their dew point. As a result, most of the moisture that was obtained during the passage of warm air masses across the Pacific precipitates as snow, which is preserved in heavy packs at middle and higher altitudes until late spring or summer. On flowing down the eastern escarpment, the descending air warms adiabaticaly and can hold more moisture than it contains. Hence, the Great Basin receives little precipitation.

The Sierra Nevada batholith is a different entity from the Sierra Nevada and originated much earlier, during the Mesozoic. It comprises the more or less continuous granitoid terrane that makes up most of the Sierra Nevada and, in this report, is considered to include granitoids in the adjacent desert ranges to the east. Isolated plutons are present both to the east and to the west of the batholith. The batholith is continuous northward with granitoids in northwestern Nevada and southward with granitoids in southern California.

Most of the studies of the batholith by the U.S. Geological Survey during the last 30 years have been concentrated in a broad belt across the middle of the batholith between 37° and 38° N. lat., and most of the data for this report are drawn from studies of this area (Figs. 2 and 3). Scattered studies to the north and south indicate this area is representative of the entire batholith.

AGES, COMPOSITIONS, AND STRUCTURES OF THE COUNTRY ROCKS

The country rocks into which the batholith was emplaced

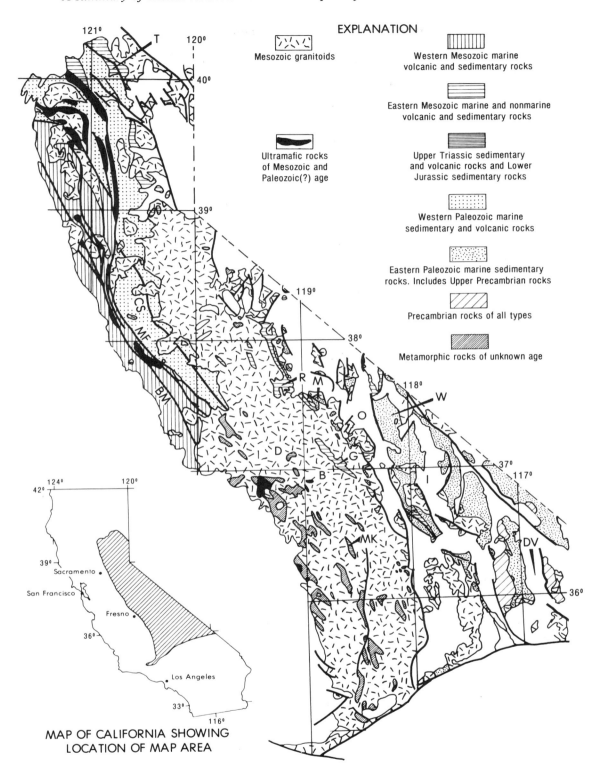

Figure 1. Sierra Nevada and adjacent areas in eastern California showing the Sierra Nevada batholith and its country rocks. Taylorsville, T; Ritter Range roof pendant, R; Mount Morrison roof pendant, M; White Mountains, W; Inyo Mountains, I; Dinkey Creek roof pendant, D; Boyden Cave roof pendant, B; Mineral King roof pendant, MK; Owens Valley, O; Death Valley, DV; Melones fault zone, MF; Calaveras-Shoe Fly fault zone, CS; Bear Mountain fault zone, BM.

are complexly deformed, weakly to moderately metamorphosed strata of Precambrian, Paleozoic, and Mesozoic ages. Precambrian rocks are exposed only along the eastern side of the batholith in the White Mountains and in the Death Valley region, and no outcrops of Precambrian rocks have been found west of Owens Valley in the Sierra Nevada (Fig. 1). The country rocks west of the batholith are mostly in the greenschist facies of regional metamorphism, and except for Precambrian rocks in the Death Valley region, those east of the batholith are virtually unmetamorphosed. Remnants within the batholith are in the hornblende hornfels or albite-epidote hornfels facies of thermal metamorphism. The pre-batholithic rocks were deposited in a wide range of environments and include a correspondingly great variety of lithologic types. Late Paleozoic and early Mesozoic ophiolite sequences and volcanogenic and sedimentary strata in the western part of the western metamorphic belt are widely believed to have been deposited adjacent to ocean ridges and island arcs and to have accreted tectonically along the continental margin.

Paleozoic country rocks occur along the eastern margin of the batholith, within the batholith as roof pendants and septa (screens), and in the eastern part in the western metamorphic belt (Fig. 1). East of the batholith, the strata trend generally toward the southwest into the batholith, whereas in roof pendants and in the western metamorphic belt, the strata strike northwest, parallel to the batholith. This regional change in strike has been interpreted as an oroflex in which, to the west, the southwest-trending strata east of the batholith bend successively west, northwest, west, and southeast as they cross the batholith (Albers, 1967; Stewart, Albers, and Pole, 1968). This oroflex is associated with right-lateral faults and is generally thought to result from cumulative right-lateral displacement adjacent to the continental margin.

The Paleozoic strata in the Inyo and White Mountains and in the roof pendants of the eastern Sierra Nevada are shelf or miogeosynclinal sedimentary rocks, predominantly carbonate, quartzite, and pelite. In the central Sierra Nevada between 37° and 38° N. lat (Fig. 2), the Paleozoic strata in the western metamorphic belt have been traditionally assigned to the Calaveras Formation of probable upper Paleozoic age. The Calaveras has been subdivided by Schweickert and others (1977) into four units. However, the most westerly unit, which consists predominantly of quartzite but also includes schist, marble, and calc-silicate rocks, is now considered by Schweickert (1978) to be a southern extension of the lower Paleozoic Shoe Fly Formation, which lies east of the Calaveras in the northern Sierra Nevada. The remaining three units in the Calaveras are, from west to east, a volcanogenic unit, a schist and phyllite unit that also contains minor chert and marble, and a chert unit. The Calaveras is separated on the east from the Shoe Fly Formation by the Calaveras-Shoe Fly fault zone and on the west from Mesozoic strata by the Melones fault zone.

The Mesozoic country rocks include three distinct, spa-

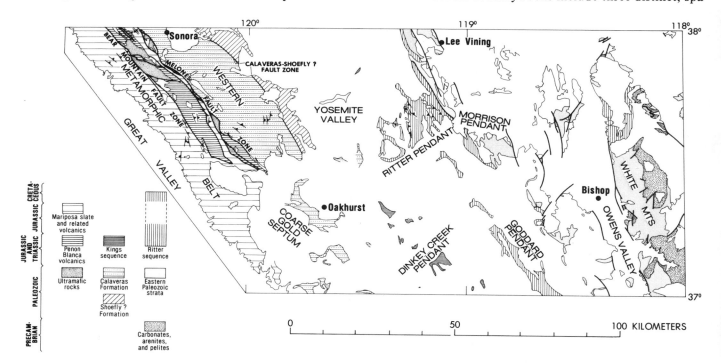

Figure 2. Distribution of country rocks in the central Sierra Nevada and White Mountains. Barbed symbols in western metamorphic belt and Ritter Range roof pendant show top facings of beds (Clark, 1964; Huber and Rinehart, 1965).

tially separated groups of strata: a group that occupies the western part of the western metamorphic belt, the Kings sequence (Bateman and Clark, 1974), which is exposed in a series of roof pendants that lie east of the Calaveras, and the Ritter Range sequence in the axial part of the batholith.

The western group of Mesozoic strata occupies the western part of the western metamorphic belt and is separated from the Calaveras Formation by the Melones fault zone. It is composed largely of slate, phyllite, and a variety of volcanic rocks, including pillow lavas. These strata are divided by the Bear Mountain fault zone into two structural blocks. Chemical analyses of composite samples (unpublished data) indicate that despite a wide range of compositions, the bulk composition of both structural blocks lies between average andesite and average dacite. Graded beds are ubiquitous, and the overall constitution of the strata indicates that they were deposited in an island-arc environment. Sparse fossils indicate that most of these strata are Late Jurassic. However, Early Jurassic U-Pb isotopic ages (Morgan and Stern, 1977), obtained from two small plutons that intrude the Penon Blanco Volcanics in the block between the Melones and Bear Mountain fault zones (Fig. 3), indicate that these volcanic rocks are at least as old as Early Jurassic and may be Triassic. In this same block, north of the area shown in Figure 2, is a melange (the so-called "western belt" of the Calaveras Formation of Clark, 1964) in which Tethyan fusilinids of Permian age are present in exotic blocks of limestone (Douglass, 1967). Although these fossils may not indicate the age of the enclosing strata, they do indicate the former presence of Permian source beds that apparently were deposited in a warm-water environment. Jones and others (1977) report that exotic blocks containing Tethyan fusilinids are sporadically distributed as far north as Alaska. They interpret these relations to indicate that Permian strata deposited in warm water moved northward tectonically during the early Mesozoic, shedding fossiliferous remnants en route. In contrast with these exotic materials, conglomerate beds that crop out between the Penon Blanco Volcanics and the Melones fault zone, within the Mariposa Slate, contain chert and granitoid boulders, suggesting a local source.

The rocks of the Kings sequence are metasedimentary in the area shown in Figure 2 — quartzite, marble, schist, calc-silicate hornfels, and biotite-andalusite hornfels (Kistler and Bateman, 1966) — but farther south, in the Mineral King and Yokohl Valley roof pendants, volcanic rocks are also present (Saleeby and others, 1978). Although the strata are strongly deformed, fossils of both Early and Late Jurassic ages have been collected from several roof pendants south of the area shown in Figure 2 (Knopf and Thelan, 1905; Christensen, 1963; Moore and Dodge, 1962; Jones and Moore, 1973; Saleeby and others, 1978).

The Ritter sequence (Bateman and Clark, 1974) is dominantly volcanogenic and consists of intermediate to felsic volcanic and volcanogenic sedimentary rocks. Fiske and Tobisch (1978) have subdivided the Ritter sequence into lower and upper sections, which are separated by an unconformity that seems to represent a significant time gap. The lower section is essentially homoclinal and dips steeply west. Common rock types are silicic ash-flow tuff, tuff breccia, bedded tuff, lapilli tuff, and thin beds of limestone. Crossbeds are common in sedimentary layers and indicate that almost all bedding tops face west. Undoubtedly, the section was deposited partly on land and partly in a shallow subaqueous environment. Although fossils collected from a single locality (Huber and Rinehart, 1965) indicate an Early Jurassic age, isotopic ages from the lower section range from Early Triassic to Middle Jurassic. The present stratigraphic thickness of the lower section is about 5 km, but it has been tectonically thinned from an original thickness approaching 11 km (Fiske and Tobisch, 1978).

The upper section of the Ritter sequence dips more gently in most places and rests unconformably on the lower section. It is less strongly deformed than the lower section and has been tectonically thinned much less — from an original thickness of about 4 km to a present thickness of about 3.4 km (Fiske and Tobisch, 1978). It consists of silicic lava flows, tuff, and tuff breccia, all of which were deposited on land and probably originated in a caldera (Fiske and others, 1977). Isotopic ages are mid-Cretaceous.

The ages of the Ritter sequence and Kings sequence overlap, but their relations to each other are uncertain; they may be in fault contact. However, along the southwestern edge of the Ritter Range roof pendant, hypabyssal rocks, compositionally similar to volcanic rocks of the Ritter sequence, intrude metasedimentary strata of the Kings sequence (Nokleberg, 1981).

All of the stratified country rocks in the central Sierra Nevada are strongly deformed, and evidence of repeated deformations has been reported from many different areas. The Late Devonian and Early Mississippian(?) Antler orogeny and the late Paleozoic Sonoma orogeny affected the eastern Paleozoic rocks, and early folds in the Mount Goddard roof pendant suggest deformation during the Triassic or Jurassic before the Late Jurassic Nevadan orogeny. The Late Jurassic Nevadan orogeny is generally presumed to account for the northwest-trending grain in the country rocks — the so-called Sierran trend. Other deformations took place as late as the mid-Cretaceous, about 100 m.y. ago.

The major structure in the country rocks of the central Sierra Nevada is defined by the preponderance of east-facing bedding tops in the western metamorphic belt and of westward-younging Paleozoic strata in the Mount Morrison roof pendant and Mesozoic strata of the Ritter range roof pendant. In some earlier reports (Bateman and Wahrhaftig, 1966; Bateman and Eaton, 1967), the present author interpreted these inward-facing strata to indicate a complex faulted synclinorium. However, no single stratigraphic unit has been identified in both limbs, and it is possible that the inward-facing bedding tops were caused by some other mechanism than folding and faulting in a synclinorium.

STRUCTURE OF THE BATHOLITH

Sierran granitoids are I types as defined by Chappell and White (1974) and Hine and others (1978). Most are diopside normative and contain hornblende. Mafic inclusions are common in hornblende-bearing granitoids, and metamorphosed sedimentary inclusions occur only adjacent to metasedimentary wall rocks. Cordierite has not been reported from the granitoids, and garnet and aluminum silicates have been found only in late felsic fractionates.

The batholith is composed of a great many different plutons that are either in sharp contact with one another or are separated by thin septa (screens) of older metamorphic or igneous rocks. Although the rocks within a pluton may vary in composition and texture from place to place, distinctive characteristics generally are present in all the variants. Contacts of granitoids with both country rocks and with other granitoids are steep or vertical. No inward-dipping contacts, as might indicate a sharp lower boundary surface, have been observed in the central Sierra Nevada.

Plutons in the same general area and having the same compositions, textures, and apparent ages are grouped into formations. Field, petrographic, and chemical criteria are interpreted as indicating that many granitoid formations are consanguineous with one another; that is, they are differentiates of a common parent magma that was produced in a single fusion event (Presnall and Bateman, 1973). Groups of consanguineous formations form granitoid sequences whose members are generally (but not invariably) more felsic with decreasing age.

The simplest kind of comagmatic granitoid sequence is a concentrically zoned pluton in which relatively mafic, high-temperature, mineral assemblages in the margins grade inward, without discontinuities, to more felsic, lower temperature, mineral assemblages. Studies of the Tuolumne Intrusive Suite (Bateman and Chappell, 1979) and the Mount Givens Granodiorite (Bateman and Nokleberg, 1978) have lead to a model involving crystal fractionation during inward solidification with falling temperature. More complex suites are attributed to movements of less-crystallized core magma, which intrudes and, in places, breaks through the solidifying carapace into the country rocks.

Unpublished studies of initial $^{87}Sr/^{86}Sr$ by R. W. Kistler and B. W. Chappell (1981, oral communication) suggest that a second mechanism also was operative in the Tuolumne Intrusive Suite. Their data show that initial $^{87}Sr/^{86}Sr$ increases generally in the younger and more felsic units of the suite. The simplest explanation for this increase in the initial ratio is that the composition of the parent magma changed with time, probably as the result of incorporating increasing amounts of low-temperature constituents of the lower crust. The relative importance of crystal fractionation and the changing composition of the parent magma in producing the range of compositions present in the Tuolumne Intrusive Suite is still uncertain.

Careful examination of the compositional and textural changes in relatively simple concentrically zoned plutons and in granitoid suites having few discontinuities leads to the following criteria for identifying the units of a granitoid suite in which a concentric arrangement is not readily apparent: (1) all the granitoid formations of a suite crop out in the same general area, and many of them are contiguous; (2) vestiges of a concentric arrangement may be recognizable, in which units are successively younger and more felsic inward; (3) intrusive relations at contacts indicate that successively younger formations are successively more felsic; (4) textural changes are in the same order as in concentrically zoned plutons having the same range of compositions; (5) consanguineous formations generally have some common mineralogical, chemical, and/or textural characteristics; (6) cataclastic zones or swarms of dikes in the granitoids of an older suite may be cut off by granitoids of a younger suite; (7) septa (screens) of older rocks generally lie between granitoid suites rather than between different units of the same suite; and (8) isotopic ages of a given suite fall in a time interval of only a few million years.

Once the formations have been assembled into a much smaller number of comagmatic suites, the magnitude of the problem of ordering the formations is much reduced. The relative ages of suites whose members are in contact generally can be established by field relations, and the relative ages of others can be determined by isotopic dating. Using these methods, most of the larger plutons in the central Sierra Nevada and many smaller ones have been assigned to formations and the formations to granitoid suites whose relative and isotopic ages are established (Fig. 3).

FABRIC

The textures of the granitoids are similar to those of mesozonal granitoids everywhere. Most are medium to coarse grained and hypidiomorphic granular to porphyritic, but the rock in some small plutons is fine grained. The porphyritic granitoids contain megacrysts of K-feldspar, which appear to have crystallized from the magma and to be phenocrysts rather than porphyroblasts (Bateman and Chappell, 1979). Bimodal size distribution of other minerals, such as occurs in porphyries, is found only in small bodies that probably lost volatiles, producing a pressure quench. Miarolitic cavities are uncommon but are found in a few small fine-grained felsic bodies.

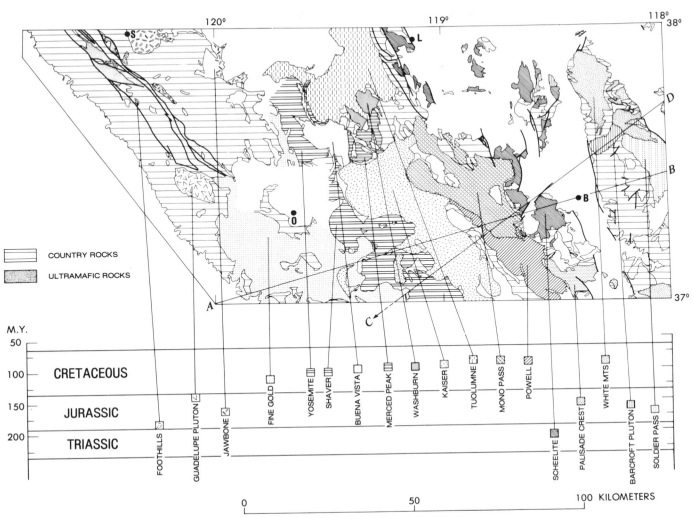

Figure 3. Provisional granitoid suites in the central Sierra Nevada and their approximate radiometric ages. Ages are based on an evaluation of U-Pb and K-Ar ages (Stern and others, 1981; Evernden and Kistler, 1970; Crowder and others, 1973; McKee and Nash, 1967; Kistler and others, 1965). Geochemical and geophysical profiles (Figures 6 and 7) are along line A-B, except the residual magnetic intensity profile, which is along line C-D.

Most bodies of quartz diorite, tonalite, granodiorite, and quartz monzonite, and some bodies of granite, are foliated. The foliation is shown both by the preferred orientation of tabular and prismatic minerals and by lens-shaped mafic inclusions. The mineral foliation generally parallels that shown by mafic inclusions but, in places, is less regular and varies around the foliation shown by mafic inclusions. Two crossing foliations, both shown by mineral alignment, have been observed in a few places but are not common. Lineation in the plane of foliation generally is difficult to identify, but in some granitoids, oriented hornblende prisms define a lineation. Although mafic inclusions appear elongate in outcrop, most are lens-shaped or only faintly triaxial. Where identifiable, the lineation generally is down the dip of the foliation.

In general, foliation is strongest close to external intrusive contacts of plutons, which it approximately parallels. With distance from the contact, the foliation is progressively weaker because of the less perfect preferred orientation of minerals and because mafic inclusions are less abundant and less elongate as seen in outcrop. Inward, the foliation may diverge from the contact, but generally it does so in broad sweeping curves that are sub-parallel to the margins of the pluton (Fig. 4). Cloos has explained foliation in igneous rocks in terms of movements in the magma during solidification (see Balk, 1937), and Mackin (1947) has identified three types of non-uniform flow that can produce foliation. He terms these deceleration flow, acceleration flow, and velocity gradient flow. Velocity changes parallel the direction of flow in deceleration and acceleration flows and are across the direction of flow in velocity gradient flow. In all three types, the primary consideration is the direction of stretching or

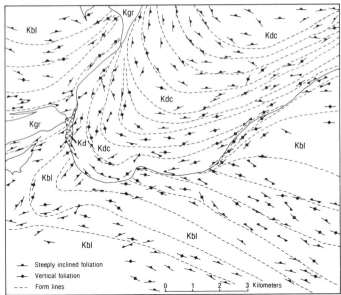

Figure 4. Original and deformed primary foliation patterns. Foliation in older tonalite of Blue Canyon (Kbl) was disturbed during the emplacement of the granodiorite of Dinkey Creek (Kdc). The arcuate patterns in the granodiorite of Dinkey Creek are original. Kgr, younger granite; Kd, older diorite. Shaver Lake quadrangle, central Sierra Nevada (Lockwood and Bateman, 1976).

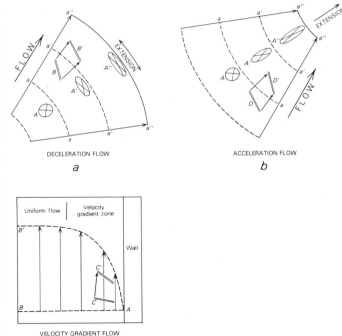

Figure 5. Three types of non-uniform intrusive flow that can produce foliation and/or lineation. Diagrams are two-dimensional and represent either horizontal or vertical sections. Figures 5a and 5c are modified from Mackin (1947).

elongation, for both mafic inclusions, which were probably only a little stiffer than the enclosing magma, and inequant crystals are aligned in this direction, which is not necessarily the flow direction.

In deceleration flow (Fig. 5a), the velocity diminishes in the direction of flow either because the magma passes through an orifice of increasing size or, more commonly, because the magma crowds the walls and roof of its chamber upward and outward. Deceleration flow offers a satisfactory explanation for foliation in plutons with subequant dimensions in which the foliation is strongest at the margins and progressively weaker inward. Magma introduced into the central part of the chamber moves with diminishing velocity toward the walls as the magma chamber expands. This movement causes concentric or arcuate elongation at a high angle to the direction of flow (Fig. 5a), so that mafic inclusions are stretched out more or less parallel with the chamber walls. Crystals are oriented approximately parallel with the mafic inclusions because the trailing ends of randomly oriented inequant crystals are carried toward the chamber walls faster than the leading ends.

Acceleration flow occurs when magma moves through an orifice of diminished size, which requires increased velocity of flow. In this type of flow, mafic inclusions are stretched out in the direction of flow rather than across it (Fig. 5b), and inequant crystals are oriented in this direction because the leading ends of crystals athwart the direction of flow move faster than their trailing ends. This type of flow occurs

in dikes and in large plutons with long, narrow shapes in which the foliations parallel the long dimension. The assumption is that the plutons narrow outward and upward, so that the velocity of flow increases toward the extremities of the pluton.

In velocity gradient flow, flow parallels a bounding crystal-liquid interface, toward which the velocity diminishes because of internal friction (Fig. 5c). This type of flow causes plastic mafic inclusions to be stretched out in a direction almost parallel to the bounding interface and randomly oriented crystals to be rotated parallel to the interface. Studies made since Mackin published his analysis of the origin of primary foliation and lineation show that velocity gradient flow also sorts crystals according to size. The smallest crystals are closest to the interface, where velocity differences (but not the velocity) are greatest. Crystals are progressively larger away from the interface to a region of uniform velocity flow (Bhattacharje and Smith, 1964; Wilshire, 1969; Komar, 1972). Velocity difference (shear stress), and not velocity itself, tends to expel crystals of all sizes, but larger crystals more effectively than smaller crystals. Schlieren are the most common features attributable to flow sorting, but flow sorting may be the true explanation for many fine-grained margins, especially on dikes, that have been attributed to chilling. Fine-grained marginal rock caused by chilling should contain a representative collection of the crystals which the magma contained when it was chilled. Any larger crystals present farther inward from the contact must have attained their larger size later, after the chilled margin had solidified.

In many plutons, especially in the youngest plutons in a given terrane, original magmatic foliation is preserved intact. However, the foliation in many older bodies wraps around younger plutons and appears to have been displaced by the younger plutons (Fig. 4). A paucity of cataclasis in such older granitoids suggests they were softened and moved plastically during the emplacement of the younger plutons.

In contrast with these primary structures are post-consolidation cataclastic structures. Cataclasis is indicated in thin section by fine-grained mortar around larger grains and in outcrops and hand specimens by reduced grain size, blurred crystal shapes, streaky layering, and ovoid or diamond-shaped porphyroclasts. Lineations are characteristic of cataclastically deformed granitoids. In such rocks, mafic inclusions commonly are drawn into elongate blade-like forms.

REGIONAL COMPOSITIONAL VARIATIONS

One of the most striking features of the Sierra Nevada batholith is eastward increase in the potassium content of the granitoids, independently of their ages. This variation is readily apparent even on casual examination of the rocks along a transect across the batholith. Tonalite and quartz diorite are the most common granitoids in the western Sierra Nevada, granodiorite predominates in the middle and higher parts of the range, and quartz monzonite and monzonite are common in the desert ranges east of the Sierra Nevada. This systematic change in the composition of the plutonic rocks across the Sierra Nevada was observed many years ago by Lindgren (1915) and has been confirmed by Moore (1959) and by Bateman and Dodge (1970). Similar compositional changes with distance from the Pacific basin have been observed along many other transects across the Mesozoic batholiths of North America.

Bateman and Dodge (1970) showed the eastward increase in the potassium content of the granitoids in terms of the potassium index 10^3 $[K_2O/(SiO_2-45)]$. However, modal data on the granitoids are far more abundant than chemical data, and in Figure 6, the variation is shown by K-feldspar/(K-feldspar + quartz). A simple plot of K-feldspar also shows eastward increase but is less regular because K-feldspar also varies within comagmatic sequences with fractionation, as does quartz.

The eastward increase in potassium is accompanied by many other compositional changes. Uranium, thorium (Wollenberg and Smith, 1968), beryllium, rubidium (Dodge, 1972a), and the oxidation ratio (Dodge, 1972b) also increase, and calcium decreases slightly (Bateman and Dodge, 1970). Initial $^{87}Sr/^{86}Sr$ increases eastward from less than 0.704 in

the western foothills to about 0.708 along the crest of the range, but levels off and even decreases toward the north and east (Fig. 6). An isopleth drawn between initial $^{87}Sr/^{86}Sr$ greater than and less than 0.706 runs northward along the western slope of the Sierra Nevada to near latitude 38° N. and then bends sharply east (Kistler and Peterman, 1973). All of these variations, like that of potassium, are independent of the age of the granitoids, making it unlikely that they were caused by eastward increasing depth of magma generation along a subduction zone, as has been proposed by Dickinson (1970). Similar variations of potassium, uranium, and thorium in the country rocks (Wollenberg and Smith, 1970) and eastward increase in K/Na in Cenozoic volcanic rocks (Moore, 1962) support the view that the lateral changes in the compositions of the granitoids across the batholith result from progressive changes with distance from the Pacific Ocean in the composition of the source materials from which the magmas were generated. The ratio of mantle-derived mafic material to crustal material also may be important.

AGE PATTERNS

Isotopic dating of biotite and hornblende by the K-Ar method (Curtis and others, 1958; Kistler and others, 1965; McKee and Nash, 1967; Evernden and Kistler, 1970; Crowder and others, 1973) and of zircon by the U-Pb method (Chen, 1977; Stern and others, 1980) gives a good picture of the ages of the granitoids across the central Sierra Nevada (Fig. 3). K-Ar ages are generally reliable for the youngest granitoids in a given area, but the K-Ar ages of many older granitoids have been reduced because of re-heating during the emplacement of nearby younger intrusions. Biotite ages are reduced more commonly, and by larger amounts, than hornblende ages, but both the horn-

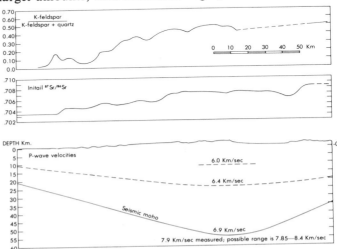

Figure 6. Geochemical and seismic profiles. Profiles are along line A-B in Figure 3. Geochemical profiles from Bateman (1979) and seismic profile from Bateman and Eaton (1967).

blende and biotite K-Ar ages on older plutons generally are widely dispersed, causing considerable uncertainty as to the reliability of a particular K-Ar age. U-Pb ages do not appear to have been modified significantly by either metamorphism or the intrusion of younger plutons, and most U-Pb ages are thought to reflect the original crystallization ages of the granitoids.

K-Ar dating has shown that a wide continuous belt of Cretaceous granitoids trending N 20° W crosses an older discontinuous belt of Jurassic granitoids trending about N 40° W in the central Sierra Nevada (Evernden and Kistler, 1970; Kistler and others, 1971). The Cretaceous belt trends northward along the axis of the Sierra Nevada into northeastern California and northwestern Nevada. South of the area shown in Figure 3, Jurassic granitoids lie east of the Cretaceous granitoids in the desert ranges of eastern California; north of the area shown in Figure 3, they lie west of the Cretaceous granitoids in the western foothills of the Sierra Nevada (Fig. 1).

In the central Sierra Nevada, where the two age belts cross (Fig. 3), the core of the batholith is occupied by Cretaceous granitoid suites that field relations and U-Pb ages indicate are progressively younger eastward over a 37 m.y. interval that extends from about 125 to 88 m.y. ago. The Cretaceous granitoids are flanked by Jurassic granitoids on both sides, and a single, but extensive, Triassic granitoid suite is also present on the eastern side (Fig. 3). The Jurassic suites range in age from 186 to 155 m.y., and the Triassic suite has an optimum age (an age based on an evaluation of all pertinent data) of 210 m.y. The distribution of Jurassic ages indicates that, prior to the emplacement of the Cretaceous granitoids, Jurassic granitoids were widely distributed across the central Sierra Nevada but were not emplaced in a west-to-east succession as were the Cretaceous granitoids.

Viewed very broadly, the U-Pb ages indicate two intervals of plutonism, one between 206 and 155 m.y. ago, during the Triassic and Jurassic, and the other between 125 and 88 m.y. ago, during the Cretaceous. Within these intervals, plutonism appears to have been episodic. Time gaps of as much as 20 m.y. separate some of the older granitoid suites, whereas some younger suites, whose different ages are established by field relations, have similar optimum U-Pb ages.

Although many plutons clearly deformed their wall rocks when they were emplaced, field relations and isotopic ages indicate that the regional deformations that affected the Sierra Nevada occurred between, rather than coincident with, granitoid intrusions. The granitoids were probably emplaced during periods of crustal relaxation, or even extension, rather than of shortening. Few granitoid ages from the central Sierra Nevada fall in the 30 m.y. interval of greatly reduced plutonic activity between 155 and 125 m.y. However, it was during this interval that the Late Jurassic Nevadan orogeny, generally considered to have been the most intense deformation to have affected the Sierra

Nevada, occurred. It was also during this time, though not necessarily coincident with the Nevadan orogeny, that the N. 40° W. Jurassic trend of plutonic activity shifted to the N. 20° W. Cretaceous trend.

The time required for the emplacement and solidification of granitoids belonging to the same intrusive suite is not yet resolved. Several millions of years are generally assumed for large plutons. However, Spera (1980) has concluded that a spherical pluton with a radius of 5 km and containing only 0.5 percent water will cool to below the solidus in about 3.3×10^5 years, and that with 4 percent water, the same pluton will solidify in 5×10^4 years. He also concluded that a pluton with a radius of 10 km and 2 percent water will cool to below the solidus in about 1.1×10^6 years. The narrow spreads of U-Pb ages within comagmatic sequences allow only a few million years at most for the emplacement and solidification of all the units in a suite and fit Spera's solidification times much better than the longer times commonly assumed. The differences among the U-Pb ages of the units within a suite generally are less than the analytical uncertainties, and the ages do not reflect the order of intrusion within suites as established by field relations.

MAFIC MASSES, INCLUSIONS, AND DIKES

The mafic rocks of the Sierra Nevada have received much less attention than the granitoids and still require more study to be understood. The mafic rocks include three separate groups: small discrete masses, inclusions in the granitoids, and dikes.

Small masses of mafic rock, chiefly hornblende gabbro and diorite, are abundant on both the eastern and the western sides of the batholith in the central Sierra Nevada. Masses of gabbro and norite are also present on the western side. Many mafic masses are associated with metamorphic rocks and, on the eastern side, also with Triassic and Jurassic granitoids, which intrude them. Textures range from fine grained and equigranular to coarse grained and splotchy, almost pegmatitic. Poikilitic (or poikiloblastic) hornblende is common in coarse-grained varieties. Whether all the textures are primary is uncertain; some coarser grained textures may be metamorphic.

The preferential association of mafic masses with metamorphic rocks in the margins of the batholith and, on the eastern side, with Triassic and Jurassic granitoids, and their paucity in the interior of the bathlith, where Cretaceous granitoids predominate, suggest that most of the mafic masses are Jurassic or older. Whether they are early differentiates of granitoid magmas or represent separate earlier magmatic pulses is uncertain. The only isotopic age that has been determined on a mafic mass is a Jurassic U-Pb age of 169 m.y. on a diorite from the eastern side of the batholith (Stern and others, 1979). However, other nearby mafic masses have been intruded by Triassic granitoids with an optimum U-Pb age of 210 m.y.

Mafic inclusions are widely distributed in such hornblende-bearing granitoids as tonalite and granodiorite. They are common in granitoids in which anhedral mafic and accessory minerals form clots of varying size but are scarce in granitoids in which hornblende and biotite are characteristically in discrete euhedral crystals. Most mafic inclusions contain the same minerals as the enclosing granitoid but in different proportions, indicating equilibrium. The attainment of equilibrium and the fact that the inclusions apparently have been shaped by magmatic movements indicate long residence in the granitoid magma. Generally, no materials suitable for forming mafic inclusions are to be found among the wall rocks of the granitoids. As a consequence of these considerations, Presnall and Bateman (1973) and Bateman and Chappell (1979) have interpreted the mafic inclusions to be residual semi-solid material that was carried in the magma from the source region of the magma. Mottled calcic cores in plagioclase and the small clots of anhedral mafic minerals are also thought to be residual material. Granitoids that lack these other residual materials, including rocks in which the hornblende and biotite occur in euhedral crystals, contain few mafic inclusions. Most euhedral hornblende and biotite crystals are believed to be precipitates from the melt phase of the magma (Bateman and Chappell, 1979). However, those present locally in mafic inclusions probably grew as porphyroblasts.

Locally, mafic inclusions are particularly abundant adjacent to mafic volcanic rocks, older diorite and gabbro, and calcareous metasedimentary rocks. This spatial association suggests that some mafic inclusions were derived from such rocks. Some locally derived mafic inclusions are rimmed by felsic minerals, indicating disequilibrium between the inclusions and the magma.

Most petrologists accept the view that most of the mafic inclusions have been carried upward in the magma from depth, but they differ as to the nature of the material from which the mafic inclusions were derived. The most common view is that the inclusions represent refractory solid mafic material in the source region of the magma (restite), such as volcanic rocks, dikes, and sills, or mafic layers in gneiss complexes. This view is supported by the local presence of elongate masses of mafic rocks, a few tens to a few hundreds of meters long, which appear to have been in the process of breaking up into smaller masses, the size of typical inclusions, when the enclosing granitoid magma solidified. Another popular view is that the mafic inclusions represent immiscible blobs of andesitic or basaltic magma that were injected into the granitoid magma as it was forming.

The mafic dikes range in composition from lamprophyre to granodiorite porphyry, the more mafic compositons predominating (Moore and Hopson, 1961). The dikes are of several ages. Dikes that intrude one pluton may be truncated by a younger pluton, which, in turn, is intruded by still younger dikes. The only dikes that have been studied in detail belong to the Independence dike swarm (Moore and Hopson, 1961). U-Pb ages show that this swarm was intruded 148 m.y. ago (Chen and Moore, 1979) into granitoids with U-Pb ages ranging from 155 to 206 m.y. during a time when few granitoids were being emplaced (Stern and others, 1979). The dikes have dilated walls and obviously were intruded during a period of relaxation rather than during a period of compression (Chen and Moore, 1979).

However, the relations between mafic dikes and the plutons they intrude are often ambiguous. In some places, dikes appear to have been intruded while the granitoid magma was still mobile enough to stretch and disrupt them. Such dikes have been termed "synplutonic" (Roddick and Armstrong, 1959). Whether mafic dikes were intruded both concurrently with granitoids and between epochs of granitoid emplacement is yet to be determined.

The possibility that the exposed mafic dikes may be merely the upper expression of more intensive diking and underplating of the lower crust merits consideration. Such deeper mafic material, emplaced somewhat earlier than the visible dikes, conceivably could have been source materials for mafic inclusions.

SUBSURFACE STRUCTURE

Interest in the deep structure beneath the Sierra Nevada began in 1936 when Lawson published "The Sierra Nevada in the light of isostasy." In a comment on Lawson's paper, Byerly (1938) inferred a root beneath the Sierra Nevada from delay in the arrival times at stations east of the range of earthquake waves that originated west and northwest of the range. Seismic and gravity studies have since shown that the Moho is depressed beneath the Sierra Nevada and forms a north-trending trough that follows the axis of Cretaceous intrusions. The position and orientation of the trough led Bateman and Eaton (1967) to conclude that it was related in origin to the batholith rather than to the Cenozoic uplift and westward tilting of the range.

Seismic refraction measurements show that the maximum depth to the Moho is 45 km in the northern Sierra Nevada (Eaton, 1963) and more than 50 km in the central part of the range (Eaton and Healy, 1963; Bateman and Eaton, 1967). Gravity studies are consistent with the seismic results (Oliver, 1977). Both the seismic and gravity data show that, in the central Sierra Nevada, the Moho defines an asymmetric trough that is centered beneath the Sierra Nevada divide (Fig. 6). From the axis of the trough, the Moho rises gently westward to less than 20 km beneath the eastern side of the Great Valley of California and more steeply eastward to less than 35 km beneath the Great Basin. The P-wave velocities beneath the central Sierra Nevada are 7.9 km/s in the upper mantle, 6.9 km/s at the base of the crust, 6.4 km/s above 24 km depth, and 6.0 km/s above 12 km depth (Bateman and Eaton, 1967).

Downward increase in P-wave velocity to 6.4 km/s in the upper crust probably reflects downward increasing abundance of settled mafic crystals and of solid residual material carried upward in the magma during its rise. The 6.9 km/s velocities in the lower crust may reflect the presence of abundant residual (high melting temperature) material left behind as the magma rose from its source region.

The trough in the Moho in conjunction with the inward-facing bedding tops in the country rocks, which more-or-less coincide with the trough, led the present author to favor the tectogene model of Griggs (1939) and Vening Meinetz (1948) for the origin of the batholith (Bateman and Wahrhaftig, 1966; Bateman and Eaton, 1967). However, the advent of plate tectonics, together with the recognition that other circum-Pacific batholiths lack this setting (see, for example, Pitcher, 1978), and additional information on the country rocks have forced the present author to abandon this model.

MAGNETIC, GRAVITY, HEAT FLOW, AND HEAT GENERATION PATTERNS

The geophysical profiles across the batholith shown in Figure 7, like the geochemical and seismic profiles (Fig. 6), are drawn along line A-B in Figure 3, except for the magnetic profile, which is drawn along line C-D in Figure 3. The negative Bouguer gravity anomaly calculated for an assumed density of 2.67 g/cm³ defines a trough whose axis coincides with the core of the batholith, as would be

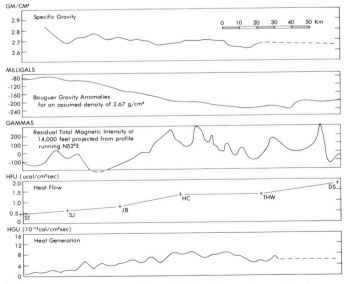

Figure 7. Geophysical profiles. All profiles except residual magnetic intensity are along line A-B in Figure 3. Residual magnetic intensity is along line C-D in Figure 3. Profiles are from Bateman (1979).

expected as a result of thickening of the relatively light rocks of the batholith (Oliver and Robins, 1973). However, measured specific gravities show that the density of the exposed granitoids actually decreases systematically eastward into the core of the batholith. If this decrease is taken into account, the Bouguer gravity anomaly would be reduced by an amount that would depend on the depth extent of the surface densities.

The gross changes in residual magnetic intensity (Oliver, 1977) appear to reflect chiefly the abundance of magnetite in the granitoids, but the dip in the eastern part of the profile reflects the structural and topographic depression of Owens Valley, inasmuch as the magnetic data were measured at a constant altitude (Fig. 7). The two features of greatest interest are the twin bumps at the western end of the profile, which reflect outcrops of olivine-bearing hornblende gabbro, and the broad depression just east of the bumps. The granitoids that correspond with the depression contain little or no magnetite, a feature that Dodge (1972b) attributed to lower oxygen fugacity contingent on lower water content in the source rocks. Preliminary petrographic studies indicate that these rocks contain rounded blebs of pyrite, suggesting the possibility that sulfur fugacity may have been important in producing the low magnetic values.

Heat generation in the surface rocks from the radioactive disintegration of isotopes of K, U, and Th increases eastward to the Sierra Crest, then diminishes farther east (Fig. 7). Heat flow measured in six borings increases steadily eastward, rising from a little more than 0.4 H.F.U. (heat-flow unit) in the west to about 1.8 H.F.U. in the White Mountains (Lachenbruch, 1968; Lachenbruch and others, 1976). All of these heat-flow values are lower than the average for continental areas, and 0.4+ H.F.U. is the lowest value yet measured and far below measurements obtained in the ocean basins. A plot of heat-flow measurements against heat generation measurements made at the drill sites yields a straight line that intersects zero heat generation at 0.4 H.F.U. Lachenbruch (1968) has interpreted this amount of heat to represent the mantle contribution beneath the Sierra Nevada, the excess above this amount coming from radioactivity in the crust. This inferred mantle contribution is low as compared to other regions and has been tentatively explained by Lachenbruch as resulting from downward flow of heat from the upper mantle, possibly to a cold subducted slab.

REFERENCES CITED

Albers, J. P., 1967, Belt of sigmoidal bending and right-lateral faulting in the western Great Basin: Geological Society of America Bulletin, v. 78, p. 143–156.

Balk, R., 1937, Structural behavior of igneous rocks: Geological Society of America Memoir 5, 177 p.

Bateman, P. C., and Chappell, B. W., 1979, Crystallization, fractionation, and solidification of the Tuolumne Intrusive Series, Yosemite National Park, California: Geological Society of America Part 1, v. 90, no. 5, p. 465–482.

Bateman, P. C., and Clark, L. C., 1974, Stratigraphic and structural setting of the Sierra Nevada batholith, California: Pacific Geology, v. 8, p. 70–89.

Bateman, P. C., and Dodge, F. C. W., 1970, Variations of major chemical constituents across the central Sierra Nevada batholith: Geological Society of America Bulletin, v. 81, no. 2, p. 409–420.

Bateman, P. C., and Eaton, J. P., 1967, Sierra Nevada batholith: Science, v. 158, p. 1407–1417.

Bateman, P. C., and Nokleberg, W. J., 1978, Solidification of the Mount Givens Granodiorite, Sierra Nevada, California: Journal of Geology, v. 86, p. 563–579.

Bateman, P. C., and Wahrhaftig, C., 1966, Geology of the Sierra Nevada: *in* Bailey, E. A., editor, Geology of Northern California: California Division of Mines and Geology, Bulletin 190, p. 107–172.

Bhattacharji, S., and Smith, C. H., 1964, Flowage differentiation: Science, v. 145, no. 3628, p. 150–153.

Byerly, P., 1938, The Sierra Nevada in the light of isostasy: Geological Society of America Bulletin, v. 48, Supplement, p. 2025–2031.

Chappell, B. W., and White, A. J. R., 1974, Two contrasting granite types: Pacific Geology, v. 8, p. 173–174.

Chen, J. H., 1977, Uranium-lead isotope ages from the southern Sierra Nevada batholith and adjacent areas, California [Ph. D. thesis]: Santa Barbara, University of California, 180 p.

Chen, J. H., and Moore, J. G., 1979, Late Jurassic Independence dike swarm in eastern California: Geology, v. 7, no. 3, p. 129–133.

Christensen, M. N., 1963, Structure of metamorphic rocks at Mineral King, California: University of California Publications in Geological Sciences, v. 42, no. 4, p. 159–198.

Clark, L. D., 1964, Stratigraphy and structure of part of the western Sierra Nevada metamorphic belt, California: U.S. Geological Survey Professional Paper 410, 70 p.

Crowder, D. F., McKee, E. H., Ross, D. C., and Krauskopf, K. B., 1973, Granitic rocks of the White Mountains Area, California-Nevada: Age and regional significance: Geological Society of American Bulletin, v. 84, no. 1, p. 385–396.

Curtis, G. H., Evernden, J. F., and Lipson, J. I., 1958, Age determination of some granitic rocks in California by the potassium-argon method: California Division of Mines and Geology Special Report 54, 16 p.

Dickinson, W. R., 1970, Relations of andesites, granites, and derivative sandstones to arc-trench tectonics: Reviews of Geophysics and Space Physics, v. 8, no. 4, p. 813–860.

Dodge, F. C. W., 1972a, Trace-element contents of some plutonic rocks of the Sierra Nevada batholith: U.S. Geological Survey Bulletin 1314-F, p. F1–F13.

——,1972b, Variation of ferrous-ferric ratios in the central Sierra Nevada batholith, U.S.A.: International Geology of Congress, 34th, Proceedings, Section 10, Montreal, 1972, p. 12–19.

Dougless, R. D., 1967, Permian Tethyan fusilinids from California: U.S. Geological Survey Professional Paper 593-A, p. 7–43.

Eaton, J. P., 1963, Crustal structure from San Francisco, California, to Eureka, Nevada, from seismic-refraction measurements: Journal of Geophysical Research, v. 68, no. 20, p. 5789–5806.

Eaton, J. P., and Healy, J. H., 1963, The root of the Sierra Nevada as determined from seismic evidence (abs.): International Union Geodesy and Geophysics, 13th General Assembly, Berkeley, California, 1963, Abstract Papers, v. 3, article B28, p. III–46.

Evernden, J. F., and Kistler, R. W., 1970, Chronology of emplacement of Mesozoic batholithic complexes in California and western Nevada: U.S. Geological Survey Professional Paper 623, 42 p.

Fiske, R. S., and Tobisch, O. T., 1978, Paleogeographic significance of volcanic rocks of the Ritter Range pendant, central Sierra Nevada, California: *in* Howell, D. G., and McDougall, K. A., editors, 1978, Mesozoic paleogeography of western United States: Pacific Section, Society of Economic Paleontologists and Mineralogists, Pacific Coast Paleogeography Symposium 2, p. 209–221.

Fiske, R. S., Tobish, O. T., Kistler, R. W., Stern, T. W., and Tatsumoto, M., 1977, Minerets caldera: A Cretaceous volcanic center in the Ritter Range roof pendant, central Sierra Nevada, California: Geological Society of America Abstracts with Programs, v. 9, no. 7, p. 975.

Griggs, D. T., 1939, A theory of mountain-building: American Journal of Science, v. 237, no. 9, p. 611–650.

Hine, R., Williams, I. S., Chappell, B. W., and White, A. J. R., 1978, Contrasts between I- and S-type granitoids of the Kosiusko batholith: Journal of Geological Society of Australia, v. 15, pt. 4, p. 219–234.

Huber, N. K., and Rinehart, C. D., 1965, Geologic map of the Devils Postpile quadrangle, Sierra Nevada, California: U.S. Geological Survey, Geological Quadrangle Map GQ-437, scale 1:62,500.

Jones, D. L., and Moore, J. G., 1973, Lower Jurassic ammonite from the south-central Sierra Nevada, Calif.: U.S. Geological Survey Journal of Research, v. 1, no. 4, p. 453–458.

Jones, D. L., Silberling, R. J., and Hillhouse, J., 1977, Wrangellia — a displaced terrane in northwestern North America: Canadian Journal of Earth Sciences, v. 14, no. 11, p. 1565–1577.

Kistler, R. W., and Bateman, P. C., 1966, Stratigraphy and structure of the Dinkey Creek roof pendant in the central Sierra Nevada, California: U.S. Geological Survey Professional Paper 524B, p. B1–B14.

Kistler, R. W., Bateman, P. C., and Brannock, W. W., 1965, Isotopic ages of minerals from granitic rocks of the central Sierra Nevada and Inyo Mountains, California: Geological Society of America Bulletin, v. 76, no. 2, p. 155–164.

Kistler, R. W., Evernden, J. R., and Shawe, H. R., 1971, Sierra Nevada plutonic cycle: Part 1, Origin of composite granitic batholiths: Geological Society of America Bulletin, v. 82, no. 4, p. 853–868.

Kistler, R. W., and Peterman, Z. E., 1973, Variations in Sr, Rb, K, Na, and initial $^{87}Sr/^{86}Sr$ in Mesozoic granitic rocks and intruded wall rocks in central California: Geological Society of America Bulletin, v. 84, no. 11, p. 3489–3512.

Knopf, A., and Thelen, P., 1905, Sketch of the geology of Mineral King, California: University of California Publications, Department of Geology Bulletin, v. 4, no. 12, p. 227–262.

Komar, P. D., 1972, Mechanical interactions of phenocrysts and flow differentiation of igneous dikes and sills: Geological Society of America Bulletin, v. 83, no. 4, p. 973–988.

Lachenbruch, A. H., 1968, Preliminary geothermal model of the Sierra Nevada: Journal of Geophysical Research, v. 73, no. 22, p. 6977–6989.

Lachenbruch, A. H., Sass, J. H., Munroe, R. J., and Moses, T. H., 1976, Geothermal setting and simple heat conduction models for Long Valley caldera: Journal of Geophysical Research, v. 81, no. 5, p. 769–784.

Lawson, A. C., 1936, The Sierra Nevada in the light of isostasy: Geological Society of America Bulletin, v. 47, no. 11, p. 1691–1712.

Lindgren, W., 1915, The igneous geology of the Cordilleras and its problems, *in* Problems of American geology: New Haven, Yale University, Silliman Foundation, p. 234–286.

Lockwood, J. P., and Bateman, P. C., 1976, Geologic map of the Shaver Lake quadrangle, central Sierra Nevada, California: U.S. Geological Survey Geologic Quadrangle Map, GQ 1271, scale 1:62,500.

Mackin, J. H., 1947, Some structural features of the intrusions in the Iron Springs district: Utah Geological Society Guidebook to the Geology of Utah, no. 2, 62 p.

McKee, E. H., and Nash, D. B., 1967, Potassium-argon ages of granitic rocks in the Inyo batholith, east-central California: Geological Society of America Bulletin, v. 78, no. 5, p. 669–680.

Moore, J. G., 1959, The quartz diorite boundary line in the western United States: Journal of Geology, v. 67, no. 2, p. 198–210.

——,1962, K/Na ratio of Cenozoic igneous rocks of the western United

States: Geochimica et Cosmochimica Acta, v. 26, p. 101–130.

Moore, J. G., and Dodge, F. C., 1962, Mesozoic age of metamorphic rocks in the Kings River area, southern Sierra Nevada, California, *in* Geological Survey research 1962: U.S. Geological Survey Professional Paper 450-B, p. B19–B21.

Moore, J. G., and Hopson, C., 1961, The Independence dike swarm in eastern California: American Journal of Science, v. 259, p. 241–259.

Morgan, B. A., and Stern, T. W., 1977, Chronology of tectonic and plutonic events in the western Sierra Nevada between Sonora and Mariposa, California: Geological Society of America, Abstracts with programs, v. 9, no. 4, p. 471–472.

Nokleberg, W. J., 1981, Stratigraphy and structure of the Strawberry Mine roof pendant, central Sierra Nevada, California: U.S. Geological Survey Professional Paper 1154, 18 p.

Oliver, H. W., 1977, Gravity and magnetic investigations of the Sierra Nevada batholith, California: Geological Society of America Bulletin, v. 88, no. 3, p. 445–461.

Oliver, H. W., and Robbins, S. L., 1973, Complete Bouguer anomaly map of the Mariposa 1° × 2° and part of the Goldfield 1° × 2° quadrangles: U.S. Geological Survey open-file map.

Pitcher, W. S., 1978, The anatomy of a batholith: Journal of the Geological Society, v. 135, pt. 2, p. 157–182.

Presnall, D. C., and Bateman, P. C., 1973, Fusion relationships in the system $NaAlSi_3O_8$-$CaAl_2Si_2O_8$-$KAlSi_3O_8$-SiO_2-H_2O and generation of granitic magmas in the Sierra Nevada batholith: Geological Society of America Bulletin, v. 84, no. 10, p. 3181–3202.

Roddick, J. A., and Armstrong, J. E., 1959, Relic dykes in the Coast Mountains near Vancouver, British Columbia: Journal of Geology, v. 67, p. 603–613.

Saleeby, J. B., Goodin, S. E., Sharp, W. D., and Busby, C. J., 1978, Early Mesozoic paleotectonic-paleogeographic reconstruction of the southern Sierra Nevada region: *in* Howell, D. G., and McDougall, K. A., editors, 1978, Mesozoic paleogeography of the western United States: Pacific Section, Society of Economic Paleontologists and Mineralogists, Pacific Coast Paleogeography Symposium 2, p. 311–336.

Schweickert, R. A., 1978, Triassic and Jurassic paleogeography of the Sierra Nevada and adjacent regions, California and western Nevada, *in* Howell, D. G., and McDougall, K. A., editors, Mesozoic Paleogeography of the western United States, Pacific Coast Paleogeography Symposium 2: Society of Economic Paleontologists and Mineralogists, p. 361–384.

Schweickert, R. A., Saleeby, J. E., Tobisch, O. T., and Wright, W. H., III, 1977, Paleotectonic and paleogeographic significance of the Calaveras complex, western Sierra Nevada: *in* Stewart, J. H., Stevens, C. H., and Fritsche, A. E., editors, 1977, Paleozoic paleogeography of the western United States: Pacific Section, Society of Economic Paleontologists and Mineralogists, Pacific Coast Paleogeography Symposium 1, p. 381–394.

Spera, F., 1980, Thermal evolution of plutons: A parameterized approach: Science, v. 207, p. 299–301.

Stern, T. W., Bateman, P. C., Morgan, B. A., Newell, M. F., and Peck, D. L., 1980, Isotopic U-Pb ages of zircon from granitoids of the central Sierra Nevada: U.S. Geological Survey Professional Paper (in press).

Stewart, J. H., Albers, J. P., and Poole, F. G., 1968, Summary of regional evidence for right-lateral displacement in the western Great Basin: Geological Society of America Bulletin, p. 1407–1414.

Vening-Meinesz, F. A., 1948, Major tectonic phenomena and the hypothesis of convection currents in the earth: Quarterly Journal of the Geological Society of London, v. 103, p. 191–207.

Wilshire, H. G., 1969, Mineral layering in the Twin Lakes Granodiorite, Colorado: Geological Society of America, Memoir 15, p. 235–261.

Wollenberg, H. A., and Smith, A. R., 1968, Radiogeologic studies in the central part of the Sierra Nevada batholith, California: Journal of Geophysical Research, v. 73, no. 4, p. 1481–1495.

——,1970, Radiogenic heat production in prebatholithic rocks of the central Sierra Nevada: Journal of Geophysical Research, v. 75, no. 2, p. 431–438.

MANUSCRIPT ACCEPTED BY THE SOCIETY JULY 12, 1982

Geological Society of America
Memoir 159
1983

The Salinian block — A structurally displaced granitic block in the California Coast Ranges

Donald C. Ross
U.S. Geological Survey
345 Middlefield Road
Menlo Park, California 94025

ABSTRACT

The Salinian block is an elongate block of granitic basement that has presumably been displaced by large lateral movements on the San Andreas fault system. Its point of origin was most likely some place farther south within the Cordilleran batholithic belt, but its present position, immersed in the Franciscan "subduction" assemblage, is a challenge to all reconstruction models for the western margin of North America. The granitic rocks generally have abundant quartz and range from quartz diorite to quartz monzonite[1]. In chemical character, they straddle the boundary between calcic and calc-alkalic with a Peacock index of about 61. The most common framework metamorphic rocks are strongly deformed, thinly layered gneiss, granofels, and impure quartzite, with lesser amounts of schist and marble, which suggest a dominantly thin-bedded silty, sandy protolith. The absence of thick pure quartzite and marble units suggests that they are not correlative with the Cordilleran miogeoclinal deposits — traditionally considered to be their parent terrane. The virtual absence of metallic mineralization is also anomalous relative to other granitic terranes of the region.

STRUCTURAL SETTING

The western margin of North America has a long and active history as a mobile tectonic region that features accretion of exotic terranes (Irwin, in press; Jones and others, 1978; Churkin and Eberlein, 1977), mind-boggling lateral movements of plate segments (Atwater, 1970), and present-day fault displacements that trigger earthquakes to remind everyone that the tectonism is still continuing. It is into this setting that the Salinian block (Fig. 1) must be accommodated. The Salinian block is a 40,000 km² fault-bounded block of granitic and metamorphic basement along California's coastline that is structurally isolated from other Cordilleran granitic terranes of the west coast by presumed large-scale, right-lateral movements of at least 500 km on the San Andreas fault system. The Salinian block now sits engulfed in the Franciscan assemblage — a subduction complex of graywacke, shale, chert, and volcanic rocks that accumulated along a convergent plate boundary during late Mesozoic and early Tertiary time. No rocks of the Franciscan assemblage have ever been found within the Salinian block, however. The original position of the Salinian block is unknown, but the chemical and petrographic characters of the granitic rocks and their metamorphic framework rocks suggest that the Salinian block belongs somewhere within the Cordilleran batholithic belt. The initial position of the Salinian block is a prime key to any reconstruction models of the western North American continental margin.

CHARACTER AND DISTRIBUTION OF BASEMENT ROCK TYPES

Basement rock outcrop areas, plus a rather generous

---

[1]The rock names used in this report follow a classification scheme (shown on Figure 3) that has been used for numerous previous reports dealing with the granitic rocks of the Salinian block. This classification can be "converted" to the nomenclature of the classification proposed by the IUGS Subcommission on the systematics of igneous rocks (Streckeisen, 1976) by replacing quartz diorite with tonalite (IUGS) and quartz monzonite with granite (IUGS). Granodiorite, as used in this paper, is also granodiorite in the IUGS classification.

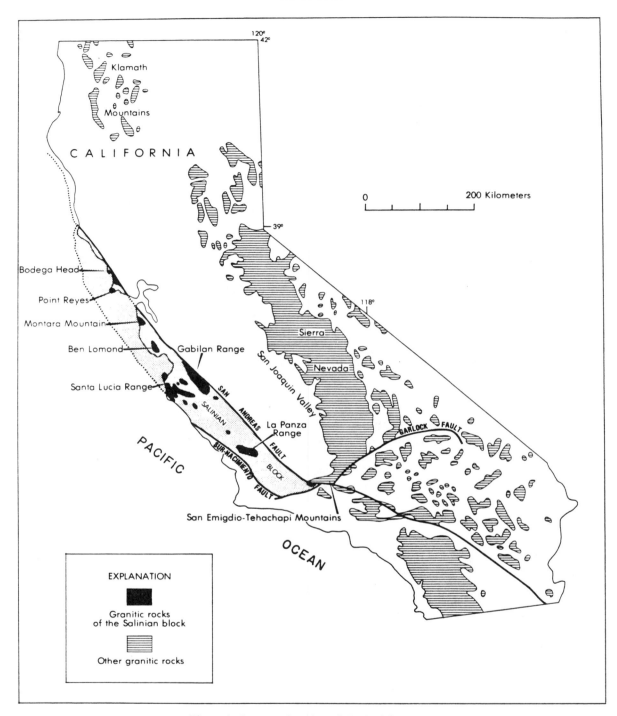

Figure 1. Structural setting of the Salinian block.

extrapolation into the subsurface based on well-core material and off-shore dredging, provide some knowledge of the basement for about 6000 km², some 15 percent, of the Salinian block. In the rest of the block, the basement is overlain by a thick pile of Mesozoic and Cenozoic clastic sedimentary rocks or is off-shore beneath the Pacific Ocean.

METAMORPHIC FRAMEWORK ROCKS

The pre-intrusive metamorphic host rocks of the Salinian block most probably are Paleozoic, but no identifiable fossils have ever been found. These rocks are dominantly strongly deformed, moderate- to high-grade gneiss, granofels, and impure quartzite, with lesser amounts of schist and marble. Pure quartzite is virtually absent, and marble, though locally prominent, is rarely a dominant rock type. Probably, the protoliths of this suite were mostly (if not wholly) sedimentary — a thin-bedded sequence of inter-bedded quartz-rich siltstone and impure sandstone, with lesser amounts of shaly, marly, and calcareous rocks.

The metamorphic rocks are almost exclusively of amphib-olite grade. Red garnet is common and widespread, and sillimanite is common in rocks of suitable argillaceous composition. Locally, particularly in the western block, hypersthene-bearing gneiss and granofels of granulite grade are found. The distribution of the granulitic rocks suggests that they reflect local hot or dry spots and not significantly deeper terranes than the amphibolite-grade rocks.

Although the foregoing describes most of the meta-morphic framework, there are two notable exceptions. In the eastern Santa Lucia Range (Sierra de Salinas) and southern Gabilan Range (Fig. 2), a homogeneous biotite quartzofeldspathic schist with minor associated quartzite, amphibolite, and marble has a chemical composition sug-gesting a graywacke protolith with minor associated volcanic rocks and marble (Ross, 1976).

The other exception is east of the Red Hills-San Juan-Chimeneas fault zone (Barrett Ridge slice), where limited outcrop areas and numerous samples from wells that reached basement suggest similarity to the gneissic, migma-titic Precambrian terrane of parts of the Transverse Ranges to the southeast. The local presence of marble, thinly layered calc-hornfels, and quartzite indicates sedimentary protoliths for parts of the section, but for most of the gneissic rocks, the degree of ultrametamorphism or granitization is intense enough to obscure original compositional layering. The similarity to the Transverse Ranges gneisses suggests that the Barrett Ridge slice may have a Precambrian framework. Framework metamorphic rocks contrast strongly across the Red Hills-San Juan-Chimeneas fault zone.

Traditionally, the pre-intrusive rocks of the Salinian block have been thought to be Paleozoic and most probably related to the Cordilleran miogeosyncline, largely on the basis of the local abundance of marble. However, the lack of thick pure quartzite and marble units in the Salinian rocks makes them unlikely candidates for correlation with the Paleozoic Cordilleran miogeosyncline terrane.

GRANITIC TERRANES

The Salinian block can be subdivided into four smaller blocks; in each, the granitic rocks are characterized by some degree of petrographic and structural coherence. The largest area, here called the central block, embraces most of the granitic outcrops of the Salinian block: the La Panza Range, the Gabilan Range, most of the Santa Lucia Range, and the Ben Lomond area. The western block includes that part of the Santa Lucia Range west of the Palo Colorado fault. The southeastern block (Barrett Ridge slice) includes the terrane east of Red Hills-San Juan-Chimeneas fault: Red Hills, Barrett Ridge, and the Mount Abel-Pinos area. The north-ern block is a catch-all for the widely separated outcrops north of the Zayante-Vergeles fault: Montara Mountain, Farallon Islands, Point Reyes, and Bodega Head. The degree of petrographic and structural coherence in the northern block is less certain because basement exposures are so few.

Central block

The granitic basement of the central block is dominated by four large plutonic units (Fig. 2). The most extensive is the granodiorite-quartz monzonite of the La Panza Range, a seriate, distinctive, and relatively homogeneous rock, peppered with small biotite flakes and much less common hornblende crystals, that is exposed over an area of some 250 km^2 in the La Panza Range. Similar rocks occur in the Adelaida area near Paso Robles and in several well cores as far north as the San Ardo oil field. Petrographically and chemically similar rocks are also exposed in the south and central Gabilan Range.

Another important granitic unit is the quartz diorite-granodiorite of Johnson Canyon in the Gabilan Range. This rock contains abundant irregular biotite and hornblende crystals that are only locally euhedral. Considerable varia-tion exists in this mass, and it is probable that some facies or separate intrusions may be present.

The other two large plutonic units are in the Santa Lucia Range: the hornblende-biotite quartz diorite of Soberanes Point and its probable correlative units, and the porphyritic granodiorite of Monterey. The contact between Soberanes and Monterey masses is a wide gradational zone of rocks intermediate in composition. The present author has inter-preted the granodiorite of Monterey as having intruded the still partially fluid Soberanes Point rocks. This interpreta-tion seems compatible with a hot, "mushy" terrane indicated by the abundant migmatitic and gneissic rocks throughout the northern Santa Lucia Range.

The granodiorite of Monterey is the most distinctive granitic unit in the central block because of its K-feldspar megacrysts, as long as 10 cm, which make remarkable "hobnail" exposures, particularly along the sea coast. Its areal extent, including submarine outcrops in Monterey Bay and its probable extent beneath Cenozoic deposits southeast of Monterey Peninsula, is about 500 km^2.

The Soberanes Point mass is also relatively dark quartz diorite that contains abundant biotite and hornblende. It underlies nearly 200 km^2 of the central part of the Santa Lucia Range and, together with two possibly correlative masses, is a major granitic unit in the central block. This unit

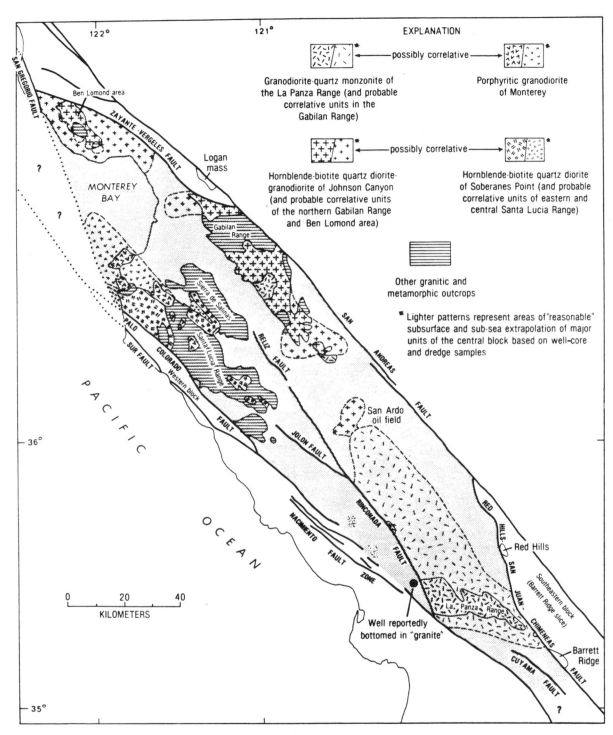

Figure 2. Central block (shaded area) of the Salinian block showing relations to the rest of the Salinian block, and the distribution of the four major granitic units.

may be related to some or all of the quartz diorites of the San Ardo oil field, the Gabilan Range, and Ben Lomond.

The occurrence of two associated major rock types (La Panza-Monterey and hornblende-biotite quartz diorite) within the central block suggests a comagmatic coherent basement framework within which later major structural dislocations seem unlikely.

Western block

The most distinctive plutonic rock type in the Salinian block is in the Santa Lucia Range west of the Palo Colorado fault zone. This rock, the "charnockitic tonalite" of Compton (1960), characteristically contains hypersthene and only about 10 percent quartz (Fig. 3). Six masses of varied size, together, cover less than 50 km[2] and are immersed in gneiss and granofels, which are also, in part, hypersthene-bearing and are generally less well-banded than the gneissic rocks east of the Palo Colorado fault zone. The metamorphic matrix is characterized by more marble than is present elsewhere in the Santa Lucia Range and by local granulite-facies gneisses bearing hypersthene.

Within the western block are aplite-alaskite-pegmatite dikes and small masses but no granitic plutons other than the charnockitic tonalite. The abrupt western termination of the large Soberanes Point mass by the Palo Colorado fault emphasizes the strong contrast of granitic basement across this fault zone.

Southeastern block

Exposures within the southeastern block are limited to the rather small areas of Red Hills, Barrett Ridge (Fig. 2), and the Mount Abel-Pinos massif (southeastern corner of the Salinian block, see Fig. 6); in addition, there are a few subsurface samples. Most of the basement is relatively high grade gneiss or ultrametamorphosed gneiss which exhibits partly- to fully- developed homogeneous granitic fabric. The only true intrusive rocks in this block are quite felsic. Several oil wells drilled in the southeastern block penetrated various kinds of gneiss, some amphibolite, and felsic granitic rocks. Thus, there is a strong basement contrast between the central and southeastern blocks across the Red Hills-San Juan-Chimeneas fault.

Northern block

North of the gabbroic outcrops of the Logan mass, the basement of the Salinian block is represented by only four isolated on-shore outcrop areas, by the tiny granitic Farallon Islands (Fig. 6), and by a few dredge samples.

The southernmost mass of the northern block, at Montara Mountain, involves some 80 km[2] of deeply weathered granitic rocks, poorly exposed in brushland and grassland.

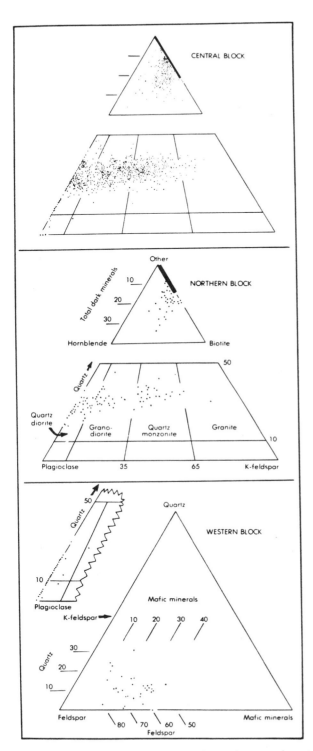

Figure 3. Modal character of the central, northern, and western blocks of the Salinian block. Central block labels same as for northern block.

Fresh rock is uncommon even in deep roadcuts and in sea cliffs, and shearing and cataclastic deformation are common. Most of the rocks are quartz diorite with abundant biotite and hornblende. The dark minerals tend to be coarse and irregular and tend to occur in splotchy aggregates. Some of these coarse foliated outcrops, particularly along the shore, are distinctly different in appearance from other Salinian block granitic rocks.

The nearest granitic outcrops to the Montara area are on the Farallon Islands, some 45 km to the northwest. Cobble beaches provide fresh granitic rocks with scattered large biotite books and minor hornblende. K-feldspar occurs solely in distinctive lacy interstitial patterns. Modal compositions cluster at the quartz diorite-granodiorite boundary.

In the Point Reyes area, about 60 km north from the Montara mass, granitic rock crops out in a small area at Point Reyes. More extensive granitic outcrops extend 30 km along the western side of Tomales Bay (Fig. 6). The northern part of this area (Tomales Point) is underlain by hornblende-biotite quartz diorite, but most of the area is underlain by a much more felsic rock rich in K-feldspar. These two rock types are separated by a metamorphic screen (or septum) at McClures Beach — on the sea side — possibly the same screen is also seen at Marshall Beach across the peninsula on Tomales Bay.

About 10 km north along the coast from the northernmost exposures of the Point Reyes area, a few square kilometers of hornblende-biotite quartz diorite crop out at Bodega Head. Although these rocks contain more hornblende than do the rocks on Tomales Point, they are otherwise similar and probably are parts of the same mass, but because of their spatial separation, they could represent separate intrusions.

Some 35 km offshore from Point Reyes, submarine granitic outcrops have been sampled by dredging. Chesterman (1952) briefly described these rocks as being composed dominantly of plagioclase and quartz with small amounts of K-feldspar and variable amounts of biotite and hornblende. The description suggests that they are much like the quartz diorite of Tomales Point and Bodega Head.

MODAL AND CHEMICAL SUMMARY OF GRANITIC BASEMENT

The bulk of the exposed granitic basement rocks is found in the central and northern blocks. They are chemically near the boundary (Fig. 4) between the calc-alkalic and calcic fields of Peacock (1931). On this same figure, plots of Alk-F-M and Or-Ab-An for all the chemically analyzed specimens show normal granitic fields and trends. The analyses on which these summaries are based can be found in Ross (1972a, 1975, and 1977).

Modally, these rocks fall in a broad swath across the quartz diorite, granodiorite, and quartz monzonite fields

(Fig. 3). The modal field plots nearly horizontal. In contrast, many granitic terranes in the West Coast batholithic belt have modal fields that trend toward lower quartz values in the quartz diorite end of the field. Horizontal modal fields of this sort characterize some trondhjemitic suites, but the rocks of the Salinian block are not chemically trondhjemitic (Ross, 1973). A general plot by area of outcrop (Fig. 5) shows that no one rock type dominates; in the past, the prevalent opinion was that quartz diorite was predominant in the Salinian block. The hypersthene-bearing, relatively quartz-poor granitic rocks of the western block (Fig. 3) present a sharp contrast to the other granitic rocks of the Salinian block. Though the charnockitic rocks have a limited area of outcrop, their distinctive mineralogy and chemistry make them noteworthy from the standpoint of correlation with potential parent terranes.

AGE OF THE GRANITIC BASEMENT

In the pioneering work on age determinations by the K/Ar method by Curtis, Evernden, and Lipson (1958), biotite from granitic samples (Fig. 6) from Point Reyes, the Farallon Islands, Montara Mountain, the Santa Lucia Range, the Gabilan Range, and the La Panza Range gave radiometric ages in the range of 82-92 m.y. More recent refinement of decay rates revises these ages to 78-88 m.y. (Evernden and Kistler, 1970). Originally interpreted as the age of emplacement, these radiometric dates are now generally considered to reflect some post-emplacement event. Compton (1966), for example, noted that biotite-bearing granitic rocks in the Santa Lucia Range yielding radiometric ages of 70 to 75 m.y. are overlain by Upper Cretaceous sedimentary rocks containing Campanian fossils, indicating that those radiometric ages could not date emplacement. Other K/Ar ages on biotite and hornblende from granitic rocks of the Salinian block have generally been in the 70 to 90 m.y. range (Evernden and Kistler, 1970; Huffman, 1972; Hart, 1976). Biotite ages as young as 55 m.y. have been reported from the La Panza Range (Hart, 1976), but mineral alteration with consequent argon loss may account for the younger age, as biotite from fresh quarry rock in the same mass yielded a radiometric age of 72 m.y.

Rubidium-strontium age determinations by R. W. Kistler (reported in Ross, 1972b) on six granitic samples in the Gabilan Range give a good isochron of 110 ± 5 m.y. Three granitic samples from the Santa Lucia Range plot on this same isochron. These data also agree with a thorium-lead age of 106 m.y. determined by Hutton (1959) on monazite from the porphyritic Monterey mass.

In contrast, rubidium-strontium data from the southeastern block (Kistler and others, 1973) suggest that the homogenized gneiss of the Mount Abel-Pinos massif is Precambrian and that the felsic rocks intruding it are compatible with a reference isochron age of about 180 m.y.

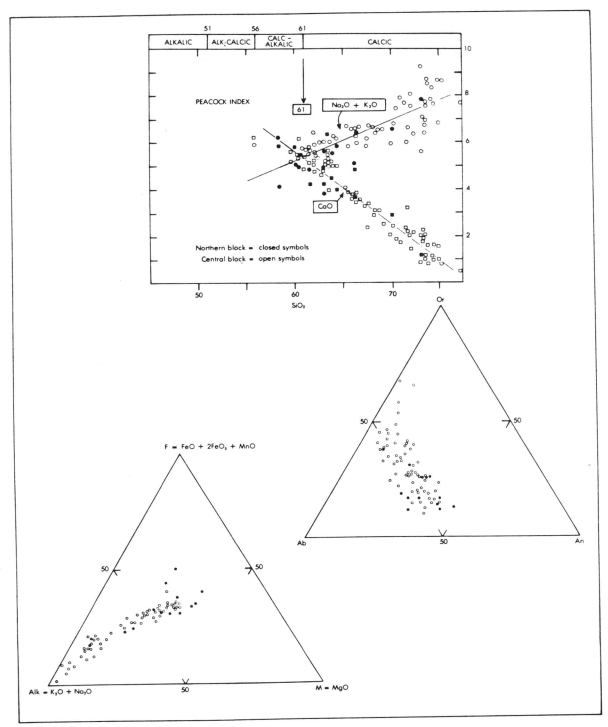

Figure 4. Chemical data plots for granitic rocks of the central and northern blocks of the Salinian block.

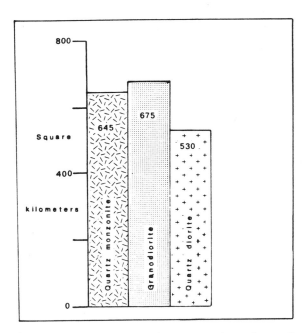

Figure 5. Area of exposed granitic basement by rock type for the Salinian block.

In addition, the basement rock at Red Hills has a much higher initial $^{87}Sr/^{86}Sr$ (0.7095) than other Salinian block rocks, which are in the range of 0.706 to 0.7080. These data, though limited, are compatible with the idea that the southeastern block is separate and distinct from the rest of the Salinian block.

Uranium-lead studies of sphene, feldspar, and apatite fractions from a granitic sample in the Point Reyes area gave an isochron of 61 m.y. (Mattinson, Davis, and Hopson, 1971). Uranium-lead studies on similar mineral fractions from the Santa Lucia Range (Mattinson, Hopson, and Davis, 1972) yielded an isochron of 79 ± 2 m.y. These isochrons are within the span of K-Ar ages for biotite and hornblende samples of the Salinian block.

Mattinson, Hopson, and Davis (1972) also studied zircon separates from Bodega head, Point Reyes, and the Santa Lucia Range. Their data suggest zircon crystallization about 100 m.y. ago for the Bodega Head and Santa Lucia Range samples and probably a similar age for the Point Reyes samples. However, the Point Reyes samples give discordant ages, which may reflect the incorporation of Paleozoic or Precambrian zircons by assimilation or anatexis of wall-rocks.

Fission-track ages have been determined for 26 sphene and 24 apatite separates from granitic rocks of the Salinian block (Naeser and Ross, 1976). The sphene determinations range from 68 to 93 m.y., virtually the same age range as the K/Ar ages for biotite and hornblende. The more temperature-sensitive apatite yielded ages ranging from 3 to 74 m.y., in part, owing to the probable influence of Tertiary volcanism. The sphene ages have an unexplained bimodal

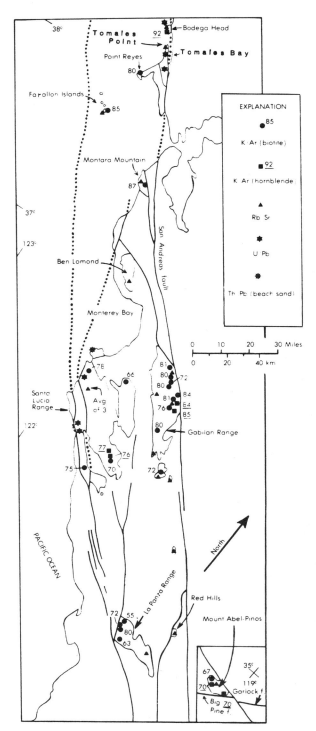

Figure 6. Location of radiometrically dated granitic samples (other than fission-track dates) for the Salinian block. Ages in m.y.

distribution, with the ages in the Gabilan and Santa Lucia Ranges slightly but consistently lower (avg. 75 m.y.) than the ages for parts of the Salinian block to the north and south (avg. 85 m.y.)

Present data (Rb/Sr and U/Pb) suggest emplacement

ages of 100-110 m.y. for much of the granitic terrane of the Salinian block. The southeastern block may have 180-m.y.-old felsic intrusions and possibly Precambrian gneiss. The charnockitic western block appears to share the 100-110-m.y. intrusive age with the strikingly different granitic terrane to the east (the central block).

MINERALIZATION IN THE SALINIAN BLOCK

The basement terrane of the Salinian block, with locally abundant marble and abundant felsic to intermediate granitic rock types, would seem to be a likely terrane for contact metamorphic and replacement mineral deposits, but surprisingly, the Salinian block is virtually barren of metallic mineralization. There is no recorded lode metal production; from 1840 to 1934, about $100,000 in placer gold were produced from stream deposits east of the La Panza Range, but the source of the placers has never been found.

Traces of scheelite, molybdenite, chalcopyrite, argentiferous galena, and arsenopyrite have been reported from the Santa Lucia and Gabilan Ranges, and minor scheelite has been reported from the Point Reyes area. Many semiquantitative spectrographic analyses of altered rocks and stream sediments in the Santa Lucia Range (Pearson, Hayes, and Fillo, 1967) revealed nothing of ore-grade, and no fluorescent tungsten minerals were seen. The absence of prospect pits and dumps, however, is the most telling piece of evidence for lack of metallic mineralization in this well-travelled area.

The barren nature of the Salinian block relative to other nearby granitic terranes along the western margin of North America is a fact, but the cause of the difference and its significance are not known. It is tempting to speculate that this difference reflects the possibility that the Salinian block is less closely related to other California granitic terranes than has previously been proposed.

REFERENCES CITED

Atwater, T., 1970, Implications of plate tectonics for the Cenozoic tectonic evolution of western North America: Geological Society of America Bulletin, v. 81, p. 3513-3536.

Chesterman, C. W., 1952, Descriptive petrography of rocks dredged off the coast of central California: California Academy of Sciences Proceedings, 4th Ser., v. 27, n. 10, p. 359-374.

Churkin, M., Jr., and Eberlein, G. D., 1977, Ancient borderland terranes of the North American Cordillera: correlation and microplate tectonics: Geological Society of America Bulletin, v. 88, p. 769-786.

Compton, R. R., 1960, Charnockitic rocks of Santa Lucia Range, California: American Journal of Science, v. 258, p. 609-636.

——,1966, Granitic and metamorphic rocks of the Salinian block, California Coast Ranges: California Division of Mines and Geology Bulletin 190, p. 277-287.

Curtis, G. H., Evernden, J. F., and Lipson, J., 1958, Age determinations of some granitic rocks in California by the Potassium-Argon method:

California Division of Mines Special Report 54, 16 p.

Evernden, J. F., and Kistler, R. W., 1970, Chronology of emplacement of Mesozoic batholithic complexes in California and western Nevada: U.S. Geological Survey Professional Paper 623, 42 p.

Hart, E. W., 1976, Basic geology of the Santa Margarita area, San Luis Obispo County, California: California Division of Mines and Geology Bulletin 199, 45 p.

Huffman, O. F., 1972, Lateral displacement of Upper Miocene rocks and the Neogene history of offset along the San Andreas fault in central California: Geological Society of America Bulletin, v. 83, p. 2913-2946.

Hutton, C. O., 1959, Mineralogy of beach sands between Halfmoon and Monterey Bays, Calif.: California Division of Mines and Geology Special Report 59, 32 p.

Irwin, W. P., 1981, Tectonic accretion of the Klamath Mountains, in W. G. Ernst, ed., The geotectonic development of California; Rubey Symposium Volume: Englewood Cliffs, N.J., Prentice-Hall, p. 29-49.

Jones, D. L., Silberling, N.J., and Hillhouse, J.W., 1978, Microplate tectonics of Alaska — significance for the Mesozoic history of the Pacific Coast of North America, in Howell, D. G., and McDougall, K. A., eds., Mesozoic paleogeography of the western United States; Society of Economic Paleontologists and Mineralogists, Pacific Sec., Pacific Coast Paleogeography Symposium 2, p. 71-74.

Kistler, R. W., Peterman, Z. E., Ross, D. C., and Gottfried, D., 1973, Strontium isotopes and the San Andreas fault, in Proceedings of the conference on tectonic problems of the San Andreas fault system: Stanford University Publication, Geological Sciences, v. 13, p. 339-47.

Mattinson, J. M., Davis, T. E., and Hopson, C. A., 1971, U-Pb studies in the Salinian block in California: Annual Report of the Director, Geophysical Laboratory Carnegie Institution Yearbook 70, p. 248-251.

Mattinson, J. M., Hopson, C. A., and Davis, T. E., 1972, U-Pb studies of plutonic rocks of the Salinian block, California: Annual Report of the Director, Geophysical Laboratory, Carnegie Institution Yearbook 71, p. 571-576.

Naeser, C. W., and Ross, D. C., 1976, Fission-track ages of sphene and apatite of granitic rocks of the Salinian block, Coast Ranges, California: U. S. Geological Survey Journal of Research, v. 4, no. 4, p. 415-420.

Peacock, M. A., 1931, Classification of igneous rock series: Journal of Petrology, v. 39, n. 1, p. 54-67.

Pearson, R. C., Hayes, P. T., and Fillo, P. U., 1967, Mineral resources of the Ventana primitive area, Monterey County, California: U.S. Geological Survey Bulletin 1261-B, 42 p.

Ross, D. C., 1972a, Petrographic and chemical reconnaissance study of some granitic and gneissic rocks near the San Andreas fault from Bodega Head to Cajon Pass, California: U.S. Geological Survey Professional Paper 698, 92 p.

——,1972b, Geologic map of the pre-Cenozoic basement rocks, Gabilan Range, Monterey and San Benito Counties, California: U.S. Geological Survey Miscellaneous Field Investigation Map MF-357.

——,1973, Are the granitic rocks of the Salinian block trondhjemitic?: U.S. Geological Survey Journal of Research, v. 1, no. 3, p. 251-254.

——,1975, Modal and chemical data for granitic rocks of the Gabilan Range, Central Coast Ranges, California: National Technical Information Service, Report PB-242 458/AS, 42 p.

——,1976, Metagraywacke in the Salinian block, central Coast Range, California — and a possible correlative across the San Andreas fault: U.S. Geological Survey Journal of Research, v. 4, no. 6, p. 683-696.

——,1977, Maps showing distribution of granitic rocks and modal and chemical data patterns, Santa Lucia Range, Salinian block California Coast Ranges: U.S. Geological Survey Miscellaneous Field Studies Map MF-799.

Streckeisen, A., 1976, To each plutonic rock its proper name: Earth Science
    Reviews, v. 12, p. 1–33.

MANUSCRIPT ACCEPTED BY THE SOCIETY JULY 12, 1982

Geological Society of America
Memoir 159
1983

Mesozoic and Cenozoic granitic rocks of southern California and western Mexico

Gordon Gastil
Department of Geological Sciences
San Diego State University
San Diego, California 92192

ABSTRACT

Plutonic and volcanic rocks of Middle Jurassic to Miocene age are widely exposed in southern California (U.S.A.), the peninsula of Baja California, and the states of Sonora, Sinaloa, and Jalisco (Mexico).

The Jurassic emplacement occurred in two belts, one in the extreme western margin of the continent, the other 500 kilometers inland in central Sonora. During early Cretaceous to Oligocene time, the axis of magmatism migrated from the peninsular ranges of southern and Baja California eastward into central Mexico. In the Miocene, magmatism moved westward to the area soon to become the Gulf of California.

In the continental margin island arc-type plutonic and volcanic magma erupted into Triassic and Jurassic ophiolite basement. In the Peninsular Ranges, the granitic rocks intruded the apron of Triassic-Jurassic clastic strata derived from the Precambrian craton to the east. Farther inland, Late Cretaceous and Early Tertiary plutons intruded Precambrian, early Paleozoic, and Cretaceous cratonic strata.

In the continental borderland and mainland Mexico, the observable plutons are shallowly emplaced into low-grade metamorphic rocks. In the Peninsular Ranges, uplift and deep erosion have exposed diapir-shaped bodies emplaced within upper amphibolite grade schists and gneisses.

In the continental margin, gabbro and low-potassium tonalite predominate; in the western Peninsular Ranges, gabbro is associated with a variety of rocks, ranging from diorite to granite. In the eastern Peninsular Ranges, leuco-tonalite is predominant, with granite becoming progressively important eastward into mainland Mexico.

The distribution of Jurassic arcs suggests that an oceanic arc was sutured onto North America in Early Cretaceous time. The tremendous volume of added sialic magma dilated the western edge of the continent by almost 300 kilometers, perhaps causing the continent to override the trench at the end of the Oligocene.

INTRODUCTION

In describing the Mesozoic-Cenozoic granitic rocks of southern California (U.S.A.) and western Mexico, this paper will palinspastically move peninsular California back to what is, in the author's opinion, the best pre-Gulf of California fit (Figs. 2-7). The original position of the different peninsular elements is conjectural.

This paper will be concerned with the granitic rocks as far north as the Los Angeles basin and the Channel Islands of California, and as far south as Barra Navidad, in western Jalisco (Fig. 1). For summaries of past work, see Gastil and others (1975) and Silver and others (1979a,b).

THE AGES OF THE GRANITIC ROCKS

Comparisons between isotopic U/Pb determinations on zircons and K/Ar determinations on various minerals (Krummenacher and others, 1975; Silver and others, 1979b) show that, for the northern, axial portion of the Peninsular

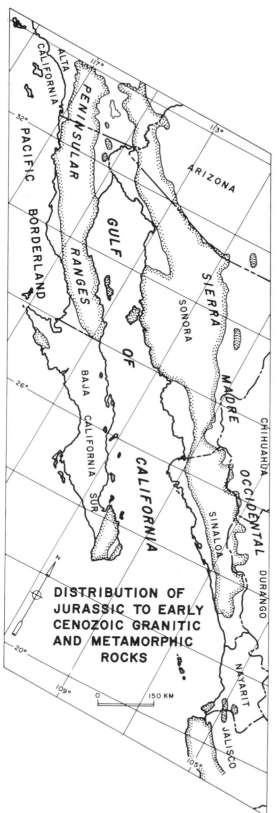

Figure 1. Index map of Jurassic to Miocene granitic and metamorphic rocks of southern California and western Mexico.

Ranges of southern California and northern Baja California, there are disparities of up to 25 m.y., almost invariably showing greater ages for zircons. Since these zircons are largely unaffected by inherited lead (Silver and others, 1979b), workers in this area have concluded that the zircon ages more closely correspond to the ages of pluton emplacement and that the K/Ar data (including many concordant hornblende-biotite pairs) give the dates at which the rocks cooled to the blocking temperatures for argon. Both sets of data are important in unraveling the regional history, and their comparison has tectonic implications.

Where the granitic rocks seen at the surface today intrude weakly metamorphosed rocks, as along the Pacific margin of the peninsula, the disparity between zircon ages and K/Ar ages tends to be small (less than 10 m.y.); where the present erosional level shows plutons intruding essentially unmetamorphosed host rocks, as in western Sonora and Sinaloa, the difference between zircon and K/Ar ages is even less (Gastil and others, 1978; Henry, 1975). Thus, whereas K/Ar ages may reveal little about the age of emplacement in deeply eroded areas, K/Ar ages of shallowly emplaced bodies may be very close to the age of emplacement.

Where no zircon data are available, K/Ar data can generally be considered as minimum ages. A possible exception may be found in minerals of extremely low potassium content, where inherited radiogenic argon can make the K/Ar age excessive.

EMPLACEMENT AGES

In the continental borderland (Santa Cruz Island, Isla Cedros, Vizcaino Peninsula), zircon data on plutonic rocks (Mattinson and Hill, 1976; Boles and Hickey, 1979) and K/Ar data on volcanic rock (Kilmer, 1979) date the first calc-alkaline magmatism at approximately 155 m.y. Biostratigraphy on Isla Cedros (Kilmer, 1979) and the Vizcaino Peninsula (Rangin, 1978; Boles and Hickey, 1979; Barnes and Berry, 1979) confirms the volcanic age and places confirming constraints on the plutonic ages.

In central Sonora, both zircon age work (Anderson and Silver, 1979) and biostratigraphy (Rangin, 1978) identify an interval of andesitic volcanic-plutonic activity of late Middle to Late Jurassic age.

The belts of Jurassic magmatism in the Pacific borderland and in central Sonora are today separated by more than 500 kilometers. Much of this separation has probably resulted from dilation caused by the emplacement of the Mesozoic granitic rocks; however, the Pacific margin andesites rest on oceanic crust, the Sonoran andesites rest on the Precambrian craton, and the two terranes are separated by a considerable basin of upper Paleozoic to Middle Jurassic marine strata. The two andesite belts must have been separated by a minimum of 250 kilometers at the time of emplacement.

Upper Jurassic (Portlandian) volcanic-volcaniclastic rocks near San Diego (Fife and others, 1967) contain clasts of granitic rocks. This was first noted by Hanna (1926) and has since been observed at many localities from Lake Hodges to Crest (unpublished field class studies, San Diego State University). These range from gabbro to granodiorite in composition. Many clasts had been stream-rounded, and some had been partly fused by the enclosing volcanic rock. In contrast, the oldest plutonic dates in the western part of the peninsula are 130 m.y. for zircon (Silver and others, 1979b) and 126-142 m.y. for four K/Ar ages on hornblende (which may reflect inherited argon).

The granitic clasts in the Santiago Peak Volcanics may have had a provenance to the west in the continental borderland belt (mentioned above). However, the andesite volcanism itself suggests the probability of co-magmatic plutons of Jurassic age, although none have as yet been identified in the Peninsular Ranges.

Zircon data (summarized by Silver and others, 1976) clearly show that the majority of relatively unmetamorphosed granitic bodies in the western part of the peninsula were emplaced between 120 and 105 million years ago. These ages are distributed across what Gastil (1975) described as the "gabbro belt" with no apparent age gradient. East of the gabbro belt, zircons give ages of less than 105 million years, with declining ages eastward (Silver and others, Fig. 3, 1979b).

The ages of 105 to 120 m.y. correspond fairly well to the well-documented biostratigraphic age of the Alisitos Formation (summarized in Gastil and others, 1975). The Alisitos is an andesite volcanic-volcaniclastic formation which appears to have received little or no detritus from the craton to the east. No direct evidence for the deposition of this age volcanic rock has as yet been found in the peninsula north of the Agua Blanca fault (Gastil and others, 1975, 1978).

In Sinaloa, volcanic rocks have been biostratigraphically dated as Albian-Cenomanian (Bonneau, 1971). The plutonic rocks in this area range in age from 102 m.y. (U-Pb) for quartz diorite to 45 m.y. (K/Ar) for granodiorite (Henry, 1975). In these rocks, zircon U-Pb ages are only slightly older than K/Ar ages for concordant biotite-hornblende. The oldest plutonic rocks may predate the biostratigraphically dated Cenomanian andesites.

In coastal Sonora, Anderson and Silver (1969) report zircon ages on plutonic rocks as young as 83 m.y., and there are similar K/Ar ages (Gastil and others, 1978). Nonmarine volcanic rocks and related shallow intrusive bodies are common through the Upper Cretaceous and Lower Tertiary strata of eastern Sonora and adjacent areas (Coney and Reynolds, 1977; Keith, 1978).

As yet, no zircon ages have been reported for granitic rocks in the tip of Baja California, Nayarit, or Jalisco. Volcanic and plutonic K/Ar ages for these areas are largely Late Cretaceous and Early Tertiary (Jensky, 1975; Gastil

and others, 1978). There may be older plutonic and volcanic rocks which have been reset. Older discordant K/Ar ages have been obtained for rocks in the southernmost part of Baja California (Gastil and others, 1976).

Magmatism of Oligocene and Miocene age was extensive in western Mexico. The present erosional level exposes primarily volcanic rocks, but small granitic plutons are exposed in several places. The dating of these rocks by K/Ar is consistent with nonmarine biostratigraphy (summarized by Coney and Reynolds, 1977; and Clark and others, 1977; Keith, 1978).

SUMMARY OF MAGMATIC HISTORY

The earliest widespread andesitic, volcanic-plutonic interval was late Middle to Early Jurassic and occurred along two belts: one in what is now the Pacific borderland (possibly including Arroya San Jose on the peninsula, Minch, 1969) and the other in central Sonora. These belts were, at the time, separated by at least 250 kilometers and may have been much farther apart.

Volcanic rocks of latest Jurassic and Early Cretaceous age were deposited along the northwestern edge of the peninsula, possibly extending into what is now northwestern Sonora (Anderson and Silver, 1969). Evidence of contemporary plutonic rocks has yet to be adequately documented.

A belt of volcanic rocks, Aptian-Albian in northern Baja California and Albian-Cenomanian in Sinaloa, extended from at least the Agua Blanca fault in northern Baja California to the vicinity of Mazatlan in Sinaloa. Gabbroic to granodioritic rocks of this age are abundant in northwestern Baja California and in southern California to the northernmost limit of the granitic province under discussion.

From Cenomanian to Oligocene time, the locus of volcanic and plutonic emplacement moved eastward across the continent, erupting on and into a nonmarine terrane. In Oligocene time, the eastward movement halted, and about 23 million years ago, the locus of magmatism jumped westward to the area now bordering the Gulf of California.

COOLING AGES

Since the blocking temperatures for argon in biotite and hornblende (Hart, 1964) are considerably below the temperatures at which zircons tend to equilibrate with their environment (Gastil, 1976), one can expect K/Ar ages to be less than zircon ages where cooling was slow and essentially the same where it was geologically fast. Where biotite and hornblende cooling ages are within experimental agreement, and still considerably less than the zircon age, the rock probably remained above the argon blocking temperature for some millions of years and then rapidly cooled (Krummenacher and others, 1975).

Figure 2 shows the location of all the K/Ar dates

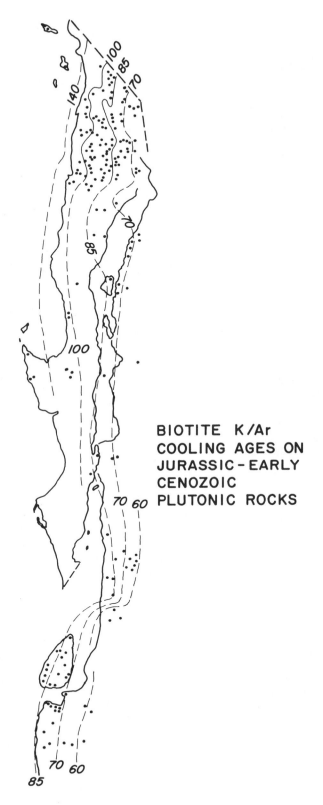

BIOTITE K/Ar
COOLING AGES ON
JURASSIC - EARLY
CENOZOIC
PLUTONIC ROCKS

Figure 2. Cooling ages for plutonic rocks in Baja California and western Mexico (peninsula placed in hypothesized pre-Miocene position).

tabulated by Krummenacher and others (1975), Gastil and others (1976), as well as Henry (1975) and miscellaneous other data. The contours are necessarily both generalized and conjectural. East of what is now the Gulf of California, cooling ages may approximate emplacement ages.

Unfortunately, the number of zircon dates is not yet sufficient to adequately evaluate the disparity on an areal basis. Present evidence indicates that, in the Laguna Mountains, the Sierra Juarez, and in the desert ranges in the northern part of the Gulf of California depression, there is commonly 25 million years difference for biotite and 15 million years difference for hornblende (see Silver and others, Fig. 3, 1979b). To the west and to the east (not shown on the above diagram), the difference is less. This belt of maximum disparity (Fig. 3) corresponds to host rocks that show the highest grade of metamorphism and, therefore, probably also the greatest uplift and erosion. Thus, one might predict that large disparities between zircon and K/Ar ages will be found in the Sierra Victoria south of La Paz and in the Cabo Corrientes area of northwestern Jalisco.

HOST ROCK TERRANES

The andesitic magmas of the Pacific borderland were emplaced through and onto a basement of ultramafic rocks, strataform gabbro, pillow basalt, chert, trench-type and deep sea fan clastic deposits, and tectonic and depositional melanges of all of these, including rocks affected by both greenschist and blueschist metamorphism. All the plutonic rocks appear to have been emplaced at shallow depths (Kilmer, 1979; Rangin, 1979; Boles and Hickey, 1979; Moore, 1979).

East of the Pacific margin is a zone in which the exposed host rocks consist almost entirely of Jurassic and Cretaceous volcanic rocks, which range in composition from basalt to rhyolite but which are predominantly calc-alkaline andesite-dacite. This zone (Fig. 4) extends from southern California and the length of northern Baja California, crossing to the eastern edge of the southern half of the peninsula and following the coast of Sinaloa, apparently continuing into Nayarit and Jalisco. The age of the predominant volcanic host rock changes from Late Jurassic in southern California, to Aptian-Albian in northern Baja California, to Albian-Cenomanian in northern Baja California, to Albian-Cenomanian in Sinaloa, and to even younger Cretaceous in Jalisco. Rangin (1978) believes that pillow basalt and pyroxenites near El Arco, Baja California, are unconformably beneath the Aptian-Albian sequence — suggesting that this volcanic "arc" is also built directly on older oceanic crust. Walawender and Smith (1979) cite petrographic and trace element data supporting considerable chemical interaction between the intrusive and host rocks.

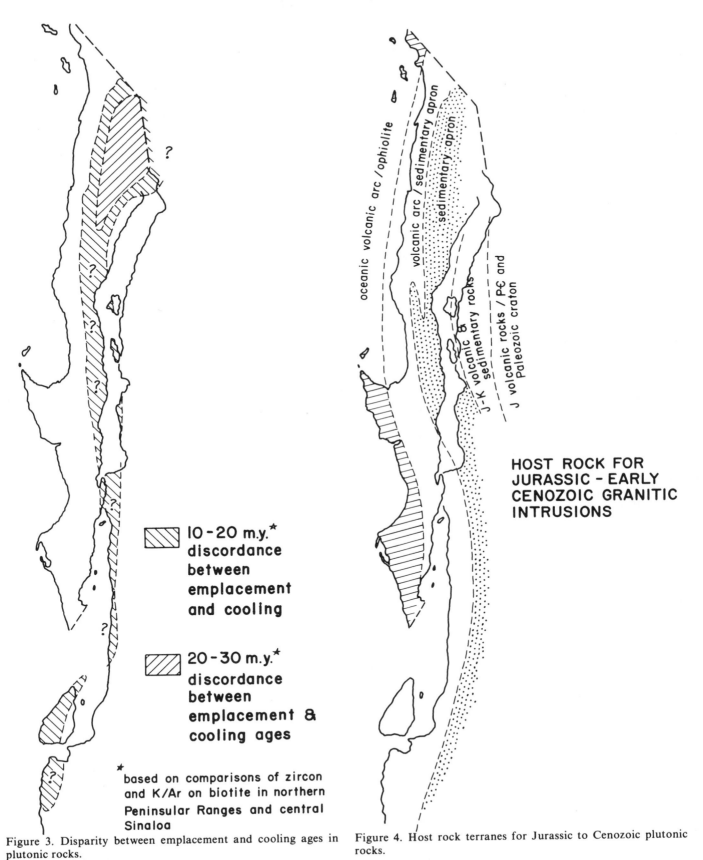

10-20 m.y.* discordance between emplacement and cooling

20-30 m.y.* discordance between emplacement & cooling ages

*based on comparisons of zircon and K/Ar on biotite in northern Peninsular Ranges and central Sinaloa

oceanic volcanic arc / ophiolite

volcanic arc / sedimentary apron

sedimentary apron

J-K volcanic & sedimentary rocks

J volcanic rocks / PЄ and Paleozoic craton

HOST ROCK FOR JURASSIC - EARLY CENOZOIC GRANITIC INTRUSIONS

Figure 3. Disparity between emplacement and cooling ages in plutonic rocks.

Figure 4. Host rock terranes for Jurassic to Cenozoic plutonic rocks.

East of the zone of volcanic host rock is a belt in which the host rock consists of Triassic and Jurassic marine clastic rocks derived from the craton to the east (Criscioni and others, 1978; Gastil and others, 1978). This host rock predominates in the western slopes of the Peninsular Ranges of Southern California and northern Baja California but, if present, has not been identified in Sinaloa, Nayarit, and Jalisco. These rocks host the eastern edge of the gabbro zone rocks and the western edge of the leucotonolite zone. The boundary with the volcanic terrane to the west is generally abrupt (Gastil and others, 1980), but the boundary to the east is indefinite. Walawender and others (1979) cite evidence for significant chemical interaction between the intrusives and the metasedimentary host rock, and Taylor and Silver (1978) conclude that the intrusions in the eastern part of the peninsula must have interacted with oxygenated shallow crustal rocks. However, Silver and others (1979b) and Gromet and Silver (1979) do not believe that the variations in strontium isotopes and rare elements can be explained by shallow contamination. Certainly, the major compositional boundaries that Silver, Taylor, Chappel and Gromet cite between what Silver calls the older and younger arcs do not clearly correspond to the boundary between arc-volcanic and craton-derived clastic host rocks.

In the eastern slopes of the Peninsular Ranges, extending into the ranges of the Gulf of California depression and onto the coast of Sonora, the host rock consists of carbonate and associated metasedimentary rocks, at least partly, of Carboniferous to Triassic age (Gastil and others, 1975, 1980). These sequences, most of which have had only cursory study, may include marginal marine sediments of the Atlantic type as well as basalt and trench-type deposits of a marginal basin. In the eastern slope of the major ranges, these rocks contain wollastonite, sillimanite, pyroxene, and garnet-rich rocks, but at many places near the Gulf of California, the metamorphism is lower green schist facies, and in the northern Sierra Pintas, fossils have been preserved.

A short distance inland from the coast of Sonora, the plutons intrude demonstrable Proterozoic-Lower Cambrian strata, and nearby these strata rest unconformably upon older Precambrian gneiss. The intrusions into this terrane are shallow.

In eastern Sonora, Chihuahua, Durango, and localities to the south, the Precambrian craton is overlain by a thick sequence of Mesozoic carbonate and volcanic strata. The Late Cretaceous and Cenozoic plutons of this region are observed largely in contact with these Mesozoic strata, emplaced at very shallow depths.

COMPOSITIONAL ZONING

In Figure 5, the Mesozoic-Cenozoic magmatic rocks are divided into six compositional belts. Published chemical

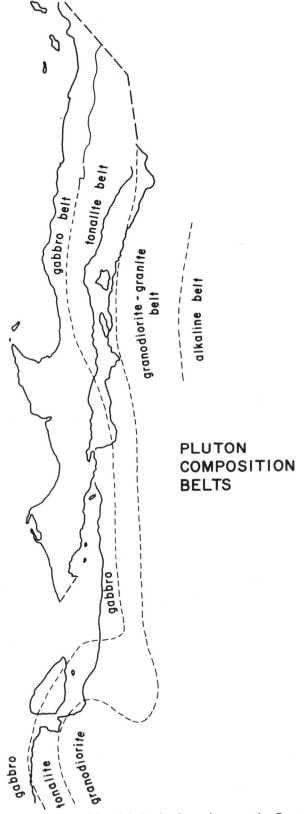

PLUTON COMPOSITION BELTS

Figure 5. Compositional belts in Jurassic to early Cenozoic magmatic rocks.

analyses (summarized in Gastil and others, 1975; Baird and Welday, 1974) give a very spotty coverage. Compositional distinctions, therefore, rely heavily on petrographic analyses.

Pacific Margin Belt

In the Channel Islands of southern California, Isla Cedros, and the western Capes of Baja California, the plutonic rocks include gabbro, tonalite, and plagiogranite (Mattinson and Hill, 1976; Moore, 1979; Boles and Hickey, 1979; Barnes and Berry, 1979). Figure 5 shows this belt swinging east across the southern part of Baja California (where there are almost no outcrops of Mesozoic rocks). This is based upon the 143 m.y. hornblende K/Ar age on a shallowly emplaced tonalite north of Loreto and blocks of ultramafic rock in a basal conglomerate near Tambobiche (latitude 25° 19'N, longitude 110° 56'W).

Gabbro Belt

The western part of the Peninsular Ranges of southern and Baja California is intruded by a combination of gabbro, tonalite, and granodiorite (Gastil, 1975; Silver and others, 1979b). The gabbro bodies have been described by Walawender (1976) and Walawender and Smith (1979). A large body of gabbroic rock in northern Sinaloa has recently been described by Mullan (1978). The location of the northern Sinaloa gabbro has caused the author to considerably revise the distribution of the gabbro belt as originally proposed by Gastil and others in 1972. Exposures of Henry (1975) have documented the continuity of the belt in Sinaloa as far south as Mazatlan. Gastil and others dated (K/Ar, hornblende) a single exposure of Cretaceous gabbro (98 ± 3 m.y.) in Nayarit. The belt is well represented in the western part of the peninsula of Baja California south of La Paz, is present on the northernmost part of the Tres Marias Islands (latitude 21° 45'N, longitude 106° 50'W; Chinas, 1963), and, although absent in extreme northwestern Jalisco, is well represented in southwestern Jalisco. Throughout this belt, gabbro constitutes no more than 20 percent of the plutonic rocks. The bodies commonly show evidence of original stratiform structure. In addition to common hornblende and clinopyroxene gabbro, they include orthopyroxene and olivine-rich phases, anorthosite, and pyroxenite. In many places, there are quartz and biotite-bearing phases, gradational into quartz diorite.

Most of the belt is intruded into the andesite arc host terrane, but the eastern edge of the belt, in the northern part of the peninsula, intrudes the craton-derived clastic host terrane.

Within portions of the gabbro belt, many of the plutonic rocks are cut by swarms of diabase to rhyolite dikes, the majority being dacite. The younger plutons, in turn, crosscut the dikes. In some places, the dike swarms occupy the greater part of the surface area (Gastil, 1975).

Near San Vicente Reservoir in southern California, Edelman (1980) has found that the oldest (foliated) granodiorite is intruded by gabbro and associated dioritic rocks. Similarly, south of Ojos Negros Valley in northern Baja California, gabbro crosscuts foliated tonalite. The gabbros, then, are not necessarily the earliest of the arc magmatic rocks.

Leucotonalite Belt

Tonalites, for example, the Green Valley and Bonsal type tonalites of Larsen (1948), are the most abundant rock type found in the gabbro belt. However, to the east of the gabbro belt, gabbros are almost entirely absent, and the tonalite bodies are larger, more homogeneous, and more leucocratic (Gastil, 1975). These tonalite bodies, characterized by the La Posta of eastern San Diego County, vary from potash, feldspar-free tonalite to granodiorite, without perceptible internal boundaries. The characterizing phase has idiomorphic hornblendes, biotite books, and idiomorphic sphene (up to 0.5 cm.). It is doubtful whether an unambiguous line can be drawn between the gabbro and leucotonalite belts.

Silver and others (1979b) see a strong correlation between the western edge of the leucotonalite zone and the ages of the plutonic rocks. From Silver and others (1979b, Fig. 3), it would appear that the average emplacement age drops about 10 million years across this boundary. These authors also see a correlation between this boundary and a sharp increase in the $w^{18}O$. Gromet and Silver (1979) see a regional contrast between the proportion of large ion lithophile elements in the leucotonalite and the gabbro belts.

Granodiorite — Granite Belt

In the gabbro belt, there are a few very small bodies of aplitic granite and some larger bodies (Woodson Mountain) that vary in composition from granodiorite to what until recently had been called adamellite (now called granite, according to Strekeisen, 1973). The potassium feldspars in these rocks are rarely large, idiomorphic, or pink. Thus, it has been local tradition that the Peninsular Ranges are, for practical purposes, devoid of "true granite."

However, just west of the Gulf of California, one finds a few large bodies of adamellite (for example, the southeastern corner of the Sierra Pintas, latitude 31° 31'N, longitude 115° 12'W) where the potassium feldspars are well formed, up to one centimeter long, and distinctly pink. Inland from the coast of Sonora, similar granodiorites and adamellites are common, and there are bodies of even coarser grained rocks, some of which are potash, feldspar-rich granite. Thus, the western boundary of the Granodiorite-Granite belt lies a little to the west of the westernmost exposures of older Precambrian basement rocks, about where the first metamorphosed remnants of Proterozoic — Cambrian carbonate rock and quartzite are encountered.

Alkali-Calcic and Alkaline Belt

In peninsular California, there are no monzodiorites, no monzonites, no syenites, and no silica undersaturated granitic rocks. The data on coastal Sonora, Sinaloa, and areas to the south are limited, but no alkali-calcic and alkaline igneous rocks have been reported there either. However, Guillermo A. Salas (personal communication, 1972) has mapped monzonitic rocks in the vicinity of Hermosillo, Sonora (about 100 kilometers northeast of the Gulf). The mid-Tertiary volcanic rocks of the Sierra Madre Occidental (McDowell and Keiser, 1977) are distinctly richer in potassium than rocks farther to the southwest. Truly alkaline volcanic rocks of early Tertiary age are widely distributed from western Texas south into central Mexico (Barker, 1977). One must, of course, question whether rocks as far inland as western Texas can be related to the Farallon — North American convergence at all. The boundary between the Granodiorite Granite belt and the Alkaline belt (Fig. 5) is diagrammatic. Keith (1978) divides a magmatic profile across southern California, Arizona, southern Mexico, and western Texas into calcic, calc-alkaline, potassic calc-alkalic, alkali-calcic, and alkaline zones.

Discussion of the Origin of the Compositional Belts

From the Pacific margin to the interior of the continent, there is a gradual change from potassium-poor to potassium-rich igneous rocks.

It may be significant that Permo-Triassic igneous rocks in Arizona and northern Mexico (see summary in Gastil and others, 1979) form a predominantly rhyolitic association, not unlike the Cretaceous-Lower Tertiary rocks of the same area; and the Lower to Middle Miocene igneous rocks around the Gulf of California (Gastil and others, 1979) constitute an andesite-dacite association similar to the granitic rocks emplaced 75 million years earlier.

There is some correlation between the host rock terranes and the compositional zones. The potassium-poor magmatic rocks of the Pacific borderland are found in a host terrane of ophiolite-type rocks. The rocks of the gabbro belt are largely in arc volcanic rocks, which probably rest on oceanic crust. The leucotonalite belt overlaps the zone of cratonal clastic sediments and the zone of Carboniferous-Triassic carbonate rocks. Rocks of the Granodiorite-Granite and Alkaline zones are emplaced largely in old sialic crust.

Possibly, the rocks of the Pacific margin were emplaced in an oceanic island arc at some distance from the more continental zones (Gastil and others, 1978; Rangin, 1978). It is also possible that the andesite arc and older (foliated) granitic rocks, host to most of the gabbro zone, originated in an oceanic arc which was later sutured to the Pacific edge of continent.

STRUCTURE OF PLUTONIC EMPLACEMENT

The host rocks can be described in terms of three metamorphic and structurally distinguishable terranes (Fig. 6). These are the supradiapiric rocks, the interdiapiric rocks, and the subdiapiric rocks. The supradiapiric rocks are those which were too rigid to allow diapiric penetration. Granitic rocks emplaced within such rocks entered via fractures and caulderic collapse. The host rocks have not been tightly folded or foliated as a result of intrusion, and the metamorphism is contact and very low-grade regional.

The interdiapiric host rocks have been deformed both by the insertion of the intruding bodies and by the inflation of these bodies. They are typically schistose with steep foliation and steep lineations (fold axes). Their regional metamorphic grade varies with their depth relative to the zone of plutonic emplacement and the duration of plutonic activity.

Subdiapiric rocks are those through which the diapirs rose to reach the level at which they came to rest. Subdiapiric terranes can be expected to have a higher grade of metamorphism and show a more complex structural history than the interdiapiric rocks (Gastil, 1979). Nowhere in the western-Mexican granitic province can the subdiapiric terrane be demonstrated as in the Idaho batholith (Hyndman and Talbot, 1975). In the axial zones of the northern and central Peninsular Ranges, however, there are schist-gneiss terranes of sillimanite grade rock with migmatite phases. Gastil (1975) has suggested that these might be belts of subdiapiric rocks which have risen isostatically along the axis of maximum crustal underplating. Figure 6 attempts to map the areas of supradiapiric, interdiapiric, and subdiapiric terranes. The synplutonic and postplutonic deformation of the granitic rocks of San Diego County, Alta California, has been described by Todd and Shaw (1979).

TECTONIC IMPLICATIONS

The tenets of plate tectonics and the realities of subduction were not discovered by studying granitic rocks. However, the weight of seismic, heat flow, magnetic, and sea floor geologic data makes it imperative to look at the present as the key to the past and discover how the history of Mesozoic-Cenozoic magmatism in southern California and western Mexico is related to the convergence between the Farallon and North American plates which was occurring during the magmatic interval.

An observation one can make today in volcanic arcs around the world is that, at any given time, the products of volcanism erupt along a very narrow belt. The erosion of such arcs (for example, Peru; Pitcher, 1978) shows that the width of contemporaneous plutonic emplacement is no

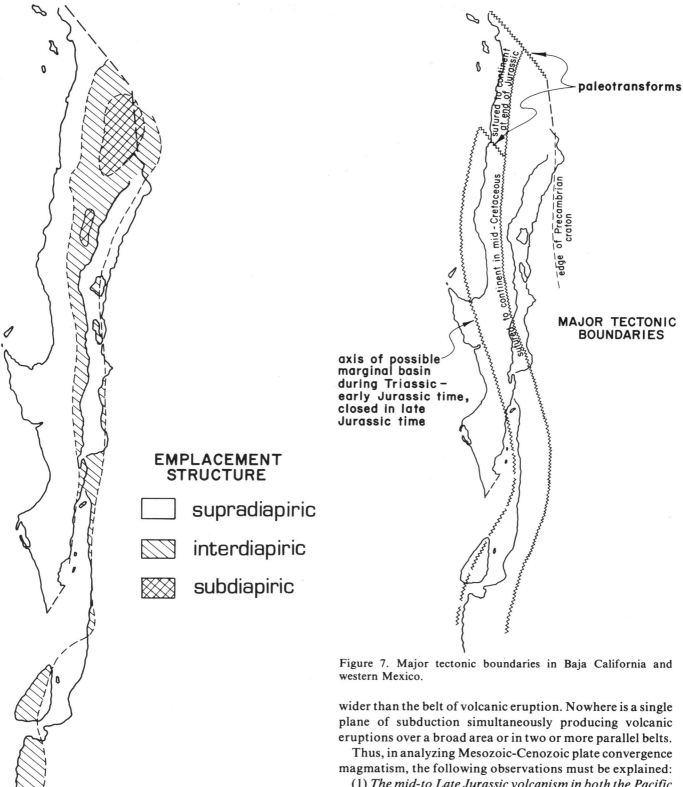

EMPLACEMENT
STRUCTURE

- ☐ supradiapiric
- ▨ interdiapiric
- ▩ subdiapiric

Figure 6. Structural zones of plutonic emplacement.

paleotransforms

MAJOR TECTONIC
BOUNDARIES

axis of possible
marginal basin
during Triassic –
early Jurassic time,
closed in late
Jurassic time

Figure 7. Major tectonic boundaries in Baja California and western Mexico.

wider than the belt of volcanic eruption. Nowhere is a single plane of subduction simultaneously producing volcanic eruptions over a broad area or in two or more parallel belts.

Thus, in analyzing Mesozoic-Cenozoic plate convergence magmatism, the following observations must be explained:

(1) *The mid-to Late Jurassic volcanism in both the Pacific borderland and west-central Sonora (separated today by 500 kilometers.* Dilation of the older continental crust by Cretaceous batholiths can account for much, but not all, of

this separation. The borderland andesites are clearly on oceanic crust, and the eastern andesites are clearly on the Precambrian craton. Gastil and others (1978, 1980) have proposed that these two belts resulted from two parallel, simultaneous planes of subduction, ultimately welding the oceanic arc onto the continent (Fig. 7).

(2) *The continentward migration of the axis of magmatism from Jurassic to Oligocene time, with a westward return during the Miocene.* Most authors suggest a flattening of the angle of subduction to account for the continentward migration of magmatism. This flattening may have resulted, at least in part, from the dilation resulting from sialic accretion. Gastil (1977) calculated that the continent grew in width 315 kilometers over an interval of 150 million years.

(3) *The reversal in the direction of magmatic migration about 35 million years ago and the abrupt southwestward jump about 23 million years ago.* By Luyendyke's hypothesis (1970), this westward reversal would require a steepening of the plane of subduction resulting from a slowing of the rate of convergence. Alternatively (Gastil and others, 1980), the continuing continental accretion may have finally caused the continent to override the Farallon/North American plate boundary. When this happened, a new subduction plane formed seaward of the old one. With the angle of subduction unchanged, magma resulting from the new subduction plane appeared to the southwest (see Fig. 7, in Gastil and others, 1980).

ACKNOWLEDGMENTS

To make this a compact summary, covering many aspects of a very large province, it was not possible to adequately reference the many published and unpublished contributions. Most of these can be found in the articles which have been referenced. Although I am the sole author, many of the ideas and the bulk of the data have been generated by other researchers. Among those who are currently active in the province, I am particularly grateful to Leon T. Silver, Daniel Krummenacher, Victoria Todd, Michael Walawender, and Claude Rangin. My part in the research over the years has been supported by the National Science Foundation and the National Geographic Society. I have been assisted by many scientists and organizations in Mexico.

REFERENCES CITED

Anderson, T. H., and Silver, L. T., 1979, Mesozoic magmatic events of the northern Sonora coastal region, Mexico; Geological Society of America Abstracts with Program, v. 1, p. 3.

Baird, K. W., and Welday, E. E., 1974, Chemical trends across Cretaceous batholithic rocks of southern California: Geology, v. 2, p. 493–496.

Barker, D. S., 1977, Northern Trans-Pecos magmatic province: Introduction and comparison with the Kenya Rift: Geological Society of America Bulletin, v. 88, p. 1421–1427.

Barnes, D. A., and Berry, K. D., 1979, Jura-Cretaceous Paleogeography —the Eugenia Group, western Vizcaino Peninsular, Baja California Sur, Mexico: *in* Abbott, P. L., and Gastil, R. G., eds., Baja California Geology: Geological Society of America Annual Meeting, San Diego, Publ. by Dept. of Geological Sciences, San Diego State University, p. 53–64.

Boles, J. R., and Hickey, J. J., 1979, Eugenia Formation (Jura-Cretaceous), Punta Eugenia Area: *in* Abbott, P. L., and Gastil, R. G., eds., Baja California Geology: Geological Society of America Field Guidebook, Annual Meeting at San Diego, publ. by Dept. of Geological Sciences, San Diego State University, p. 65–72.

Bonneau, M., 1971, Una nueva area Cretacica fosilifera en el Estado de Sinaloa: Societad Geológica Mexicana Boletin, v. 32, n. 2, p. 159–167.

Chiñas, R., 1963, Geologia de las Tres Maria: tesis profesional, Instituto Politecnico Nacional, Mexico.

Clark, K. F., Carrasco, C., Damon, P. E., and Sandoval, S. H., 1977, Posición estratigraphia en tiempo y espacio de mineralización en la provincia de la Sierra Madre Occidental en Durango, México: Associación Ingeneria Minera y Metalogia, y Geologia de México, 12 Con. Nac., Memoria, p. 197–244.

Coney, P. J., and Reynolds, S. J., 1977, Cordilleran Benioff zones: Nature, v. 270, p. 403–406.

Criscioni, J. J., David, T. E., and Ehlig, P., 1978, The age and sedimentation diagenesis for the Bedford Canyon Formation and the Santa Monica Formation in Southern California: A Rb/Sr evaluation: *in* Howell, D. G., and McDougall, K. A., eds., Pacific Coast Paleogeography Symposium 2, p. 385–396.

Edelman, S., 1980, The origin of foliation in gneissic granite rocks near San Vicente Reservoir, San Diego County, California: [M.S. thesis] San Diego State University, San Diego, California.

Fife, D. L., Minch, J. A., and Crampton, P. J., 1967, Late Jurassic age of the Santiago Peak Volcanics, California: Geological Society of America Bulletin, v. 78, p. 299–303.

Gastil, R. G., 1975, Plutonic zones in the Peninsular Ranges of southern California and northern Baja California: Geology, v. 3, no. 6, p. 361–363.

——, 1976, Some effects of progressive metamorphism on zircons: Geological Society of America Bulletin, v. 78, p. 879–906.

——, 1977, Subduction, accretion, and batholith emplacement, *in* Plutonism in relation to volcanism in relation to volcanism and metamorphism: International Geological Correlations Program, Circum-Pacific Plutonism Project, 7th Meeting, Toyama, Japan, p. 62–73.

——, 1979, A conceptual hypothesis for the relation of differing tectonic terranes to plutonic emplacement: Geology, v. 7, p. 542–544.

Gastil, R. G., and Krummenacher, D., 1978, A reconnaissance geologic map of the west-central part of the state of Nayarit, Mexico: Geological Society of America Map and Chart Series MC-24 with text.

Gastil, R. G., Krummenacher, D., Doupont, J., and others 1976, La zona batolitica del sur de California y el occidente de México: Sociedad Geológica Mexicana Boletin, v. 37, p. 84–90.

Gastil, R. G., Krummenacher, D., and Morgan, G., 1978, Mesozoic history of peninsular California and related areas east of the Gulf of California, *in* Howell, D. G., and McDougall, K. A., eds., Mesozoic Paleogeography of the Western United States: Pacific Section Society of Economic Paleontologists and Mineralogists, California, U.S.A., p. 107–116.

——, 1981, The tectonic history of peninsular California and adjacent Mexico, *in* Ernst, W. G., ed., Ruby Symposium Volume 1, The Geotectonic Development of California; Englewood Cliffs, N.J., Prentice-Hall Inc., 710 p.

Gastil, R. G., Phillips, R. P., and Allison, E. C., 1975, A reconnaissance geology of the State of Baja California: Geological Society of America

Memoir 140, 170 p.

Gastil, R. G., Phillips, R. P., and Rodriguez-Torres, R., 1972, The reconstruction of Mesozoic California: 24th International Geological Congress, Section 3, p. 217–229.

Gromet, P. L., and Silver, L. T., 1979, Profile of rare earth element characteristics across the Peninsular Ranges batholith near the international border, southern California, U.S.A. and Baja California, Mexico: Mesozoic Crystalline Rocks, Field Guidebook, Geological Society of America Meeting, San Diego, California, published by Dept. Of Geological Sciences, San Diego State University, San Diego California, p. 133–142.

Hanna, M. A., 1926, Geology of the La Jolla quadrangle, California: University of California, Dept. of Geological Sciences Bulletin, v. 16, p. 187–246.

Hart, S. R., 1964, The petrology and isotopic-mineral age relations of a contact zone in the Front Range, Colorado: Journal of Geology, v. 72, p. 493–525.

Henry, C. D., 1975, Geology and geochronology of the granitic batholithic complex, Sinaloa, Mexico: Geological Society of America Abstracts with program, v. 7, p. 172; doctoral dissertation, University of Texas, Austin, 158 p.

Hyndman, D. W., and Talbot, J. L., 1975, Boulder batholith: a result of emplacement of a block detached from the Idaho batholith infrastructure?: Geology, v. 3, n. 7, p. 401–404.

Jensky, W. A., 1975, Reconnaissance geology and geochronology of the Bahia de Banderas area, Nayarit and Jalisco, Mexico: [M.S. thesis] University of California, Santa Barbara.

Keith, S. B., 1978, Paleosubduction geometrics inferred from Cretaceous and Tertiary magmatic patterns in southwestern North America: Geology, v. 6, p. 516–521.

Kilmer, F. H., 1979, A geological sketch of Cedros Island, Baja California, Mexico, *in* Abbott, P. L., and Gastil, R. G., eds., Baja California Geology: Geological Society of America Field Trip Guidebook, Annual Meeting at San Diego, published by Dept. Of Geological Sciences, San Diego State, p. 11–28.

Krummenacher, D. Gastil, R. G., Bushee, J., and Doupont, J., 1975, K-Ar apparent ages, Peninsular Ranges batholith, southern California: Geological Society of America Bulletin, v. 86, p. 760–768.

Larsen, E. S., 1948, Batholith and associated rocks of Corona, Elsinore, and San Luis Rey quadrangles, southern California: Geological Society of America Memoir 29, 182 p.

Luyendyk, B. P., 1970, Dip of the downgoing lithospheric plate beneath island arcs: Geological Society of America Bulletin, v. 81, p. 3411–3416.

Mattinson, J. M., and Hill, D. J., 1976, Age of plutonic basement rocks, Santa Cruz Island, California, *in* Aspects of the geologic history of the California continental borderland: Pacific Section American Association of Petroleum and Geology, Miscellaneous Publication 24, p. 53–59.

McDowell, F. W., and Keizer, R. P., 1977, Timing of mid-Tertiary volcanism in the Sierra Madre Occidental between Durango City and Mazatlan, Mexico; Geological Society of America Bulletin, v. 88, p. 1479–1487.

Minch, J. A., 1969, A depositional contact between the pre-batholithic Jurassic and Cretaceous rocks in Baja California, Mexico: Geological Society of America Abstracts with Program for 1969, pt. 3, p. 42–43.

Moore, T. E., 1979, Geologic summary of the Sierra de San Andres ophiolite, *in* Abbott, P. L., and Gastil, R. G., eds., Baja California Geology: Fieldtrip Guidebook for Geological Society of America annual Meeting, San Diego, published by Dept. of Geological Sciences, San Diego State University, p. 95–106.

Mullan, H. S., 1978, Evolution of part of the Nevadan orogeny in northwestern Mexico: Geological Society of America Bulletin, v. 89, p. 1175–1188.

Pitcher, W. S., 1978, The anatomy of a batholith: Journal of the Geological Society of London, v. 135, p. 157–182.

Rangin, C., 1978, Speculative model of Mesozoic geodynamics, central Baja California to northeastern Sonora, Mexico, *in* Howell, D. G., and McDougall, K. A., eds., Mesozoic Paleogeography of the western United States: Pacific Section Society of Economic Paleontologists and Mineralogists, California, U.S.A., p. 85–106.

——, 1979, Evidence for superimposed subduction and collision processes during Jurassic-Cretaceous time along Baja California continental borderland, *in* Abbott, P. L., and Gastil, R. G., eds., Baja California Geology: Fieldtrip Guidebook for Geological Society of America Annual Meeting, San Diego, published by Dept. of Geological Sciences, San Diego State University, p. 95–106.

Silver, L. T., Taylor, H. P., Jr., and Chappell, B., 1979a, Some petrological geochemical and geochronological observations of the Peninsular Ranges batholith near the International border of the U.S.A. and Mexico: U.S. Geological Survey open-file report 78-701, p. 423–426.

——, 1979b, Peninsular Ranges Batholith, San Diego and Imperial Counties: Mesozoic Crystalline Rocks, Guidebook for Geological Society of America meeting, San Diego, published by Dept. of Geological Sciences, San Diego State University, San Diego, California, p. 83–110.

Strekeisen, A. L., 1973, Plutonic rocks-classification and nomenclature: Geotimes, v. 18, p. 26–30.

Taylor, H. P., and Silver, L. T., 1978, Oxygen isotope relationships in plutonic igneous rocks of the Peninsular Ranges batholith, southern and Baja California: Short papers of the Fourth International Conference, Geochronology, Cosmochronology, Isotope Geology, Zartman, R. E., ed., U.S. Geological Survey Open File Report 78-701, p. 423–426.

Todd, V. R., and Shaw, S. E., 1979, Metamorphic, intrusive, and structural framework of the Peninsular Ranges batholith in southern San Diego County, California: Mesozoic crystalline rocks, Field Guidebook Geological Society of America meeting, San Diego, Published by San Diego Department of Geological Sciences, p. 177–232.

Walawender, M. J., 1976, Petrology and emplacement of the Los Pinos pluton, southern California: Canadian Journal of Earth Sciences, v. 13, p. 1288–1300.

Walawender, M. J., Hoppler, H., Smith, T. E., and Riddle, C., 1979, Trace element evidence for contamination in a gabbro-quartz diorite sequence in the Peninsular Ranges batholith: Journal of Geology, v. 87, p. 87–97.

Walawender, M. J., and Smith, T. E., 1980, Geochemical and petrologic evolution of the basic plutons of the Peninsular Ranges batholith, southern California: Journal of Geology, v. 88, p. 233–242.

MANUSCRIPT ACCEPTED BY THE SOCIETY JULY 12, 1982

PRINTED IN U.S.A.

Geological Society of America
Memoir 159
1983

Andean plutonism in Peru and its relationship to volcanism and metallogenesis at a segmented plate edge

E. J. Cobbing
Institute of Geological Sciences
Keyworth Nottingham NG1 25GG, England

W. S. Pitcher
University of Liverpool
Department of Geology P.O. Box 147
Liverpool L69 3BX, England

ABSTRACT

Segmentation of both the Lower Paleozoic and the Mesozoic orogenic belts of Peru is indicated by rapid variation in thickness and lithology of stratigraphic units and by the presence of lateral structures. Some of these segmental boundaries are common to both the Paleozoic and the Mesozoic fold belts and are probably old structures which have been episodically rejuvenated.

The Coastal Batholith is also segmented compositionally, and some of the segment boundaries coincide with the orogenic segmentation. Others, however, do not. Those which do are probably deep-reaching, crustal-mantle faults which have penetrated sufficiently deeply into the source region of the granitoid magmas to affect both the composition of the magmas and the timing of magma generation within adjacent segments.

A suite of early gabbros is common to the batholith as a whole, but the batholithic segments comprise granitoid super-units which are specific to the segments. Each super-unit may range from diorite to monzogranite, and the differences are established on the basis of textural features, the order of emplacement, and geochronology. Two batholithic segments have been mapped in detail. These two segments are compared, and it is shown that the Lima segment is the more complex, having no more than ten units and super-units which were emplaced between 100 and 30 Ma. The Arequipa segment consists of only 5 super-units which were emplaced between 100 and 80 Ma.

A belt of Tertiary intrusives has also been distinguished to the east of the batholith which consists of numerous, small, scattered, plutons.

The mainly Cretaceous plutonism of the batholith is spatially associated with thick sequences of pre-orogenic submarine lavas and volcaniclastic rocks which are the products of a volcanic arc constructed on thinned continental crust, whereas the Tertiary intrusives are associated with post-orogenic terrestrial plateau volcanics which were deposited on thickened crust. There is some evidence that the Cretaceous and Tertiary lavas are isotopically different.

In terms of metal association, the Cretaceous belt is a copper, molybdenum, and gold metallogenic province, whereas the Tertiary belt is polymetallic, producing mainly copper, lead, zinc, and silver.

INTRODUCTION

The purpose of this paper is to summarize the Coastal Batholith of Peru in terms of age, structure, and composition so that it may be viewed as a whole and in relationship to other tectonic and magmatic elements which have combined to produce the Andean mountain chain at the boundary between the Pacific Ocean and the South American continent. Sillitoe (1974) suggested that tectonic segmentation of the Andes was reflected in the metallogenetic provinces and mineral occurrences. In this contribution, special attention will be drawn to segmentation and its influence on sedimentation, structure, volcanism, plutonism, and metallogenesis.

REGIONAL TECTONIC SETTING

The Andes of Peru comprise two distinct tectonic units, the Eastern Cordillera and the Western Cordillera.

The Eastern Cordillera comprises an old fold belt, which has been extensively studied by French geologists (Mégard and others, 1971; Laubacher, 1978; Marocco 1978). It is divided into two parts, a southern part, consisting of folded Lower Paleozoic slates and quartzites, and a northern part of Precambrian schists. Intrusives of Permian age pierce these folded rocks (Dalmayrac, 1977) and also an overlying volcanic and conglomeratic Upper Permian molasse, the Mitu Formation, which rests discordantly upon the folded structures. These Permian intrusions mark the final stabilization of the Eastern Cordillera as an orogenic belt; henceforward, it acted as a stable block. Thus, the red beds of the Mitu Formation, which truncate the Paleozoic structures, form the basal unit of a stratigraphic sequence which was structurally coherent up to the Incaic orogeny in the early Paleocene.

The rocks of the Western Cordillera developed in a zone of crustal weakness immediately to the west. They are almost entirely of Mesozoic and Tertiary ages and can be divided, in terms of environment, into an eastern miogeosynclinal basin of clastics in the lower part and carbonates in the upper and a western eugeosynclinal basin consisting of marine volcanic rocks. This volcanic belt itself varies from flysch-like volcaniclastics and andesitic or basaltic andesite pillow lavas on the western side, to fragmental submarine and subaerial flows and pyroclastics of andesite and dacite composition on the eastern side (Webb, 1976).

Two episodes of deformation have affected the rocks of the Western Cordillera: a pre-batholithic Subhercynian orogeny, at the end of the Albian, which affected only the eugeosynclinal zone; and the Incaic orogeny, of Paleocene age, which affected the miogeosynclinal zone in particular, but also the whole of the Western Cordillera.

The Incaic orogeny was manifested by folding of the Cretaceous strata, by crustal thickening, and by uplift. An erosion surface bevelled all the Cretaceous structures, cutting across the sedimentary rocks of the miogeosyncline, the volcanic rocks of the eugeosyncline, and the Cretaceous intrusives of the Coastal Batholith. The plateau volcanics of the Calipuy Group of Eocene to Late Miocene ages were subsequently deposited on this erosion surface. These volcanics are mainly fragmental andesite and dacite of terrestrial type with intercalated lacustrine sedimentary rocks.

The Ancient Floor

A glimpse of the floor upon which the Mesozoic rocks were formed is seen in southern Peru, where Upper Paleozoic clastics overlie extensive outcrops of gneiss, for which an age of 2000 Ma has been obtained (Cobbing and others, 1977a; Dalmayrac, 1977; Schackleton and others, 1979). This cratonic basement is actually quite varied, as is shown by the fact that in southern Peru the Mesozoic cover rests directly on ancient gneisses or on their Paleozoic cover, whereas in northwestern Peru, it lies on schists of late Precambrian age (Bellido, 1969). Nevertheless, the pre-Mesozoic basement, in spite of its complexity, is everywhere clearly of continental crustal character.

The Mesozoic sedimentary and volcanic basins of the Western Cordillera were fault-controlled, and their sediments are all cratonic in character. They are well-sorted, compositionally well-differentiated clastics which show evidence of having been derived from a mature land surface and deposited in shallow seas (Wilson, 1963). These characters are also known for the Upper Paleozoic geosyncline (Dalmayrac, 1977). The flysch-like fragmental volcanics which are intercalated with some of these sediments were probably derived from highlands represented by the Cretaceous volcanic arc.

Tectonic Segments

Both the Eastern and the Western Cordillera are tectonically divided (Sillitoe, 1974) into segments having distinctive characteristics. The Eastern Cordillera is affected by three breaks. The most southerly, the Abancay deflection (Marocco, 1971), causes both the sedimentary basins forming the Lower Paleozoic geosyncline and the structures within it to follow an easterly trend for about 200 km (Fig. 1). In contrast, north and south of this deflection, the structures follow the northwesterly Andean trend. The second break is near the latitude of Lima and is marked by the transition from Lower Paleozoic strata in the south to late Precambrian schists in the north, both belts following the Andean trend. The third break, the Huancabamba deflection in northern Peru (Ham and Herrera, 1963), occupies a rather ill-defined area underlain by a broad outcrop of late Precambrian schists which extends as far west as the coast. This zone also marks the change in strike

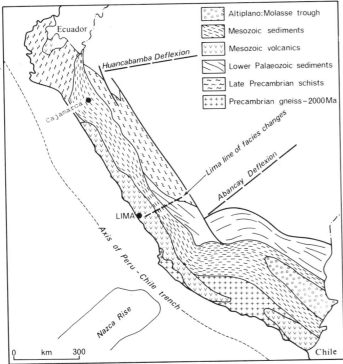

Figure 1. The main tectonic elements of Peru, compiled from the 1:1,000,000 map, 1975, and from Dalmayrac and others (1977).

within the schist belt from northwesterly in Peru to north-easterly in Ecuador; it also divides the Central Andes from the Northern Andes.

The same three breaks also affect the Mesozoic formations of the Western Cordillera. South of the Abancay deflection, the Andean fold belt was uplifted earlier, and for a longer period than to the north, with the result that the formations exposed to the south are Jurassic rather than Cretaceous (Fig. 6a), though Cretaceous rocks may well have been deposited. The Jurassic formations are contained in faulted grabens in the Precambrian basement, and the division into a western eugeosyncline and an eastern miogeosyncline, although discernible, is not so well defined as in the Cretaceous formations to the north. The break near the latitude of Lima occurs where the outcrop of the miogeosyncline is restricted, whereas that of the eugeosyncline is increased (Cobbing, 1978). The third break, at the Huancabamba deflection, is again rather ill-defined but is indicated by the local thinness of the Cretaceous strata, forming a bridge between that of the Peruvian Andes and the Ecuadorian Andes. A fourth break, which is specific to the Western Cordillera, occurs at the latitude of Cajamarca, where the formations which comprise the miogeosyncline die out rapidly to the north, and where the sediments and structures within the miogeosyncline follow westerly trends for 140 km in a manner reminiscent of the Abancay deflection.

Most of these breaks, which result in a tectonic segmentation of the fold belts, are evidently long-lived structures

which have affected both the Eastern and the Western Cordillera; further evidence that the present Andes resulted from recurrent orogenic activity in the same zone from the late Precambrian to the present.

DISTRIBUTION OF THE INTRUSIVES OF THE COASTAL BATHOLITH: UNITS, SUPER-UNITS, AND SEGMENTS

Within the Lima segment (Fig. 2), the Coastal Batholith of Peru consists mainly of Cretaceous intrusions, but intrusions of Paleocene (60 Ma) and Oligocene (33 Ma) ages also occur. In spite of the exceptions, and in order to distinguish between the mainly Cretaceous rocks of the Coastal Batholith and the much younger Tertiary intrusives further east, the batholith has been shown as entirely Cretaceous on Figure 6.

All Cretaceous intrusives in Peru are confined to the narrow batholith lineament, which has an average width of 65 km. This plutonic lineament, which is 2000 km long, traverses the several segmental boundaries discussed above and is not deflected by them; nevertheless, it is itself segmented internally in a compositional way. Cobbing and others (1977b) showed that families of rock types could be distinguished within the batholith in chains of separated plutons, which they called super-units. Each super-unit consists of a consanguineous magma series which ranges

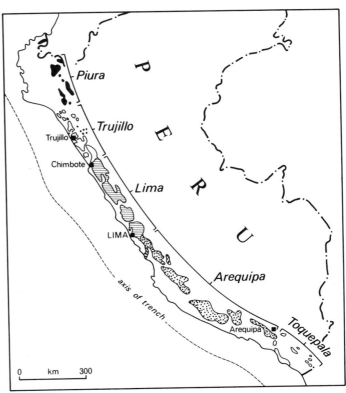

Figure 2. The major segments of the Coastal Batholith of Peru.

from basic to acid both in composition and in order of emplacement, and each is the surface expression of a deep-seated melt-cell shaped as a long, narrow, podded cylinder, the pods probably corresponding in location to the physically separated outcrop of the plutons and complexes. Similar assemblages of granitoids have been described from the Sierra Nevada by Bateman and Dodge (1970), who called them sequences, and from the Tasman syncline of eastern Australia by Chappell and White (1974), who called them suites.

Each of the segments described by Cobbing and others (1977b) comprises a limited number of super-units which occur within the segment and nowhere else. Two of these segments (Fig. 2) have been mapped in reasonable detail: the Lima segment and the Arequipa segment. The remaining segments are only poorly known. The Lima segment has been shown to consist of seven super-units (and three units), whereas the Arequipa segment consists of five super-units (Table 1). Four batholith segments were recognized by Cobbing and others (1977b), but subsequently this number has been increased to five. In each segment mapped, and everywhere else in the batholith, the earliest component of the batholith is gabbro. These gabbro precursors, now much dissected by later arrivals, vary somewhat in size and importance, but along the whole plutonic lineament, they form the earliest constituent.

Some of the breaks between the segments correspond to those affecting the Eastern and Western Cordilleras as noted above, but for some there is no correspondence. Thus, the Arequipa segment of the batholith stretches from Arequipa to Lima and crosses the Abancay deflection with no apparent change (Figs. 1,2). The southern end of the Lima segment, south of Lima, corresponds approximately to the breaks in the Eastern and Western Cordilleras noted at the same latitude, but the northern termination of the Lima segment of the batholith, at Chimbote, corresponds to no recognized regional structural change. On the other hand, the termination of the Trujillo segment near Chiclayo does correspond to the beginning of the Huancabamba deflection. The tectonic segmentation of the cordilleras was apparently caused by structures that were capable of influencing the generation of the magmas but which did not invariably do so, possibly because the structures were only intermittently active and were quiescent for long periods.

The following descriptive section on the Lima and Arequipa segments will focus on the areal distribution of the super-units, their relative importance, and their sequence in time. For questions of emplacement and structure and for details of petrography, the reader is referred to Pitcher (1978) and Cobbing and others (1981). Here it will only be noted briefly that all the plutons were emplaced at a high level in the crust as sharply defined plutons. The constituents of the super-units range from gabbro to monzogranite, but, overall, tonalite-granodiorite is the most important lithology.

TABLE 1. AGES OF THE SUPER-UNITS

LIMA SEGMENT			AREQUIPA SEGMENT		
Super-unit and units	Lithology	Preferred Age Ma	Super-unit	Lithology	Preferred Age Ma
Pativilca	Monzogranite	33			
Cañas-Sayán	Monzogranite	61			
San Jerónimo	Monzogranite	62			
Puscao	Granodiorite-monzogranite	61			
La Mina	Tonalite	66			
Dyke Swarm	Andesite	72			
Humaya	Granodiorite	73			
Santa Rosa	Tonalite-granodiorite	90-75	Tiabaya	Tonalite-granodiorite	81
Paccho	Diorite-tonalite	95*	Incahuasi	Diorite-tonalite	83
Jecuan	Granodiorite	102*	Linga	Monzodiorite-monzogranite	97
			Pampahuasi	Tonalite	97
Patap	Gabbro	102*	Patap	Gabbro	102*

* Estimated age from a combination of geological and radiometric considerations

THE LIMA SEGMENT OF THE COASTAL BATHOLITH

The Units and Super-units

The oldest intrusives are the gabbros of the Patap Super-unit (Figs. 3 and 4) with a probable age of 102 Ma. The largest outcrops occur on the western side of the batholith, though outcrops may occur over its entire width. The gabbro plutons are cut into and dissected by later granitoids so that the area of their original outcrop is now much reduced. Petrologically, they are two-pyroxene cumulate gabbros of tholeiitic affinity (Cobbing and others, 1981) with linear zones of recrystallization producing rocks of pyroxene hornfels-facies within the plutons — which suggests a complex emplacement history (Regan, 1976). The gabbros have been much affected by the growth of amphibole which crystallized directly from the magma during the later stages and also occurs as over-growths and replacements of pyroxene in the solid state. These changes are thought to be the result of the build-up of P_{H_2O} during the consolidation of the magma. The gabbros have sharp discordant contacts against volcanic and sedimentary rocks. These display only limited effects of contact megamorphism, and it is considered that, in spite of their complex internal history, they were finally emplaced as crystal mushes in the form of relatively simple plutons along the entire length of the segment.

Closely associated with the gabbros on the western side of the batholith is the Jecuan granodiorite (Fig. 3), which is present as a chain of small plutons, one of which, the Atocongo adamellite, has been dated by Stewart and others (1974) at 102 Ma.

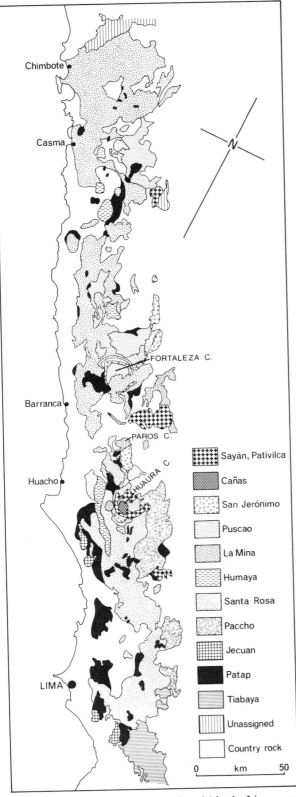

Figure 3. The distribution of super-units within the Lima segment. Field work in 1980 established that the plutons of Tiabaya, which are shown to cut the Santa Rosa east of Lima in Figure 3, are in fact variants of Santa Rosa. There is now no known locality where the field relationships between Tiabaya and Santa Rosa may be directly observed.

Legend for Figure 3:
- Sayán, Pativilca
- Cañas
- San Jerónimo
- Puscao
- La Mina
- Humaya
- Santa Rosa
- Paccho
- Jecuan
- Patap
- Tiabaya
- Unassigned
- Country rock

0   km   50

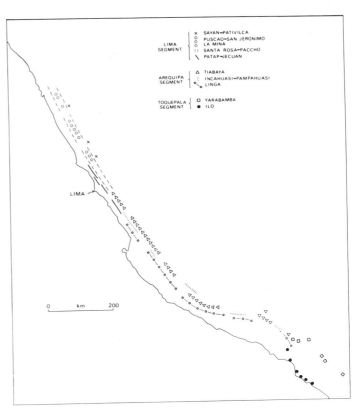

Figure 4. The distribution of the super-units within the Lima, Arequipa, and Toquepala segments.

Legend for Figure 4:

LIMA SEGMENT
- × SAYAN–PATIVILCA
- ○ PUSCAO–SAN JERONIMO
- ○ LA MINA
- ‖ SANTA ROSA–PACCHO
- \ PATAP–JECUAN

AREQUIPA SEGMENT
- △ TIABAYA
- · INCAHUASI–PAMPAHUASI
- · LINGA

TOQUEPALA SEGMENT
- □ YARABAMBA
- ● ILO

The Patap gabbros were invaded by the Santa Rosa Super-unit (Fig. 3), which, in this segment, is the most important member of the batholith. It consists of a wide range of rock types which range from basic diorite to monzogranite. Typically, it consists of two principal types: an early quartz diorite which in some areas may be as silicic as tonalite, and a later, paler type which ranges from tonalite to granodiorite. The darker, older type is commonly marginal to the later, paler one, but large, discrete plutons of one or the other type also occur.

Small monzogranitic plutons of 5 - 10 km diameter with similar textural characteristics were emplaced as single pulses within the earlier granodiorites, tonalites, and diorites. These may themselves, in places, show an inward zonation to monzogranite.

However, for the most part, the Santa Rosa intrusives are arranged in four large lenticular complexes of 100 to 130 km in length (Fig. 3), each comprising many plutons. These complexes are cut by large plutons of a distinctive pale granodiorite with well-formed books of biotite, the Humaya granodiorite, now regarded as a separate unit. Both this granodiorite and the various members of the Santa Rosa Super-unit are cut by a swarm of dikes which range from andesite to rhyolite in composition but are mainly andesitic. This dike swarm is particularly characteristic of those sectors of the batholith characterized by the Santa Rosa Super-unit.

On the eastern side of the batholith is a large complex of quartz diorites and tonalites which is older than the Santa Rosa. This is the Paccho Super-unit, and it is less widespread than the Santa Rosa, providing discontinuous outcrops of rather small plutons except northeast of Lima, where outcrops are more extensive.

Following emplacement of the main dike swarm, one super-unit of tonalite and two of more acid composition were emplaced in the central zone of the batholith. They are restricted to a strike-length of 250 km, as opposed to the 400 km of the whole Lima segment (Figs. 3, 4), which may indicate reduction of the zone of magma generation with time (Pitcher, 1978). Both the field relationships and the radiometric evidence on the timing of these two silicic super-units are equivocal, but they seem to be more or less coeval.

The largest of these in outcrop area is the Puscao Super-unit, which forms elongate, rather rectangular plutons composed of coarse monzogranite. The main body of the Puscao plutons show cryptic layering (Taylor, 1976), and the upper parts of the plutons are characterized by flat-lying sheets of granophyric aplogranite, whereas the outer marginal zones consist of a darker member which is cut by the main monzogranite.

A San Jerónimo Super-unit, which consists of porphyritic monzogranites, is temporally and spatially associated with the Puscao. The rocks are undoubtedly different, however, in petrology and geochemistry.

The most striking feature of both the Puscao and the San Jerónimo Super-units is that typical members contribute to the formation of three well-defined centered complexes, (Bussell et al., 1976), Rio Huara, Quebrada Paros, and Rio Fortaleza. In all three, the ring dikes comprise both Puscao and San Jerónimo members in a most intimate mixture.

These equally spaced ring complexes occupy a central position in the southern half of the Lima segment. The reason for the regularity of their occurrences is not known, though it may be related to the intersection of major fracture systems (Bussell and others, 1976). Although Puscao and San Jerónimo super-units are the main constituents of the ring complexes, both older and younger intrusives may be locally important. The ring complexes are considered to be the eroded remnants of calderas which may have contributed to the volcanic cover (Pitcher, 1978).

The La Mina tonalite is one of the super-units which is also associated with the ring complexes. It is present in both the Huaura and the Quebrada Paros complexes but not in that of the Rio Fortaleza. The medium-grained tonalite forms generally well-defined plutons 5 to 10 km across.

Two acid intrusives also contribute to the Rio Huara ring complex but not to the others. These are the monzogranites of Cañas and Sayán. The Sayán pluton, which consists of coarse monzogranite with K-feldspar megacrysts,

cuts across the earlier ring structures and is itself cut by the well-defined Cañas pluton (Fig. 3), which is also a coarse monzogranite but lacks the K-feldspar megacrysts.

The relative intrusion ages of all the intrusive members of the ring complexes are well established by cross-cutting relationships. Nevertheless, the radiometric determinations indicate that they were all close in time, at about 60 Ma (Wilson, 1975).

Megacrystic monzogranites also form large, but widely spaced, plutons, which outcrop mainly along the eastern side of the batholith at Pariacoto in the valley of the Rio Casma, the Rio Pativilca valley itself, and at Vilca in the valley of the Rio Chancay. Although these rocks and the Sayán monzogranite are petrographically similar and could be grouped together, the Sayán and Pativilca plutons show an age discrepancy (K-Ar) of about 30 Ma. The Sayán pluton gives an age of 60 Ma, which is similar to that of all the components of the ring complexes into which it is emplaced, whereas the Pativilca gives an age of 34 Ma (Wilson, 1975). There is at present no isotopic data on the remaining plutons.

Plutonic phases

The history of construction of the Coastal Batholith within this segment can be divided into four or five well-defined periods, each characterized by its particular set of super-units (Table 1). The earliest phase, from 102 to 95 Ma, was marked by the emplacement of the Patap gabbros and the Jecuan granodiorites; in the second phase from 95 to 75 Ma, the most important volumetrically, the Paccho and Santa Rosa Super-units were emplaced; the third phase, at about 72 Ma, saw the intrusion of the Humaya granodiorite and the main dike swarm; and the fourth phase, from 66 to 55 Ma, was marked by a generally more acid magmatism and the construction of the ring complexes, and involved the super-units and units of Puscao, San Jerónimo, La Mina, Sayán, and Cañas. The final late granites of Pativilca were emplaced after a long hiatus at 34 Ma (Fig. 5).

These five phases of plutonic activity may be compared with tectonic events affecting the envelope. The Patap phase

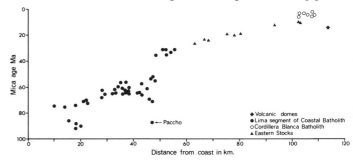

Figure 5. K/Ar ages from a composite traverse across the Western Cordillera from the coast to the Cordillera Blanca. (From Wilson, 1975).

is geochemically related to the marine volcanics of the Casma Group, and emplacement took place after folding had affected the Albian volcanics, but not long after, as is clear from the close geological and petrological association and the radiometric ages. This folding is referred to the "Subhercynian" orogeny (Cobbing, 1978). The Santa Rosa-Paccho phase was emplaced during a period when no marked tectonism affected the envelope; in fact, the Jumasha and Celendin limestones of Cenomanian to Santonian age were being deposited in the miogeosyncline to the east during this interval. The emplacement of the Humaya granodiorite and the regional dike swarm may correspond to uplift of the eugeosynclinal zone from which debris was shed to the east to form the Casaplaca Red-bed Formation of Upper Cretaceous to Lower Tertiary age in the miogeosyncline (Bellido, 1969). The Puscao, San Jerónimo, La Mina, Sayán, and Cañas phases may also correspond, at least in part, to the main Incaic orogeny of early Paleocene age which affected mainly the miogeosyncline. During this orogenic phase, the Cretaceous and Paleocene strata were folded and uplifted, and an erosion surface was formed upon which the Calipuy Group, with an age range of 52 Ma to 14 Ma, was deposited. It is possible, but not yet certain, that the granites of the ring complexes contributed silicic volcanic material to the lower part of this group. The final phase, marked by the Pativilca intrusion, took place after a long interval which probably corresponded to the time of formation of the erosion surface and the deposition of the older members of the Calipuy Group.

THE AREQUIPA SEGMENT OF THE COASTAL BATHOLITH

Units and Super-units

The Arequipa segment of the batholith, at 900 km, is over twice as long as the Lima segment, but it is much less complex, comprising only four granitoid super-units in addition to the gabbro precursor, as opposed to seven in the Lima segment. These four super-units are Linga, Pampahuasi, Incahuasi, and Tiabaya (Table 1). Three of these super-units outcrop along the length of the Arequipa segment, but the Pampahuasi is restricted to a relatively small area near the town of Ica.

The relative field relationships are less readily established in the Arequipa segment than in the Lima segment because only the youngest super-unit, the Tiabaya, is abundant everywhere along the length of the segment. The other super-units are somewhat patchy in their distribution, and although they may be extremely important or even dominant in some traverses, there are large areas in which they are absent. For this reason, representative plutons of these super-units may not come into contact with those of others, and their relative order of emplacement may be difficult to

determine from field evidence alone. Nevertheless, the order of emplacement suggested by field evidence is Pampahuasi, Linga, Incahuasi, and Tiabaya (Moore, pers. comm., 1979).

In this segment, the Patap gabbros do not normally form large plutons but are present as small remnants strung out along the length of the segment. They are the earliest intrusives emplaced and are similar in most respects to the gabbros in the Lima segment.

The Pampahuasi Super-unit is fairly restricted, outcropping as moderately sized plutons in the section between the Rio Canete and the Rio Huambo. It consists of dark tonalite and diorite and is cut by representative plutons of the Tiabaya and Incahuasi Super-units. The Pampahuasi Super-unit is not found in contact with the Linga Super-unit.

The Linga Super-unit was first recognized and described by Stewart (1968). It outcrops on the western side of the batholith throughout the Arequipa segment as far north as Canete. It is a very complex group of rocks which consist principally of monzogabbros and monzodiorites, yet individual types within the group may be mapped over long distances. The Linga Super-unit may be divided into two types, one of monzogabbros and monzodiorites, and another of monzonites ranging to monzogranites, but it is not yet known how the two groups are genetically related. The plutons of both types intrude, with sharp discordant relationships, the volcanics of Jurassic and Cretaceous ages which form the envelope.

The Incahuasi Super-unit consists principally of coarse-grained diorite and tonalite with some local variation to monzogranite. This super-unit is characterized by transitional variations of one type to another which make it difficult to map the various phases as discrete pulses. In places, it outcrops across the entire width of the batholith but is somewhat discontinuous along strike. Furthermore, it is absent from substantial parts of the Arequipa segment, but where present, it forms large areas. Unlike many of the other super-units, its rocks are commonly well foliated along Andean trends.

The Tiabaya Super-unit is the major constituent of the Arequipa segment, along which it is almost continuous, forming a number of relatively simple plutons of great size, together with a host of smaller ones. The super-unit is composed of two principal rock types: a dark, earlier variety of dioritic composition is normally present in marginal situations but may also form complete plutons; and a pale, later variety of tonalitic or granodioritic composition which cuts the earlier diorite and forms the bulk of the plutons. Many of the smaller plutons of Tiabaya display a concentric zoning to monzogranitic cores, but the larger plutons are remarkably homogeneous.

An important suite of massive porphyritic dikes, ranging from 5 to 30 m in individual width, is associated with Tiabaya and frequently cuts the earlier granitoid hosts to this super-unit, particularly those of Incahuasi. Mafic dikes are also present in small number in all phases of all super-units in the Arequipa segment, but they are nowhere so abundant as the main post-Humaya dike swarm in the Lima segment.

Plutonic Phases

There are few radiometric determinations covering the Arequipa segment, and since most of these were made before the detailed mapping was done, their interpretation is difficult. However, fifteen determinations using the Potassium/Argon method from mapped plutons of the four super-units in the valleys of the Rio Pisco and the Rio Ica give ages which range between 97 and 77 Ma (Moore, 1979). The Pampahuasi and Linga Super-units group within an age-span 97 - 94 Ma; the Incahuasi, 83 - 78 Ma; and the Tiabaya within 85 - 77 Ma.

Ages quoted by Stewart and others (1974) from near Canete and Chincha of 92 and 85 Ma are similar to those obtained by Moore, but ages of 66, 65, and 55 Ma from the upper Pisco valley, also quoted by Stewart and others (1974), are very different. Stewart and others (1974) also recognized two main events in the batholith at Arequipa and Toquepala, an earlier one at 77 Ma and a younger at 58 Ma; the latter age is similar to those for the upper Pisco valley. An age of 57 ± 2.0 Ma was recorded by Stewart and others (1974) from the Tiabaya pluton at Tiabaya, which provides the type lithology for the Tiabaya Super-unit, but this is now interpreted as a reset age, both because of the systematic dating of Moore, recorded above, and an age of 77 Ma determined by Le Bel (pers. comm., 1979) from the same Tiabaya Super-unit. In addition, at the transition from the Lima to the Arequipa segment, just south of Lima, the plutons of Tiabaya are thought to be cut by the main Santa Rosa dike swarm, which has been dated at about 72 Ma (Wilson, 1975).

Plutonic activity within the Arequipa segment seems to correspond approximately to the first and second phases of activity in the Lima segment. It immediately followed the "Subhercynian" deformation and continued through the Late Cretaceous, during which time limestones were being deposited in the miogeosyncline to the east (Bellido, 1969).

COMPARATIVE SUMMARY OF THE LIMA AND THE AREQUIPA BATHOLITHIC SEGMENTS

The granitoids of the Arequipa segment are all of Cretaceous age and were, moreover, emplaced during a relatively brief phase of activity which lasted for about 20 Ma. This is in marked contrast to plutonism in the Lima segment, which endured for 70 Ma. Nevertheless, the great bulk of plutonic activity in the Lima segment took place at about the same time as that in the Arequipa segment. The most striking difference between the two segments is the

lack, in the Arequipa segment, of the later, more acid super-units with their attendant ring complexes. The Arequipa segment may thus be considered to be more primitive in batholithic terms, whereas the Lima segment is more evolved.

The concentration in the Late Cretaceous of the greatest volume of plutonic activity within the two segments lends credence to the hypothesis that magmatic activity may be correlated with the maximum rates of sea-floor spreading (Larson & Pitman, 1972; Charrier, 1973).

Preliminary geochemical investigations indicate that the gabbro precursors are related to the Cretaceous volcanic rocks and that both represent primitive magmas probably derived from the mantle. The granitoids themselves are typically calc-alkaline and correspond to the I-types of Chappell and White (1974). The initial $^{87}Sr/^{86}Sr$ ratio of the super-units is uniformly very near 0.7042 (W. P. Taylor, pers. comm., 1979), which suggests a rather primitive origin for all the components, though not necessarily as direct derivatives of the mantle (see Atherton et al., 1979, for full discussion). All of these factors together indicate that the best model for the generation of the granitoid magmas is one related to the subduction of oceanic lithosphere.

METALLOGENESIS IN RELATION TO THE CRETACEOUS INTRUSIVES

The eugeosynclinal belt into which the batholith is emplaced is a copper-molybdenum geochemical province in which gold is also developed in certain sectors. Mineral deposits occur both in the volcanic country rocks and in the granitoids, and there are several paragenetic associations, some syngenetic and others epigenetic. Many, however, are clearly spatially associated with the intrusives (Hudson, 1979).

It is a striking fact that the intrusives of the Lima segment are less abundantly mineralized than those of the Arequipa segment. Copper is the principal metal in both segments, but whereas there are many producing mines in the Arequipa segment, there are few in the Lima segment.

Within the Arequipa segment, copper mineralization takes three main forms:

a)   Veins and disseminations in contact aureoles within the volcanic country rocks of plutons of the Linga and Tiabaya Super-units.

b)   Dike-like bodies of chalcopyrite, magnetite, and apatite within plutons of the Linga Super-unit, e.g., at Cobrepampa.

c)   Dissemination of chalcopyrite in alteration zones in plutons of the Tiabaya and Incahuasi Super-units; these may be related to porphyry-copper type deposits.

In addition to these, however, are the known porphyry-copper deposits of Toquepala, Cerro Verde, Quellaveco,

and Cuajones, which are located in the Yarabamba granodiorite of the Toquepala segment of the batholith immediately south of Arequipa; the Yarabamba granodiorite
bears a close, but so far unresolved, relationship to the Linga
Super-unit of the Arequipa segment.

Of these different kinds of mineralization, only that of the
chalcopyrite "dikes" of Cobrepampa is unequivocally linked
to the magmas of the batholith. The "dikes" follow fractures
in the intrusive which have also been channels for substantial potash metasomatism. In the contact-zone deposits, the
action of the intrusives may have concentrated metals which
were already present in lower concentration in the volcanics.
However, the disseminated sulphides in the alteration zones
clearly occur within fractured granitoid intrusives. Possibly,
the latter simply host a mineralization which may have had
its source in some other intrusive (unexposed or unrecognized). Nevertheless, the spatial association of the copper
mineralization with the intrusives of the batholith is striking
and suggestive of a primary connection.

The Cerro Verde porphyry-copper deposit immediately
south of Arequipa may be taken as representative of all true
porphyry coppers in the southern sector. The main focus of
mineralization is provided by a small stock of plagioclase
porphyry emplaced into the Yarabamba granodiorite. Breccia pipes and breccia dikes are associated with the porphyry
and are, in places, mineralized. In addition, tourmaline is
abundant in intrusives throughout the area and is particularly associated with aplite stocks and veins which
penetrate the intrusive. These are also the locus of copper
mineralization.

Finally, an important gold mineralization affects the
western side of the batholith in the Arequipa segment
between Jaqui and Camaná. The mineralization takes the
form of auriferous quartz veins which cut the plutons and
locally cut the volcanic rock of the envelope. As plutons of
all the super-units are affected, the mineralization is clearly
later than emplacement of the plutons of the batholith. The
source of the gold is uncertain; it may have originated at
depth and be unrelated to the intrusives of the batholith, or
it may have been scavenged from the intrusives by mineral-
izing fluids.

The metallogenesis of the batholith is being investigated,
and preliminary work confirms a degree of difference in the
batholithic segments. As an example, C. Vidal (pers. comm.,
1980) finds that fluid inclusions from the two segments of
Lima and Arequipa differ in size and composition; those
from the Arequipa segment are larger, more abundant, and
richer in salines.

DISTRIBUTION OF THE TERTIARY INTRUSIVES

As noted above, the eastern part of the Coastal Batholith
in the Lima segment contains some large plutons of Tertiary
age. Farther east, scattered stocks, 1 to 10 km across, spread

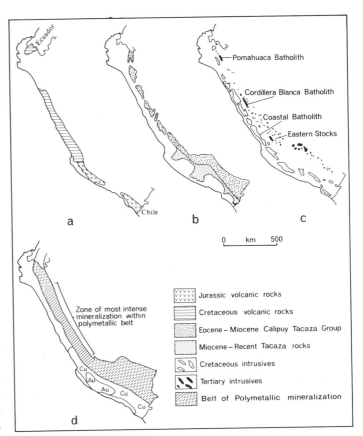

Figure 6. Distribution of: (a) Jurassic and Cretaceous volcanic
rocks, (b) Upper Cretaceous and younger volcanic rocks, (c) Mesozoic and Tertiary plutonic rocks, (d) Metallogenic belts.

inland for up to 200 km from the coast (Fig. 6c). Also inland
are the substantial batholiths of the Cordillera Blanca,
Pomahuaca, and Tupe which lie in well-defined, linear
zones. These scattered stocks and batholiths form a more or
less continuous outcrop from the Abancay deflection to the
Ecuador border.

In the Arequipa segment, Tertiary plutonic rocks have
not been recognized within the batholith. However, south of
the Abancay deflection, Tertiary intrusives occupy a well-
defined linear zone east of the Coastal Batholith, but now
about 300 km from the coast, where they form both small
stocks and the irregularly shaped Abancay Batholith of
Miocene age (Marocco, 1971).

Because the Tertiary intrusives outcrop over such a great
distance, but only as discontinuous scattered stocks, relatively little is known about them. Their discontinuous
nature would inhibit the recognition of super-units such as
has been achieved in the investigation of the Coastal
Batholith.

An exception is the Cordillera Blanca Batholith, which is
fairly well known (Egeler and de Booy, 1956). It consists
mainly of granodiorite and monzogranite, and the main
rock type is coarse-grained, with megacrysts of grey K-

feldspar, and contains both biotite and muscovite (the only two-mica granite known from the Peruvian Andes). Isolated stocks of similar lithology extend as far south as the Rio Huaura.

Otherwise, the small scattered stocks to the east of the Coastal Batholith are, for the most part, porphyritic with phenocrysts of plagioclase, corroded quartz, biotite, and hornblende set in a very fine-grained matrix of quartz and alkali feldspar. Most of these rocks are of rhyodacitic composition, but those which outcrop nearer to the Coastal Batholith, and which also tend to be somewhat larger and are coarse grained, have a granitoid rather than a porphyritic texture. This is also true for those Tertiary plutons within the Lima segment of the Coastal Batholith, some of which, however, are coarse monzogranites.

Radiometric determinations are few and scattered and range (Fig. 5) from 34 Ma to 4.7 Ma (Stewart and others, 1974). They indicate that all the Tertiary intrusives were emplaced after the Paleocene deformation, which affected the Cretaceous sedimentary and volcanic formations, and after the development of an erosion surface which truncates both the Cretaceous strata and intrusives of the Coastal Batholith.

A major time break of about 30 Ma between the mainly Cretaceous plutonism of the Coastal Batholith and the mid-Tertiary plutonism corresponds to the period of folding, uplift, and erosion known as the Incaic orogeny. Whereas the axis of Cretaceous plutonism was fixed in a belt within 70 km of the coast, the axis of Tertiary plutonism extended from 50 to 200 km from the coast.

RELATIONSHIP BETWEEN PLUTONISM AND VOLCANISM

These two axes of plutonic activity also coincide with the outcrop of the Cretaceous and Tertiary volcanics (Figs. 6 A,B,C), and this spatial association suggests a temporal relationship. This, however, is only partly true. Cretaceous volcanism occurred mainly in the Albian, whereas plutonism lagged until Late Cretaceous. Similarly, in the Tertiary, whereas volcanism ranges from 52.5 Ma (Wilson, 1975) to 14.5 Ma (Farrar and Noble, 1976), the plutonism was delayed until between 34 Ma to 4.7 Ma (Stewart and others, 1974). The overlap in time suggests that a connection exists but that it is not a direct one.

Cretaceous volcanism was marine and took place in structurally well-defined, probably fault-bounded, troughs, and the structures which controlled the troughs also guided the emplacement of the Coastal Batholith, producing its linearity.

Tertiary volcanism, by contrast, was terrestrial, more widespread, and deposited flows and pyroclastic sheets on a Paleocene erosion surface. The spatially associated pluto-nism was scattered, but a linear structural control was in

some places important, as in the case of the Cordillera Blanca Batholith.

Furthermore, the area of Tertiary volcanism and pluto-nism is divided into two well-marked sectors; that north of the Abancay deflection and that to the south (Fig. 6b). The volcanic sequence to the north is of Eocene to Middle Miocene age, but to the south the volcanicity, which began in the Eocene, has continued to the present. The northern sequence corresponds to the Calipuy Group and the southern sequence to the Tacaza-Sencca and Barroso Formations (Bellido, 1969). The axis of outcrop of the southern group of volcanics and intrusives lies farther east (at about 200 km) than that of the northern group. The authors conclude that the Abancay deflection has influenced the generation of magma through the Upper Neogene to the present, whereas during the Cretaceous and Paleocene, its influence was minimal.

METALLOGENESIS AND THE TERTIARY INTRUSIVES

The mineral wealth of Peru resides principally in those mineral deposits which pertain to the belt of polymetallic mineralization which runs throughout the Andes of Ecua-dor, Peru (Fig. 6D), Bolivia, and Chile. The metals obtained are principally copper, lead, zinc, silver, and tungsten, though molybdenum, bismuth, antimony, and mercury are locally important. Many authors, e.g., McLaughlin (1924), have stressed the importance of the Tertiary intrusives in mineral genesis, and it is certainly true that the majority of mineral deposits occur as skarns, replacements, and veins in or around these small intrusive bodies. The host rock may be intrusive, sedimentary, or volcanic, but, in the main, volcanic rocks and limestones are the most commonly mineralized lithologies. Local factors, such as faulting and host rock lithology, have been decisive in determining the mineralization pattern in individual deposits, but all recent work confirms that the polymetallic belt is linked with the Tertiary plutonic and volcanic activity. It is, moreover, important to note that, whereas the polymetallic belt (Bellido and De Montreuil, 1972) corresponds closely with the outcrop of all the Tertiary igneous rocks, the zone of most intense mineralization coincides with the outcrop of the Calipuy Group of volcanic rocks, which lies to the north of the Abancay deflection and which seems, therefore, to have also affected the genesis of the metallic minerals.

DISCUSSION

Subduction and Magma Genesis

The geological and geochronological evidence presented above shows two distinct and separate episodes of volcanism and plutonism in the Peruvian Andes. Each of these episodes was lengthy, and the second was generated farther inland than the first. Moreover, each was characterized by its own distinct pattern of mineralization, the copper, molybdenum, and gold belt being linked to Cretaceous magmatism, whereas the polymetallic belt is related to the Tertiary magmatism. These are pertinent factors in a discussion of the generation of granitoid magmas at a convergent plate margin, such as the Andes. A subduction model for magma generation is more acceptable than other models for reasons mentioned above, and also because subduction of the Nazca plate below the continental margin is modelled as taking place at present, the associated magma generation being visible in the form of the chain of recent volcanoes south of the Abancay deflection. Interestingly, the absence of such volcanoes north of the Abancay deflection has been interpreted by Mégard and Philip (1976) as being due to the dip of the subduction zone being too shallow to permit the generation there of magma. Directly relating magma generation to subduction implies that, as no magma was generated during the Paleocene orogeny and possibly not during the most active phase of the Sub-hercynian, no subduction took place at these times. This possibility has already been indicated by Charrier (1973), who correlated tectonic activity in the Andes with interruptions of sea-floor spreading.

The fact that Tertiary magmas were generated farther inland than Cretaceous magmas may indicate either that the dip of the Tertiary subduction zone was different from that of the Cretaceous, or that the location of the trench relative to the present continental margin was different during the two eras. This could have been achieved if the Cretaceous subduction zone developed within the oceanic crust, as may have been the case in Ecuador (Kennerley, 1980), or if the continental margin was tectonically eroded by subduction, as envisaged by Rutland (1971).

There are sufficient chemical and petrological differences between the Cretaceous and Tertiary magmatisms to suggest that a comparative study might illuminate the question of magma generation in the two provinces. Thus, whilst both sequences are calc-alkaline, the Tertiary magmatism is probably more silicic overall. These differences are further indicated by the different kinds of metallogenesis associated with the two magmatic provinces.

James and others (1974), from southern Peru, suggested that there are indeed significant differences in the initial $^{87}Sr/^{86}Sr$ ratios for the Mesozoic and Neogene volcanic rocks. They reported initial ratios of 0.7040 from the volcanic rocks of the Chocolate Formation of Jurassic age and 0.7042 for volcanic rocks of the Toquepala Formation of Cretaceous age. In contrast, the initial ratios for Miocene and Recent volcanic formations in the same area are: Huavlillas dacite, 0.7062; Sencca dacite, 0.7085; Arequipa andesite, 0.7080; and Barroso dacite, 0.7068.

Initial ratios for the Cretaceous intrusives (about 0.7042) are uniform throughout the batholith and are evidently near those for the Chocolate and Toquepala volcanics; however, no data of this kind are available for the Tertiary intrusives.

In order to explain these isotopic differences, James (1978) appealed to a model of subduction of Cretaceous trench sediments and their subsequent return to the crust as recycled magmas. The necessity for the subduction of Mesozoic continental waste and its possible contribution to subsequent magmas had also been previously suggested by Gilluly (1971) and Cobbing (1976).

Accepting the subduction model, several complicating factors exist which could affect magma generation. The simple case of partial melting of the down-going slab is possibly sufficient to account for some of the magmatism (Ringwood, 1974), and it is likely that the early magmas of the Cretaceous volcanics were derived by partial melting within the subduction zone or in the overlying mantle wedge. However, the later arrival of the plutons of the Coastal Batholith suggests that the delay in their emplacement relative to the spatially associated mid-Cretaceous volcanics was the result of a change in the process of magma formation, which is also, perhaps, indicated by their overall more silicic composition with respect to the mid-Cretaceous volcanics.

This difference is at its most extreme for the circa 60 Ma Puscao and San Jerónimo Super-units of the Lima segment, which form the spectacular ring complexes. This contrast in source may be due to the generation of granitoid magmas by partial melting of the lower continental crust, a process triggered by the introduction of heat via the basic magmas derived from an underlying subduction zone. The timing of emplacement and the ubiquity of the gabbro precursor suggest that it could have played such a role. A similar mechanism for such crustal fusion was suggested for the Sierra Nevada by Presnall and Bateman (1973). The geological evidence for the minor role of crustal shortening in Andean tectonics (Cobbing and others, 1981) supports a hypothesis of underplating as a means of attaining the great thickness of the crust in this area, which the geophysical evidence also requires (James, 1971). Such underplating may have its origin in material derived from a mantle wedge above the subduction zone. The underplate would have produced a zone of more readily fusible material with $^{87}Sr/^{86}Sr$ initial ratios similar to that of the mantle itself, and it is likely that this underplate material was supplied by magmas related to the generation of the volcanic arc. This could explain the apparent mantle characteristics of the initial ratios with the association of granitoid material and pre-existing continental crust (Fig. 7).

It seems likely that a model based on subduction of oceanic crust beneath the continent, leading to the generation of volcanic and plutonic magmas in a two-stage cycle, could eventually account for most of the magmatic phe-

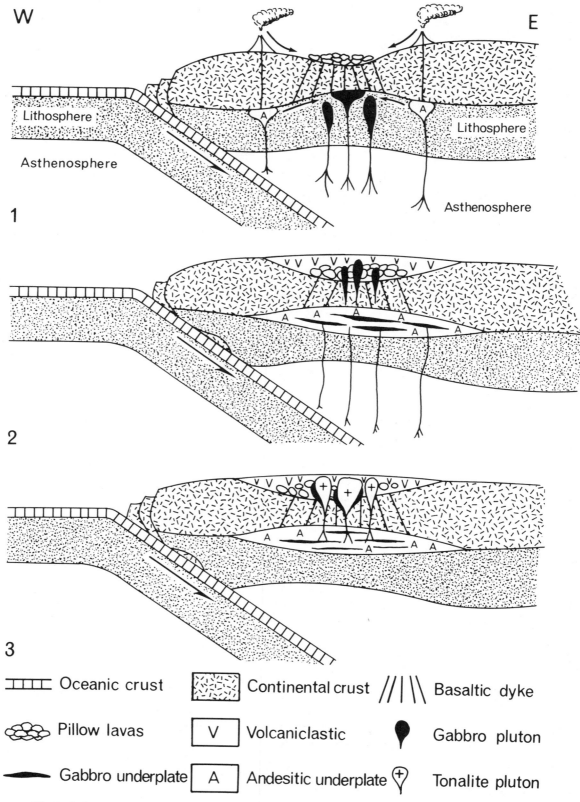

Figure 7. Cartoon to illustrate the formation of a crustal underplate at an Andinotype margin. The underplate provides the source area for the granitoid magmas.

nomena actually observed. From the geological point of view, however, it is more rewarding to comment on the relationship of magma generation to crustal structure.

Crustal Structure and Batholith Emplacement

The highly linear nature of the Coastal Batholith has been commonly noted (Pitcher, 1972) and is indeed impressive: on the geological map of Peru, the plutons of the batholith can be contained within two ruled lines (see Myers, 1975). It has been shown that the formation of the Mesozoic sedimentary and volcanic troughs was controlled by block faulting parallel to the continental margin (Wilson and others, 1967; Myers, 1974; Cobbing, 1978). The block faulting penetrated the crust and, in combination with pull-apart tectonics appropriate to a back-arc basin, resulted in crustal thinning and trough formation. It has also been suggested that these same faults penetrated into the mantle and provided channels for the emplacement of volcanic material and later for the granitoids (Cobbing, 1976; Pitcher, 1978). The confining of the plutons of the Coastal Batholith within a narrow zone implies that the deep-reaching, crustal-mantle faults were active over the same extended period as batholith emplacement (Bussell, 1976).

In contrast, the Tertiary intrusives, although occupying a broad band parallel to the continental margin, are diffusely spread over the band. It is true that highly linear zones are discernible within this diffuse spectrum, but, in comparison to the Coastal Batholith, they are generally relatively short in length and short lived. The plutonic lineaments represented by the Cordillera Blanca Batholith and the Pomahuaca Batholith are examples of such zones which are obviously fault controlled, and the Tertiary intrusives in southeastern Peru may well be another.

It is not difficult to account for the diffuse pattern of Tertiary plutonism as opposed to the linear pattern of mainly Cretaceous plutonism. The Tertiary plutonism followed the Paleocene orogeny, and the constituent plutons were emplaced into terrestrial volcanic sequences laid down upon the erosion surfaces produced by that orogeny. Thus, by the time Tertiary magmatism was initiated, substantial thickening of the crust had occurred. This may well have healed deep penetrating fractures so that only some persisted. Those few, such as that on the eastern side of the Coastal Batholith in the Lima segment and that now represented by the Cordillera Blanca Fault, continued to permit the ascent of major plutons. Other plutons which had to traverse a wide area of thickened crust were eventually emplaced in a diffuse pattern. The thickened crust also may well have been important in the genesis of the Tertiary magmas and may account for any differences between them and the Cretaceous magmas.

Lateral Structures and Batholith Segmentation

The transverse structures which affect plutonism are more enigmatic. Several of these have been noted above, namely, the Abancay and Huancabamba deflections, the stratigraphic and plutonic change at the latitude of Lima, and the plutonic change at the northern end of the Lima segment at Chimbote. It is most significant that three of these structures, the Abancay deflection, the Lima "change," and the Huancabamba deflection, are reflected in the Paleozoic Eastern Cordillera, as well as in the distribution of Mesozoic-Cenozoic intrusions, and must, accordingly, have a long history. The nature of these deflection zones is not known, but each appears to be different from the rest.

The most clearly defined is the Abancay deflection, which, superficially, could be a sinistral wrench-fault cutting across the entire Eastern Cordillera. Its easterly extension could form the southern margin to the Amazon depression (Dalmayrac and others, 1977). Also perhaps significant is the presence of the Quiros fracture zone, which bounds the Nazca rise and which continues the trend of the Abancay deflection into the Pacific as a possible transform fault. This particular structure may thus be a fracture of continental dimensions. It is evidently very deep reaching, for it affected the generation of magma during the Tertiary era and must have penetrated as deep as the subduction zone.

The other structures are probably similar, if less well defined, and together suggest that an ancient cross-fracture pattern contributed to the development of the Andean fold belt. One of the most puzzling features of the Abancay deflection, however, is the fact that it did not affect magma generation during the Cretaceous, because the line of plutons of the Arequipa segment runs straight across it. The structure at Lima, however, did affect Cretaceous magmatism, the evidence being that plutonism in the Lima segment has a much longer history than that within the Arequipa segment.

If present day subduction is used as a model to explain how this may have occurred, it is apparent in the chain of modern volcanoes which extends through southern Peru as far north as the Abancay deflection (noted above). As already stated, Mégard and Philip (1976) suggested that the abrupt change in the distribution of the volcanoes is related to the fact that to the south the Benioff zone dips beneath the continent at about 30°, whereas to the north it dips at 10-15°. This strongly indicates that the transverse structures do penetrate sufficiently deep to affect the subduction zone and that they thus define segments within each of which the Benioff zone has a different but constant dip. This would allow the conditions of magma generation for adjacent segments to be different and favor the formation of a segmented batholith.

ACKNOWLEDGMENTS

This research project was financed by the Overseas Development Administration and the Natural Environment Research Council of Great Britain. The work was carried out in co-operation with the Servicio de Geología y Minería del Perú over a lengthy period, and the invaluable help rendered by that organization under two directors, Ing E. Bellido and Ing B. Morales, is warmly acknowledged. We thank our colleagues W. P. Taylor, C. Vidal, and N. Moore for permission to quote their unpublished analytical results. Dr. E. J. Cobbing writes by permission of the Director of the Institute of Geological Sciences, Natural Environment Research Council.

REFERENCES CITED

Atherton, M. P., McCourt, W. J., Sanderson, L. M., and Taylor, W. P., 1979, The geochemical character of the segmented Peruvian Coastal Batholith and associated volcanics: *in* Atherton, M. P., and Tarney, J., eds., Origin of Granite Batholiths: Shiva Publishing, Orpington, Kent, p. 65-75.

Bateman, P. C., and Dodge, F. C. W., 1970, Variations of major chemical constituents across the central Sierra Nevada batholith: Geological Society of America Bulletin, v. 81, no. 2, p. 409-420.

Bellido, E., 1969, Sinopsis de la geología del Perú: Servicio de geología y minería del Perú Boletin, No. 22, 54 p.

Bellido, E., and De Monteuil, L., 1972, Aspectos generales de la metalogenia del Perú. Servicio de geología y minería del Perú: Geología Economica, No. 1, 149 p.

Bussell, M. A., 1976, Fracture control of high-level plutonic contacts in the Coastal Batholith of Peru: Proceedings of the Geological Association, v. 87, p. 237-246.

Bussell, M. A., Pitcher, W. S., and Wilson, P. A., 1976, Ring complexes of the Peruvian Coastal Batholith, a long-standing sub-volcanic regime: Canadian Journal of Earth Sciences, v. 13, p. 1020-1030.

Chappell, B. W., and White, A. J. R., 1974, Two contrasting granite types: Pacific Geology, v. 8, p. 173-174.

Charrier, R., 1973, Interruption of spreading and the compressive tectonic phases of the meridional Andes: Earth and Planetary Sciences Letters, v. 20, p. 242-249.

Cobbing, E. J., 1976, The geosynclinal pair at the continental margin of Peru: Tectonophysics, v. 36, p. 157-165.

Cobbing, E. J., 1978, The Andean geosyncline in Peru and its distinction from Alpine geosynclines: Journal of the Geological Society of London, v. 135, p. 207-218.

Cobbing, E. J., Baldock, J., McCourt, W., and others, 1981, The geology of the Western Cordillera of Northern Peru: Institute of Geological Sciences, Overseas Memoir 5, London, 143 p.

Cobbing, E. J., Ozard, J. M., and Snelling, N. J., 1977a, Reconnaissance geochronology of the crystalline basement rocks of the Coastal Cordillera of southern Peru: Geological Society of America Bulletin, v. 88, p. 241-246.

Cobbing, E. J., Pitcher, W. S., and Taylor, W. P., 1977b, Segments and super-units in the Coastal Batholith of Peru: Journal of Geology, v. 85, p. 625-631.

Dalmayrac, B., 1977, Gèologie de la Cordillère Orientale de la région de Huanuco. Sa place dans une transversale des Andes du Pérou Central (19° - 10° 30′ S): Thesis, University of Montpellier.

Egeler, C. G., and De Booy, T., 1956, Geology and petrology of part of the southern Cordillera Blanca, Peru: Verhandelingen Van Het Koninklijk Nederlandsch Geologische-Mijnbouwkundig Genootschap Geologische Serie. v. 17, p. 11-86.

Farrar, E., and Noble, D. C., 1976, Timing of late Tertiary deformation in the Andes of Peru: Geological Society of America Bulletin, v. 87, p. 1247-1259.

Gilluly, J., 1971, Plate tectonics and magmatic evolution: Geological Society of America Bulletin, v. 82, p. 2383-2396.

Ham, C. K., and Herrera, L. J., 1963, Role of Subandean fault systems in tectonics of eastern Peru and Ecuador: *in* Childs, D., and Beebe, B. W., eds., Backbone of the Americas: American Association of Petroleum Geologists Memoir, No. 2, p. 47-61.

Hudson, C., 1979, Zoneamiento de la metalogenia Andina del Perú: Bol. Soc. geol. Perú., v. 60, p. 61-72.

James, D. E., 1978, On the origin of the calc-alkaline volcanics of the Central Andes: a revised interpretation: Carnegie Institute of Washington Yearbook, p. 562-590.

James, D. E., Brooks, C., and Cuyubamba, A., 1974, Andean Cenozoic volcanism: magma genesis in the light of strontium isotope composition and trace element geochemistry: Carnegie Institute of Washington Yearbook, p. 983-997.

Kennerley, J. B., 1980, Outline of the geology of Ecuador: Overseas Geology and Mineral Resources, No. 55.

Larson, R. L., and Pitman, W. G., III, 1972, World-wide correlation of Mesozoic magnetic anomalies and its implications: Geological Society of America Bulletin, v. 83, p. 3645-3662.

Laubacher, G., 1978, Estudio geologico de la region norte del Lago Titicaca: Instituto de Geología y Minería del Perú Boletin, No. 5, p. 120.

Marocco, R., 1971, Etude Géologique de la Chaîne andine au niveau de La déflexion d'Abancy (Pérou): Cahiers *Orstom*, Serie Géologique, v. 3, p. 45-58.

—, 1978, Estudio geologico de la Cordillera de Vilcabamba: Institute de Geología y Minería del Perú Boletin, No. 4, 157 p.

McLaughlin, D. H., 1924, Geology and physiography of the Peruvian Cordillera, Departments of Junin and Lima: Geological Society of America Bulletin, v. 35, p. 591-632.

Mégard, F., and Philip, H., 1976, Plio-Quaternary tectono-magmatic zonation and plate tectonics in the central Andes: Earth and Planetary Science Letters, v. 33, p. 231-238.

Mégard, F., Dalmayrac, B., Laubacher, G., and others, 1971, La Chaîne hercynienne au Pérou et en Bolivie; premiers resultats: Cahiers Orstom Serie Géologique, v. 3, p. 5-43.

Moore, N., 1979, The geology and geochronology of the Coastal Batholith of southern Peru: University of Liverpool, Ph. D. thesis, unpublished.

Myers, J.S., 1974, Cretaceous stratigraphy and structure, western Andes of Peru between latitudes 10° and 10° 30′ S: Geological Society of America Bulletin, v. 58, p. 474-487.

—, 1975, Vertical crustal movements of the Andes in Peru: Nature, v. 254, p. 672-674.

Pitcher, W. S., 1972, The Coastal Batholith of Peru: some structural aspects. 24th International Geological Congress, Section 2, No. 24, p. 156-163.

—, 1978, The anatomy of a batholith: Journal of the Geological Society of London, v. 135, p. 157-182.

Presnall, D. C., and Bateman, P. C., 1973, Fusion relations in the system $NaAlSi_3O_8$-$CaAl_2Si_2O_8$-$KAlSi_3O_8$-SiO_2-H_2O and generation of granitic magmas in the Sierra Nevada Batholith: Geological Society of America Bulletin, v. 84, p. 3181-3202.

Regan, P. F., 1976, The genesis and emplacement of mafic plutonic rocks of the coastal Andean batholith, Lima province, Peru: University of Liverpool, Ph. D. thesis, unpublished.

Ringwood, A. E., 1974, The petrological evolution of island arc systems: Journal of the Geological Society of London, v. 130, p. 183-204.

Rutland, R. W. R., 1971, Andean orogeny and ocean floor spreading: Nature, v. 233, p. 252-255.

Shackleton, R. M., Ries, A. C., Coward, M. P., and Cobbold, P. R., 1979, Structure, metamorphism, and geochronology of the Arequipa Massif of coastal Peru: Journal of the Geological Society of London, v. 136, p. 195–214.

Sillitoe, R. H., 1974, Tectonic segmentation of the Andes: implications for magmatism and metallogeny: Nature, v. 250, p. 542–545.

—, 1976, Andean mineralization: model for the metallogeny of convergent plate margins: Geological Association of Canada, Special Paper No. 14, p. 59–100.

Stewart, J. W., 1968, Rocas intrusivas del cuadrangulo de la Joya: Servicio de Geología y Minería del Perú Boletin, No. 19, p. 43–78.

Stewart, J. W., Evernden, J. F., and Snelling, N. J., 1974, Age determinations from Andean Peru: a reconnaissance survey: Geological Society of America Bulletin, v. 85, p. 1107–1116.

Taylor, W. P., 1976, Intrusion and differentiation of granitic magma at a high level in the crust: the Puscao pluton, Lima Province, Peru: Journal of Petrology, v. 17, p. 194–218,

Webb, S., 1976, The volcanic envelope of the Coastal Batholith in Lima and Ancash, Peru: University of Liverpool, Ph. D. thesis, unpublished.

Wilson, J. J., 1963, Cretaceous stratigraphy of the Central Andes of Peru: American Association of Petroleum Geologists Bulletin, v. 47, p. 1–34.

Wilson, J. J., Reyes, L., and Garayar, J., 1967, Geología de los cuadrangulos de Mollebamba, Tayabamba, Huaylas, Pomabamba Carhuaz, y Huari: Servicio de Geología y Minería del Perú Boletin, No. 16, p. 95.

Wilson, P. A., 1975, K-Ar age studies in Peru with special reference to the emplacement of the Coastal Batholith: University of Liverpool, Ph. D. thesis, unpublished.

MANUSCRIPT ACCEPTED BY THE SOCIETY JULY 12, 1982

Geological Society of America
Memoir 159
1983

Granitoids in Chile

Luis Aguirre
Department of Geology
University of Liverpool P.O. Box 147
Liverpool L69 3BX, England

ABSTRACT

Granitoid rocks are exposed in nearly forty per cent of Chile. The main periods of emplacement, indicated by isotopic data and stratigraphic relations, are Carboniferous, Jurassic, Cretaceous, and Tertiary, and they approximately coincide with the ages of the principal orogenic phases recorded in the country. The commonest rock types, regardless of age, are granodiorite and tonalite, but the compositional range is broad, and diorite, granite, quartz diorite, and gabbro are common.

The Paleozoic granitoids intrude stratified formations from Late Ordovician to latest Permian in age and are intruded by Cretaceous and Tertiary plutons. Their contacts are sharply discordant in some localities and transitional, accompanied by widespread migmatization, in others. A close relationship between the Paleozoic granitoids and high-temperature, low-pressure metamorphic series is well known in Central Chile. The Paleozoic granitoids are closely related in space with rhyolitic volcanic rocks of Permo-Triassic age in the Andes of north Central Chile.

The Mesozoic and Cenozoic (Andean) granitoids cut rocks ranging in age from Paleozoic to Late Tertiary. Their transgressive contact relations indicate a post-kinematic character, and thermal aureoles are present in several regions. Mineral deposits, especially copper, iron, molybdenum, and zinc are genetically related to these granitoids. The Tertiary granitoids are richer in alkalis than those of the Mesozoic, and some are directly associated with porphyry copper deposits. A genetic relation between the Tertiary granitoids and Pliocene andesitic flows has been suggested on the basic of Rb/Sr ratios. The age of the Andean granitoids decreases from west to east, and the initial Sr^{87}/Sr^{86} ratio of granitoids, with ages ranging from mid-Cretaceous to Quaternary, shows a systematic west to east increase from 0.7022 to 0.7077, which indicates a change in the locus of melting.

INTRODUCTION

Granitoid outcrops occupy nearly 40% of the surface of Chile, and they form part of the circum-Pacific batholiths. A long term, systematic study of the Chilean granitoids, such as those carried out in the Sierra Nevada, Baja California, the Peruvian Coastal Batholith, and other regions of the Pacific belt, has not yet been undertaken. However, regional mapping, prospecting, geochronological work, and research on the petrological and chemical character of some specific plutonic complexes provide abundant information which broadly characterizes plutonism in Chile.

Geochronologic and stratigraphic information, especially in the last twenty years, shows that granitoid generation and emplacement range in age from the Paleozoic to Neogene. The following account has been chronologically structured within three main parts: Paleozoic Granitoids; Granitoids of the Andean Orogenic Cycle (Mesozoic/Cenozoic); and Essential Features of Chilean Plutonism.

PALEOZOIC GRANITOIDS

General Statement

Until the end of the 1950's the names "Andean Batholith" or "Andean Diorite" were used in Chilean geological literature to designate all the granitoids in the country. These rocks were assigned to the plutonic activity connected with the Andean Orogenic Cycle during the Mesozoic and Cenozoic.

In certain parts of the territory, such as in the Andes of Coquimbo (latitude 30° S) and Atacama, farther north, stratigraphic relationships had early revealed a pre-Liassic age for some of the batholiths (Domeyko, 1845; Willis, 1929). Moreover, Steinmann (1929) had suggested that some of the granitoids present along the coast and the Coastal Ranges of Central Chile (region west of Santiago) could actually be older than the "Andean" Mesozoic types exposed farther east. Granitoids in this coastal region are closely related in space to the metamorphic rocks of the Chilean basement which were then considered to be of Precambrian age. Muñoz Cristi (1960) proposed a division of the Andean Batholith of Central Chile into the Coastal Granite and the Central Granite, although a Paleozoic age for the Coastal Granite was not implied in its definition. In the same year, the first Paleozoic radiometric age of 265 ± 30 Ma (lead-alpha method on zircon) for a Chilean intrusive, a granite from Juntas, 85 km southeast from Copiapó, was reported (Ruiz et al., 1960). Two years later, Muñoz Cristi (1962) obtained a K/Ar, biotite age of 287 ± 20 Ma for a tonalite from his Coastal Granite west of Santiago. Since then, the existence of large volumes of Paleozoic granitoids has been identified in several parts of the country. About the same time, geochronological studies on the metamorphic basement (González-Bonorino, 1967; Munizaga et al., 1973; Hervé et al., 1974b, 1976) led to the establishment of a Paleozoic age (mainly Carboniferous) for this complex. The term Crystalline Basement has been used in the last few years (González-Bonorino and Aguirre, 1970; Aguirre et al., 1972) to embrace the metamorphic basement and the Paleozoic granitoids in Central Chile (33° S to 40° S, approximately) where these two units appear closely related spatially and genetically (González-Bonorino, 1970, 1971).

Main characteristics of the Paleozoic granitoids

Paleozoic batholiths are found both along the Coastal and Andean Ranges between latitudes 23° S and 42° S (Fig. 1). The general features of the Paleozoic batholiths are mainly based on information from the publications listed in Table 1.

The main phases are granodiorite, quartz diorite, granite, and tonalite, with subordinate diorite and gabbro. Minor amounts of porphyries are present in places. In some parts of the country, these petrographic phases are considered to be associated units belonging to a cogenetic, composite batholith. According to Tobar (1977) and Mpodozis et al. (1976), this would be the case for the El Salvador - Potrerillos area and the Andes of Coquimbo, respectively.

The Paleozoic granitoids cut across stratified formations

TABLE 1. MAIN AREAS OF PALEOZOIC BATHOLITHIC EXPOSURES, AND REFERENCES

Area	References
East of Antofagasta	Halpern, 1978
El Salvador - Potrerillos	Tobar, 1977; Halpern, 1978
Region between lat. 26° S and 29° S	Clark et al., 1976; Farrar et al., 1970; Levi et al., 1963; McBride et al., 1976; McNutt et al., 1975; Segerstrom, 1967; Zentilli, 1974
Andes of Coquimbo	Dediós, 1967; Mpodozis et al., 1976
Coastal Range, Central Chile, between lat. 33° S and 38° 30′S	Muñoz Cristi, 1964; Corvalán and Munizaga, 1972; Corvalán and Dávila, 1964; Hervé, 1977; Hervé et al., 1976; Hervé et al., 1974b; Levi et al., 1963; Muñoz Cristi, 1962; Munizaga et al., 1973; Nishimura, 1971; Hervé, 1976; González-Bonorino, 1970
Andean region east of Valdivia	Hervé et al., 1974a; Groeber, 1963; Parada, 1975

ranging in age from Late Ordovician (Caradoc) (McBride et al., 1976) to the latest Permian (Mpodozis et al., 1976), including the basement rocks of Central Chile, for which the main metamorphic event has been demonstrated to be of Carboniferous age (Munizaga et al., 1973). The Paleozoic intrusives are, in turn, unconformably covered in several localities by sedimentary marine rocks of Early Jurassic (Liassic) and also probable Triassic age. In places, they are covered by terrestrial rocks of Late Triassic (Keuper) age, as in the Andes of Coquimbo (Mpodozis et al., 1976). Stratified volcanic and sedimentary rocks of Early Cretaceous (Neocomian) and Late Cretaceous age unconformably rest on the Paleozoic intrusives west of Santiago (Corvalán and Dávila, 1964) and in the Valdivia region (Hervé et al., 1974a; Parada, 1975), respectively. Granitoids of Cretaceous and Tertiary age intrude the Paleozoic batholiths in several areas.

The nature of the contacts varies from one locality to another. In the Andes of Coquimbo, Mpodozis et al. (1976) described the Paleozoic granitoids as epizonal, disharmonic plutons showing sharp discordant boundaries, devoid of migmatites and dyke swarms. Their contact aureoles are narrow or absent. According to these authors, such features are characteristic of the late to post-tectonic batholiths of the geosynclinal belts. Rather similar features were described by González-Bonorino (1970) for the segment of the Coastal Range between latitudes 34° S and 38° S, where migmatitic zones are restricted to a few metres or tens of metres, and the contacts are generally neat and remarkably regular in shape, being essentially parallel to the schistosity and the fold axes of the metamorphic rocks of the basement, with which they

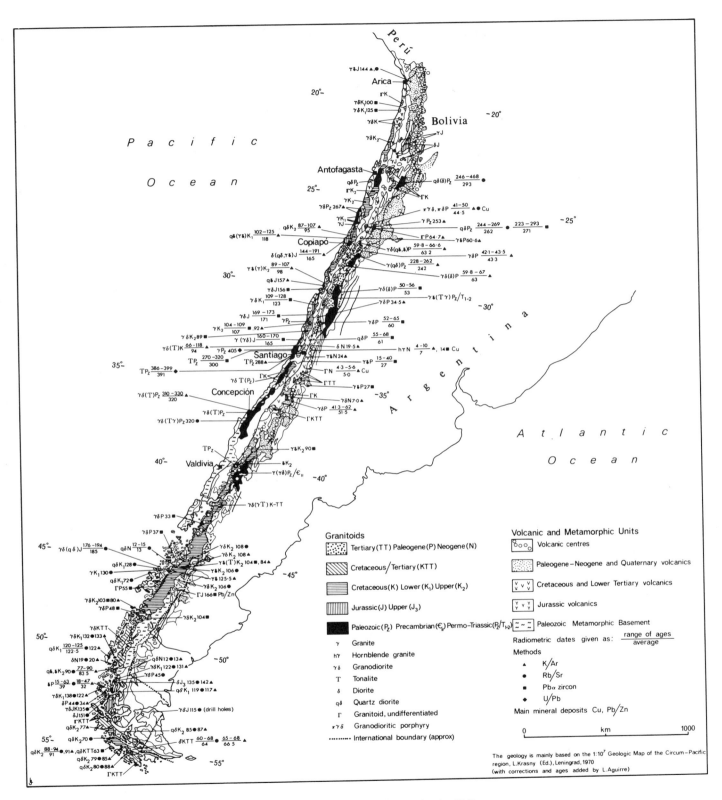

Figure 1. Distribution of granitoids in Chile.

are closely associated. This situation clearly contrasts with the contact relations observed west of Santiago and in the Nahuelbuta Mountains (38°S latitude). In the former region, broad migmatitic zones are associated with high-grade metamorphic facies (amphibolite-granulite transition) of the basement. In the latter region, migmatites are common at the contact zone, and their paleosome consists of gneisses similar to those found in the metamorphic basement surrounding the intrusions (Hervé, 1977). Abundant pegmatitic veins are associated with these mixed rocks. Moreover, mineralogical and structural changes in the granitoids can be detected as far as about 5 km from the contact with the metamorphic basement. These changes are represented by the appearance of muscovite, abundant garnet, and occasional andalusite and sillimanite, as well as by the disappearance of hornblende. Contamination of the Paleozoic granitoids by admixture with the surrounding metamorphic rocks of the basement was also described by Tobar (1977) for the El Salvador - Potrerillos region, where diorite is typically found close to the contacts with the metamorphic rocks from which numerous xenolithic inclusions have been derived.

A direct role of the Coastal Range Batholith of Central Chile in the generation of the high-temperature, low-pressure Eastern Series of the metamorphic basement is evident. According to González-Bonorino (1970), the Nirivilo and Pichilemu Series (Eastern Series of Aguirre et al., 1972) would have been partly developed during a late-kinematic phase of the batholithic emplacement, partly under the influence of post-kinematic magma movements which superposed thermal aureoles on the already regionally metamorphosed rocks. This rather long-standing interaction would also explain the changeable nature of the contacts as observed in regions described above.

Thirty-two radiometric ages for Paleozoic granitoids from different regions have been compiled, mainly from the papers referred to in Table 1. Thirteen of them were obtained by the Pbα method on zircon, eight by K/Ar, and four by Pb/U (Appendix A). These dates are supported in most cases by stratigraphic evidence and show that plutonic activity culminated in the Early Devonian and Early Permian, with most values between the Early Carboniferous and the end of the Paleozoic (Fig. 2).

The compositional range of the late Hercynian granitoids, modal analyses (Appendix B) and their corresponding QAP diagram (Fig. 3) have been taken from the comprehensive work of Mpodozis et al. (1976) covering the Andes of Coquimbo and adjacent Argentina between latitudes 30°S and 33°S.

The Paleozoic granitoids in Chile appear to be almost unrelated to metallogenesis; one exception would be the copper mineralization in the Imilac area, Antofagasta region, which Halpern (1978), on the basis of geochronology, reported as Paleozoic. A small deposit of fluorite and topaz,

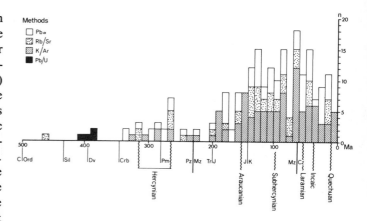

Figure 2. Histogram showing number of radiometric determinations (n) versus age in millions of years (Ma) for the Chilean granitoids listed in Appendix A. Major orogenic phases documented in the geological evolution of the country are shown. Age boundaries taken from the Geological Time Scale (Van Eysinga, 1975).

the only one of this type known in Chile, is present in Paleozoic granites southeast from the town of Vicuña in Coquimbo province (Ruiz et al., 1965).

GRANITOIDS OF THE ANDEAN OROGENIC CYCLE (Mesozoic/Cenozoic)

General Statement

Large volumes of granitoids were generated during the Andean Orogenic Cycle. Comprehensive accounts of the global geological evolution during this cycle can be found mainly in Aubouin et al., (1973) and Aguirre et al., (1974).

Mesozoic and Cenozoic granitoids crop out all along the Chilean territory as extensive batholiths, massifs, minor stocks, and scattered bodies (Fig. 1). Several petrological phases are represented, the most common being granodiorite, tonalite, and granite. Quartz monzonite, adamellite, gabbro, and various types of porphyry are also present.

These plutons intrude rocks ranging from the Paleozoic Crystalline Basement to the Late Tertiary series of the Andean Range.

Contact relationships indicate a post-kinematic character for the granitoids. In most places, they clearly cut across the structures; transitional margins and migmatitic borders are absent or very rare. Contact-metamorphic aureoles are well developed in several bodies, and hydrothermal alteration is common in some. Mineral deposits of economic value are closely related to these intrusives, especially copper, iron, molybdenum, and zinc.

Geochronological studies carried out in the last fifteen years have led to the identification of at least three main cycles of emplacement, dated approximately as Jurassic, Cretaceous, and Tertiary. Almost one third of the ages

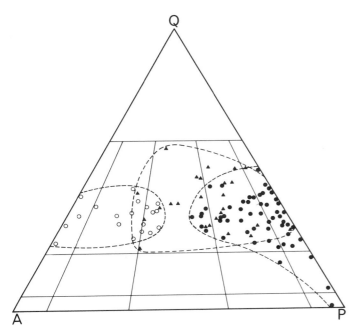

Figure 3. QAP plot of the late-Hercynian granitoids from the western flank of the Cordillera Frontal between 30° S and 33° S (Appendix B, 1, 2 and 3). Full circles = biotite/hornblende granodiorite-Tonalite Association; Open circles = hololeucocratic granites; Triangles = muscovite granodiorites. (Figure and petrographic naming of groups taken from Mpodozis et al., 1976).

available have been obtained by the Pbα method, a fact which precludes a refined correlation with other geotectonic events in the territory. Nevertheless, a broad agreement between the plutonic activity and the deformational phases known in the Andean Orogenic Cycle is apparent.

The granitoids referred to in this section are mainly within the domain of the Andean Geosyncline, between the northern boundary of the country at latitude 18° S and the southern extreme of "continental Chile" at latitude 41° S. However, some are south of this latitude and belong to the domain of the Magellan Geosyncline.

Main characteristics of the Mesozoic and Cenozoic granitoids

Jurassic Granitoids. Most outcrops of these rocks are found along the Coastal Range from latitude 18° 30'S to latitude 32° S. Others are known on Victoria Island (45° S), around Lake General Carrera (46° 33'S, 72° 25'W), and in a few localities in the fjord region, south of latitude 51° S (Fig. 1).

At Quebrada Camarones, near the Peruvian border, a granodiorite with an age of 144 Ma (K/Ar in biotite and hornblende) is covered depositionally by a terrestrial clastic formation of inferred Early Cretaceous age. Outcrops in this northern-most region are scattered and small; the main and more continuous exposures are found farther south, between

latitudes 26° S and 32° S. The Jurassic granitoids cut across the Paleozoic metamorphic basement on Damas Island (29° S) (Aguirre, 1967) and are, in turn, intruded by Cretaceous granitoids along most of their eastern boundary south of latitude 26° S (Farrar et al., 1970; Thomas, 1967; McNutt et al., 1975; Zentilli, 1974). A tectonic contact with the Paleozoic metamorphic basement has been established along most of the western boundary of the Jurassic intrusives south of latitude 30° S (Thomas, 1967).

Petrographic types include, from north to south: granodiorite at Quebrada Camarones (Levi et al., 1963); gabbro in the Coast Range of Tarapacá province (Tobar et al., 1968); diorite, quartz diorite, quartz monzonite, granodiorite, and granite in the Coast Range around latitude 27° S in Atacama province (Farrar et al., 1970; Zentilli, 1974); gneissic granite on Damas Island (Aguirre, 1967); quartz monzodiorite at latitude 30° S, north of La Serena (Aguirre and Egert, 1970; Curtis, written com., 1972); and dark gray tonalite and granodiorite cut by a great number of andesitic dykes in the coastal region at Punta Lengua de Vaca, between latitudes 30° S and 32° S in Coquimbo province (Thomas, 1967). Granitoids of Jurassic age from the fjord region south of latitude 51° S range from diorite to granite (Halpern, 1973). Two representative modal analyses of Jurassic granitoids are given in Table 2 (samples 1 and 2) and shown on Figure 4.

No regional metamorphic events related to the Jurassic granitoids are known. A thermal aureole produced by a small body (0.1 km²) of gneissic granite on Damas Island has been described. There, greenschists belonging to the Paleozoic metamorphic basement have been transformed into andalusite-bearing hornfels in a narrow zone around the intrusive (Aguirre, 1967).

The eleven Pbα, twenty K/Ar, and five Rb/Sr dates available (Appendix A) are clustered in the Early Jurassic and latest Jurassic. Possible culminations are suggested between 170 and 190 Ma, and between 140 and 160 Ma, although plutonic activity maintained its intensity into the Cretaceous (Fig. 2).

According to Ruiz et al. (1965), mineral deposits genetically related to Late Jurassic intrusive activity would be represented by copper veins which are accompanied by small quantities of cobalt, molybdenum and uranium. The distribution of these deposits coincides with the described belt of Jurassic granitoids (see Metallogenic Map of Chile, 1962, *in* Ruiz et al., 1965). At Lake General Carrera (46° 33'S, 72° 25'W), granitoids with a reported age of 166 Ma (Pbα) are related to lead/zinc deposits.

Cretaceous Granitoids. These rocks outcrop throughout the country and constitute the most important plutonic episode of the Andean Orogenic Cycle, both in volume and areal extent (Fig. 1). They cut across stratified formations ranging in age from Early Jurassic to mid-Cretaceous and, in several places, they intrude Paleozoic and Jurassic granitic masses. Large roof pendants of Mesozoic forma-

TABLE 2. MODAL ANALYSES OF MESOZOIC AND TERTIARY GRANITOIDS BETWEEN 30° AND 35° S LATITUDE.*

Sample No.	Locality †	Qtz	K-feld	Plag	Biot	Amph	Pyrox	Opaque Min.	Accessories	Ser Clor Ep	Name of Rock (after Streckeisen
1	Lambert	12.3	13.9	52.6	6.9	11.3		2.2			Qtz Monzodiorite
2	Papudo	26.0	29.0	29.0	6.0	7.0	—	—	—	—	Monzogranite
3	La Higuera	21.8	24.4	32.5	1.6	3.8	12.3	4.0	0.6	0.6	Monzogranite
4	Sta. Gracia	27.3	32.0	33.6	3.0	2.5	—	1.0	—	—	Monzogranite
5	Sta. Gracia	3.6	5.0	62.9	1.0	22.9	0.9	2.6	0.7	—	Diorite
6	Lambert	—	—	61.3	30.6	—	8.0	—	—	—	Diorite
7	LaLigua River	28.7	37.8	31.0	—	1.3	0.3	0.4	0.5	—	Monzogranite
8	LaLigua River	7.9	3.8	64.4	—	22.9	—		0.3	0.7	Qtz Diorite
9	La Campana	1.0	—	65.2	—	24.1	3.7	2.7	6.3	2.8	Gabbro
10	La Campana	14.9	7.2	46.0	15.0	8.5	1.9	0.9	0.4	5.0	Granodiorite
11	La Campana	15.2	9.6	54.3	6.8	7.9	2.7	1.5	0.3	1.6	Qtz Monzodiorite
12	La Campana	7.1	—	64.0	7.0	14.1	—	2.0	0.5	3.1	Qtz Diorite
13	La Campana	16.6	21.9	42.3	8.4	1.9	4.1	1.5	0.7	2.5	Granodiorite
14	La Campana	10.8	7.1	52.3	4.9	14.8	3.6	1.7	0.5	4.0	Qtz Monzodiorite
15	La Campana	8.5	19.1	64.1	—	5.5	1.3	0.6	0.7	0.1	Qtz Monzodiorite
16	La Campana	7.0	—	67.9	4.6	2.3	14.1	2.7	1.2	0.1	Qtz Diorite
17	La Campana	4.0	10.2	64.1	7.6	1.5	10.0	2.3	—	—	Monzodiorite
18	La Campana	24.0	17.6	45.0	0.7	5.3	—	1.4	—	6.0	Granodiorite
19	La Campana	22.5	24.1	39.6	—	—	—	2.1	0.1	11.1	Monzogranite
20	Valpo San Antonio	20.0	—	60.0	5.0	15.0	—	—	—	—	Tonalite
21	Valpo San Antonio	15.0	28.0	45.0	—	12.0	—	—	—	—	Qtz Monzonite
22	Valpo San Antonio	9.0	29.0	39.0	22.0	—	—	—	—	—	Qtz Monzonite
23	Valpo San Antonio	24.0	11.0	51.0	7.0	5.0	—	—	—	—	Granodiorite
24	Peñaflor	27.4	—	59.5	—	5.2	—	0.8	—	7.1	Tonalite
25	High Limarí	16.7	18.1	45.9	4.5	14.8	—	—	—	—	Granodiorite
26	High Limarí	10.2	3.1	62.3	11.5	8.1	—	2.0	1.5	—	Qtz Diorite
27	High Limarí	29.3	—	52.7	7.4	8.0	—	0.6	—	2.0	Tonalite
28	Tascadero	3.1	1.9	53.9	—	30.6	4.1	—	6.3	—	Qtz Diorite
29	Tascadero	41.7	36.8	18.9	0.5	—	—	—	1.0	—	Syenogranite
30	Campo deAhumada	15.0	2.0	64.7	4.0	10.5	0.2	4.0	0.2	—	Qtz Diorite
31	Putaendo River	27.4	24.5	41.1	—	6.3	—	0.5	0.2	—	Monzogranite
32	Putaendo River	2.9	0.1	83.9	—	10.2	0.2	2.3	0.5	—	Diorite
33	Rocín River	8.1	13.1	54.9	—	19.8	0.6	3.0	0.5	—	Qtz Monzodiorite
34	El Toro Brook	2.9	—	68.0	—	—	24.2	4.9	—	—	Diorite
35	Pan de Azúcar	27.8	—	65.3	6.2	—	—	0.7	—	—	Tonalite
36	Juncal River	4.4	—	71.3	—	—	19.4	3.9	1.0	—	Qtz Diorite
37	Est. San José	2.4	—	72.7	—	—	21.5	0.1	3.3	—	Diorite
38	Riecillos	13.0	20.5	55.0	3.5	7.5	—	—	0.5	—	Qtz Monzodiorite
39	Riecillos	24.0	32.0	38.0	3.0	4.0	—	—	—	—	Monzogranite
40	Riecillos	13.0	26.0	46.0	—	6.0	—	—	—	2.1	Monzogranite
41	La Obra	30.0	15.6	44.4	6.7	2.8	—	0.7	—	—	Granodiorite
42	San Gabriel	7.6	27.2	39.2	—	19.5	—	6.0	0.5	—	Qtz Monzonite
43	Rio Colorado	28.6	28.0	34.3	4.1	3.8	—	—	1.2	—	Monzogranite
44	Portillo	20.8	11.2	59.1	3.7	1.8	—	—	3.4	—	Granodiorite

* From Aguirre et al., 1974, p. 21.

† Geographical coordinates for the localities can be read from Appendix A with the exception of the following: samples 7 and 8, La Ligua River (32° 22′S, 71° 03′W); samples 20 to 24, Valparaiso-San Antonio, Peñaflor (between 33° 22′S - 33° 34′S and 71° 14′W -70° 58′W); samples 25 to 27, High Limari (30° 46′S - 70° 25′W); samples 28 and 29, Tascadero (31° 07′S, 70° 30′W); samples 30 to 37 (between 32° 11′S - 32° 30′S and 70° 18′W - 70° 30′W); sample 42, San Gabriel (33° 47′S, 70° 11′W); sample 43, Rio Colorado (33° 30′S, 70° 11′W); sample 44, Portillo (32° 53′S, 70° 12′W); and sample 3, La Higuera (29° 37′S, 71° 70′W).

tions are enclosed in the Cretaceous batholiths. Lack of detailed geological knowledge has made impossible the separation of Jurassic, Cretaceous, and Tertiary granitoids in several parts of the territory, resulting in the designation "Mesozoic and/or Cenozoic granitoids undifferentiated" on the 1 : 1,000,000 Geological Map of Chile (*Instituto de Investigaciones Geológicas*, 1962). In Figure 1 (this paper), Cretaceous/Tertiary undifferentiated granitoids (KTT) are designated in some places.

Different petrographic phases are present in these batho-

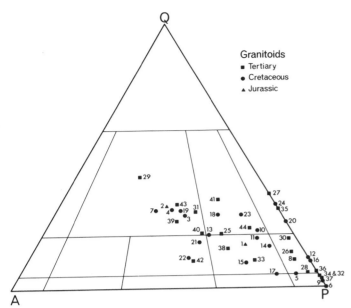

Figure 4. QAP plot of Mesozoic and Tertiary granitoids from Central Chile between latitudes 30°S and 35°S. Numbers correspond to the modal analyses in Table 2. (Taken from Aguirre et al., 1974).

liths. Granodiorite, diorite, granite, and tonalite are the most common types; quartz monzonite and gabbro are more restricted. Twenty-two modal analyses of representative Cretaceous granitoids from Central Chile are given in Table 2 (samples 3 to 24) and are shown on Figure 4.

Thirty-five K/Ar, twenty-three Rb/Sr, and twenty-one Pbα dates show that plutonism was intense and continuous during the whole Cretaceous period, with slight culminations suggested between 120 and 140 Ma and between 80 and 90 Ma (Fig. 2).

Thermal metamorphic aureoles produced by the Cretaceous granitoids are known in several regions of the country. Most of them vary in thickness from 0.3 km to a maximum of 2.0 km and have been generated in volcanic and sedimentary rocks of Early Cretaceous age in north Cental Chile (Muñoz Cristi, 1950; Aguirre and Egert, 1970; Chávez, 1972, 1975; Münchmayer, 1972; Tidy, 1970; Tilling, 1962). Contact metamorphic facies described by these authors correspond mainly to albite-epidote hornfels and hornblende hornfels; pyroxene hornfels is rarely present.

Mineral deposits genetically related to the Cretaceous granitoids (Ruiz et al, 1965; Oyarzún, 1971) include iron and apatite, mainly developed by contact metamorphism in andesitic flows, and copper, gold, and silver as veins in limestone, shale, and andesite.

Tertiary Granitoids. They form a discontinuous belt located in the east-central part of the territory and crop out mainly in the highest areas of the Andean Range. Compared with the Cretaceous bodies, the Tertiary granitoids are smaller and more scattered.

The Tertiary plutons cut across stratified formations with ages ranging from the Paleozoic to the Early Tertiary, but most intrude the terrestrial volcano-sedimentary formations of the Upper Cretaceous and Lower Tertiary of Central Chile (Abanico and Farellones formations). Their contacts with the stratified units are sharply defined, and most of the bodies present typical features corresponding to a shallow emplacement.

The most common petrographic types are granodiorite, diorite, tonalite, and porphyry. Modal analyses of some representative Tertiary granitoids from Central Chile are given in Table 2 (samples 25 to 44) and are shown on Figure 4. Thirty-seven K/Ar, twenty-five Pbα, and nine Rb/Sr dates show a marked tendency to concentrate in the intervals 60 to 70 Ma, 40 to 50 Ma, and less than 20 Ma (Fig. 2).

Contact-metamorphic effects from the Tertiary granitoids are narrow, but no particular aureole has been described. However, hydrothermal alteration related to these bodies is widespread and is closely connected to the genesis of porphyry copper deposits. This alteration is characterized mainly by strong silicification and the development of sericitic, argillic, and chloritic zones, commonly arranged in a concentric pattern. A typical example of this arrangement is found at Los Pelambres (31°45′S, 70°28′W), where potassic, sericitic, argillic, and propylitic zones are developed sequentially from the centre to the periphery (United Nations Report, 1970). K/Ar dates of 9.96 ± 0.18 Ma and 9.74 ± 0.16 Ma have been obtained for hydrothermal biotite from this locality (Quirt et al., 1971).

The most important copper and molybdenum deposits of the country, such as Chuquicamata, El Salvador, Potrerillos, and El Teniente, are genetically associated with the Tertiary granitoids, especially with porphyry types. Dykes, mineralized with silver, zinc, mercury, antimony, and gold, are present in many parts of the periphery of the main porphyry copper deposits. Tourmaline breccia-pipe deposits containing copper, tungsten, and gold are also genetically related to Tertiary shallow intrusive bodies.

ESSENTIAL FEATURES OF PLUTONISM IN CHILE

The Paleozoic

The lack of comprehensive and detailed studies concerning Paleozoic plutonism in Chile precludes an extensive analysis of its characteristic features. However, some general trends can be pointed out.

The distribution of available granitoid ages shows that plutonic activity in the Paleozoic was concentrated mainly in the interval from the Early Carboniferous to the end of the era. An older event seems to have taken place during the Early Devonian (Fig. 2). These culminations of plutonism, closely connected in space and time with regional metamorphic complexes, are interpreted as pulses of the Hercynian and Caledonian orogenies, respectively. Stratigraphic relations in different parts of the country also indicate an important Hercynian event.

In the region between latitudes 26° S and 29° S in northern Chile, and in the contiguous Argentinian territory to the east, a general westward migration of Paleozoic plutonic, volcanic and metamorphic activity, together with a displacement of the main basins of sedimentation, is indicated by geological mapping and geochronological information (McBride et al., 1976; Zentilli, 1974; Clark et al., 1976). However, considerable overlap of plutonic centres of different ages is observed. According to these authors, each pulse of batholithic emplacement affected areas with wide east-west extension (up to 400 km), which is in marked contrast with the much narrower, longitudinal, plutonic complexes intruded during the Andean Orogenic Cycle (Clark et al., 1976). In this same region, mineralization is poor, especially in the western part of the Paleozoic crystalline basement, where most of the granitic bodies are of mid-Carboniferous to Late Permian age. Clark et al. (1976) suggested that the intensity of mineralization associated with the late Paleozoic orogenic activity decreased drastically with time as the locus of plutonism shifted to the west.

A general calc-alkaline nature seems to characterize the Chilean Paleozoic batholiths (Table 3). In the Andes of Coquimbo, the chemical features of a part of the Composite Batholith of the Frontal Cordillera have been investigated by Mpodozis et al. (1976). They conclude that the calc-alkaline plutonism in the region tends towards alkalinity at the end of the cycle, as indicated by the late appearance of hypersolvus granites. According to the previous authors, this type of evolution would indicate the transition, for the Hercynian belt, from the late-geosynclinal period to the post-geosynclinal period, this last being typified by pluto-

nism subsequent to final volcanism. A close relationship between the late Paleozoic granites and acidic volcanics of Permo-Triassic age exists in the Coquimbo Andean region.

In several areas (Coastal Range of Central Chile west of Santiago, Nahuelbuta Mountains), the Paleozoic batholiths are associated with high-grade sillimanite gneisses and widespread border migmatization. Hervé (1977) suggested that in the Nahuelbuta Mountains the intrusion of granodiorites and tonalites, which represent the main phases present, took place in a water-rich environment produced by dehydration reactions taking place in the surrounding metamorphic rocks. These metamorphic reactions occurred within the stability field of sillimanite, near to or within the melting field of rocks of granitic composition. The crystallization of muscovite and garnet in the Nahuelbuta Mountain granitoids, as a result of contamination with the alumina-rich country rock, would have resulted in this way according to Hervé (1977). Moreover, the presence of muscovite in this temperature range would support the existence of high P_{H_2O} at that stage.

The Andean Orogenic Cycle

The Andean granitoids show decreasing ages from the Pacific Ocean towards the east (Ruiz in Segerstrom, 1968; Farrar et al., 1970; Aubouin et al., 1973; Zentilli, 1974; Aguirre et al., 1974; Clark et al., 1976; Vergara and Drake, 1979). The Jurassic granitoids occupy the westernmost position; the Cretaceous intrusives are exposed as a central belt; and the Tertiary granitoids are confined to the main Andean Range to the east. This arrangement is best observed between latitudes 26° S and 32° S (Fig. 1). A plot of ages of the granitoids against longitude for the region between latitudes 30° S and 35° S was presented by Aguirre et al. (1974, Fig. 8), and similar diagrams for the region between latitudes 26° S and 19° S were given by Zentilli (1974, Fig. 2. 7) and by Clark et al. (1976, Fig. 2).

The isotopic ages shown by Figure 2 indicate a rather continuous process of granitoid production and emplacement throughout the whole of the Andean Cycle. However, a rather close relationship between the main plutonic culminations and the deformational phases can be observed. The predominant early post-kinematic nature of the main intrusive pulses is also made apparent, this being especially true for the plutonism related to the Araucanian (Late Jurassic), Subhercynian (mid-Cretaceous, Albian-Cenomanian), Laramide (Paleocene), Incaic (Late Eocene-Early Oligocene) and Quechuan (Late Miocene) phases.

South of latitude 42° S, in the domain of the Magellan geosynclinal basin, Halpern (1973) suggested the existence of three igneous episodes, each lasting for about 25 to 40 Ma: Late Jurassic to Early Cretaceous (155 to 120 Ma), Late Cretaceous (100 to 75 Ma), and mid- to Late Tertiary (50 to 10 Ma). The same author pointed out that no evidence in

TABLE 3. AVERAGE CHEMICAL ANALYSES OF PALEOZOIC, MESOZOIC, AND TERTIARY GRANITOIDS FROM NORTH CENTRAL CHILE (26° S - 35° S)

	PALEOZOIC *	MESOZOIC †	CENOZOIC §
SiO_2	69.59	61.20	62.12
TiO_2	0.38	0.57	0.60
Al_2O_3	14.83	15.83	15.84
Fe_2O_3	1.00	2.32	1.86
FeO	2.32	4.01	1.57
MgO	1.15	2.96	1.66
CaO	2.37	8.35	4.64
Na_2O	3.50	3.48	4.38
K_2O	3.34	1.93	2.63
Na_2O - K_2O	6.84	5.41	7.18
Fe^3/Fe^2	.55	0.52	1.33

* Based on 39 analyses, one of which is an average of 7 gabbros.

† Based on 18 analyses, one of which is an average of 9 diorites.

§ Based on 22 analyses.

this part of the territory suggests that the plutonic bodies become progressively younger or older from east to west (Halpern, 1973; Halpern and Fuenzalida, 1978).

When Mesozoic and Cenozoic granitoids and the volcanic rocks of a corresponding age are compared, several relevant features become apparent. Jurassic plutonic and volcanic rocks occupy a westernmost position, generally in the Coastal Range, although areas of common outcrop are scarce. The Jurassic volcanics are present along the coast north of latitude 26° S where no, or only minor, intrusives of Jurassic age have been identified up to the present, whereas between latitudes 26° S and 32° S, where the belt of Jurassic granitoids is mainly displayed, the volcanic formations of this age are almost completely absent. In contrast, Cretaceous and Tertiary granitoids and volcanic products of the same age are usually in close spatial association, occupying the central and eastern belt of the territory, respectively. Cretaceous intrusives are mainly associated with Lower Cretaceous volcanoclastic formations, whereas Tertiary intrusives are coincident with Upper Cretaceous and Lower Tertiary volcanoclastic formations. All these spatial relations show a coincidence in the direction of migration for the plutonic and the volcanic foci during the Andean Orogenic Cycle and would suggest a comagmatic origin for some of the granitoids and volcanic products broadly corresponding in age. However, much more work would be needed to substantiate the "comagmatic origin" view.

Regional deformative metamorphic events are not known to have occurred during the Andean Orogenic Cycle in the domain of the Andean Geosyncline. However, granitoid generation on a batholithic scale is thought to have played an important part in controlling the thermal gradients active during the generation of the regional, non-deformative, episodic, burial-metamorphic facies-series developed in the Mesozoic and Cenozoic strata (Levi, 1968, 1969, 1970; Thompson, 1971; Aguirre et al., 1978).

Chemical information shows that the plutonic and volcanic rocks generated during the Andean Orogenic Cycle are oversaturated and calc-alkaline and that both ultrabasic and alkaline rocks are absent. The volcanic rocks have been described as being mainly rhyolitic in composition in the early phase and andesitic in the final phase of each stratigraphic-structural stage (Aguirre et al., 1974, Fig. 5). Cretaceous and Tertiary granitoids range from diorite to granite with major concentration in the granodiorite and quartz monzodiorite field (Fig. 4). Information about Jurassic plutons is scarce.

Based on a restricted number (40) of chemical analyses from Andean granitoids of the north-central part of the country (between latitudes 26° S and 35° S), which cover a wide compositional range, the Tertiary intrusives appear to be richer in alkalis ($Na_2O + K_2O$) than those of the Mesozoic and lower in lime, ferric and ferrous iron, and magnesia. Silica and alumina are slightly more abundant and the

Fe^3/Fe^2 ratio (Table 3) is higher in the Tertiary granitoids. However, Montecinos (1979) found the reverse relationship for the alkalis, total iron, and MgO contents when comparing Tertiary and Mesozoic granitoids from the region between 22° and 29° S (30 analyses).

Major and trace element contents of 35 granitoids belonging to the Patagonian Batholith south of Tierra del Fuego were studied by Suárez (1977, 1978), who suggested that the plutons from western Canal Beagle can be distinguished from those in the Cape Horn area. According to him, the chemical trends observed in the variation diagrams might be explained in terms of magmatic differentiation involving predominantly hornblende and plagioclase fractionation. Assuming a similar source for the two groups, their chemical differences may be explained by the removal of more plagioclase in the Cape Horn magma than in that of the Canal Beagle, where hornblende fractionation was possibly predominant. A probable depletion in heavy REE, as indicated by the Y content, also suggests, according to Suárez, that hornblende and/or garnet took part in the genesis of these rocks, either in fractional crystallization process or as a residual phase during partial melting. Based on the low Ba and Sr contents of these granitoids, he finds it difficult to attribute an origin by anatexis of continental crust for the Patagonian Batholith. However, the few Sr^{87}/Sr^{86} initial ratios known for some of these granitoids give values of 0.7049 (two samples of Late Jurassic-Early Cretaceous age), 0.707 (one sample of Tertiary age; Halpern, 1973) and 0.7092 (Middle Jurassic foliated granitoid; Hervé et al., 1979) and would rather suggest that crustal melting processes could be related to the genesis of some of the granitoids of the Patagonian Batholith.

Rubidium and strontium contents and Rb/Sr ratios for the Andean granitoids and volcanics have been determined by McNutt et al. (1975) in the region between latitudes 26° S and 29° S, and by Oyarzún (1971) between latitudes 30° S and 35° S (Table 4).

The rubidium-strontium data listed in Table 4 show a high Rb/Sr ratio for the Jurassic granitoids, owing mainly to a very low Sr content. Low Rb and closely similar Rb/Sr ratios characterize both the plutonic and extrusive rocks of Cretaceous age in the northernmost area. The Rb/Sr ratio for extrusives of Tertiary age, in general, and Pliocene age, in particular, shows a broad similarity between the two areas. Oyarzún (1971) pointed out the close relationship in the Sr and Rb content between the Tertiary granitoids and the Pliocene andesitic flows and inferred a genetic relation between these magmatic products. Along this same line. López-Escobar et al. (1979) concluded that Upper Tertiary granodiorites between latitudes 33° S and 34° S have important geochemical similarities, such as highly fractionated REE distributions relative to chondrites, to some modern andesites in this region (e.g., Marmolejo volcano).

Initial Sr^{87}/Sr^{86} ratios were determined by McNutt et al.

TABLE 4. RUBIDIUM AND STRONTIUM IN ANDEAN GRANITOID AND VOLCANIC ROCKS

Age	Rock	Between Lat. 30°S and 35°S (Oyarzún, 1971)			Between Lat. 26°S and 29°S (McNutt et al., 1975)		
		Rb(ppm)	Sr(ppm)	Rb/Sr	Rb(ppm)	Sr(ppm)	Rb/Sr
Jur.	Granitoids				75.9	261.3	0.290
Cret.	Granitoids	54	425	0.127	47.3	458.9	0.103
Cret.	Andesites	70	395	0.180	40.5	476.0	0.090
Tert.	Granitoids	63	700	0.090	93.3	519.0	0.180
Tert.	Andesites	63	340	0.185			
Tert.	Andesites & rhyolites				87.5	430.2	0.203
Tert. (Plioc.)	Andesites	69	630	0.109	71.0	604.0	0.118
Quat.	Andesites				107.0	524.0	0.204

(1975) for the same set of Andean plutonic and volcanic rocks of Table 4. These authors established that the initial ratios of Jurassic plutons vary from 0.7043 to 0.7059 and do not correlate with age, whereas mid-Cretaceous to Quaternary rocks (125 Ma to 0 Ma) exhibit a systematic west to east increase in mean strontium isotope ratio from 0.7022 to 0.7077. Very low initial ratios (e.g., 0.7022, 0.7023) were obtained by McNutt et al. (1975) for several Mesozoic plutonic rocks and were thought to strongly imply a sub-crustal source for these magmas. These authors pointed out that "the time-dependent, post-Jurassic increase in initial ratio is considered to reflect a systematic change in the composition of partial melts generated in response to the progressive subduction of a lithospheric slab." They suggest that the locus of melting changes systematically "from along or close to the upper surface of the subduction slab into hanging-wall mantle peridotite as subduction continues."

López-Escobar et al. (1979) studied the geochemical characteristics of granitoids from Central Chile (33° - 34°S). In this region, three main batholiths — Paleozoic, Cretaceous, and Tertiary — are elongated in a north-south direction and arranged with a west to east sense of decreasing age. Based on a limited sampling of the Cretaceous and Tertiary batholiths, the above referred authors suggest a west to east increase in light REE/heavy REE ratio and in Sr content.

Halpern (1978) reported Sr^{87}/Sr^{86} values from ten samples of middle Tertiary porphyries related to the copper mineralization of the El Salvador mine. The mean value obtained is 0.7033 ± 0.0004, similar to the figure of 0.7041 ± 0.0003 found by Gustafson and Hunt (1975). Halpern thought the low Sr^{87}/Sr^{86} ratio, similar to those reported by Subbarao (1972) for basaltic rocks from the East Pacific and Chile Rises and the Nazca plate (0.7033 in average), would support the hypothesis (Sillitoe, 1972; Mitchell and Garson, 1972; Oyarzún and Frutos, 1974; and Dymond et al., 1973) relating the genesis of the circum-Pacific porphyry copper deposits to subduction processes and suggested that the mineralization at El Salvador did not originate from sialic crustal source rocks. The same author (Halpern, 1979) reported Sr^{87}/Sr^{86} ratios of total rock specimens from granitoids of the Disputada copper mine located northeast of Santiago. The values obtained vary from 0.7037 to 0.7044, and, according to Halpern, they suggest that the rocks involved are probable melting products of subcontinental lithosphere at a subduction zone or melting of a downgoing oceanic plate rather than products of simple partial melting of old sialic crust.

The Pb-isotopic ratios of thirteen samples of Andean plutonic and volcanic rocks from the area 26°S - 29°S, varying in age from 125 Ma to Recent (McNutt et al., 1976), show broad correlation between the initial Sr^{87}/Sr^{86} and Pb^{206}/Pb^{204} ratios in that both increase with decreasing age up to maximum values of 0.7079 and 18.85, respectively.

APPENDIX A
RADIOMETRIC AGES OF CHILEAN GRANITOIDS

N°	Locality	Rock Type	Age (Ma)	Method	Reference
1	Salar Punta Negra 24° 17'S, 68° 30'W	quartz diorite adamellite	468 ± 100	Rb/Sr, w.r.	Halpern (1978)
2	Quintay, Caleta 33° 12'S, 71° 43'W	adamellite	405	Pb/U	Corvalán & Munizaga (1972)
3	Lo Orrego Arriba 33° 25'S, 71° 26'W	tonalite	399	Pb/U	Corvalán & Munizaga (1972)
4	Cuesta Ibacache	tonalite	389	Pb/U	Corvalán & Munizaga (1972)
5	(west side) 33° 26'S, 71° 23'W	tonalite	386	Pb/U	Corvalán & Munizaga (1972)
6	Paihuano 30° 02'S, 70° 31'W	tonalite granodiorite	339 Av.	Pbα	Dediós (1967)
7	Posada Los Hidal-gos-Cifuncho. 25° 39'S, 70° 38'W	adamellite	340 ± 40	Pbα	Ruiz et al. (1960)
8	Concepción 36° 50'S, 73° 00'W	granodiorite	330	K/Ar	Nishimura (1971)
9	Nahuelbuta Mts. 37°–38°S, 73°W	tonalite granodiorite	320 ± 24	Rb/Sr, w.r.	Hervé et al. (1976)
10	Coastal Ranges, Central Chile 34°–41°S	mainly granodiorite	320	Rb/Sr, w.r.	Munizaga (unpl.)
11	Concepción 36° 50'S, 73° 00'W	granite	310	K/Ar	Nishimura (1971)
12	El Quisco 33° 23'S, 71° 43'W	tonalite	310 ± 35	Pbα	Levi et al. (1963)
13	Sierra Castillo 26° 22'S, 69° 17'W	granite	293 ± 30	Pbα	(IIG, unpl.)
14	Montandón Station 26° 24'S, 69° 17'W	tonalite	293 ± 30	Pbα	(IIG, unpl.)
15	Isla Negra 33° 25'S, 71° 44'W		288 ± 7	K/Ar, amph.	Hervé et al. (1976)
16	El Quisco 33° 23'S, 71° 43'W	tonalite	287 ± 20	K/Ar, biot.	Muñoz Cristi (1962)
17	Montandón Station 26° 24'S, 69° 17'W	diorite	287 ± 30	Pbα	IIG (unpubl.)
18	Sierra Castillo 26° 22'S, 69° 17'W	diorite	276 ± 30	Pbα	IIG (unpubl.)
19	Sierra Castillo	diorite	276 ± 30	Pbα	IIG (unpubl.)
20	El Quisco 33° 23'S, 71° 43'W	granite	270 ± 30	Pbα	Levi et al. (1963)
21	East of El Salvador Mine 24° 14'S, 69° 23'W	tonalite	269 ± 4	Rb/Sr, w.r.	Halpern (1978)
22	Sierra Bórax 26° 28'S, 69° 11'W	quartz diorite	268	Rb/Sr, w.r.	Halpern, *fide* Tobar (1977)
23	Posada Los Hidal-gos-Cifuncho 25° 39'S, 70° 38'W	granodiorite	267 ± 8	K/Ar, biot.	McBride et al. (1976)

APPENDIX A (continued)

N°	Locality	Rock Type	Age (Ma)	Method	Reference
24	Juntas 28° 10'S, 69° 50'W	granite	265 ± 30	Pbα	Ruiz et al. (1960)
25	Juntas 28° 00'S, 69° 52'W	quartz diorite	262 ± 10	K/Ar, biot.	McBride et al. (1976)
26	Imilac 24° 15'S, 68° 45'W	quartz diorite quartz porphyry	262 (Av.)	Rb/Sr, w.r.	Halpern (1978)
27	Sierra Bórax 26° 28'S, 69° 11'W	adamellite	250 ± 30	Pbα	IIG (unpubl.)
28	Montandón Station 26° 24'S, 69° 17'W	tonalite	244	Rb/Sr, biot.	Halpern, *fide* Tobar (1977).
29	Juntas 28° 10'S, 69° 53'W	quartz diorite	236 ± 7	K/Ar, biot.	Farrar et al. (1970)
30	Juntas 28° 09'S, 69° 49'W	granite porphyry	228 ± 7	K/Ar, biot.	McBride et al. (1976)
31	Sierra Bórax 26° 28'S, 69° 11'W	granite	223 ± 25	Pbα	IIG (unpubl.)
32	Paihuano 30° 02'S, 70° 31'W	granite	200	Pbα	Dediós (1967)
33	Victoria Island 45° 11'S, 73° 51'W	granodiorite	194 ± 67	Rb,Sr, w.r.	Halpern & Fuenzalida (1978)
34	Puerto Flamenco 26° 34'S, 70° 41'W	granite porphyry	191 ± 6	K/Ar, biot.	Farrar et al. (1970)
35	Qda. La Cachina 25° 54'S, 70° 34'W	diorite	189 ± 9	K/Ar, biot.	Farrar et al. (1970)
36	Chañaral 26° 22'S, 70° 40'W	granodiorite	188 ± 6	K/Ar, biot.	Farrar et al. (1970)
37	Caldera 27° 00'S, 70° 47'W	quartz diorite	187 ± 6	K/Ar, biot.	Farrar et al. (1970)
38	Morro de Copiapó 27° 11'S, 70° 59'W	quartz diorite	182 ± 6	K/Ar, biot.	Farrar et al. (1970)
39	Puerto Viejo 27° 21'S, 70° 56'W	quartz diorite	182 ± 6	K/Ar, biot.	Farrar et al. (1970)
40	Qda. Flamenco 26° 35'S, 70° 34'W	quartz diorite	177 ± 5	K/Ar, biot.	McNutt et al. (1976)
41	Punta Zenteno 26° 49'S, 70° 47'W	quartz diorite	176 ± 5	K/Ar, biot.	Zentilli, (1974)
42	Victoria Island 45° 11'S, 73° 51'W	quartz diorite granodiorite	176	Rb/Sr, biot.	Halpern & Fuenzalida (1978)
43	Huentelauquén 31° 19'S, 71° 35'W	granodiorite	173 ± 20	Pbα	Munizaga (1972)
44	Mincha 31° 35'S, 71° 24'W	tonalite	173 ± 20	Pbα	Munizaga (1972)
45	Damas Island 29° 13'S, 71° 31'W	gneissic granite	171 ± 20	Pbα	Aguirre (1967)
46	Fray Jorge 30° 26'S, 71° 32'W	granodiorite	171 ± 20	Pbα	Thomas (1967)
47	Limarí River 30° 40'S, 71° 32'W	diorite	171 ± 20	Pbα	Thomas (1967)
48	Catapilco 32° 36'S, 71° 23'W	granodiorite	170 ± 20	Pbα	Levi et al. (1963)

APPENDIX A (continued)

N°	Locality	Rock Type	Age (Ma)	Method	Reference
49	Peña Blanca 30° 55'S, 71° 35'W	diorite	169 ± 20	Pbα	Thomas (1967)
50	Papudo 32° 29'S, 71° 25'W	monzogranite	160 ± 20	Pbα	Levi et al. (1963)
51	Lambert 29° 47'S, 71° 09'W	quartz monzodiorite	157	K/Ar, biot.	Curtis (written comm. 1972)
52	Herradura Bay 29° 57'S, 71° 32'W	granodiorite	156 ± 15	Pbα	Munizaga (1972)
53	Añañucal 27° 22'S, 70° 55'W	greisen	156 ± 5	K/Ar, musc.	Zentilli (1974)
54	Magellan Basin 52° 20'S to 53° S 68° 20'W to 69° 40 (drill holes)	granodiorite	155 ± 20	Rb/Sr, w.r.	Halpern (1973)
55	Mal Paso 18° 40'S, 70° 23'W	gabbro	153 ± 20 (Av.)	Pbα	Tobar et al. (1968)
56	Qda. Los Apestados 28° 06'S, 70° 56'W	diorite	151 ± 5	K/Ar, biot.	Zentilli (1974)
57	Puerto Año Nuevo 52° 12'S, 73° 38'W	granite	151 ± 10	Rb/Sr, w.r.	Halpern (1973)
58	Cerro Bajo 28° 05'S, 70° 54'W	diorite	148 ± 5	K/Ar, biot.	Zentilli (1974)
59	Canto del Agua 28° 08'S, 70° 56'W	diorite	148 ± 9	K/Ar, biot.	Zentilli (1974)
60	Río Salado 26° 33'S, 70° 23'W	quartz monzonite	148 ± 7	K/Ar, biot.	Zentilli (1974)
61	Río Salado 26° 33'S, 70° 23'W	granodiorite	147 ± 5	K/Ar, biot.	McNutt et al. (1975)
62	Mincha 31° 35'S, 71° 24'W	tonalite	147	Pbα	Munizaga (1972)
63	Camarones 19° 00'S, 70° 00'W	granodiorite	144 ± 5 (Av.)	K/Ar, biot. K/Ar, hornb.	Levi et al. (1963)
64	Bahía Stewart 51° 46'S, 73° 41'W	diorite	142 ± 7	K/Ar, biot.	Halpern (1973)
65	Cerro Vetado 26° 72'S, 70° 29'W	quartz monzonite	144 ± 4	K/Ar, biot. K/Ar, hornb.	Farrar et al. (1970)
66	Camarones 19° 00'S, 70° 00'W	granodiorite	140 ± 5	Rb/Sr, biot.	Levi et al. (1963)
67	Islas Daroch 51° 28'S, 74° 44'W	adamellite	138 ± 20	Rb/Sr, w.r.	Halpern (1973)
68	Qda. Huamanga 26° 33'S, 70° 24'W	quartz diorite	137 ± 4	K/Ar, hornb.	Zentilli (1974)
69	Las Rojas 29° 59'S, 71° 21'W	granodiorite	136 ± 15	Pbα	Munizaga (1972)
70	Camarones 19° 00'S, 70° 00'W	granodiorite	135 ± 20 (Av.)	Pbα	Levi et al. (1963)
71	Bahía Stewart 51° 46'S, 73° 41'W	diorite	135 ± 7	Rb/Sr, biot.	Halpern (1973)
72	Bahía Isthmus 52° 10'S, 73° 38'W	granodiorite	135 ± 8	Rb/Sr, biot.	Halpern (1973)

APPENDIX A (continued)

N°	Locality	Rock Type	Age (Ma)	Method	Reference
73	Tamaya 30° 04′S, 71° 20′W	granodiorite	134 ± 20	Pbα	Munizaga (1972)
74	Isla Froilán 50° 38′S, 74° 40′W	granodiorite	133 ± 6	K/Ar, biot.	Halpern (1973)
75	Isla Froilán 50° 38′S, 74° 40′W	granodiorite	132 ± 6	Rb/Sr, biot.	Halpern (1973)
76	Peninsula Staines 51° 28′S, 73° 56′W	adamellite	131 ± 7	K/Ar, biot.	Halpern (1973)
77	Qda. Huamanga 26° 33′S, 70° 24′W	quartz diorite granodiorite	131 ± 4 (Av.)	K/Ar, biot. K/Ar, hornb.	McNutt et al. (1975)
78	Carrizal Alto 28° 06′S, 70° 55′W	diorite	130 ± 20	Pbα	Levi et al. (1963)
79	Carrizal Alto 28° 06′S, 70° 55′W	tonalite	130 ± 20	Pbα	Levi et al. (1963)
80	Locality 23 45° 24′S, 74° 03′W	monzonite	130	Rb/Sr, biot.	Halpern & Fuenzalida (1978)
81	Punitaqui 30° 56′S, 71° 19′W	granodiorite	128 ± 15	Pbα	Munizaga (1972)
82	Boquerón Chañar 28° 04′S, 70° 56′W	biotite vein	128 ± 4	K/Ar, biot.	Zentilli (1974)
83	Locality 18 45° 44′S, 74° 00′W	quartz diorite	128	Rb/Sr, w.r.	Halpern & Fuenzalida (1978)
84	Montepatria 30° 41′S, 71° 03′W	granodiorite	125 ± 15	Pbα	Thomas (1967)
85	Isla Duque de York 50° 32′S, 75° 05′W	quartz diorite	125 ± 8	Rb/Sr, biot.	Halpern (1973)
86	Iquique (Punta Negra). 20° 10′S, 70° 08′W	granodiorite	125 ± 15	Pbα	Ruiz et al. (1960)
87	Pedegua 32° 19′S, 71° 05′W	granodiorite	123 ± 20	Pbα	Munizaga (1972)
88	Río Salado 26° 25′S, 70° 20′W	granodiorite	123 ± 4	K/Ar, biot.	Zentilli (1974)
89	Peninsula Staines 51° 28′S, 73° 56′W	adamellite	122 ± 5	Rb/Sr, biot.	Halpern (1973)
90	Islas Daroch 51° 28′S, 74° 44′W	adamellite	122 ± 6	K/Ar, biot.	Halpern (1973)
91	Isla Caracciolo 50° 25′S, 74° 40′W	quartz diorite	122 ± 6	K/Ar, biot.	Halpern (1973)
92	Sierra Monardes 27° 20′S, 70° 31′W	quartz diorite	122 ± 4 (Av.)	K/Ar, biot.	Zentilli (1974)
93	Pozo Almonte 20° 22′S, 69° 47′W	adamellite	120 ± 15	Pbα	Ruiz et al. (1960)
94	Cuesta El Melón 32° 35′S, 71° 16′W	monzogranite	120 ± 20	Pbα	Levi et al. (1963)
95	Isla Caracciolo 50° 25′S, 74° 40′W	quartz diorite granodiorite	120 ± 8	Rb/Sr, biot. feld. w.r.	Halpern (1973)
96	Punta del Nicho 28° 13′S, 70° 50′W	quartz diorite	120 ± 4	K/Ar, biot.	Zentilli (1974)
97	Bahía Sandy 52° 08′S, 73° 43′W	quartz diorite	119 ± 6	Rb/Sr, biot.	Halpern (1973)

APPENDIX A (continued)

N°	Locality	Rock Type	Age (Ma)	Method	Reference
98	Puangue 33°20′S, 71°08′W	tonalite	118	K/Ar, biot.	Corvalán & Munizaga (1972)
99	Bahía Sandy 52°08′S, 73°43′W	quartz diorite	117 ± 6	K/Ar, biot.	Halpern (1973)
100	Puesco 39°32′S, 71°34′W	granite	117	K/Ar	Aguirre & Levi (1964)
101	Qda. Huamanga 26°37′S, 70°15′W	quartz diorite	115 ± 4 (Av.)	K/Ar, biot. K/Ar, hornb.	Zentilli (1974)
102	Rungue 33°00′S, 70°50′W	granodiorite	110 ± 1	K/Ar plag.	Drake et al. (1976)
103	Santa Inés Mine 29°48′S, 71°10′W	diorite	109	K/Ar, biot	Curtis (written comm. 1972)
104	Pan de Azúcar 29°59′S, 71°21′W	granodiorite	109 ± 10	Pbα	Munizaga (1972)
105	Salamanca 31°45′S, 70°57′W	granite	109 ± 10	Pbα	Munizaga (1972)
106	Puerto Aysén 45°24′S, 72°28′W	adamellite	108	Rb/Sr, w.r.	Halpern & Fuenzalida (1978)
107	Localities 10, 11 and 12 45°03′S, 72°10′W	granodiorite	108	Rb/Sr, w.r.	Halpern & Fuenzalida (1978)
108	Pampa Travesía 27°36′S, 70°24′W	quartz diorite	107 ± 3	K/Ar, hornb.	Farrar et al. (1970)
109	Balboa Creek 29°43′S, 71°10′W	granodiorite	106 (Av.)	K/Ar, biot. K/Ar, hornb.	Curtis (written comm. 1972)
110	La Paloma Lake 45°56′S, 72°12′W	adamellite	106 ± 9	Rb/Sr, w.r.	Halpern & Fuenzalida (1978)
111	Illapel 31°41′S, 71°09′W	granite	104 ± 10	Pbα	Munizaga (1972)
112	Cerro Imán 27°16′S, 70°26′W	greisen	102 ± 3	K/Ar, musc.	Zentilli (1974)
113	Lago Todos Los Santos 41°05′S, 72°17′W	diorite	102 ± 10	Pbα	Aguirre & Levi (1964)
114	Huara, Tarapacá 20°00′S, 69°49′W	granodiorite	100 ± 10	Pbα	Ruiz et al. (1960)
115	Río Simpson 45°29′S, 72°15′W	adamellite	110 ± 6	Rb/Sr, w.r.	Halpern & Fuenzalida (1978)
116	Melipeuco (E of) 38°52′S, 71°34′W	granite	98 ± 10	Pbα	Aguirre & Levi (1964)
117	Pampa Travesía 27°40′S, 70°28′W	quartz diorite	96 ± 3	K/Ar, biot.	Farrar et al. (1970)
118	Cerro Juan de Morales, Tarapacá 20°08′S, 69°23′W	adamellite	95 ± 10	Pbα	Ruiz et al. (1960)
119	Santa Gracia 29°45′S, 71°05′W	monzogranite	94 ± 1 (Av.)	K/Ar, biot.	Curtis (written comm. 1972)
120	Punta de Piedra 29°57′S, 71°06′W	granodiorite	94 (Av.)	K/Ar, biot.	Curtis (written comm. 1972)
121	Isla Londonderry 54°55′S, 70°47′W	quartz diorite	94 ± 6	Rb/Sr, biot.	Halpern (1973)

APPENDIX A (continued)

N°	Locality	Rock Type	Age (Ma)	Method	Reference
122	Illapel 31°41'S, 71°09'W	granite	92	K/Ar, biot.	Munizaga (1972)
123	Isla Londonderry 54°55'S, 70°47'W	quartz diorite	91 ± 5	K/Ar, biot.	Halpern (1973)
124	Lambert Quadr. 29°47'S, 71°09'W	hornb. gabbro	90	K/Ar, biot.	Curtis (written comm. 1972)
125	Paso Mallín de Icalma 38°51'S, 71°17'W	granodiorite	90	Pbα	Aguirre & Levi (1964)
126	East of Lake Colico 39°03'S, 71°50'W	adamellite	90	Pbα	Aguirre & Levi (1964)
127	Qda. Pacunto 38°34'S, 71°16'W	granodiorite	90 ± 10	Pbα	Levi et al. (1963)
128	Puerto Bueno 50°59'S, 74°13'W	quartz diorite	90 ± 5	Rb/Sr, biot.	Halpern (1973)
129	Puerto Bueno 50°59'S, 74°13'W	quartz diorite	90 ± 5	K/Ar, biot.	Halpern (1973)
130	Cerro Bodega 27°22'S, 70°23'W	mica vein	90 ± 3	K/Ar, biot.	Zentilli (1974)
131	Salamanca 31°49'S, 70°54'W	granodiorite	89 ± 10	Pbα	Munizaga (1972)
132	Isla Londonderry 54°55'S, 70°47'W	quartz diorite	88 ± 5	Rb,Sr, biot.	Halpern (1973)
133	Bahía Tres Brazos 54°58'S, 69°48'W	quartz diorite	88 ± 5	K/Ar, hornb.	Halpern (1973)
134	Sierra de Varas 28°42'S, 70°10'W	quartz diorite	87 ± 3	K/Ar, biot.	Farrar et al. (1970)
135	Punta Hope 54°07'S, 71°00'W	quartz diorite	87 ± 5	K/Ar, biot.	Halpern (1973)
136	Punta Hope 54°07'S, 71°00'W	quartz diorite	85 ± 10	Rb/Sr, w.r. Rb/Sr, biot.	Halpern (1973)
137	Seno Ventisquero 54°50'S, 70°19'W	quartz diorite	85 ± 5	K/Ar, biot.	Halpern (1973)
138	La Campana 33°05'S, 71°01'W	granodiorite	83	K/Ar, biot.	Tidy (1970)
139	Bahía Tres Brazos 54°58'S, 69°48'W	quartz diorite	80 ± 10	Rb/Sr, w.r.	Halpern (1973)
140	Seno Ventisquero 54°50'S, 70°19'W	quartz diorite	79 ± 5	Rb/Sr, biot.	Halpern (1973)
141	Cabo San Antonio 50°52'S, 74°12'W	diorite	77 ± 7	K/Ar, hornb.	Halpern (1973)
142	Locality 21 45°50'S, 74°42'W	quartz diorite	72	Rb/Sr, biot.	Halpern & Fuenzalida (1978)
143	Punta Pirando 54°26'S, 71°07'W	quartz diorite	70 ± 8	Rb/Sr, biot.	Halpern & Fuenzalida (1978)
144	Ventisquero Alemania. 54°53'S, 69°25'W	diorite	68 ± 6	Rb/Sr, w.r. Rb/Sr, biot.	Halpern & Fuenzalida (1978)
145	Putaendo River 32°30'S, 70°37'W	granodiorite	68 ± 10	Pbα	Munizaga (1972)
146	Qda. Algarrobal 28°20'S, 70°21'W	diorite	67 ± 2	K/Ar, hornb.	Zentilli (1974)

APPENDIX A (continued)

N°	Locality	Rock Type	Age (Ma)	Method	Reference
147	La Campana 33°05'S, 71°01'W	gabbro	66	K/Ar, plag.	Tidy (1970)
148	Inca de Oro 26°45'S, 69°50'W	hydrothermal alteration	65 – 2	K/Ar, seric.	Zentilli (1974)
149	Salamanca 31°42'S, 70°47'W	granite	65 ± 10	Pbα	Munizaga (1972)
150	Salamanca 31°43'S, 70°48'W	granodiorite	64 ± 10	Pbα	Munizaga (1972)
151	Hacienda Manflas 28°09'S, 70°02'W	quartz monzonite	63 ± 2 (Av.)	K/Ar, biot.	Farrar et al. (1970)
152	Pampa Larga 27°35'S, 70°10'W	diorite	63 ± 2	K/Ar, hornb.	Farrar et al. (1970)
153	Qda. Algarrobal 28°20'S, 70°28'W	granodiorite	63 ± 2 (Av.)	K/Ar, biot.	Farrar et al. (1970)
154	Isla Tabara 51°48'S, 73°57'W	diorite	63 ± 5	Rb/Sr, biot.	Halpern (1973)
155	Los Loros 27°49'S, 70°08'W	quartz monzonite	62 ± 2	K/Ar, biot.	Farrar et al. (1970)
156	El Melado 35°40'S, 70°26'W	granodiorite	62 ± 1	K/Ar, w.r.	Drake (1976)
157	Cachiyuyo de Oro 27°07'S, 70°01'W	granodiorite	61 ± 2	K/Ar, biot.	Farrar et al. (1970)
158	Qda. Los Cóndores 27°29'S, 70°03'W	granodiorite	61 ± 2	K/Ar, biot.	Farrar et al. (1970)
159	Qda. San Miguel 27°24'S, 69°35'W	quartz diorite	60 ± 2	K/Ar, biot.	Farrar et al. (1970)
160	Qda. Algarrobal 28°18'S, 70°30'W	granodiorite	60 ± 2	K/Ar, biot.	Farrar et al. (1970)
161	Aguada Maray 27°36'S, 70°05'W	granodiorite	59 ± 2	K/Ar, biot.	Farrar et al. (1970)
162	Cabeza de Vaca 27°37'S, 70°03'W	granodiorite	59	K/Ar	Ruiz et al. (1965)
163	Vicuña Quadr. 30°12'S, 70°43'W	diorite	56 ± 10	Pbα	Dediós (1967)
164	Putaendo River 32°30'S, 70°34'W	dioritic prophyry	55 ± 10	Pbα	Munizaga (1972)
165	Quebrada de Tarapacá 19°52'S, 69°23'W	granodiorite	54	K/Ar	Ruiz et al. (1965)
166	El Salvador 26°15'S, 69°34'W	granite porphyry	54 ± 10 (Av.)	Pbα	Ruiz et al. (1965)
167	Campanani 18°19'S, 69°42'W	granodiorite	53 ± 10	Pbα	Ruiz et al. (1965)
168	Cabeza de Vaca 27°37'S, 70°03'W	granodiorite	53 ± 10 (Av.)	Pbα	Ruiz et al. (1965)
169	Cuncumén 31°53'S, 70°42'W	granodiorite	52 ± 10	Pbα	Munizaga (1972)
170	Las Ñipas 30°06'S, 70°41'W	granodiorite	50 ± 10	Pbα	Dediós (1967)
171	Riecillo Brook 32°59'S, 70°21'W	quartz monzodiorite	50 ± 20	Pbα	Levi et al. (1963)

L. Aguirre

APPENDIX A (continued)

N°	Locality	Rock Type	Age (Ma)	Method	Reference
172	Hacienda Manflas 28°09'S, 70°02'W	pegmatite	50 ± 4 (Av.)	K/Ar, biot. K/Ar, chlor.	Zentilli (1974)
173	El Salvador 26°17'S, 69°34'W	rhyolite (intr.)	50 ± 3	Rb/Sr, w.r.	Gustafson & Hunt (1975)
174	El Salvador 26°17'S, 69°34'W	rhyolite (intr.)	46 ± 1	Rb/Sr, w.r. K/Ar, seric.	Gustafson & Hunt (1975)
175	Seno Yussef 51°40'S, 73°39'W	adamellite	45 ± 6	Rb/Sr, w.r.	Halpern (1973)
176	Isla Rennell 51°59'S, 73°53'W	diorite	44 ± 6	Rb/Sr, biot.	Halpern (1973)
177	Qda. Carrizalillo 27°38'S, 69°39'W	granodiorite	44 ± 10	Pbα	Ruiz et al. (1965)
178	Aguada Cortadera 27°48'S, 69°56'W	granite	44 ± 1	K/Ar, biot.	Zentilli (1974)
179	Qda. Carrizalillo 27°39'S, 69°40'W	granodiorite	43 ± 1 (Av.)	K/Ar, biot.	Zentilli (1974)
180	Chuquicamata 22°17'S, 68°54'W	granodiorite	43 ± 10	Pbα	Ruiz et al. (1965)
181	Cerro Vicuñita 26°47'S, 69°18'W	quartz porphyry	42 ± 1	K/Ar, w.r.	Zentilli (1974)
182	Chuquicamata 22°17'S, 68°54'W	porphyry	41 ± 10	Pbα	Ruiz et al. (1965)
183	El Indio 35°50'S, 70°50'W	granodiorite	41 ± 1	K/Ar, hornb.	Drake (1976)
184	El Salvador 26°17'S, 69°34'W	porphyries	41 ± 1	K/Ar, Rb/Sr	Gustafson and Hunt (1975)
185	El Salvador 26°17'S, 69°34'W	quartz diorite porphyry	40 ± 1	K/Ar, biot.	Farrar et al. (1970)
186	Yeso River 33°46'S, 70°13'W	granodiorite	40 ± 10	Pbα	Munizaga (1972)
187	Potrerillos 26°30'S, 69°25'W	feldspar porphyry	36 ± 1 (Av.)	K/Ar, biot.	Zentilli (1974)
188	Chuquicamata 22°17'S, 68°54'W	granodiorite	35	K/Ar	Ruiz et al. (1965)
189	Loica 31°04'S, 70°41'W	?	35 ± 5	K/Ar, hydr. biot.	Quirt et al. (1971)
190	Isla Rennell 51°59'S, 73°53'W	diorite	34 ± 4	K/Ar, biot.	Halpern (1973)
191	Isla Tabara 51°48'S, 73°57'W	diorite	33 ± 5 (Av.)	K/Ar, biot. K/Ar, hornb. K/Ar, biot. plus horn.	Halpern (1973)
192	Potrerillos 26°30'S, 69°25'W	porphyry	30 ± 10	Pbα	Ruiz et al. (1965)
193	Riecillo Brook 32°59'S, 70°21'W	monzogranite	30 ± 20	Pbα	Levi et al. (1963)
194	Yeso River 33°46'S, 70°13'W	granite	30 ± 10	Pbα	Ruiz et al. (1965)
195	Isla Tabara 51°48'S, 73°57'W	diorite	29 ± 3	K/Ar, biot.	Halpern (1973)

APPENDIX A (continued)

N°	Locality	Rock Type	Age (Ma)	Method	Reference
196	Las Lajas 35° 03'S, 70° 38'W	granodiorite	27 ± 10	Pbα	Munizaga (1972)
197	La Obra 33° 35'S, 70° 29'W	granodiorite	24	K/Ar, biot.	Ravich, (written comm., 1967)
198	Disputada Mine 33° 09'S, 70° 18'W	granite	23 ± 10	Pbα	Ruiz et al. (1965)
199	Yeso River 33° 46'S, 70° 13'W	granodiorite	23 ± 10	Pbα	Ruiz et al. (1965)
200	La Coipa 26° 49'S, 69° 15'W	granodioritic porphyry	23 ± 1 (Av.)	K/Ar, biot.	McNutt et al. (1975)
201	Salto del Soldado 32° 52'S, 70° 20'W	granodiorite	20 ± 1	K/Ar, w.r.	Drake (1976)
202	Bahía Inservible 50° 45'S, 74° 24'W	diorite	20 ± 2	K/Ar, biot.	Halpern (1973)
203	Bahía Inservible 50° 45'S, 74° 24'W	diorite	19 ± 3	Rb/Sr, biot.	Halpern (1973)
204	Pérez-Caldera 33° 12'S, 70° 18'W	granodiorite	18 ± 10	Pbα	Munizaga (1972)
205	El Tatio 22° 26'S, 68° 01'W	?	17 ± 10	Pbα	Ruiz et al. (1965)
206	Yeso River 33° 46'S, 70° 13'W	granodiorite	15 ± 10	Pbα	Ruiz et al. (1965)
207	Locality 25 45° 18'S, 73° 24'W	quartz diorite	15	Rb/Sr, biot.	Halpern & Fuenzalida (1978)
208	Isla Tabara 51° 48'S, 73° 57'W	diorite	15 ± 10	Rb/Sr (biot. - hb.)	Halpern (1973)
209	Colorado River 33° 30'S, 70° 13'W	tonalite	14 ± 10	Pbα	Ruiz et al. (1965)
210	Localities 6, 7 and 14 45° 24'S, 72° 50'W	quartz diorite	13 (Av.)	Rb/Sr, biot.	Halpern & Fuenzalida (1978)
211	Cerro Paine 51° 00'S, 73° 00'W	quartz diorite	13 ± 1	K/Ar, biot.	Halpern (1973)
212	Cerro Paine 51° 00'S, 73° 00'W	quartz diorite	12 ± 2	Rb/Sr, w.r.	Halpern (1973)
213	Los Pelambres Mine 31° 45'S, 70° 28'W	?	10 ± 0.17	K/Ar, hydr. biot.	Quirt et al. (1971)
214	Disputada Mine 33° 09'S, 70° 18'W	granite	10	K/Ar, biot.	Ruiz et al. (1965)
215	Río Blanco Mine 33° 09'S, 70° 18'W	quartz porphyry	5 ± 0.5 (Av.)	K/Ar, biot.	Quirt et al. (1971)
216	Laguna Invernada 35° 42'S, 70° 48'W	granodiorite & granite	7 ± 0.3 (Av.)	K/Ar, ort. K/Ar, biot. K/Ar, hornb.	Drake (1976)
217	El Teniente Mine 34° 55'S, 70° 20'W	?	5 ± 0.1 (Av.)	K/Ar, w.r. K/Ar, hydr. seric.	Quirt et al. (1971)

APPENDIX B
MODAL ANALYSES OF GRANITOIDS FROM THE UPPER MENDOZA AND SAN JUAN RIVERS.*

BIOTITE/HORNBLENDE GRANODIORITE-TONALITE ASSOCIATION

N°	Qtz.	K.f.	Pl	Bt	Hb	Op	Acc	Q	A	P	Field N°	Locality	Rock Type
1	28.1	6.6	48.2	10.5	4.1	2.5	—	33.8	7.9	58.3	(A.3042)	Stock Punta de Vacas-Q. de Vargas	Granodiorite
2	26.8	16.0	38.3	14.2	3.2	1.4	0.1	33.0	19.7	47.2	(A.3728)	Stock de Punta de Vacas	Granodiorite
3	26.8	12.8	45.7	8.5	4.4	1.5	0.4	31.4	15.0	53.5	(A.3729)	Stock de Punta de Vacas	Granodiorite
4	26.3	15.5	51.3	5.1	0.6	0.9	0.3	28.2	27.3	44.5	(A.1551)	Rio de los Patos, Vega del Burro	Granodiorite
5	0.2	2.0	52.0	—	45.2	0.4	0.2	0.4	3.6	95.6	(A.3590)	Rio Verde	Diorite
6	12.0	0.5	53.5	12.4	20.2	1.4	—	18.2	0.7	81.1	(A.3586)	Rio Verde	Tonalite
7	24.0	1.1	55.7	12.0	7.0	0.2	—	29.7	1.3	68.9	(A.3326)	Rio Carnicería	Tonalite
8	29.9	3.9	48.2	14.4	3.3	0.1	0.2	36.7	4.7	58.6	(A.3369)	Confluencia Rio Yeso-Rio Colorado	Tonalite
9	28.9	2.2	53.4	11.2	4.2	0.1	—	34.2	2.6	63.2	(A.3549)	Rio Yeso	Tonalite
10	17.4	0.4	53.4	15.2	12.6	1.0	—	24.4	0.6	75.0	(A.3587)	Juntas Rios Verde y Salinas	Tonalite
11	21.8	0.5	52.4	24.3	—	1.0	—	29.2	0.6	70.2	(A.3325)	Rio Carnicería	Tonalite
12	25.8	2.0	51.1	17.3	2.7	—	0.1	32.7	2.5	64.8	(A.3327)	Rio Carnicería	Tonalite
13	34.3	4.2	46.4	15.0	—	0.1	—	40.4	4.9	54.6	(A.3550)	Rio Yeso	Tonalite
14	25.2	7.2	55.2	9.2	1.6	1.0	0.4	28.7	8.2	63.1	(A.3308)	Rio Pachón	Granodiorite
15	28.3	6.2	56.5	0.3	—	8.4	0.3	31.0	6.8	62.2	(A.3309)	Rio Pachón	Tonalite
16	32.7	6.6	42.8	12.2	5.1	0.6	—	39.8	8.0	52.1	(A.3547)	Rio Yeso	Granodiorite
17	11.8	7.4	30.4	23.7	25.4	1.3	—	23.7	14.9	61.3	(A.3548)	Rio Yeso	Granodiorite
18	18.9	8.0	51.6	4.0	17.2	0.3	—	24.1	10.2	65.7	(A.3584)	Juantas Rios Verde y Salinas	Granodiorite
19	32.4	21.4	36.5	7.1	1.2	1.3	0.1	35.8	23.7	40.5	(A.3280)	Rio Mondaca	Monzogranite
20	30.3	23.7	36.5	7.3	0.8	1.4	—	33.4	26.1	40.5	(A.3288)	Rio Mondaca	Monzogranite
21	24.8	23.2	43.9	6.2	0.8	0.9	0.2	27.2	25.3	57.5	(A.3329)	Rio Carnicería	Monzogranite
22	30.6	27.1	35.1	5.5	1.4	—	0.2	32.9	29.1	38.0	(A.3349)	Arroyo Coipa	Monzogranite
23	30.7	23.6	36.6	6.2	2.7	—	0.2	33.6	25.9	40.5	(A.3352)	Arroyo Coipa	Monzogranite

HOLOLEUCOCRATIC GRANITES

N°	Qtz.	K.f.	Pl	Bt	Musc	Op	Acc	Q	A	P	Field N°	Locality	Rock Type
24	38.9	41.7	18.2	1.8	—	—	—	38.8	42.6	18.5	(A.1325)	Rio de las Leñas	Syenogranite
25	32.6	49.8	14.6	2.1	—	0.9	—	33.6	51.3	15.0	(A.1350)	Rio de los Patos	Syenogranite
26	29.7	43.8	23.6	1.8	—	1.1	—	30.6	45.1	24.3	(A.3596)	Pircas - La Junta	Monzogranite (graphic)
27	33.1	45.0	18.2	1.5	—	2.2	—	34.4	46.7	18.9	(A.3598)	Pircas - La Junta	Syenogranite (perthitic)
28	39.7	57.8	0.5	0.9	0.8	0.3	—	40.5	59.0	0.5	(A.3591)	Rio Verde	Alaskite (perthitic)
29	36.2	40.4	22.3	0.6	—	0.5	0.1	36.6	40.9	22.5	(A.3328)	Rio Carnicería	Monzogranite (perthitic)
30	34.3	37.6	25.5	0.8	—	1.8	—	35.2	38.6	26.2	(A.3559)	Juntas Santa Cruz - Salinas	Monzogranite (graphic)
31	24.9	49.5	23.1	1.5	—	1.0	—	25.5	50.8	23.7	(A.3567)	Juntas Santa Cruz - Salinas	Syenogranite (graphic)
32	36.8	55.9	6.7	—	0.6	—	—	37.0	56.2	6.7	(A.3557)	Juntas Santa Cruz - Salinas	Syenogranite
33	42.5	41.0	14.7	—	1.1	0.7	—	43.3	41.7	15.0	(A.3560)	Juntas Santa Cruz - Salinas	Syenogranite
34	29.7	67.5	0.9	0.2	—	1.7	—	30.3	68.8	0.9	(A.3581)	Cañón del Rio Salinas	Alaskite (graphic)
35	32.9	61.9	3.7	—	—	1.5	—	33.4	62.8	3.8	(A.3582)	Cañón del Rio Salinas	Alaskite (graphic)
36	23.7	74.4	0.9	—	—	1.0	—	23.9	75.2	0.9	(A.3583)	Rio Salinas	Alaskite (graphic)

APPENDIX B2
MODAL ANALYSES OF GRANITOIDS FROM THE UPPER LIMARI RIVER.*

BIOTITE/HORNBLENDE GRANODIORITE-TONALITE ASSOCIATION

N°	Qtz.	K.f.	Pl	Bt	Amph.	Op	Acc	Q	A	P	Field N°	Locality	Rock Type
37	30.9	1.0	38.6	25.4	2.7	—	1.4	43.8	1.4	54.8	(517)	Quebrada Carachas	Tonalite
38	37.3	1.0	51.1	8.8	—	—	1.8	42.4	1.1	57.6	(514)	Quebrada Carachas	Tonalite
39	35.6	8.9	49.6	5.0	0.9	—	—	37.8	9.5	52.7	(512-A)	Quebrada Carachas	Granodiorite
40	31.9	5.4	48.6	9.8	1.6	2.7	—	37.1	6.3	56.6	(506)	Quebrada Carachas	Granodiorite
41	30.2	0.7	50.0	19.1	—	—	—	37.3	0.9	61.8	(505)	Quebrada Carachas	Tonalite
42	17.4	3.4	58.9	12.5	—	6.9	0.9	21.8	4.3	73.9	(498)	Quebrada Carachas	Tonalite
43	28.1	12.2	37.5	20.9	—	0.7	0.6	36.1	15.9	48.0	(497-B)	Quebrada Carachas	Granodiorite
44	38.1	17.0	33.0	11.2	—	0.2	0.5	43.2	18.2	38.6	(406)	Quebrada El Toro	Granodiorite
45	12.9	8.6	57.4	12.2	—	8.4	0.5	16.3	10.9	72.8	(401)	Quebrada El Toro	Monzodiorite
46	23.6	—	45.7	24.9	—	5.8	—	34.1	—	65.9	(399)	Quebrada El Toro	Tonalite
47	14.7	4.6	46.4	16.0	14.9	2.6	0.8	22.4	7.0	70.6	(169-R)	Rio Patillos	Tonalite
48	25.1	—	39.1	18.3	14.9	0.2	0.5	39.1	—	60.9	(154-R)	Quebrada El Toro	Tonalite
49	24.5	2.9	55.6	16.3	—	—	0.6	29.5	3.5	67.0	(P-16)	Quebrada La Lunca	Granodiorite
50	26.8	18.5	40.7	12.5	0.9	0.3	1.1	31.2	21.5	47.3	(SM-77)	Rio Divisadero	Granodiorite
51	18.0	11.8	47.1	17.7	4.0	—	1.6	23.4	15.3	61.3	(SM-68)	Portezuelo Soldado	Granodiorite
52	26.2	17.1	38.5	15.1	1.5	0.1	0.7	32.0	20.9	47.1	(SM-67)	Portezuelo Soldado	Tonalite
53	29.5	—	41.2	26.5	2.0	0.1	0.7	41.7	—	58.3	(SM-66)	Rio Colorado	Tonalite
54	44.1	—	44.5	3.6	7.6	0.2	—	49.8	—	50.2	(SM-65)	Rio Colorado	Tonalite
55	31.9	3.1	43.6	11.2	8.7	0.8	0.7	40.6	3.9	55.5	(SM-41)	Rio San Miguel	Tonalite
56	24.4	2.4	54.0	3.6	14.9	0.3	0.4	30.2	3.0	66.3	(T-20)	Quebrada Mala	Tonalite
57	20.9	2.8	56.8	16.3	—	2.8	0.4	26.0	3.5	70.5	(T-19)	Quebrada Talca	Tonalite

HOLOLEUCOCRATIC GRANITES

N°	Qtz.	K.f.	Pl	Bt	Amph.	Op	Acc	Q	A	P	Field N°	Locality	Rock Type
58	26.6	44.0	27.4	1.6	0.2	0.2	—	27.1	44.9	28.0	(C-3)	Rio Los Molles	Monzogranite
59	28.2	39.5	28.7	3.6	—	—	2.6	29.2	41.0	29.8	(C-57)	Rio Los Molles	Monzogranite
60	29.2	50.5	16.4	2.1	0.8	1.0	—	30.4	52.6	17.0	(C-65)	Rio Los Molles	Monzogranite
61	37.2	36.7	25.5	0.4	0.1	0.1	—	37.4	36.9	25.7	(C-66)	Rio Los Molles	Monzogranite

MUSCOVITE GRANODIORITES

N°	Qtz.	K.f.	Pl	Bt	Musc	Amph.	Op	Acc	Q	A	P	Field N°	Locality	Rock Type
62	38.6	9.4	38.2	8.5	5.3	—	—	—	44.8	10.9	44.3	(60-R)	El Toro Chico	Granodiorite
63	19.6	10.6	43.8	11.3	0.2	—	—	2.6	26.0	14.3	59.2	(490)	El Toro Chico	Granodiorite
64	30.3	13.5	45.1	4.1	7.0	—	—	—	34.1	15.2	50.7	(557)	El Toro Chico	Granodiorite
65	43.9	13.4	30.2	5.6	6.7	—	—	—	50.2	15.3	34.5	(558)	El Toro Chico	Granodiorite
66	31.5	16.2	40.9	6.2	4.0	—	1.2	—	35.3	18.3	46.2	(549-B)	Rio Patillos	Granodiorite
67	37.8	11.6	35.7	10.3	4.0	—	0.6	—	44.4	13.6	42.0	(549-A)	Rio Patillos	Granodiorite
68	19.7	46.8	25.0	7.5	0.6	—	0.4	—	21.5	51.1	27.4	(544-A)	Rio Patillos	Monzogranite
69	33.1	34.5	24.2	3.2	3.8	—	1.1	0.1	36.0	37.6	26.4	(544-B)	Rio Patillos	Monzogranite
70	25.2	23.8	38.9	10.1	1.8	—	0.2	—	28.7	27.0	44.3	(146-R)	Quebrada El Toro	Monzogranite
71	15.0	1.0	37.0	5.0	—	38.0	—	4.0	27.7	2.0	70.3	(144-R)	Quebrada El Toro	Tonalite
72	35.0	35.0	15.0	—	—	10.0	0.5	5.5	41.2	41.2	17.6	(109-R)	Quebrada El Toro	Syenogranite
73	37.0	21.1	34.6	1.2	6.1	—	—	—	39.9	22.8	37.3	(P-98)	Quebrada La Lunca	Monzogranite
74	29.6	41.1	22.3	1.7	4.0	—	0.6	0.7	31.8	44.2	24.0	(SM-47)	Quebrada Relojero	Monzogranite
75	34.7	31.2	27.5	5.2	0.6	—	0.3	0.5	37.2	33.4	29.4	(SM-2)	Rio San Miguel	Monzogranite

APPENDIX B2 (continued)

MUSCOVITE GRANODIORITES (continued)

N°	Qtz.	K.f.	Pl	Bt	Musc	Op	Acc	Q	A	P	Field N°	Locality	Rock Type
76	39.3	6.1	33.6	9.9	10.6	0.3	0.2	49.8	7.7	42.5	(SM-51)	Rio San Miguel	Granodiorite
77	43.2	18.9	30.1	4.8	2.7	—	0.3	46.9	20.4	32.7	(SM-50)	Quebrada Relojero	Monzogranite
78	51.4	21.9	16.4	5.7	4.2	—	0.4	57.3	24.4	18.3	(SM-46)	Quebrada Relojero	Monzogranite
79	24.9	22.0	38.2	7.7	6.5	—	0.7	29.6	25.5	44.9	(SM-1)	Rio San Miguel	Monzogranite
80	45.6	14.2	32.2	5.1	2.6	—	0.3	49.6	16.4	35.0	(SM-55)	Quebrada Navio	Granodiorite
81	20.4	16.5	49.2	17.5	4.8	0.1	0.6	26.5	21.4	52.1	(SM-82)	Rio Mostazal	Granodiorite
82	39.8	15.5	43.3	0.9	—	—	0.4	40.4	15.7	43.9	(M-3)	Rio Mostazal	Granodiorite
83	41.3	17.6	30.4	5.3	5.3	—	0.4	40.4	19.8	33.8	(M-3)	Rio Mostazal	Monzogranite
84	29.3	20.0	40.4	7.8	1.4	—	0.3	32.0	22.0	45.8	(O-67)	Rio Los Malles	Granodiorite

APPENDIX B3

MODAL ANALYSES OF GRANITOIDS FROM THE UPPER ELQUI RIVER.*

BIOTITE/HORNBLENDE GRANODIORITE-TONALITE ASSOCIATION

N°	Qtz.	K.f.	Pl	Bt	Amph.	Op	Q	A	P	Field N°	Locality	Rock Type
85	42.0	13.0	40.0	5.0	—	—	44.2	13.7	42.1	(2)	Rio Turbio	Granodiorite
86	28.3	17.0	44.7	7.0	3.0	—	31.4	18.9	49.7	(18)	Rio Turbio-Qda. Pucalume	Granodiorite
87	34.7	4.4	47.9	7.2	5.8	—	39.8	5.1	55.1	(3)	Quebrada Las Lechuzas	Tonalite
88	5.5	1.2	76.0	3.5	13.9	—	6.7	1.3	92.0	(4-A)	Guanta	Quartz diorite
89	32.5	12.2	48.2	—	7.1	—	35.0	13.1	51.9	(5-C)	Guanta	Granodiorite
90	29.0	9.5	48.0	11.1	2.4	—	33.5	11.1	55.4	(8-B)	El Milagro	Granodiorite
91	24.6	23.1	50.8	1.5	—	—	25.0	23.5	51.6	(9)	Cordón Los Tilos	Granodiorite
92	14.2	5.4	45.3	—	35.1	—	21.9	8.3	69.8	(15)	Quebrada Calvario	Tonalite
93	30.4	4.9	60.0	—	4.4	0.1	31.9	5.1	63.0	(1-A)	Quebrada Pucalume	Granodiorite
94	36.7	6.6	50.5	5.9	—	0.3	39.1	7.1	53.8	(18-B)	Rio Turbio	Monzogranite
95	27.3	25.9	41.5	—	4.7	0.6	28.8	29.4	43.8	(12)	Rio Turbio	Granodiorite
96	22.4	10.0	46.8	5.0	12.9	2.9	28.3	12.6	59.1	(5-B)	Rio Turbio	Granodiorite
97	33.4	9.2	45.7	4.6	6.2	0.9	37.8	10.4	51.8	(10)	Cordón de Los Tilos	Granodiorite

HOLOLEUCOCRATIC GRANITES

N°	Qtz.	K.f.	Pl	Bt	Amph.	Op	Q	A	P	Field H°	Locality	Rock Type
98	34.5	40.0	25.5	—	—	—	34.5	40.0	25.5	(16-B)	Juntas R. Toro-R. La Laguna	Monzogranite
99	25.8	44.1	24.1	1.9	3.4	0.7	27.5	46.9	25.6	(19)	Cerro Colorado	Monzogranite

MUSCOVITE GRANODIORITES

N°	Qtz.	K.f.	Pl	Bt	Musc.	Op	Q	A	P	Field H°	Locality	Rock Type
100	43.0	3.5	39.3	7.1	7.1	—	50.1	4.1	45.8	(7—A)	Rio Turbio, E de Guanta	Tonalite
101	35.1	29.0	29.7	3.2	3.0	—	37.4	30.9	31.7	(8-A)	El Milagro	Monzogranite
102	26.0	9.8	52.5	5.1	5.8	0.8	29.4	11.1	59.5	(7-B)	Rio Turbio, E de Guanta	Granodiorite

* From Mpodozis et al, 1976

REFERENCES CITED

Aguirre, L., 1967, Geología de las islas Choros y Damas y de Punta Choros, provincia de Coquimbo: Revista Minerales, v. 22, p. 73–83.

Aguirre, L. Charrier, R., Davidson, J., and others, 1974, Andean Magmatism: Its paleogeographic and structural setting in the central part (30° S - 35° S) of the Southern Andes: Pacific Geology, v. 8, p. 1–38.

Aguirre, L., and Egert, E., 1970, Cuadrángulo Lambert (La Serena), provincia de Coquimbo. Instituto de Investigaciunes Geológicas, Carta 23, 14 p. Santiago, Chile.

Aguirre, L., Hervé, F., and Godoy, E., 1972, Distribution of metamorphic facies in Chile, an outline: Krystalinikum, v. 9, p. 7–19.

Aguirre, L., and Levi, B., 1964, Geología de la Cordillera de los Andes de las provincias de Cautín, Valdivia, Osorno y Llanquihue: Instituto de Investigacioner Geológicas, v. 17, 37 p. Santiago, Chile.

Aguirre, L., Levi, B., and Offler, R., 1978, Unconformities as mineralogical breaks in the burial metamorphism of the Andes: Contributions to Mineralogy and Petrology, v. 66, p. 361–366.

Aubouin, J., Borrello, A. V., Cecioni, G., and others, 1973, Esquisse paléogéographique et structurale des Andes Meridionales: Revue de Geographie Physique et de Geologie Dynamique, v. 15, N° 1-2, p. 11–72.

Chávez, L., 1972, Metamorfismo de contacto en la serie volcánico-sedimentaria de Santa Gracia, provincia de Coquimbo: Department of Geology, University of Chile, [unpub. thesis], 90 p.

——,1975, Contact metamorphism and regional alteration of basic volcanic rocks in the Santa Gracia area, Lambert quadrangle, Coquimbo province, Chile: Krystalinikum, v. 11, p. 25–52.

Clark, A. H., Farrar, E., Caelles, J. C., and others, 1976, Longitudinal variations in the metallogenetic evolution of the Central Andes; A progress report: Geological Association of Canada Special Paper 14, p. 23–58.

Corvalán, J., and Dávila, A., 1964, Observaciones geológicas en la Cordillera de la Costa entre los rios Aconcagua y Mataquito: Sociedad Geológica de Chile, Resúmenes, v. 9, p. 1–4.

Corvalán, J., and Munizaga, Fdo., 1972, Edades radiométricas de rocas intrusivas y metamórficas de la Hoja Valparaiso-San Antonio: Instituto Investigaciones Geológicas, Santiago, Chile, v. 28, p. 1–40.

Dediós, P. 1967, Cuadrángulo Vicuña, provincia de Coquimbo: Instituto Investigaciones Geológicas, Santiago, Chile, Carta 16, p 1–65.

Domeyko, I., 1845, Memoria sobre la constitución jeolójica de Chile. Constitución jeolójica del sistema andino i de los terrenos que atraviesan en la latitud de Coquimbo: Mineralojía (Jeolojía), t. 5, p. 173–294. Cervantes, ed. (1903), Santiago, Chile.

Drake, R. E., 1976, Chronology of Cenozoic igneous and tectonic events in the Central Chilean Andes - Latitudes 35° 30'S to 36° S: Journal of Volcanology and Geothermal Research, v. 1, p. 265–284.

Drake, R. E., Curtis, G., and Vergara, M., 1976, Potassium-Argon dating of igneous activity in the Central Chilean Andes - Latitude 33° S: Journal of Volcanology and Geothermal Research, v. 1, p. 285–295.

Dymond, J., Corliss, J. B., Heath, G. R., and others, 1973, Origin of metalliferous sediments from the Pacific Ocean: Geological Society of America Bulletin, v. 84, p. 3355–3372.

Farrar, E., Clark, A. H., Haynes, S. J., and others 1970, K-Ar evidence for the post-Paleozoic migration of granitic intrusion foci in the Andes of Northern Chile: Earth and Planetary Science Letters, v. 10, p. 60–66.

González-Bonorino, F., 1967, Nuevos datos de edad absoluta del Basamento Cristalino de la Cordillera de la Costa de Chile Central: Notas y Comunicaciones, Department of Geology, University of Chile, v. 1, p. 1–7.

——,1970, Series metamórficas del Basamento Cristalino de la Cordillera de la Costa, Chile Central: Department of Geology, University of Chile, Publicación 37, p. 1–68.

——,1971, Metamorphism of the crystalline basement of Central Chile: Journal of Petrology, v. 12, p. 149–175.

González-Bonorino, F., and Aguirre, L., 1970, Metamorphic facies-series of the Crystalline Basement of Chile: Geologische Rundschau, v. 59, p. 979–994.

Groeber, P., 1963, La Cordillera entre las latitudes 22° 20' y 40° S: Boletín Academia Nacional de Ciencias, Córdoba, Argentina, v. 43.

Gustafson, L. B., and Hunt, J. P., 1975, The porphyry copper deposit at El Salvador, Chile: Economic Geology, v. 70, p. 857–912.

Halpern, M., 1973, Regional geochronology of Chile south of 50° latitude: Geological Society of America Bulletin, v. 84, p. 2407–2422.

——,1978, Geological significance of Rb-Sr isotopic data of Northern Chile crystalline rocks of the Andean orogen between latitudes 23° and 27° South: Geological Society of America Bulletin, v. 89, p. 522–532.

——,1979, Strontium isotope composition of rocks from the Disputada copper mine, Chile: Economic Geology, v. 74, p. 129–130.

Halpern, M., and Fuenzalida, R., 1978, Rubidium-Strontium geochronology of a transect of the Chilean Andes between latitudes 45° and 46° S: Earth and Planetary Science Letters, v. 41, p. 60–66.

Hervé, F., 1976, Petrografía del Basamento Cristalino en el área Laguna Verde-Quintay, provincia de Valparaiso, Chile: Actas Primer Congreso Geológico Chileno, Tomo II, F125–F144.

——,1977, Petrology of the Crystalline Basement of the Nahuelbuta Mountains, South Central Chile: In Comparative Studies on the Geology of the Circum-Pacific Orogenic Belt in Japan and Chile, 1st Report, Japan Society of the Promotion of Science, T. Ishikawa and L. Aguirre (eds.), Tokyo, p. 1–51.

Hervé, F., Moreno, H., and Parada, M. A., 1974a, Granitoids of the Andean Range of Valdivia province, Chile: Pacific Geology, v. 8, p. 39–45.

Hervé, F., Munizaga F., Godoy, E., and Aguirre, L., 1974b, Late Paleozoic K-Ar ages of blueschists from Pichilemu, Central Chile: Earth and Planetary Science Letters, v. 23, p. 261–264.

Hervé, F., Munizaga, F., Mantovani, M., and Hervé, M., 1976, Edades Rb/Sr neopaleozoicas del Basamento Cristalino de la Cordillera de Nahuelbuta: Actas Primer Congreso Geológico Chileno, Tomo II, F19–F26.

Hervé, F., Nelson, E., and Suárez, M., 1979, Edades radiométricas de granitoides y metamorfitas provenientes de Cordillera Darwin, XII Región, Chile: Revista Geológica de Chile, N° 7, p. 31–40.

IIG (Instituto de Investigaciones Geológicas, Chile), 1962, Mapa Metalogénico de Chile, 1 : 1,500,000, in Ruiz, C. et al., 1965, Geología y Yacimientos Metalíferos de Chile, Instituto de Investigaciones Geológicas Santiago.

Levi, B., 1968, Cretaceous volcanic rocks from a part of the Coast Range west from Santiago, Chile: [PhD Thesis], Department of Geology, Univ. of California, Berkeley.

——,1969, Burial metamorphism of a Cretaceous volcanic sequence West from Santiago, Chile: Contributions to Mineralogy and Petrology, v. 24, p. 30–49.

——,1970, Burial metamorphic episodes in the Andean Geosyncline, Central Chile: Geologische Rundschau, v. 59, p. 994–1013.

Levi, B., Mehech, S., and Munizaga, Fdo., 1963, Edades radiométricas y petrografia de granitos Chilenos: Instituto de Investigaciones Geológicas, Bol. 12, p. 1–40.

López-Escobar, L., Frey, F. A., and Oyarzún, J., 1979, Geochemical characteristics of Central Chile (33° - 34° S) granitoids: Contributions to Mineralogy and Petrology, v. 70, p. 439–450.

McBride, S. L., Caelles, J. C., Clark, A. H., and Farrar, E., 1976, Paleozoic radiometric age provinces in the Andean basement, latitudes 25° -30° S: Earth and Planetary Science Letters, v. 29, p. 373–383.

McNutt, R. H., Crocket, J. H., Clark, A. H., and others, 1975, Initial Sr[87]/Sr[86] ratios of plutonic and volcanic rocks of the Central Andes between latitudes 26° and 29° S: Earth and Planetary Science Letters, v. 27, p. 305–313.

McNutt, R. H., Crocket, J. H., and Clark, A. H., 1976, Pb isotopic ratios of plutonic and volcanic rocks of the Central Andes between latitudes 26° and 29° South: Geological Society of Canada Annual Meeting Program with Abstracts, v. 1, p. 40.

Mitchell, A. H. G., and Garson, M. S., 1972, Relationship of porphyry copper and circum-Pacific tin deposits to paleo-Benioff zones: Institute of Mining and Metallurgy Transactions (Sect. B Applied Earth Science), v. 81, B10–B25.

Montecinos, P., 1979, Plutonismo durante el Ciclo Tectónico Andino en el Norte de Chile, entre los 18° - 29° lat. Sur: Actas Segundo Congreso Geológico Chileno, Tomo 3, E89–E108.

Mpodozis, C., Parada, M. A., Rivano, S., and Vicente, J-C., 1976, Acerca del plutonismo tardi-Hercínico de la Cordillera Frontal entre los 30° y 33° S (provincias de Mendoza y San Juan - Argentina; Coquimbo -Chile): Actas 6° Congreso Geológico Argentino (1975), Bahí Blanca, p. 143–166.

Münchmeyer, C., 1972, Geología general del distrito minero de Tambillos, provincia de Coquimbo, y estudio geológico del yacimiento Florida del distrito minero de Tambillos: [Unpub. Thesis], Department of Geology, University of Chile.

Munizaga, Fdo., 1972, Edades radiométricas de rocas chilenas: Inst. Invest. Geológicas, Jornadas de Trabajo, v. 2, Santiago, Chile.

Munizaga, Fdo., Aguirre, L., and Hervé, F., 1973, Rb-Sr ages of rocks of the Chilean Metamorphic Basement: Earth and Planetary Science Letters, v. 18, p. 87–92.

Muñoz Cristi, J., 1950, Geología del distrito minero de La Higuera, provincia de Coquimbo: Department of Geology University of Chile, Publicación 1, p. 1–107.

——, 1960, Contribución al conocimiento geológico de la Cordillera de la Costa de la zona central del país: Revista Minerales, v. 15, p. 28–46.

——, 1962, Comentarios sobre los granitos chilenos: Revista Minerales, v. 78.

——, 1964, Estudios petrográficos y petrológicos sobre el Batolito de la Costa de las provincias de Santiago y Valparaíso: Dept. Geol. Univ. of Chile, Publicación 25, p. 1–93.

Nishimura, T., 1971, On the geology of the Magellan geosyncline: Report of the Geological Survey of Hokkaido, v. 44, p. 45–53 (in Japanese).

Oyarzún, J., 1971, Contribution à l'étude geochimique des roches volcaniques et plutoniques du Chili: Doctoral Thesis, University of Paris, Sud, 114 p.

Oyarzún, J., and Frutos, J., 1974, Porphyry copper and tin-bearing porphyries. A discussion of genetic models: Physics of the Earth and Planetary Interiors, v. 9, p. 259–263.

Parada, M. A., 1975, Estudio geológico de los alrededores de los lagos Calafquén, Panguipulli y Riñihue, provincia de Valdivia: [Unpub. Thesis], Department of Geology University of Chile.

Quirt, G. S., Stewart, J., Clark, A. H., and Farrar, E., 1971, Potassium-Argon ages of porphyry copper deposits in Northern and Central Chile: Geological Society of America Abstracts with Programs.

Ruiz, C., Aguirre, L., Corvalán, J., and others, 1965, Geología y yacimientos metalíferos de Chile: Instituto de Investigaciones Geológicas, 305 p.

Ruiz, C., Segerstrom, K., Aguirre, L., and others, 1960, Edades plomo-alfa y marco estratigráfico de granitos chilenos con una discusión acerca de su relación con la orogénesis: Instituto de Investigaciones Geológicas, Bol. 7.

Segerstrom, K., 1967, Geology and ore deposits of Central Atacama province, Chile: Geological Society of America Bulletin, v. 78, p. 305–318.

——, 1968, Geología de las Hojas Copiapó y Ojos del Salado, provincia de Atacama: Instituto de Investigaciones Geológicas, Santiago, Chile, Bol. 24, 58 p.

Sillitoe, R. H., 1972, A plate tectonic model for the origin of porphyry copper deposits: Economic Geology, v. 67, p. 184–197.

Steinmann, G., 1929, Geologie von Peru: Heidelberg, C. Winters, 448 p.

Suárez, M., 1977, Notas geoquímicas preliminares del Batolito Patagnico al Sur de Tierra del Fuego, Chile: Revista Geológica de Chile, N° 4, p. 15–33.

——, 1978, Región al Sur del Canal Beagle: Carta Geológica de Chile N° 36, Esc; 1 : 500,000, Instituto de Investigaciones Geológicas, Santiago, Chile, 48 p.

Subbarao, K. V., 1972, The strontium isotopic composition of basalts from the East Pacific and Chile Rises and abyssal hills in the eastern Pacific Ocean: Contributions to Mineralogy and Petrology, v. 37, p. 111–120.

Thomas, H., 1967, Geología de la Hoja Ovalle, provincia de Coquimbo: Instituto de Investigaciones Geológicas, Santiago, Chile, Bol. 23, 56 p.

Thompson, A. B., 1971, Studies in low-grade metamorphism in Central Chile, South America: [PhD Thesis], Department of Geology, University of Manchester.

Tidy, E., 1970, Geología del distrito minero La Campana, provincia de Valparaíso: [Unpub. Thesis], Department of Geology University of Chile.

Tilling, R. I., 1962, Batholith emplacement and contact metamorphism in the Paipote-Tierra Amarilla area, Atacama province, Chile: [PhD Thesis], Department of Geology, Yale University, 202 p.

Tobar, A., 1977, Stratigraphy and structure of the El Salvador-Potrerillos region, Atacama, Chile: [PhD Thesis], Department of Geology University of California, Berkeley, 117 p.

Tobar, A., Salas, I., and Kast, R., 1968, Cuadrángulos Camaraca y Azapa, provincia de Tarapacá: Instituto de Investigaciunes Geológicas, Cartas 19, 20; Santiago, Chile, 20 p.

United Nations Report, 1970, Plan Chile 28. Technical Report 21. Santiago, Chile.

Van Eysinga, F. W. B. (Compiler), 1975 Geological Time Scale: 3rd Edition, Elsevier, Amsterdam.

Vergara, M., and Drake, R., 1979, Eventos magmatico-plutonicos en los Andes de Chile Central: Actas Segundo Congreso Geologico Chileno, Tomo I, F19–F30.

Willis, B., 1929, Earthquake conditions in Chile: Carnegie Institution of Washington, Publ. 382, 178 p.

Zentilli, M., 1974, Geological evolution and metallogenic relationships in the Andes of Northern Chile between 26° and 29° South: [PhD. thesis], Queen's University, Kingston, Ontario, Canada.

MANUSCRIPT ACCEPTED BY THE SOCIETY JULY 12, 1982

Typeset by TLC Type Company, Boulder, Colorado
Printed in U.S.A. by Malloy Lithographing, Inc., Ann Arbor, Michigan